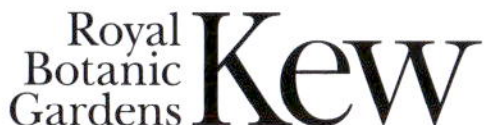

THE GOLDEN AGE OF

BOTANICAL ART

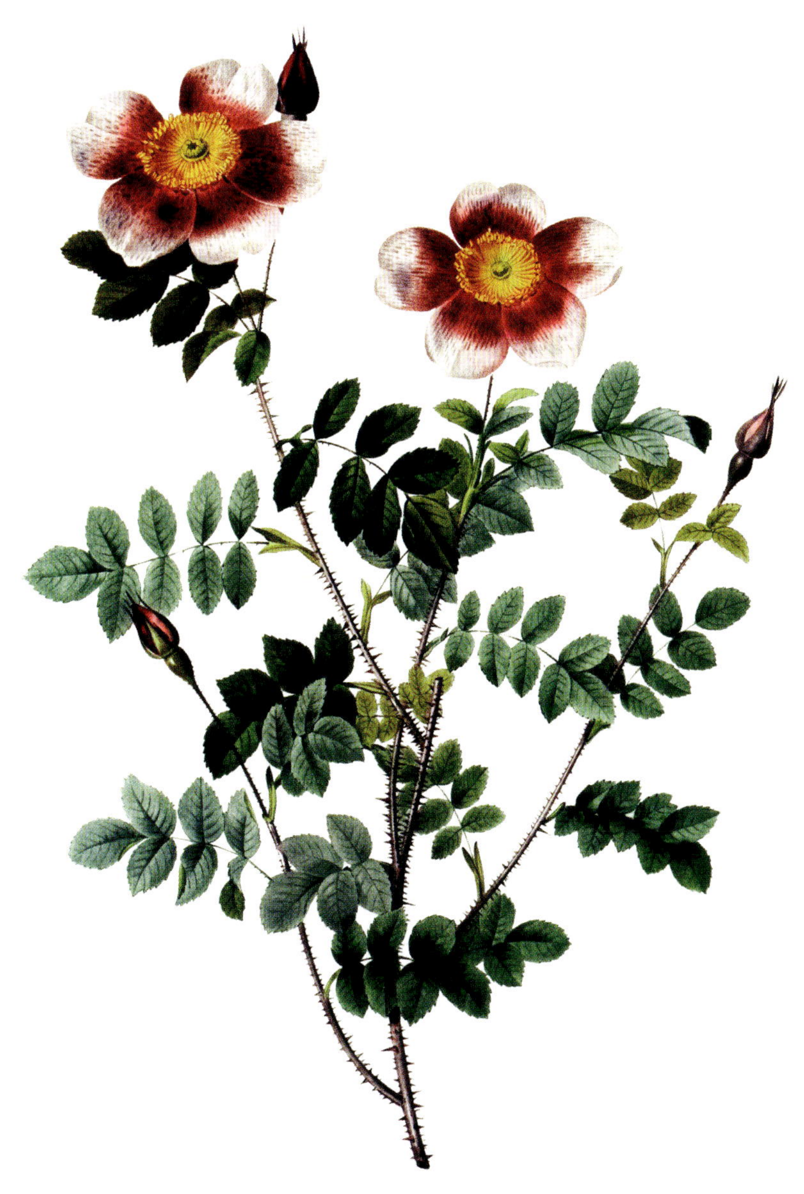

PREVIOUS PAGE: Rosa pimpinellifolia flore variegato by Pierre-Joseph Redouté from Les Roses.

RIGHT: Yellow horned poppy, Chelidonium pedunculis unifloris ... (now Glaucium flavum) by Georg Dionysius Ehret.

OPPOSITE: Bombax pentandrum (now Bombax ceiba) by General J Eyre.

Author's Acknowledgements

The author wishes to thank the staff of the Library, Art and Archives at The Royal Botanic Gardens, Kew for their help with the illustrations; Gina Fullerlove, Head of Kew Publishing for her advice, and Alison Rix for her help with research and assistance throughout the project.

First published in 2012 by Andre Deutsch.
This edition published in 2024 by Welbeck
An Imprint of HEADLINE PUBLISHING GROUP

6

Text © Martyn Rix 2012
Design © Carlton Books Limited 2012
All images unless otherwise stated in the Picture Credits on page 256 © The Board and Trustees of the Royal Botanic Gardens, Kew

The right of Martyn Rix to be identified as the author of work has been asserted by him in accordance with the Copyright, Designs and Patents Act 1988.

Apart from any use permitted under UK copyright law, this publication may only be reproduced, stored, or transmitted, in any form, or by any means, with prior permission in writing of the publishers or, in the case of reprographic production, in accordance with the terms of licences issued by the Copyright Licensing Agency.

Cataloguing in Publication Data is available from the British Library

ISBN 978 0 233 00542 3

Printed in Dubai

Headline's policy is to use papers that are natural, renewable and recyclable products and made from wood grown in well-managed forests and other controlled sources. The logging and manufacturing processes are expected to conform to the environmental regulations of the country of origin.

HEADLINE PUBLISHING GROUP
An Hachette UK Company, Carmelite House
50 Victoria Embankment, London EC4Y 0DZ

www.headline.co.uk
www.hachette.co.uk

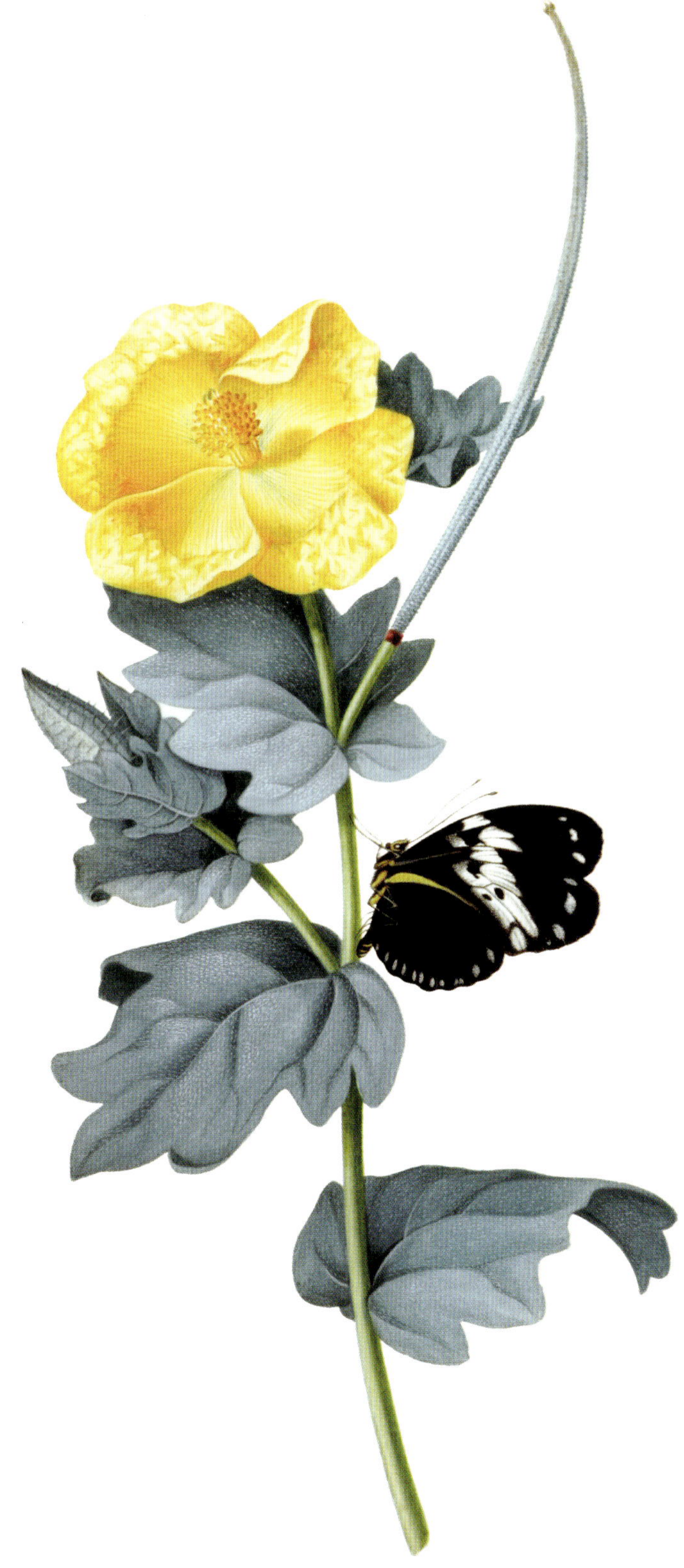

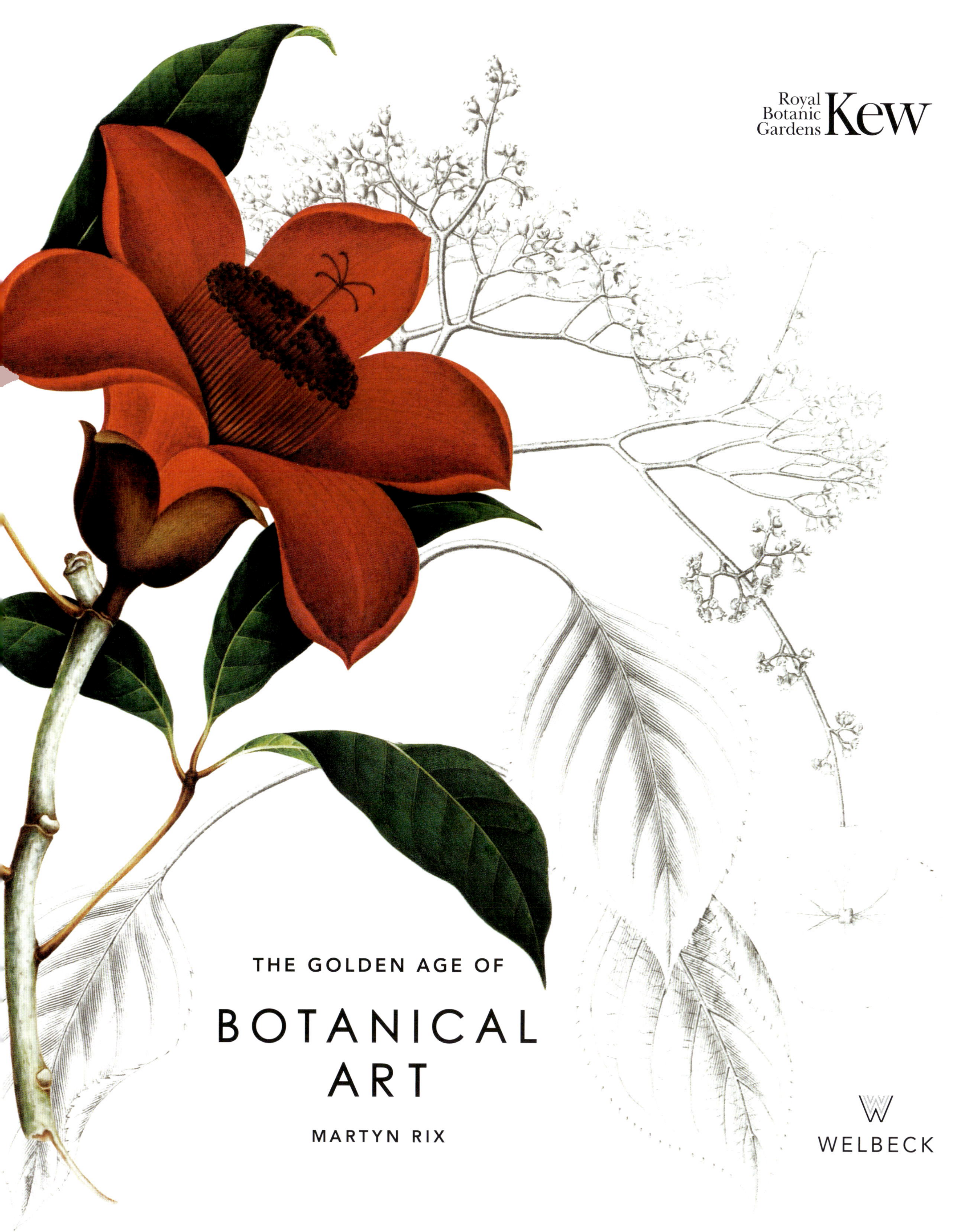

Royal Botanic Gardens Kew

THE GOLDEN AGE OF

BOTANICAL ART

MARTYN RIX

WELBECK

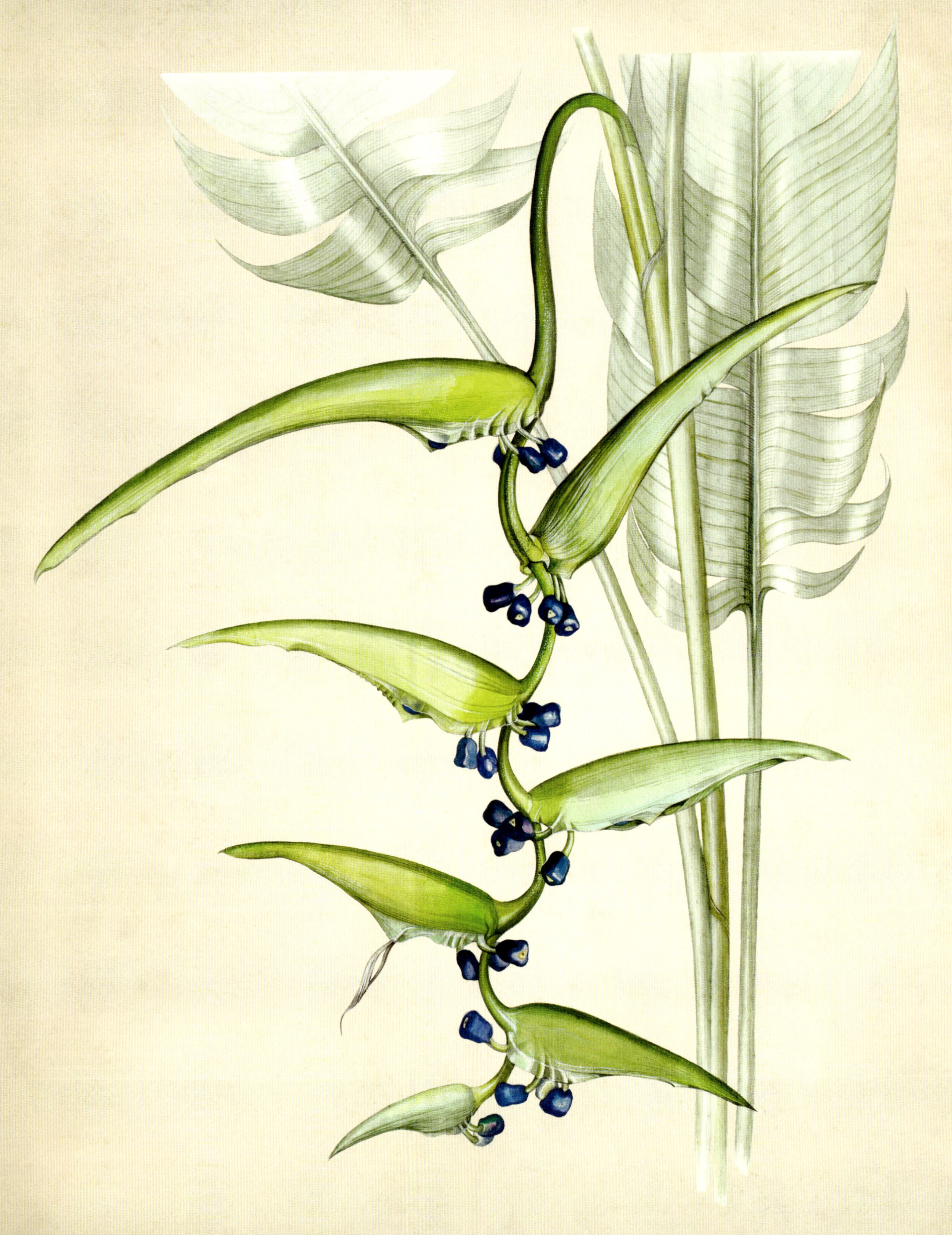

CONTENTS

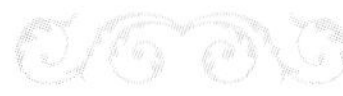

Introduction 8

1 The Origins of Botanical Art 10
Leonardo da Vinci 20
2 Early Works of the Sixteenth Century 22
Jacopo Ligozzi 32
3 Seventeenth-Century Florilegia 34
Dutch Flower Paintings 46
4 North American Plants 48
Linnaeus and Plant Classification 60
5 Travellers to the Levant 62
Maria Sybilla Merian 74
6 The Exploration of Russia and Japan 76
Les Vélins du Muséum 86
7 Botany Bay and Beyond 88
Sir Joseph Banks 98
8 The Golden Age in England 100
Mrs Delany and her Paper Mosaicks 112
9 South American Adventures 114
Thornton's The Temple of Flora, or Garden of Nature 126
10 The Golden Age in France 128
Empress Joséphine 144
11 Botanical and Horticultural Illustrated Journals 146
Henry C. Andrews 156
12 Early Chinese Plant Drawings 158
Père David and the French Missionaries 168
13 The Company School in India 170
The Story of Flora Danica 1761–1883 186
14 A New Era at Kew 188
George Maw 204
15 Victorian Travellers 206
Elwes and the Genus Lilium 216
16 Bringing China to Europe 218
Modern Florilegia 228
17 The Flowers of War and Beyond 230
Exhibiting Botanical Watercolours 240
18 Carrying on the Tradition 242

Index 250
Bibliography 255
Publishers' Credits 256

OPPOSITE: Heliconia "uaupensis", *an unamed species, by Margaret Mee.*

FOLLOWING PAGES: Victoria amazonica *from Lindley's* Victoria Regia.

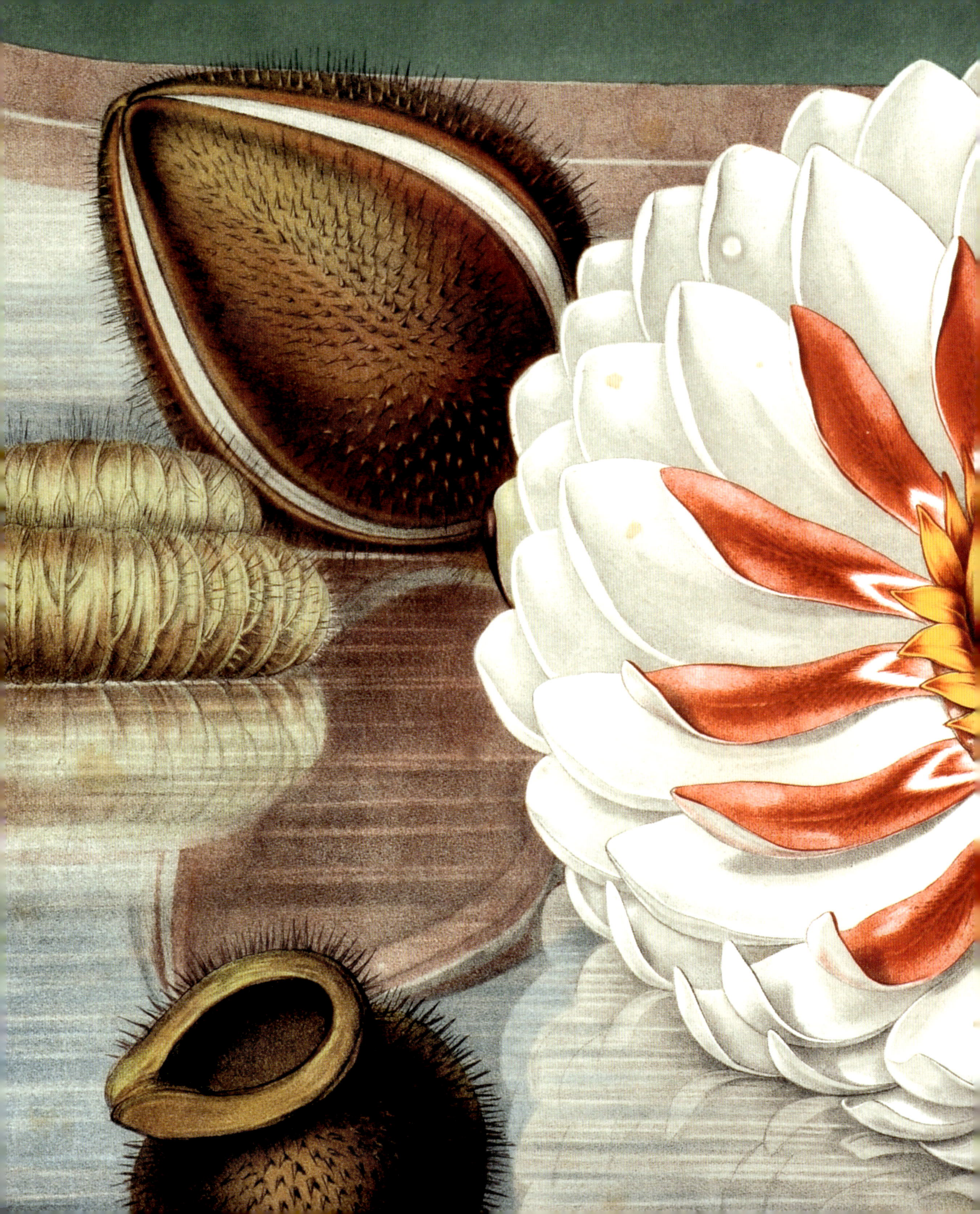

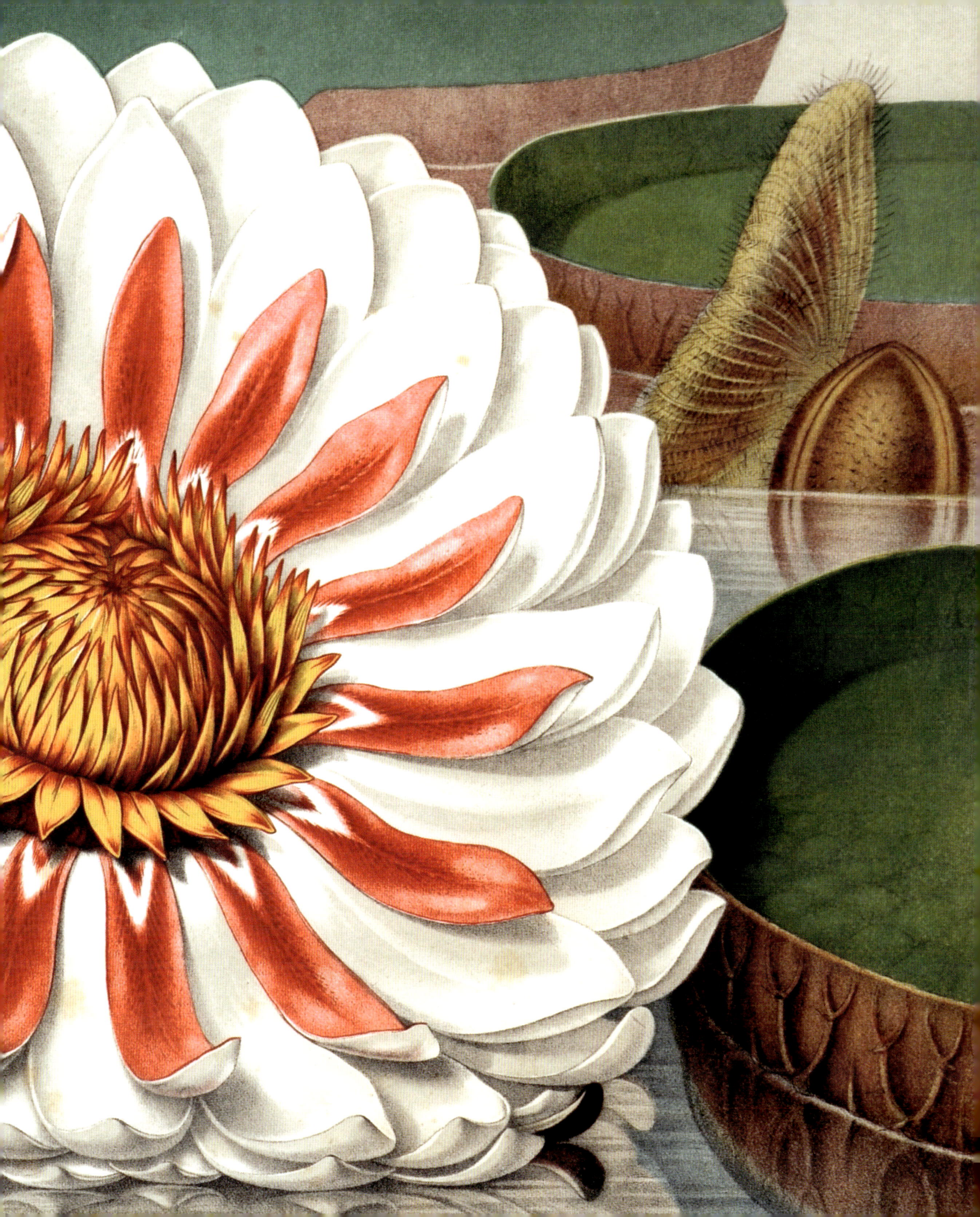

INTRODUCTION

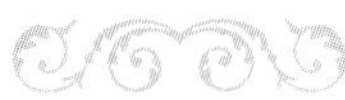

Botanical illustration reached its first peak of sophistication and achievement in the hundred or so years from 1750 to around 1850. This was a period of great discoveries in biology and other sciences, of advances in printing techniques, and of increasing wealth in Europe, so that beautiful books could be produced and find sponsors and buyers. Artistic paintings of flowers had been made from the late seventeenth century, but without publication they remained the private property of royal collections in France, England, Germany, Russia or Austria, or of a few, usually noble, patrons and collectors, such as the Earl of Derby and the Duchess of Portland in England.

With improvements in printing, initially in engraving and later in lithography, beautiful illustrated botanical books were produced. Botanical art, discovery and science were combined in Sibthorp's *Flora Graeca*, published in parts between 1806 and 1840, and containing nearly a thousand hand-coloured, copper-plate engravings of plants collected by the author in Greece and Turkey; financed by Sibthorp's estate, it is still one of the most expensive botanical books ever printed, and still one of the most beautiful.

In France, the artist Pierre-Joseph Redouté, financed by Napoléon's wife, the Empress Joséphine, produced illustrated books about her collection of garden plants, especially lilies and roses; these were printed by a novel method, stipple engraving, in which thousands of small specks were engraved onto the plate, and then filled with ink of different colours before being put through the press. This produced a soft, translucent effect and required less hand-colouring than did line engraving. *Les Liliacées* was published between 1802 and 1816 and *Les Roses* between 1817 and 1824. Redouté continued to use this method until 1833.

Meanwhile, in England, W. H. Fitch had mastered the art of plant illustration by lithography, which produces a very delicate outline and shading as a basis for hand-colouring with watercolour. Fitch worked for William Hooker and his son Joseph, at a time when they were developing and expanding the collections and herbarium at the Royal Botanic Gardens, Kew. He drew with great fluidity and confidence, but at the same time with total botanical accuracy, and he was adept at lithographing and improving the paintings of others. One of his finest works is *Illustrations of Himalayan Plants* (1855) with text by Joseph Hooker, in which he reproduced paintings made by Indian artists. Hand-coloured lithography continued to be used for books until the advent of colour printing at the end of the nineteenth century.

What is the difference between botanical art and botanical illustration? In art, the finished painting is the whole object of the artist, and it has no further purpose than to be admired. A botanical illustration has a scientific purpose, to illustrate a book or act as a record of a plant species or plant part. The illustration should have a generality that ignores the imperfections of the individual specimen, and so can represent a species or particular form of a species. This assumes some botanical knowledge on the part of the artist, in selecting as typical a specimen as possible and knowing which imperfections to ignore. In the best botanical illustration the artistic aspect is not lessened by the scientific purpose. This book traces the development of botanical art and illustration from the earliest times until the present day, particularly through the book collections in the library and the art collection of the Royal Botanic Gardens, Kew.

Will there be a second golden age of botanical illustration, and is that age with us now? We are continuing to explore and discover plants around the world, and the destruction of so many vulnerable habitats, particularly in the tropics, has given a new sense of urgency to this work. Though we have fewer botanists, we have more botanical artists than ever before, and they have ever-increasing opportunities to show their work in international exhibitions and sell their work through the Internet. Through the illustrations in *Curtis's Botanical Magazine*, our present-day paintings can be compared directly with those of artists from the late eighteenth century, and can be seen to be their equals. Printing of delicate watercolour paintings is still imperfect, but reproduction by electronic means is fast becoming so accurate and cheap that perhaps the time will come soon when large, illustrated botanical books will be published again, on demand, yet now at a price which the ordinary, enthusiastic plant-lover can afford.

Martyn Rix

OPPOSITE: Camellia reticulata *from* The Botanist.

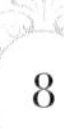

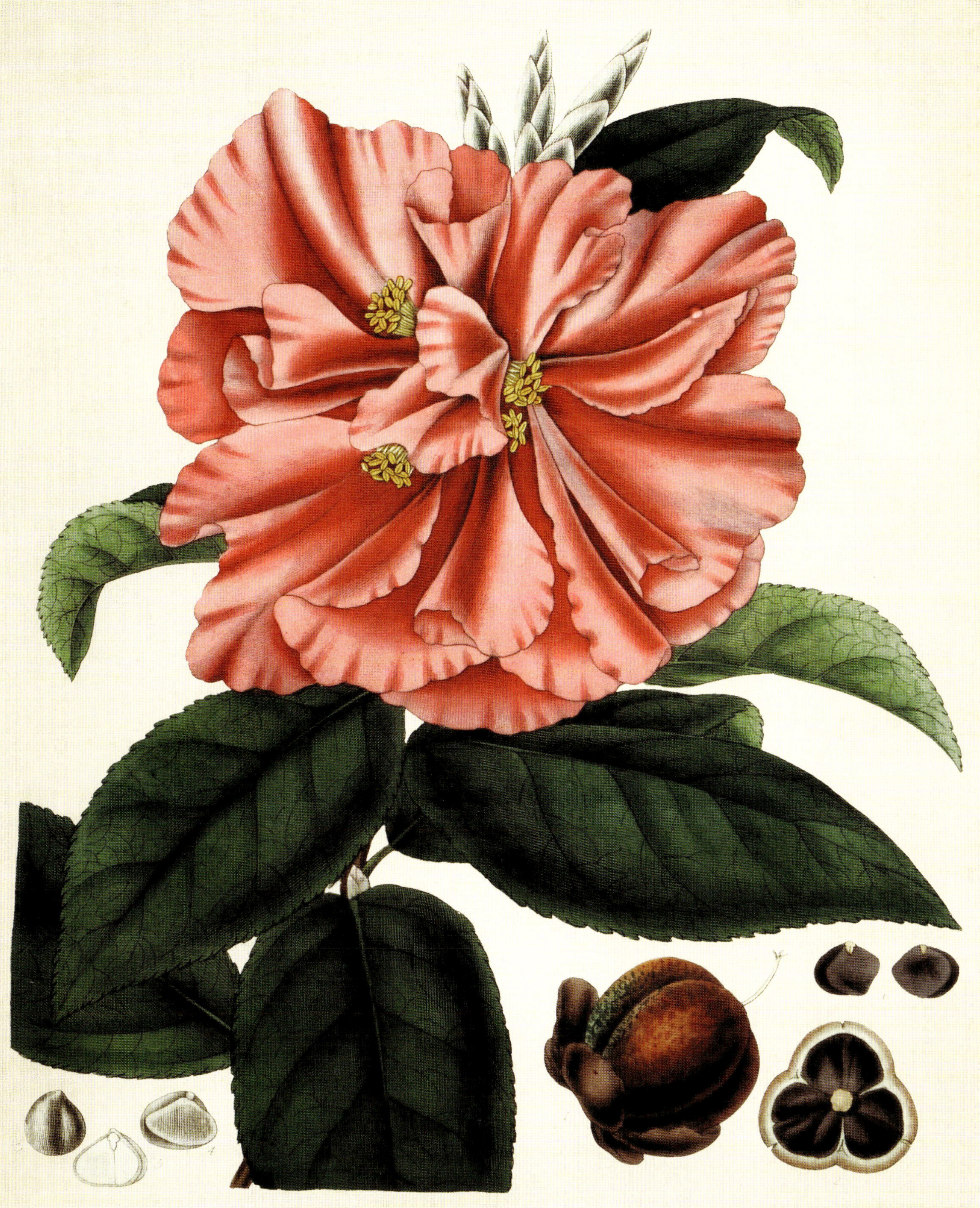

1 THE ORIGINS OF BOTANICAL ART

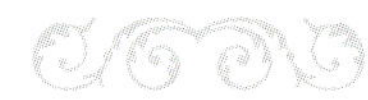

BOTANICAL ART

The representation of flowers in art has a long history. Flower paintings were originally made for two main reasons: as decoration, or as a means of identifying plants used for medicine. The earliest surviving good representations of flowers date from the late Minoan period in Crete and the eastern Mediterranean, around 1700 years ago, and were the product of a rich palace culture, with lilies, saffron and other flowers used to decorate pots and wonderful painted rooms. These depictions are not as old as paintings of animals, where the earliest cave paintings of bulls, aurochs and humans date from the end of the last Ice Age around 20,000 years ago.

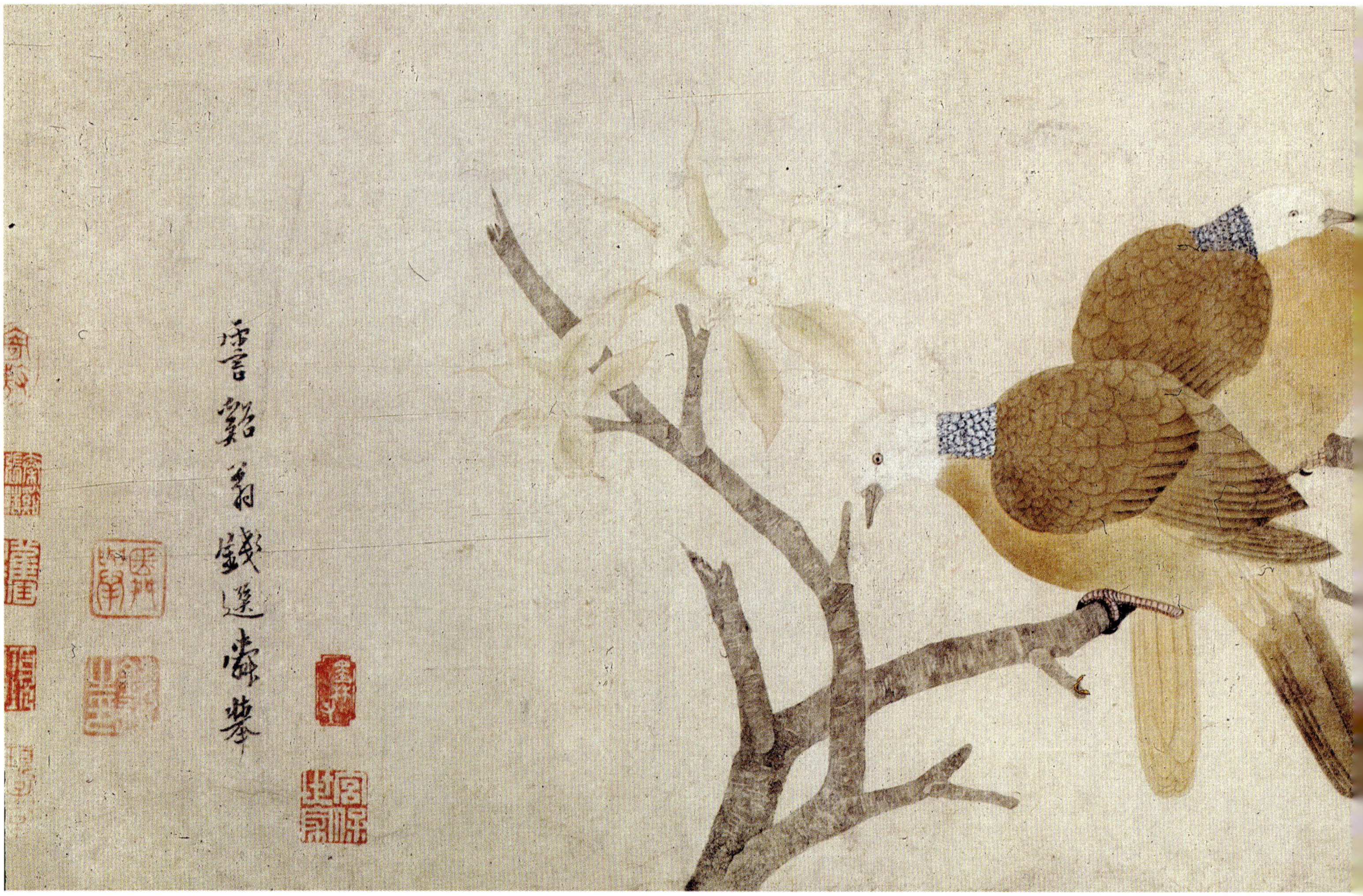

In China there are records of herbs used in medicine around 4500 years ago, but their knowledge did not spread to the west. During the classical period, in Greece, Asia Minor and Babylon, medicine began to be used by specialists, and herbals were produced, with drawings and descriptions of plants used for treatment. The first herbals that survive date from around 2000 years ago, but were based on the works of previous writers, such as Theophrastus who was working circa 300 years earlier. These classical drawings of plants, such as the great herbal of Dioscorides, continued to be copied until the Middle Ages. In the sixteenth and seventeenth centuries, herbals merged into books of flowers grown for decoration, so-called "florilegia". Pictures of the earliest European garden flowers can also be seen in paintings from the early fifteenth century, but are usually shown for their religious significance; lilies, roses, irises and carnations are commonly shown.

The earliest flower paintings to survive in which the flower itself is the main object are probably those done by Chinese painters in the thirteenth century; these were usually as hand scrolls, and show flowers and often fruit and birds as well. The scented *Narcissus tazetta*, which is early flowering and was a treasured import from Mediterranean Europe, is a frequent subject; its leaves and flowers appear to dance across the page.

In Europe, the earliest pure flower paintings to survive are by Albrecht Dürer; his famous depiction of *Das großer Rasenstück*, the big bit of turf, and a painting of peonies date from circa 1503. Two generations later, in about 1580 in Florence, Jacopo Ligozzi painted a series of realistic flowers, which are as accurate and artistic as anything depicted until the eighteenth century.

The period between the mid-eighteenth and the mid-nineteenth centuries was the golden age of botanical art, when rich patrons, skilled artists, new plant discoveries and a scientific desire for knowledge coincided. The flower paintings and wonderfully illustrated botanical books created during this century, are the main theme of this present book.

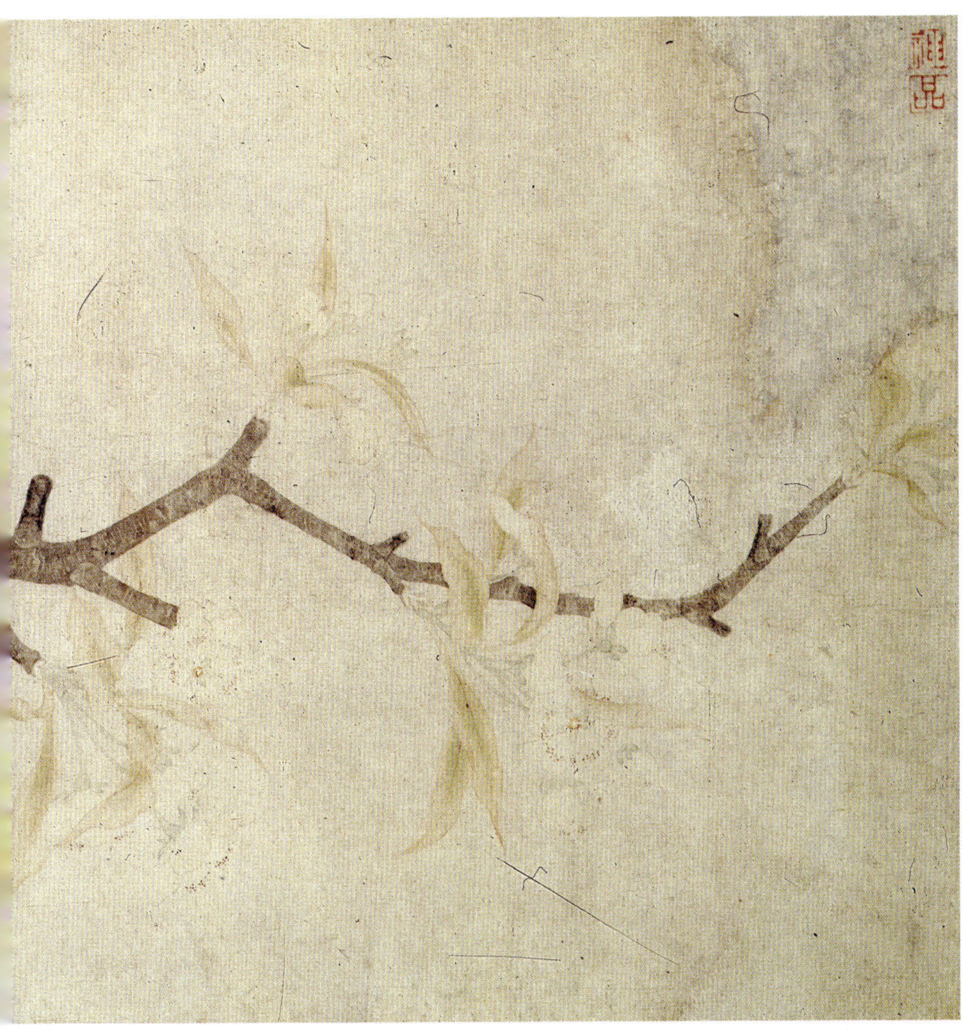

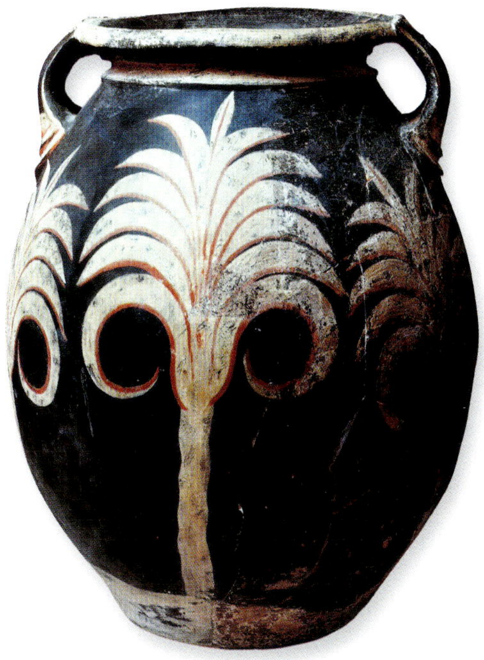

LEFT: A depiction made in ink and colour wash on paper of doves and pear blossom after rain. Produced during the Yuan Dynasty by Qian Xuan (1235–1305).

ABOVE: A vase from the Minoan Civilization (thirteenth century BC). Decorated with palm trees, it was discovered in the ruins of Knossos on the Greek island of Crete. Artist unknown.

THE PRE-CLASSICAL ERA

The frightful volcanic eruptions and earthquakes that destroyed the Minoan civilization in Crete and the eastern Mediterranean have allowed us a remarkable insight into both their culture and the decoration of their houses. When Sir Arthur Evans excavated the palace of King Minos at Knossos in Crete in around 1900, he found the remains of wall paintings, and was able to re-erect and restore some of them. Apart from the famous painting of the man leaping over a bull, there are others that show the flowers which might have been brought back to the palace or grown in the palace gardens; a fine clump of Madonna lilies (*Lilium candidum*) can be seen in the museum at Heraklion, taken from a villa in Amnissos, the port for Knossos: it dates from around 1600 BC. The saffron crocus is also shown in a Cretan wall painting, with the long styles that are the source of the flavouring and the dye, clearly protruding from the flower. The night-scented sea daffodil (*Pancratium maritimum*), familiar to anyone who uses a sandy Mediterranean beach in the summer, is commonly depicted, on frescoes, on a Mycenean sword blade and on a Cretan sarcophagus. Roses are shown, too, and more flowers can be seen on decorated Minoan pottery, showing that a love for of plants was a feature of their civilization.

The surviving wall paintings on Santorini (formerly called Thera) are even more remarkable; the city of Akrotiri was buried by a cloud of pumice in a violent eruption that is now thought to have occurred between 1627 and 1600 BC. Since 1967 the city has been excavated, and many frescoes have been discovered in a remarkable state of preservation. *The Saffron Gatherers* shows two girls picking the flowers from a rock, and in the background clumps of saffron plants seem to have been planted in rows, an early example of the cultivation of this valuable herb. In the *House of the Ladies* there is a fine representation of two clumps of sea daffodils, showing the stamens attached to the corona.

The most spectacular of all ancient flower paintings is now in the National Museum in Athens. A whole room is painted with a single scene: large clumps of lilies emerge from a rocky, weather-worn limestone landscape, and swallows swoop around the sky; the lily flowers are painted bright red, and there has been some discussion that these are the red-flowered *Lilium chalcedonicum*, from mainland Greece, rather than the white-flowered *Lilium candidum* which is the one usually shown in Minoan art. The same red lilies are shown elsewhere arranged in a large decorated vase.

A much later volcanic disaster has given us an insight into Roman painting. Pliny the Elder died during the eruption of Mount Vesuvius in 79 AD, but his nephew escaped and described the destruction of Pompeii, which was buried, like Thera, in pumice-like ash. The House of the Vettii, with its numerous frescoes therefore survived, and one of these shows a garden scene, with a red *Rosa gallica*, daisies and poppies.

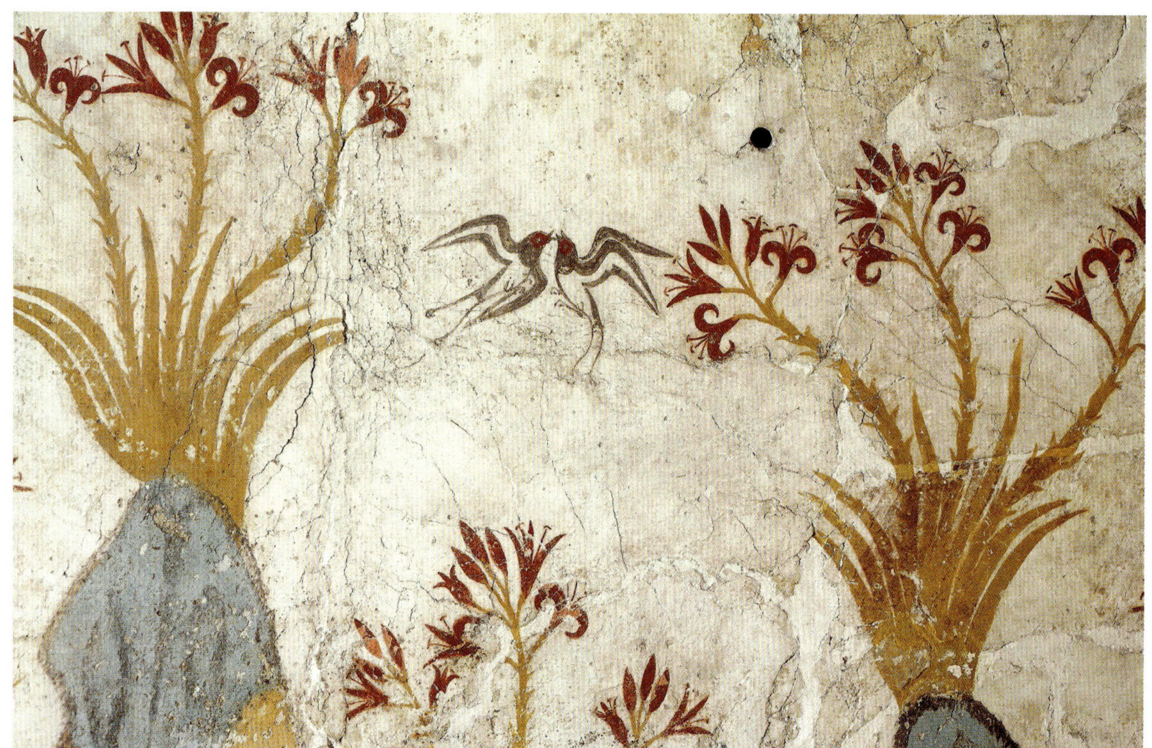

OPPOSITE: A fresco of the "Lily Prince", as he has been called, from the Palace of King Minos in Knossos (thirteenth century BC). Lilies, or possibly irises, can be seen to the left and right of him. Artist unknown.

LEFT: Part of the fresco decorating a room discovered at the Bronze Age settlement of Akrotiri on the Greek island of Santorini. It dates from the sixteenth century BC. Artist unknown.

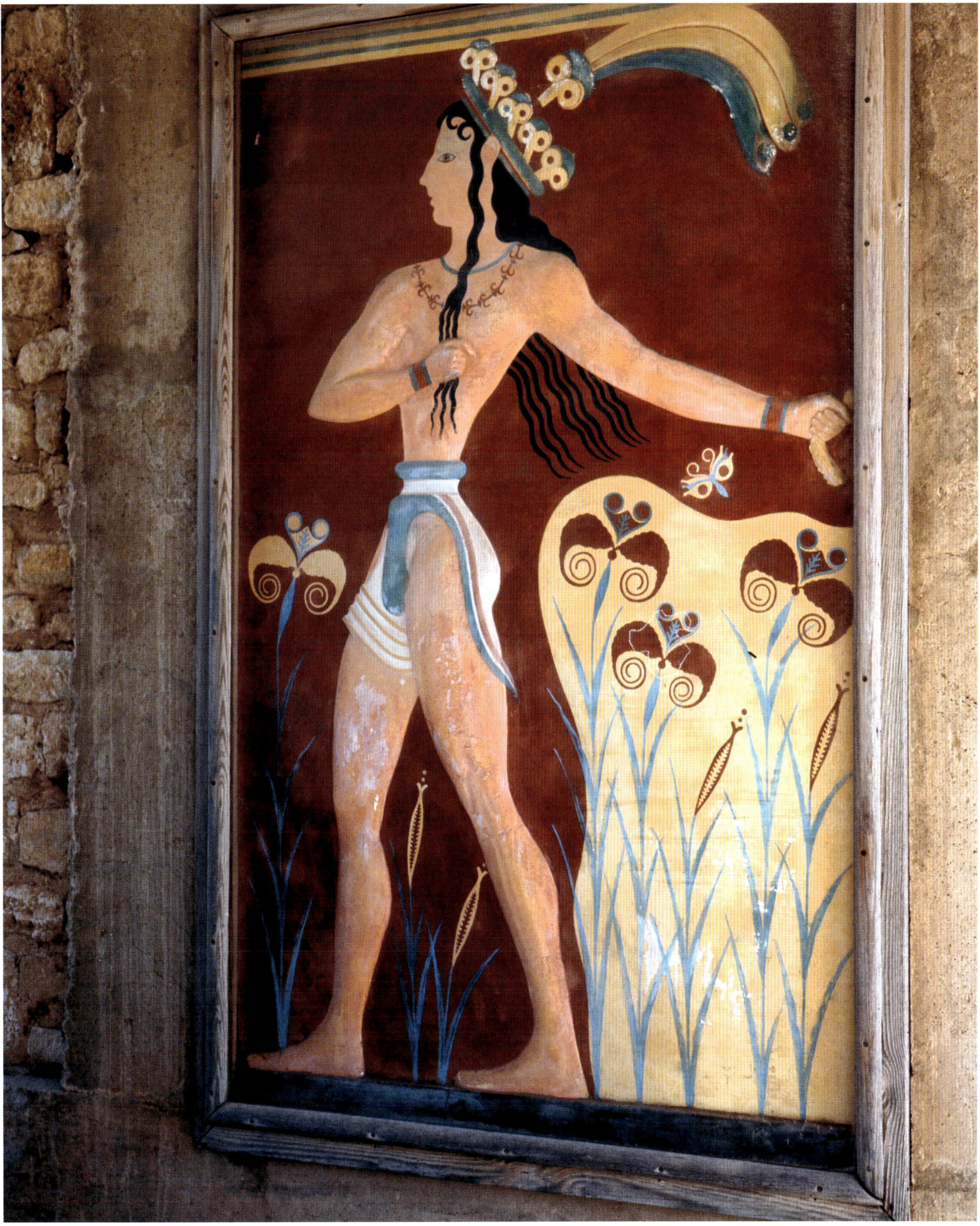

ANCIENT HERBALS

The earliest illustrated botanical books were herbals, designed to help the reader identify medicinal plants and understand their uses. Pliny the Elder mentions illustrated herbals, and the name of one of the most famous botanists, herbalists and painters of the first century BC, Cratevus, whose work became connected with the most famous herbal in existence, the *Codex Aniciae Iulianae picturis illustratus, nunc Vindobonensis Med. Gr. I*, to give it a full title, shortened to *Aniciae Iulianae Codex*. The text is mostly by Pedanios Dioscorides who was born in Anazarbus, north of present-day Çeyhan in Turkey in around 40 AD. He travelled widely in Asia Minor, as a doctor in the Roman army, and based his work on his own experience and the writings of others such as Cratevus, who was also physician to Mithridates VI of Pontus. The Vienna copy of the manuscript was made in 512, for Anicia Juliana, daughter of the Emperor Olybrius, who donated a church to the citizens of the Honorata district of Constantinople (now Istanbul); the copy was probably based on manuscripts owned by her great-grandfather, Theodosius II, Byzantine Emperor in 425, who was renowned for his learning and knowledge of herbal medicine. There are 385 plant pages on parchment, rather brown but otherwise in good condition, and it is known that the manuscript was restored and rebound in 1406 in a monastery in Constantinople. The illustrations are often very well painted and most are instantly recognizable; many of the names used are still correct today. The text of Dioscorides is written in Greek around the painting, and has been annotated in Arabic, probably by Turkish physicians who used it between the fall of Constantinople in 1485 and its purchase for the Holy Roman Emperor in Vienna in 1569.

The influence of this herbal was remarkable: it was copied again and again, initially by Greeks in Constantinople, and later by Arabic scholars who were often Nestorians, working in Baghdad for the Caliph: a good Arabic example is in Leiden (UNIV. Cod. Or. 289). Even later it was translated into Latin, but, with each copying the illustrations tended to become cruder and less accurate.

In the Early Renaissance a few Italian herbals were illustrated by new paintings; some are beautiful and good representations of the flowers described, even if the text is still basically that of Dioscorides. One example is the Egerton MS 747 in the British Library, which was produced in Salerno in around 1300. Most of the wild flowers, which the artist would have known personally, are well painted, but exotic ones, such as Dracunculus, the dragon arum, are stylized and the painter cannot resist showing a snake twining up its stem. The Carrara herbal (Sloane MS 2020), whose text is an Italian translation of an Arabic medical treatise, painted in Padua in around 1400, has elegant, original paintings swirling across the page, but these two examples are the exceptions, and even early printed herbals had debased illustrations, derived from repeated, careless copying.

LEFT: A depiction of a fumitory from the Codex Vindobonenis.

OPPOSITE: A page from the Codex Vindobonenis, *showing a sea squill bulb (*Drimia maritima*) surrounded by Greek writing.*

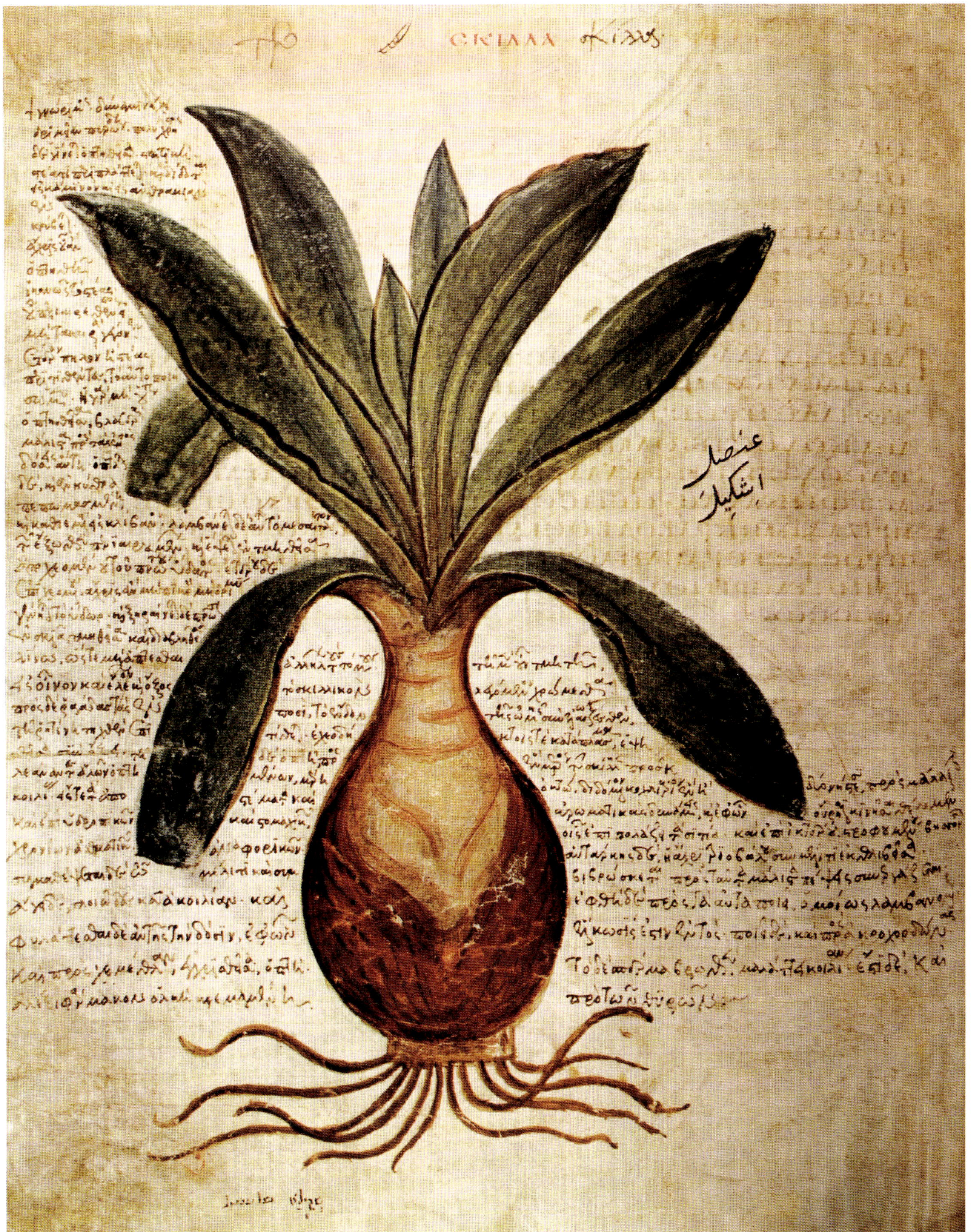
CKIΛΛΑ

FLOWERS IN RENAISSANCE PAINTING

During the early Renaissance, in the fifteenth century, painting became more realistic, discarding the stylized figures that are such a feature of Byzantine art. The figures, though still mainly of religious subjects, became more naturalistic, and gave the artist scope for scenery and other objects to be placed in the background of the painting. Two main branches of art from this period show flowers: ordinary paintings, often produced as altarpieces, and books of hours, delicately illuminated manuscripts of the psalms or prayer books. The flowers are clearly painted from nature, and are the common cottage garden flowers that the artist would have known.

One of the earliest floral altarpieces is in St Bavo's Cathedral in Ghent. It was painted in about 1430 by Hubert and Jan van Eyck (fl. c. 1420–85), and depicts the adoration of the Lamb: part of the background shows the corner of a rather unkempt garden, with a fig tree, and hedges of red gallica roses and trailing vines. The lawn is planted as a flower meadow with dandelions, daisies, wild strawberries, lady's smock, solomon's seal, rue, valerian and rose campion. The cathedral itself is shown in the background. Other parts of the altarpiece show Madonna lilies, irises, a peony, columbines, woodruff, primrose and ivy.

A second famous altarpiece, by the Flemish painter Hugo van der Goes, (c. 1440–82), is now in Florence. It was commissioned for the church of the hospital of Santa Maria Nuova in Florence by the Italian banker Tommaso Portinari, a descendent of the hospital's founder and agent of the Medici in Bruges in around 1475. The Virgin kneels, looking at the infant Jesus, and surrounded by angels and groups of ladies and friars. The vases of flowers in the foreground have special significance: the blue iris is a royal flower, often associated with the Queen of Heaven, and the white Florentine iris was also associated with the Virgin. The columbine, so-called because the petals were thought to look like a dove, symbolizes the Holy Ghost, while the orange lily is also a royal flower. Elsewhere, violets are strewn on the ground.

The painter Martin Schongauer (1469–91), a contemporary of Leonardo da Vinci, is known for only one major painting with flowers, but several other are attributed to his pupils or followers. His altarpiece, *The Madonna of the Rose Garden* of 1473, in the Église de St Martin, in Colmar, shows very natural red peonies and red *Rosa gallica*, which would have been popular garden flowers at that time.

A remarkable survival is a study of the peonies made for this painting, which is now in the Jean Paul Getty Museum in Malibu. It is on paper, watermarked around 1470 and was bought by the museum in 1992. At first it was thought to be by a follower of Dürer, but Fritz Korney, the curator of drawings at the Albertina in Vienna recognized it as the peony flowers in Schongauer's painting. Dürer went to visit Schongauer in Colmar but by the time he arrived, Schongauer had died. Dürer is said to have bought several paintings from Schongauer's brother, and it may be that this peony inspired Dürer's very similar peony painting, dated 1503, now in Bremen. As W. T. Stearn has pointed out, Schongauer's painting is remarkable for showing, with perfect botanical accuracy, the progression from leaves to sepals at the back of the flower.

BELOW LEFT: An orange lily, irises, columbines and carnations, with scattered violets, depicted in the central panel of the Portinari Altarpiece, which was painted in oil on panels by Hugo van der Goes c. 1479.

FOLLOWING PAGES: The Adoration of the Mystic Lamb, *which forms the lower half of the Ghent Altarpiece by Hubert and Jan van Eyck. Painted in oil on panels, it was completed in 1432.*

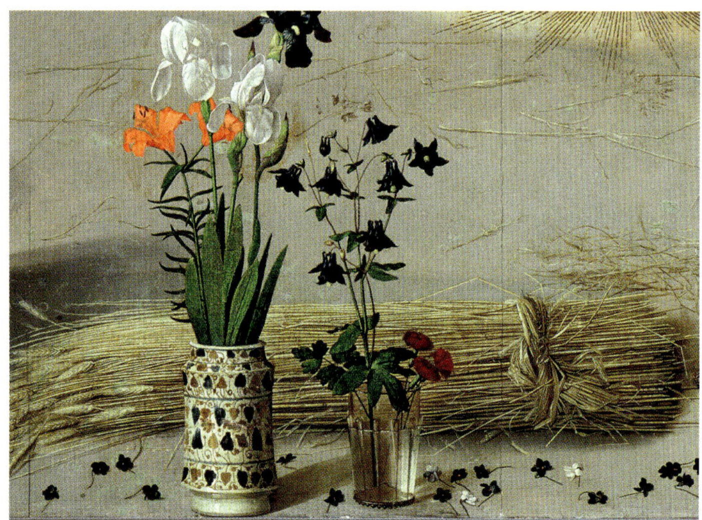

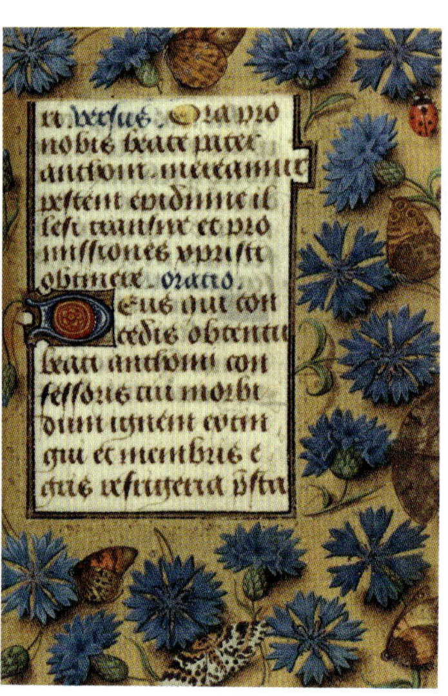

et versus. Ora pro
nobis beate pater
anthoni. mereamur
pestem condigne il
lesi transire et pro
missiones christi
obtinere. Oratio.
Deus qui con
cedis obtentu
beati anthonii con
fessoris tui morbi
dum ignem extin
gui et membris e
gris refrigeria presta

LEFT: A page featuring cornflowers from an illuminated book of hours, which formed part of the library of Francis Douce (1757–1834), who was at one time Keeper of Manuscripts at the British Museum.

*OPPOSITE: Samples of Old English lavender (*Lavandula × intermedia*) painted by Jacques Le Moyne c. 1568.*

A painting of *Mary in a Garden with Cherry Tree*, now in the National Gallery in London, is catalogued as school of Martin Schongauer. It is one of the earliest garden scenes, showing a picket fence surrounding a flowery meadow of irises, carnations, strawberries, lily of the valley, stocks, soapwort, ragged robin and plantains, all of which might be expected in a garden in northern Europe in the fifteenth century.

Another altarpiece in the National Gallery, dated to circa 1500, and probably French, shows the story of St Giles and the Hind. The King of France and a bishop kneel before St Giles, who is holding a pet hind, after a member of the royal hunt had shot at the deer, but wounded the saint. The flowers in this painting are brilliantly drawn, and perhaps show the hand of Schongauer as well. The woolly mullein on the right has its leaves tapering into the stem, and the other flowers – greater celandine, irises, mallow and a wild rose in the background – are easy to identify.

Illuminated books of hours are the most charming miniature manuscripts, and give another insight into the flowers grown in gardens in northern Europe in the fifteenth and early sixteenth centuries. The Hastings Hours, painted by the Master of Mary of Burgundy, now in the British Library, dates from around 1480; it has several pages with floral borders, showing wild roses, pinks, garden peas, speedwell and forget-me-not, while another page has irises, heartsease and a variety of insects. Some pages have text, others illustrations such as the Flight into Egypt. A beautiful book of hours in the Bodleian Library in Oxford (Douce) contains many pages where the main image is surrounded by borders strewn with all kinds of flowers.

One of the most famous artists in this period was Jean Bourdichon (1457–1521), who made illuminated books for Louis XII, King of France from 1482 to 1515. Some of the borders of his manuscripts, now in the British Library, show very detailed drawings of flowers, which appear, by the clever use of shading, to be strewn around the page.

An interesting album of flower drawings by another French artist, Jacques Le Moyne (c. 1533–88), is now in the British Museum, with others in the Victoria and Albert Museum. Le Moyne was a Huguenot, and was sent on a disastrous expedition to survey Florida in 1564; the Spanish overran the colony and massacred most of the French, though Le Moyne escaped, to spend his later years in England. His flower paintings, mostly done before and after his visit to America, are detailed and accurate; some are formal, in the tradition of the miniatures of Bourdichon, others, in delicate watercolour, are true to nature, like those of Dürer's pupil Hans Weiditz.

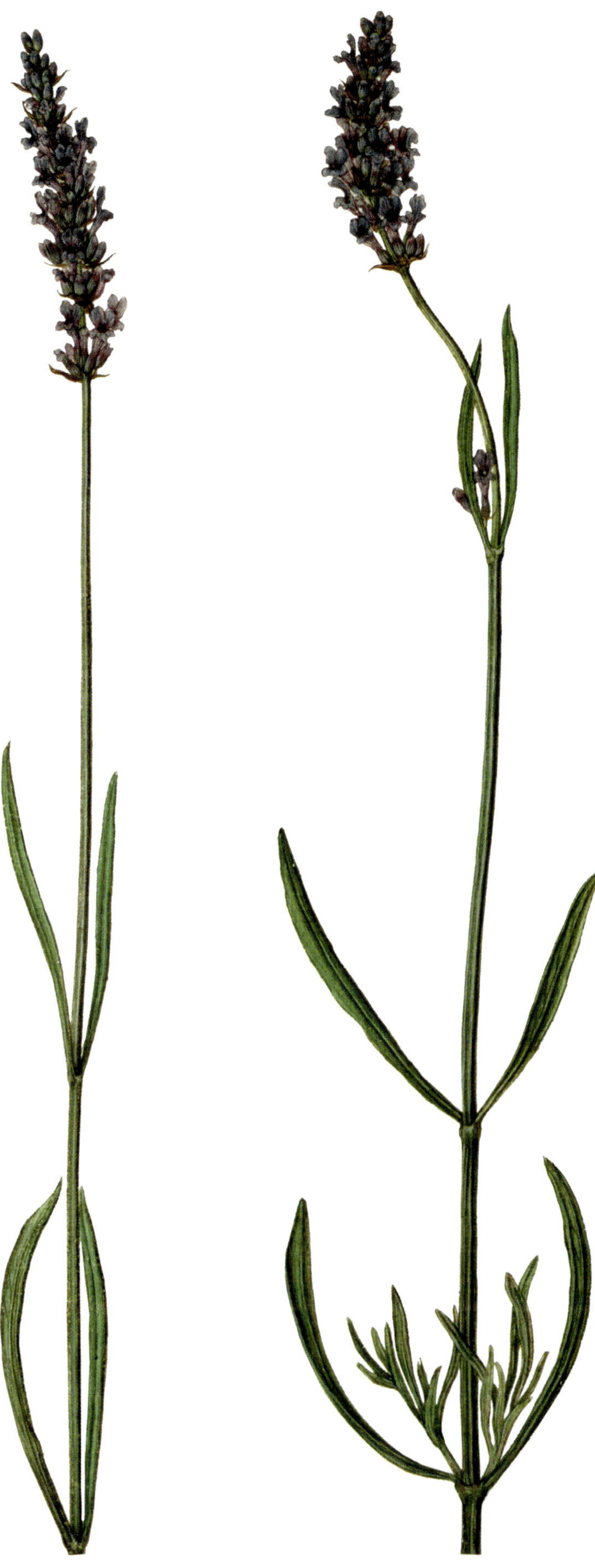

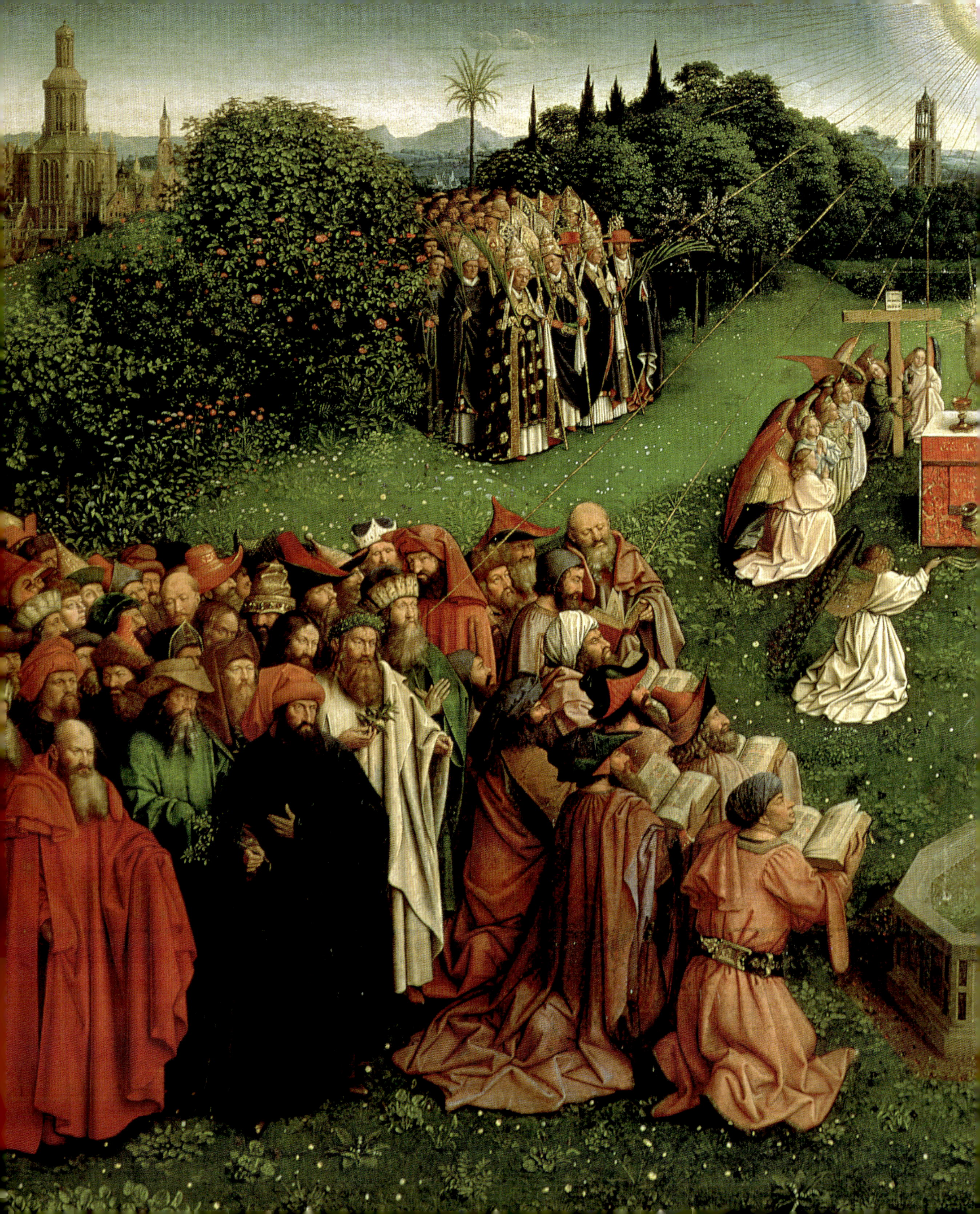

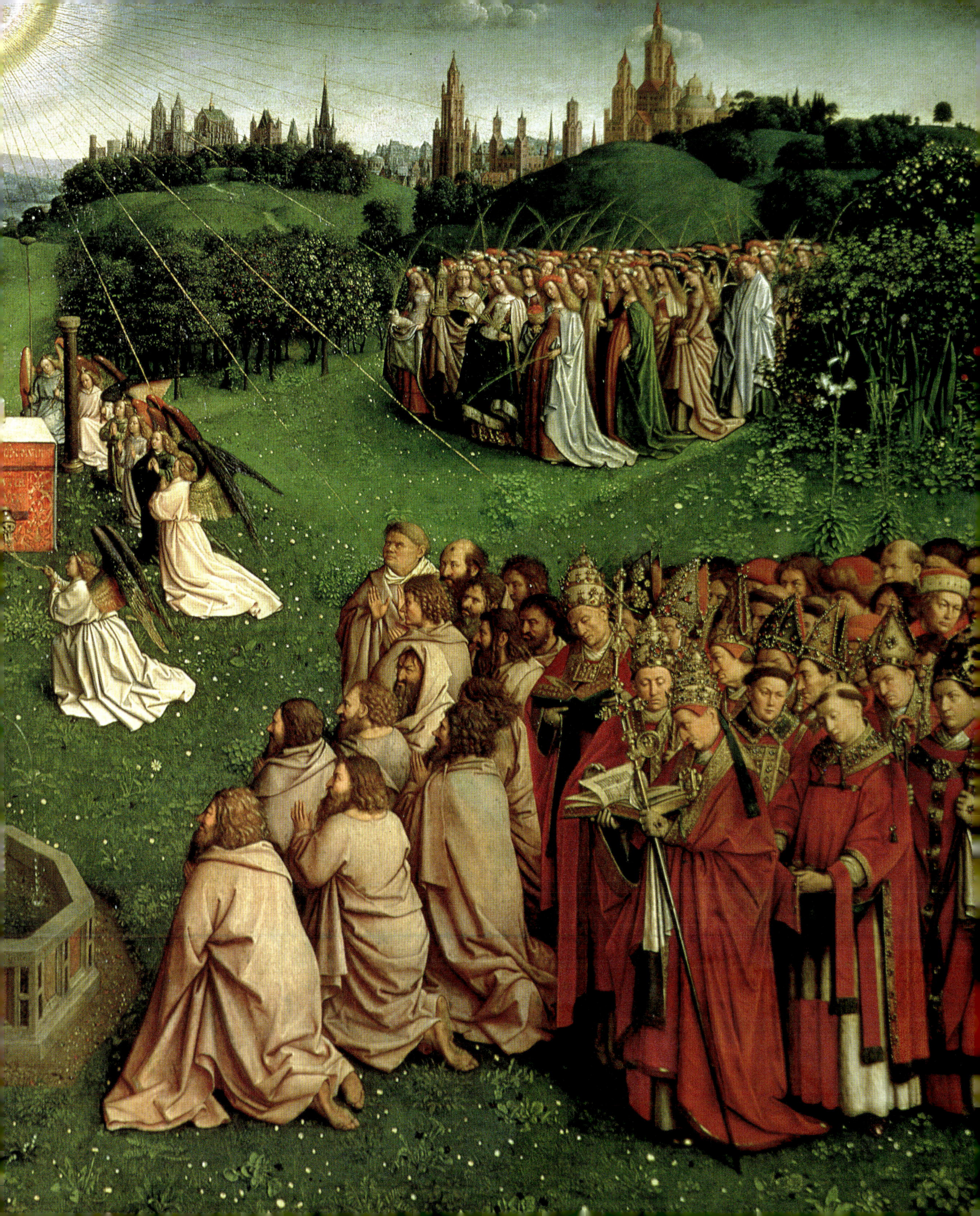

Leonardo da Vinci

Leonardo da Vinci was born in April 1452 on the outskirts of the small town of Vinci near Florence, and was apprenticed to the painter and sculptor Verrocchio in the city. His earliest known work is *The Annunciation* and was painted circa 1472–73; it remains in Florence (in the Uffizi Gallery) and shows the angel Gabriel kneeling on a flowery lawn, greeting Mary with a raised hand; a Madonna lily is growing in the background.

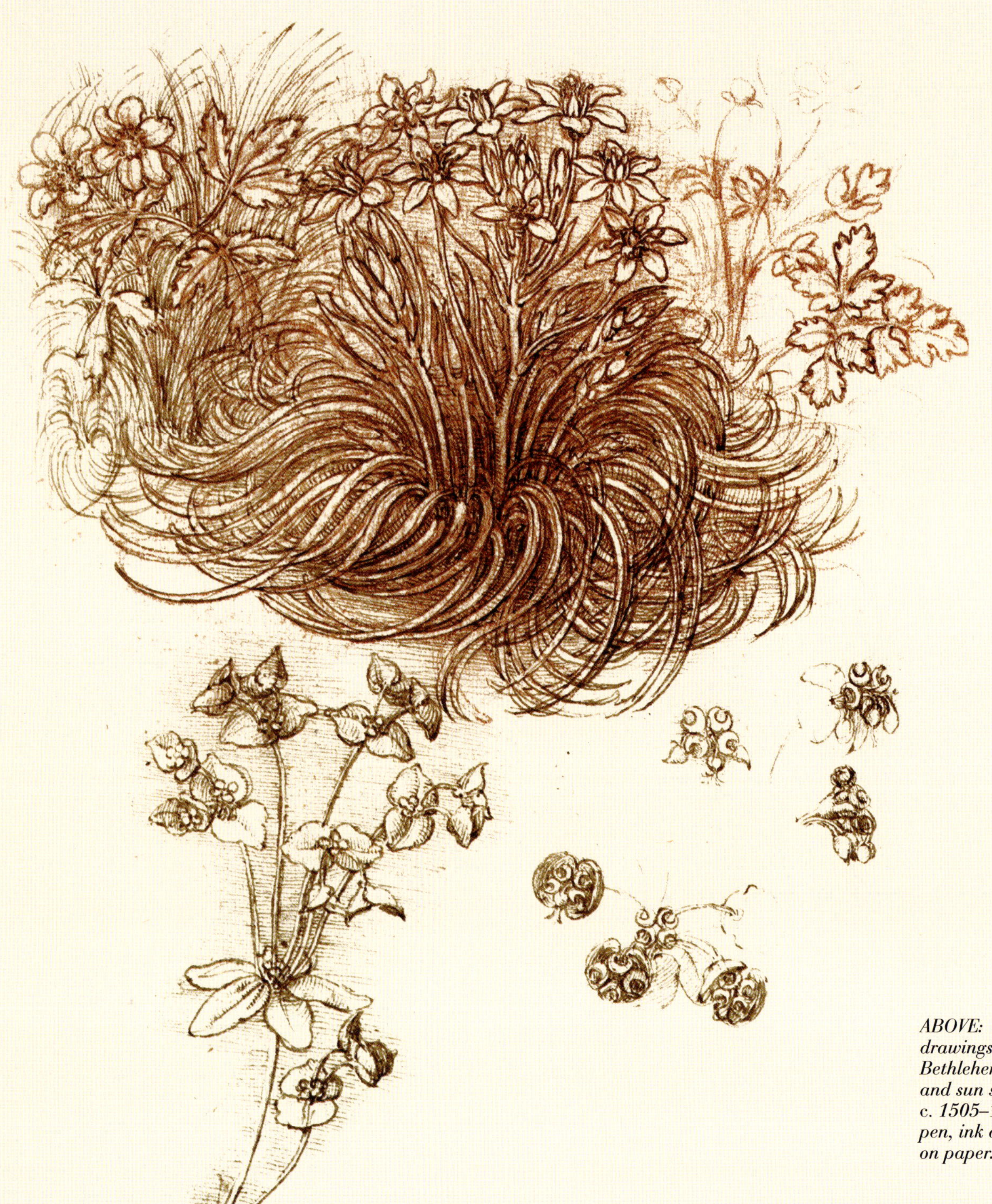

ABOVE: Leonardo's drawings of star of Bethlehem, wood anemone and sun spurge from c. 1505–10, made in pen, ink and red chalk on paper.

The custom of painting flowers in religious paintings started in the fifteenth century, and continued throughout the Renaissance. The flowers were not only for decoration, but had a deeper sybolism. The Madonna was commonly shown with a blue iris, as well as a white lily, and was also associated with both red and white roses. In *The Madonna of the Carnation* (1477–78), now in Munich, Leonardo shows Mary holding a dark red double carnation. Another Virgin and Child, now in St Petersburg, shows Mary holding a four-petalled flower, probably a scented stock.

Leonardo moved to Milan circa 1483, and it was here that he painted his two famous versions of *The Virgin of the Rocks* for an altarpiece. The painting in the Louvre was probably begun in 1483, and the National Gallery version was probably finished in 1508.

Many of Leonardo's drawings from this period survive: a sketch of a group of violets (c. 1487–90) and a group of drawings of different flowers (c. 1505) are now in the Royal Library at Windsor. These are elegant and accurately observed, and show simple flowers such as wood anemones, kingcups, star of Bethlehem (*Ornithogalum angustifolium*), a flower of spurge, an oak branch, several brambles and bulrushes (*Schoenoplectus* and *Cyperus*), grasses such as Job's tears (*Coix*), as well as the usual Madonna lily.

Though the figures in the two versions of *The Virgin of the Rocks* are very similar, the flowers in the foreground are quite different. In the Louvre version is a clump of irises, always associated with the Virgin, but in the National Gallery version the irises are replaced by a clump of *Narcissus tazetta* whose leaves are much too broad, and three other imaginary flowers. It seems inconceivable that the same hand that sketched the delicate flower drawings could have painted the flowers so badly. Perhaps the answer is that the painting of the foreground was done by an assistant. Leonardo finally received payment for the picture in August 1508, 25 years after it had been commissioned. From 1513 to 1516 he was in Rome in the papal service, where he was described by a courtier as one "of the world's finest painters, ... despises the art for which he has so rare a talent, and has set himself to learn philosophy; and in this has such strange ideas and novel fancies that for all his skill in painting, he could not depict them".

If we were to judge Leonardo only by his main paintings, he would not deserve mention in the history of botanical art, but his drawings at Windsor show that the careful study of flowers was one of the many aspects of his artistic life.

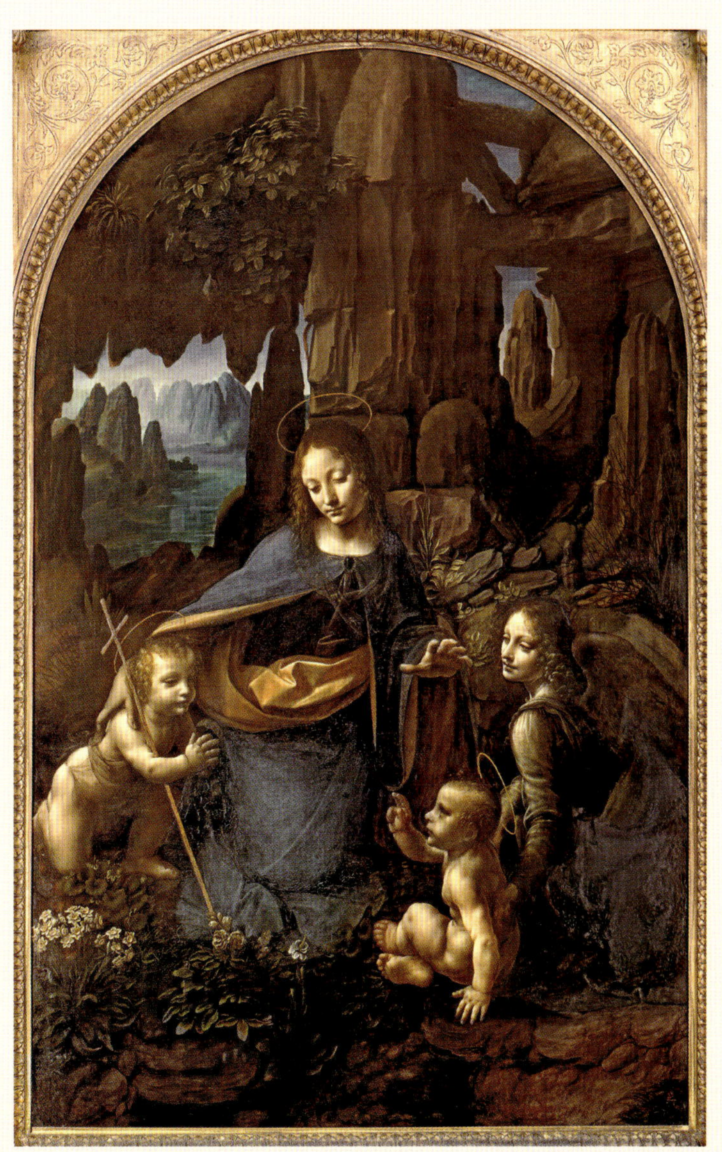

RIGHT (ABOVE): Leonardo's The Virgin of the Rocks *(c. 1508), which is held at the National Gallery in London.*

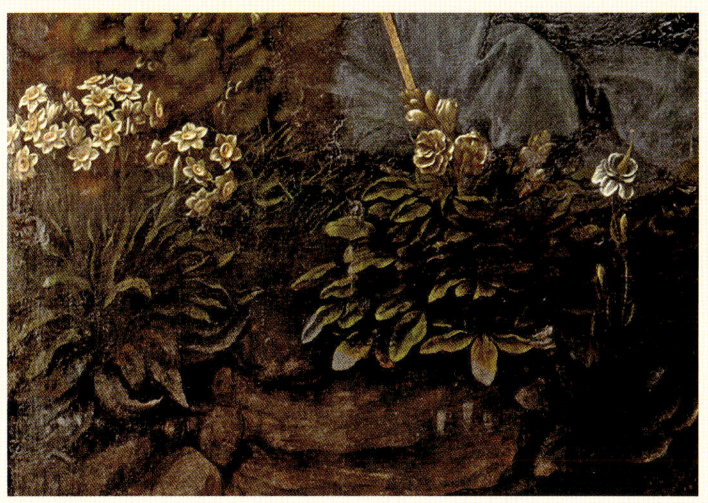

RIGHT (BELOW): A close-up of the plants in the foreground of the painting above, showing the Narcissus tazetta *and the other imaginary flowers.*

2

EARLY WORKS OF THE SIXTEENTH CENTURY

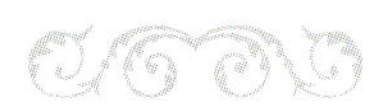

ALBRECHT DÜRER

If Leonardo da Vinci is the most famous Renaissance painter and polymath from southern Europe, Albrecht Dürer is his counterpart in the north. Dürer was born in Nürnberg in 1571, and died there in 1628. His father was a master goldsmith, and his godfather, Anton Koberger, a successful printer and publisher.

Dürer was initially destined to become a goldsmith, but his youthful talent for drawing persuaded his father to send him at the age of 15 to be apprenticed to Michael Wolgemut, an artist and printmaker, who specialized in the woodcuts then used for illustrating books. In 1590, having finished his apprenticeship, Dürer started on his *Wanderjahre*, which involved travelling to visit the workshops of other artists; he went to Colmar to meet Martin Schongauer, then famous as both a painter and engraver. By the time he arrived, Schongauer had just died, and Dürer was greeted by his brothers, from whom he bought several paintings. From there Dürer went to Luxembourg, the Netherlands, and eventually to Venice where he was influenced by Giovanni Bellini, and may have seen the superbly naturalistic painting of Jacopo Ligozzi. He returned to Nürnberg to set up his own atelier in 1495, devoted to the production of woodcuts for printing.

During this period Dürer made a few major paintings and numerous drawings, both of human anatomy and of animals and plants, some of which have survived. *Saint Jerome in the Wilderness*, now in the National Gallery in London, dates from this early period, and shows some weeds, a dock, shepherd's purse and flowering grasses. His famous watercolour of a hare, and *Das große Rasenstück* date from around 1503. In both Dürer exhibits his own dictum: "... study nature diligently. Be guided by nature and do not depart from it, thinking that you can do better yourself. You will be misguided, for truly art is hidden in nature and he who can draw it out possesses it." It is a worm's eye view, showing a sparse turf with meadow grass, salad burnet, dandelions with the flowers closed, greater plantain, milfoil and daisy leaves, painted against a plain background on a cloudy day. Other studies of flowers include a painting of peonies, which suggests that he had seen Schongauer's peonies; a piece of turf with a single and a double aquilegia, a grass and buttercup leaves; and another clump, which shows buttercups, clover and plantain. A large painting of *Iris germanica* can be identified as a study for the painting *The Madonna with the Iris* that is now in the National Gallery in London and catalogued as from the workshop of Albrecht Dürer; it also has very accurate grape vines and a pink peony. Dürer's output of flower paintings may be small, but his influence was great, particularly through the work of his likely pupil, Hans Weiditz.

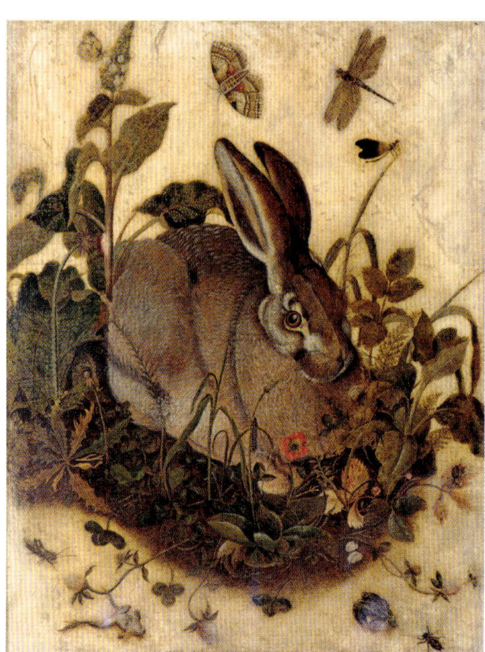

LEFT: Hare *painted by Albrecht Dürer in watercolour and bodycolour on varnished velum (c. 1503).*

OPPOSITE: Madonna with the Iris *by Albrecht Dürer, 1508, tempera on panel.*

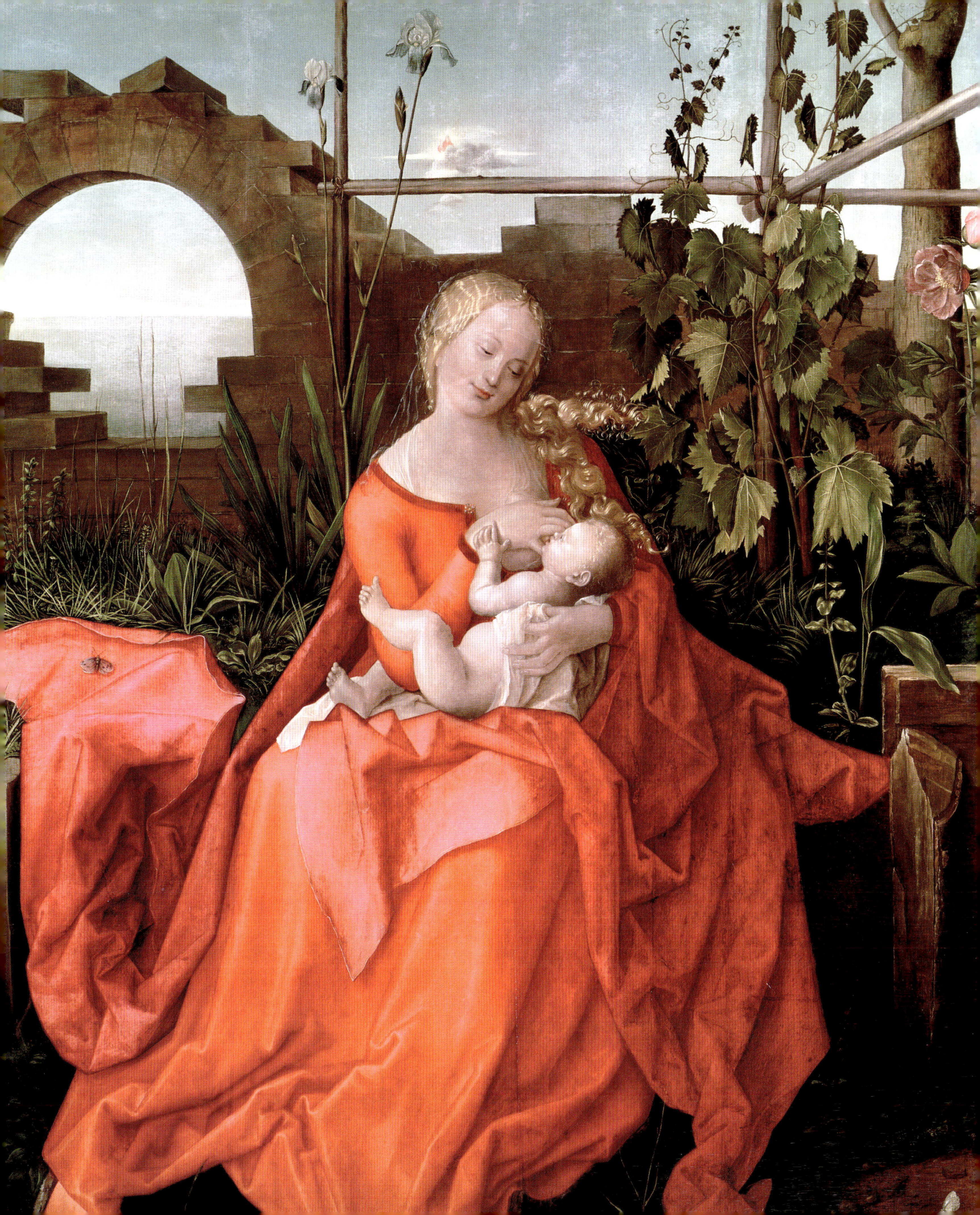

WOODCUTS

The earliest printed flower books used woodcuts for illustration; they were usually quite small, and inserted on the page with blocks of text around them. To make a woodcut, the image is drawn onto a smooth block of wood, usually beech or boxwood, and the surrounding area cut away, leaving the area to be inked raised above the surface. Printing with moveable metal type began in Germany in around 1439, and Gutenberg's famous Bible was printed in 1455 in Mainz. The first crude flower illustrations to be printed are found in the *Herbarium Apulei*, published in 1481 in Rome, and were probably copied from a Latin manuscript herbal in the library at Monte Cassino. A slightly less crude herbal was printed in Mainz in 1484, but the first to have new, and naturalistic, illustrations was the *Herbarum Vivae Eicones*, living pictures of plants, published by Otto Brunfels in 1530. It must have been a revelation for the reader, because the flowers were now recognizable and could be identified from the illustrations. Brunfels was first a Carthusian monk, but later became a Lutheran and was a physician in the city of Bern. The text was relatively standard for the time, based on Dioscorides and Pliny. It was the illustrations that were revolutionary, showing the plants with flowers and roots, with leaves and even hairs in the right places. They were the work of Hans Weiditz; because there were other engravers with this name, his details are uncertain, but he was probably born in Mainz and moved in the circle of Dürer. Until 1930, Weiditz was known only as illustrator of Brunfels's herbal, but then a number of his watercolours were found in the herbarium

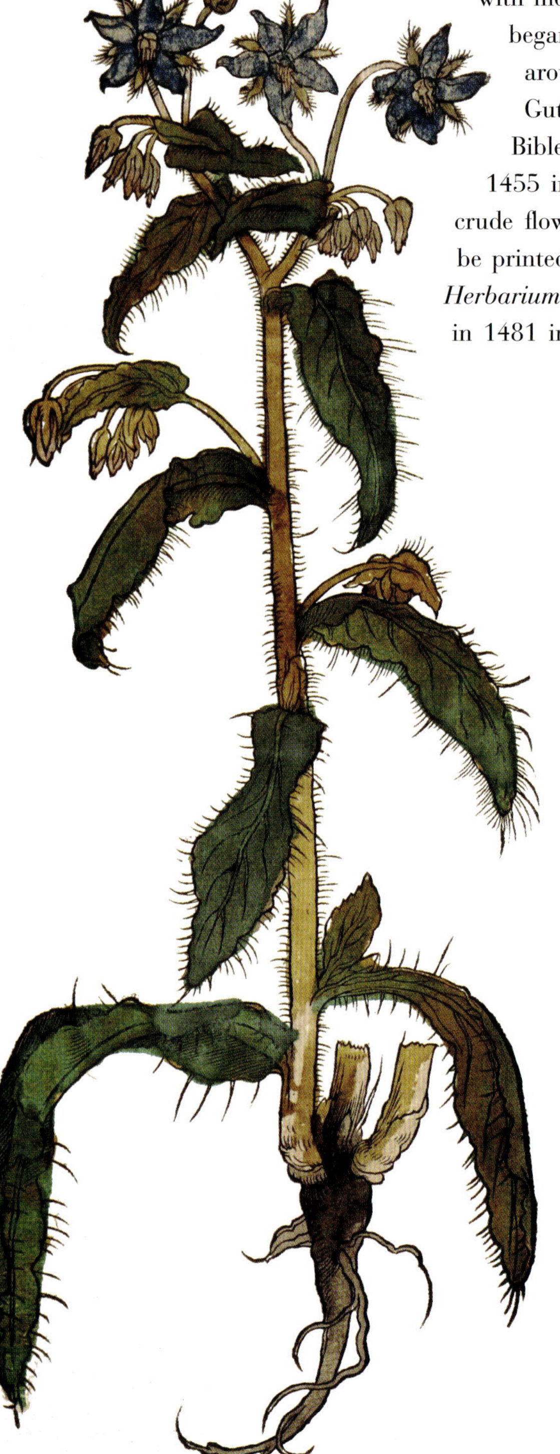

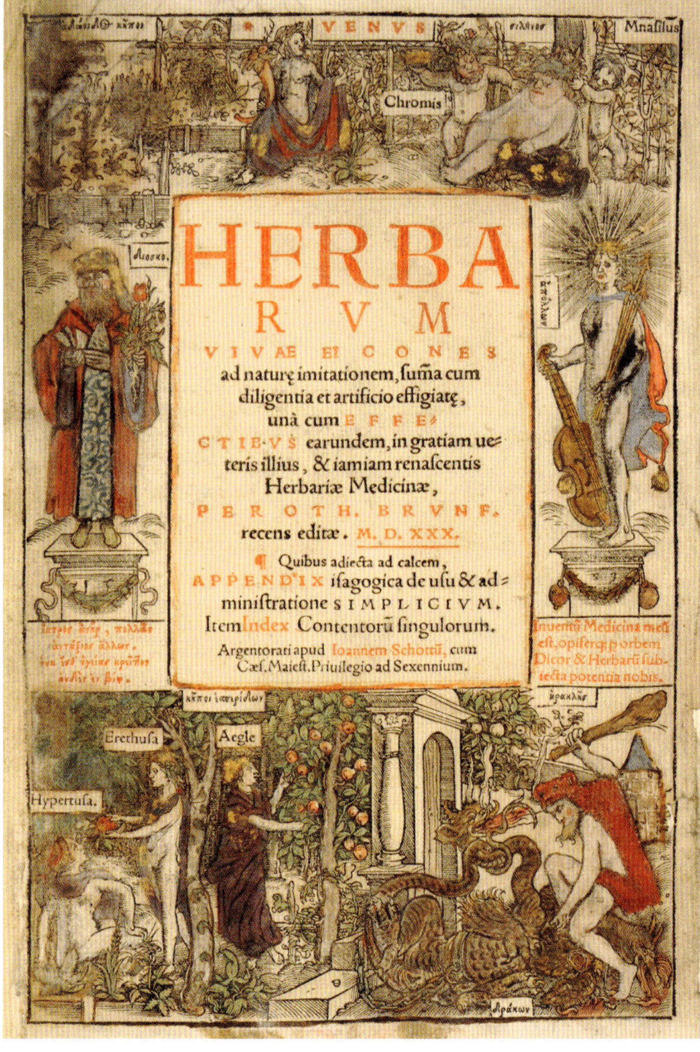

HERBA
RVM
VIVAE EICONES
ad naturę imitationem, suma cum
diligentia et artificio effigiatę,
unà cum EFFE-
CTIBVS earundem, in gratiam ue-
teris illius, & iamiam renascentis
Herbariæ Medicinæ,
PER OTH. BRVNF.
recens editæ. M. D. XXX.
¶ Quibus adiecta ad calcem,
APPENDIX isagogica de usu & ad-
ministratione SIMPLICIVM.
Item Index Contentorũ singulorum.
Argentorati apud Ioannem Schottũ, cum
Cæs. Maiest. Priuilegio ad Sexennium.

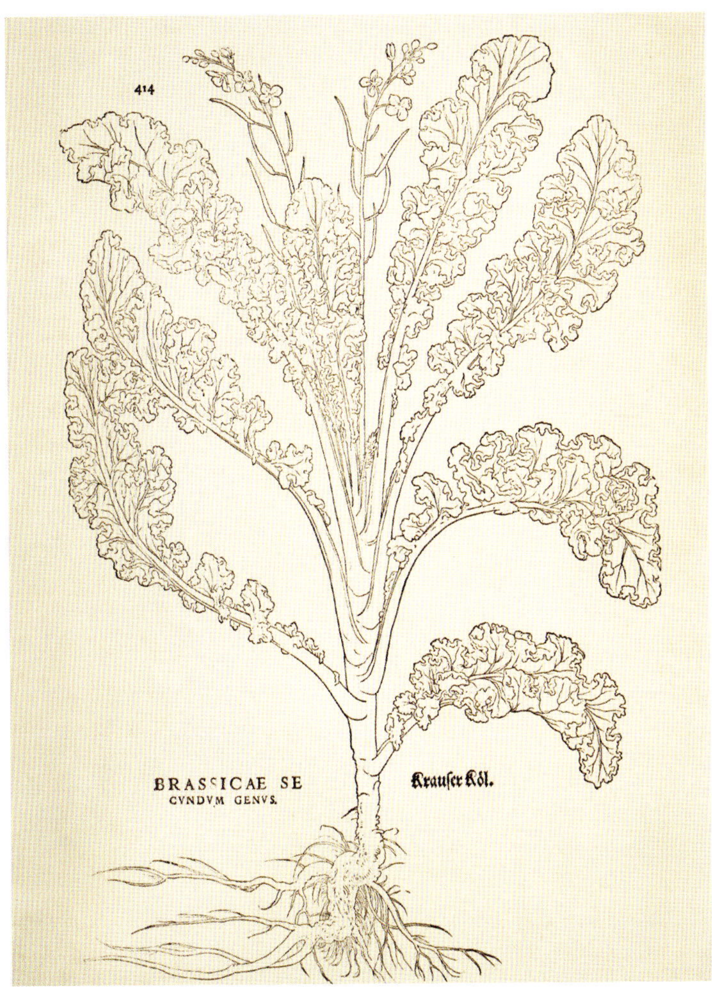

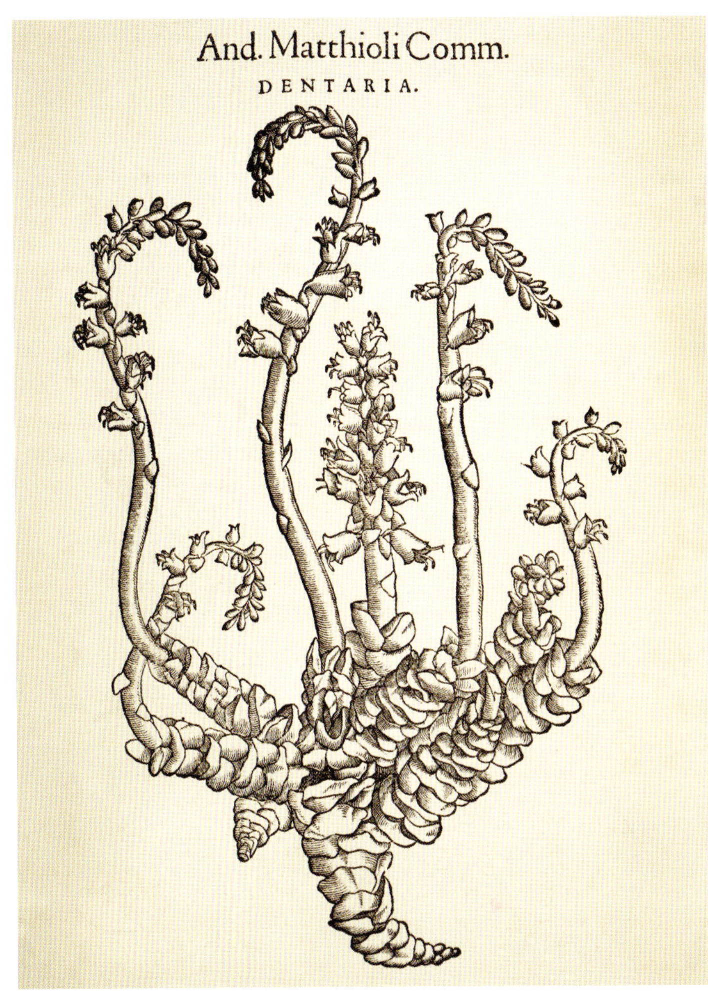

in Bern, crudely cut out and attached to large herbarium sheets. These are the earliest surviving watercolours intended primarily for plant identification, and were probably painted around 1529.

Soon after Brunfels's *Herbarum Vivae Eicones*, a second and even more beautiful herbal was published. *De Historia Stirpium*, about the history of plants, appeared in Basel in 1542; the text was by Leonhart Fuchs, (1501–66) a physician, born in Bavaria and later Professor of Medicine in the University of Ingolstadt, which was very influential in the sixteenth and seventeenth centuries but closed in 1800. Fuchs successfully treated many who were struck with the plague in 1529, so his fame spread across Europe, and he later moved to Tubingen. This is a big book, and the illustrations, which were drawn by Albrecht Meyer, are in thin outline and designed to be hand-coloured. The text gives the plants' names, habits, time of flowering, localities and their uses according to Dioscorides, Pliny and Galen. A German edition, the *New Kreüterbuch* appeared the following year. Fuchs's herbal was one of the first to show plants from the Americas, and includes a good illustration of a *Tagetes* and a *Capsicum*, as well as European vegetables, such as cabbages and curly kale. Linnaeus honoured him by naming the American genus *Fuchsia* after him.

The Italian Pier Andrea Matthioli (1501–77) was a botanist and physician, born in Siena and brought up in Venice; he spent much of his later life in Prague as physician to the Grand Dukes there and his last years in Vienna. The first edition of his herbal, *Commentarii, in libros sex Pedacii Dioscoridis de medica materia*, was published in Venice in 1554. It was botanically a great advance on earlier plant books and immensely successful, going through more than forty editions and selling, in the small, early editions, over 30,000 copies. Later, from 1562 onwards, a larger edition with new woodcuts was published. In addition to his other interests, Matthioli was a keen gardener, and received plants from Turkey, sent back by Ogier Busbecq. He also included, in the 1563 edition, the first illustration of the horse chestnut, *Aesculus hippocastanum* L., which he received from Busbecq's physician in Istanbul.

*OPPOSITE LEFT: Borage (*Borago officinalis*) from Brunfels's* Herbarum Vivae Eicones.

OPPOSITE RIGHT: The title page of Brunfels's Herbarum Vivae Eicones.

ABOVE LEFT: A brassica from Leonhart Fuchs's De Historia Stirpium.

*ABOVE RIGHT: "Dentaria" or toothwort (*Lathraea squamaria*) from Pier Andrea Matthioli's* Commentarii, in libros sex Pedacii Dioscoridis de medica materia.

ISLAM AND PLANTS FROM THE TURKISH EMPIRE

The fall of Constantinople to the army of Mehmet II, the Conqueror, in 1453 marked the end of the slow decline of the Byzantine Empire. Western tradition portrayed the Turks as morally degenerate barbarians, but their sophistication is shown by the paintings of Mehmet himself by Giovanni Bellini (c. 1430–1516) and by the miniature of him, now in the Topkapi Palace Museum, with a red rose held up to his aquiline nose. This is the red damask rose, *Rosa gallica* "Officinalis", which had probably been cultivated since pre-Roman times. A period of even greater prosperity and a high point in the culture of Ottoman Turkey coincided with the reign of Suleiman I, the Magnificent, from 1520 to 1566, who extended the empire into North Africa, and eastwards through Anatolia, as far as Erzurum. Furthermore he continued the conquest of the Balkans, only to be halted at the gates of Vienna in 1529.

The letters of Ogier Ghislen de Busbecq, who was appointed ambassador from the Holy Roman Emperor to the Ottoman Empire in late 1554, gave readers in Europe the first detailed account of Ottoman life. Busbecq reached Istanbul in January 1555, and continued across Anatolia to Amasya, where the Sultan was staying. His four long letters were sent back from Istanbul, and later embellished and edited after his return to Vienna; his two first Turkish letters were printed by Plantin in Antwerp in 1581 and 1582. In them he describes the social structures of Ottoman life, the harem, the hammams and the Turks' love of tulips and other bulbs. His first visit lasted about a year and he reported seeing narcissi, hyacinths and "what the Turks call Tulipam" flowering in the fields around Edirne. Busbecq is credited with introducing the first ornamental bulbs to Vienna and later to Antwerp, where they thrived and became the foundation of the Dutch bulb trade. Tulips, hyacinths, black irises, narcissi, Turban buttercups and crown imperials were all grown in quantity by the Turks and soon exported.

Under Suleiman, the ceramic workshops at Iznik became recognized officially as the finest in the Ottoman empire, at first copying the blue-and-white china imported from Jingdezhen in Jiangxi via the Persian gulf, and then, from 1560 onwards, developing their own many-coloured style; the tiles and plates show the long, narrow pointed tulips favoured by the Turks, as well as hyacinths and roses, red *Lilium chalcedonicum* and carnations. Some of their more fanciful flowers can be traced back to Chinese models, and a fine collection of early Chinese ceramics, sent to the Sultans, is preserved in the Topkapi.

The first illustrations of these Turkish flowers are found in the work of Charles d'Escluse, known as Clusius, born in Arras, but studying in Flanders, and based at the University of Leiden. He visited Spain in 1564, and in 1573, moved to Vienna at the invitation of the Emperor, where he remained for about 14 years. His collected works were published by the Antwerp publisher Christophe Plantin in 1601, as *Rariorum Plantarum Historia*, and showed for the first time many of the new plants that were flooding into Europe from the east and from the Americas.

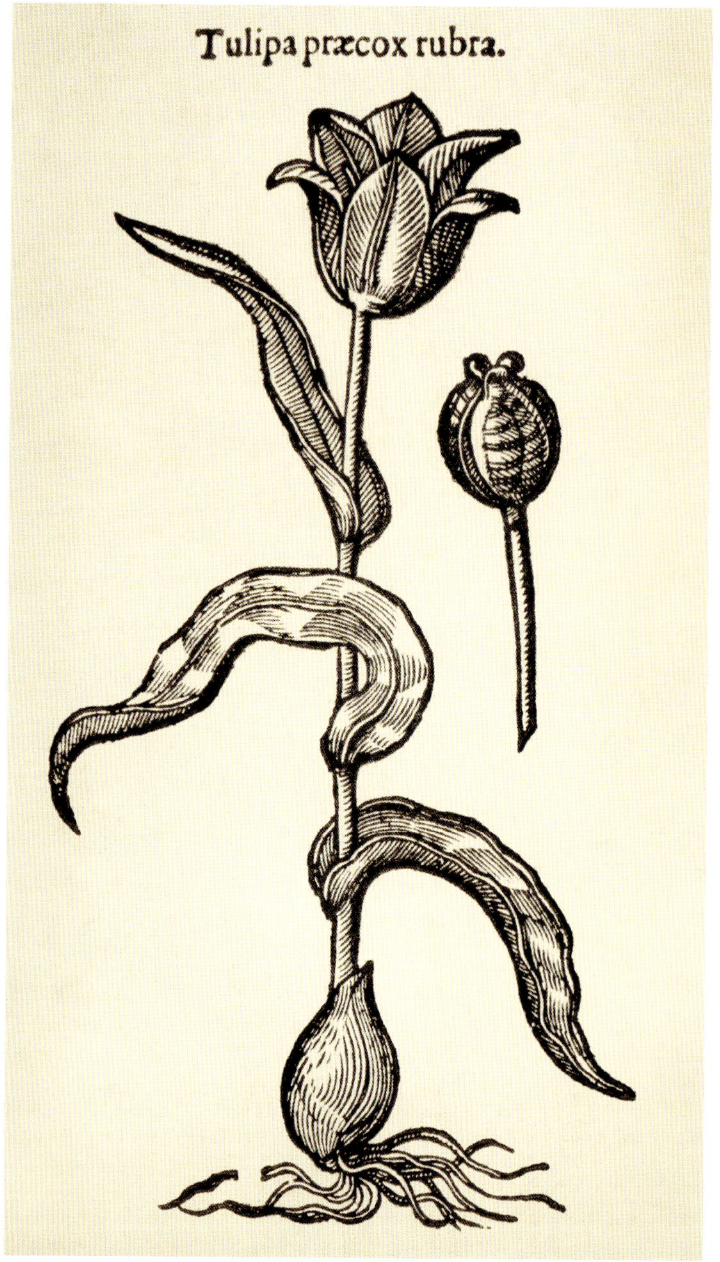

ABOVE: Tulipa praecox rubra *from Clusius's* Rariorum Plantarum Historia.

OPPOSITE: An example of the ceramic work produced at Iznik during the sixteenth century. The plate shows various flowers, including tulips.

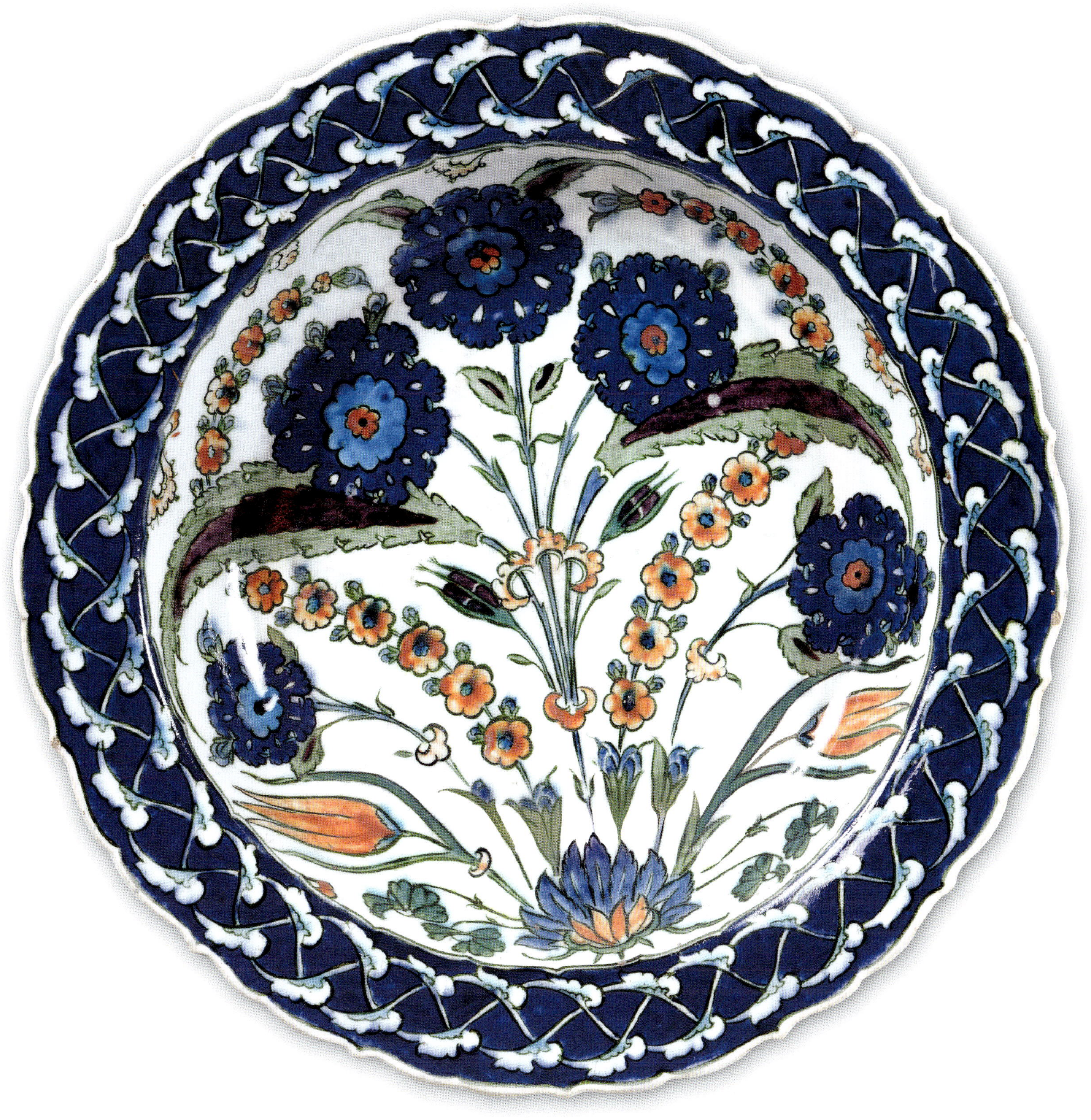

ENGLISH HERBALS

The first English illustrated Herbal, *A new Herball*, was published in parts, starting in 1551, and the complete work was printed again in 1568. Smaller herbals had been published before, and *The grete herball* of 1526, which was a translation of the French *Le Grand Herbier*, and contained some small and crude woodcuts, derived from earlier continental books.

William Turner's *A new Herball* was in a different league, and Turner himself was an interesting character, one of the first clerical botanists. He was Dean of Wells, and a protestant, who was taught at Pembroke Hall, in Cambridge by Nicholas Ridley and Hugh Latimer, both Protestants who were burnt at the stake under Mary in 1555. In 1540, Turner escaped abroad for the first time, and returned to the interests of his youth, taking a degree in botany and medicine in Italy, and visiting Switzerland, Germany and Holland; in his absence his books were banned by Henry VIII. Turner returned to England after Henry's death, and in 1548 published a checklist: *The names of plants in Greek, Latin, Englishe, Duche & Frenche, with the commune names that herbaries and apothecaries use*. It contains 105 plant records new to England, some with their localities given as well.

This was the forerunner of *A new Herball*, which, unusually for the time was published in English, but it still included synonyms in the other languages, and quotations from Dioscorides; here we can find some of the familiar English names of wild flowers, such as Cockoupynt for *Arum maculatum*, and mugwurt which "is called both of the Greeks and latines, Artemisia". The illustrations were mostly taken from the octavo version of Fuchs' *New Kreüterbuch*.

Two other important herbals were published in England in the 1570s. Mathias de L'Obel and his friend Pierre Pena, wrote *Stirpium adversaria nova*, which was published in 1570. They

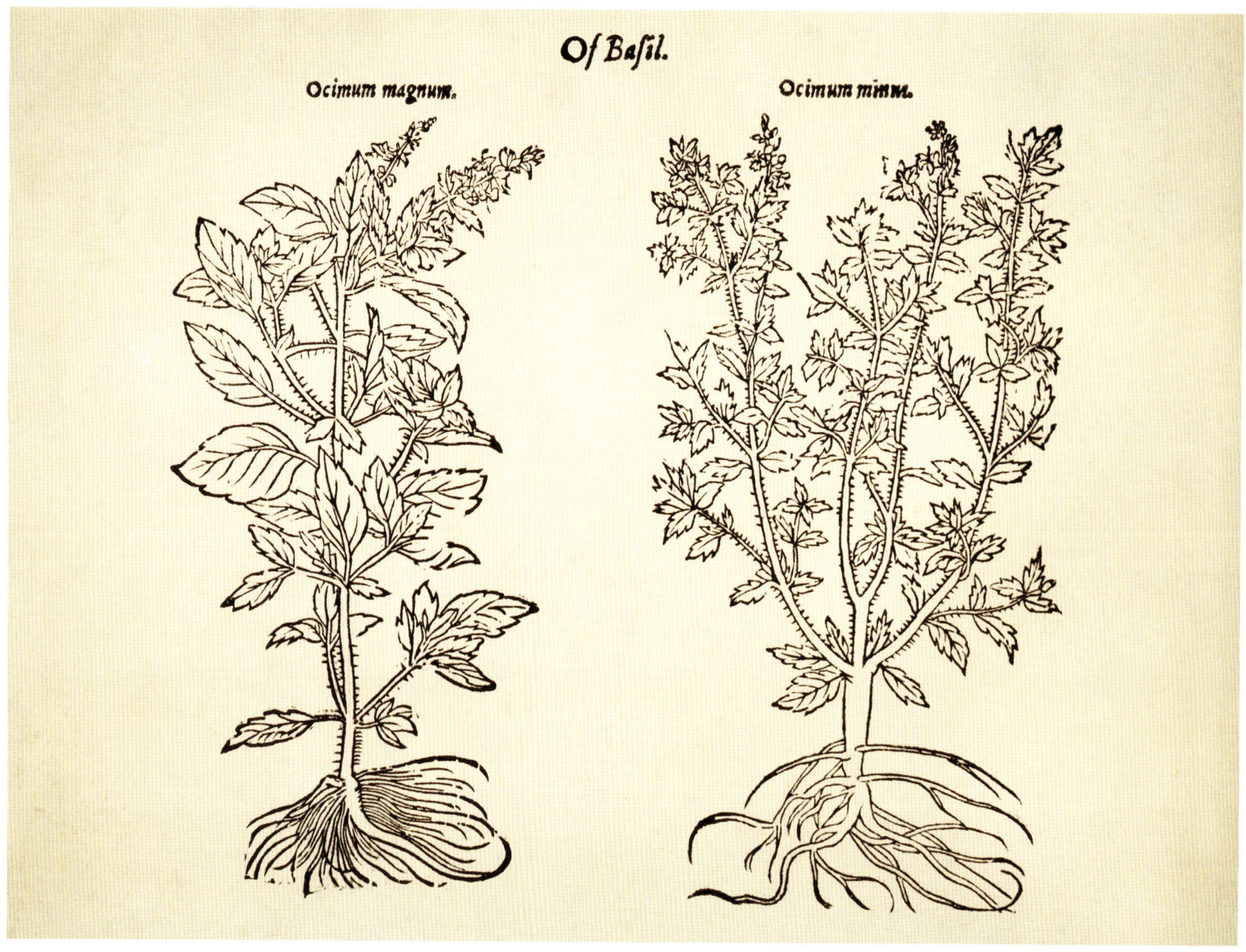

were both born in France, and this was the fruit of their botanical studies in Montpellier, their travels around Europe and a stay in England, which they visited during the reign of Elizabeth I. They published the book in England, in Latin, and dedicated it to the Queen. Botanically, this was a big advance on earlier flower books, and was based in part on leaf shape, so all the bulbs, grasses et cetera. (monocotyledons) were put together, and other plants tended to be grouped botanically. The Antwerp publisher Christophe Plantin bought 800 copies, unbound, and published them under his own imprint: he also bought 250 of the original woodblocks of the illustrations, which formed part of Plantin's famous collection of botanical images. One shows a tobacco plant, and an Indian's head smoking a giant cigar. Pena returned to the continent to practise medicine, but de l'Obel stayed in England until he died in 1616. Some of writings were published as late as 1655, others are unpublished and are now in the library at Magdalen College, Oxford.

Henry Lyte's *A Nievve Herball* was printed in Antwerp and published in London in 1578. It was not an original work, but was a translation of Clusius' French version of Rembert Dodoens' *Cruÿdeboeck* of 1554. The artist of the title page was Pierre van de Borcht, who had painted many flowers for Clusius: the woodcuts are Plantin's and after so many printings, some are beginning to show signs of wear.

OPPOSITE: Two varieties of basil from Turner's A new Herball.

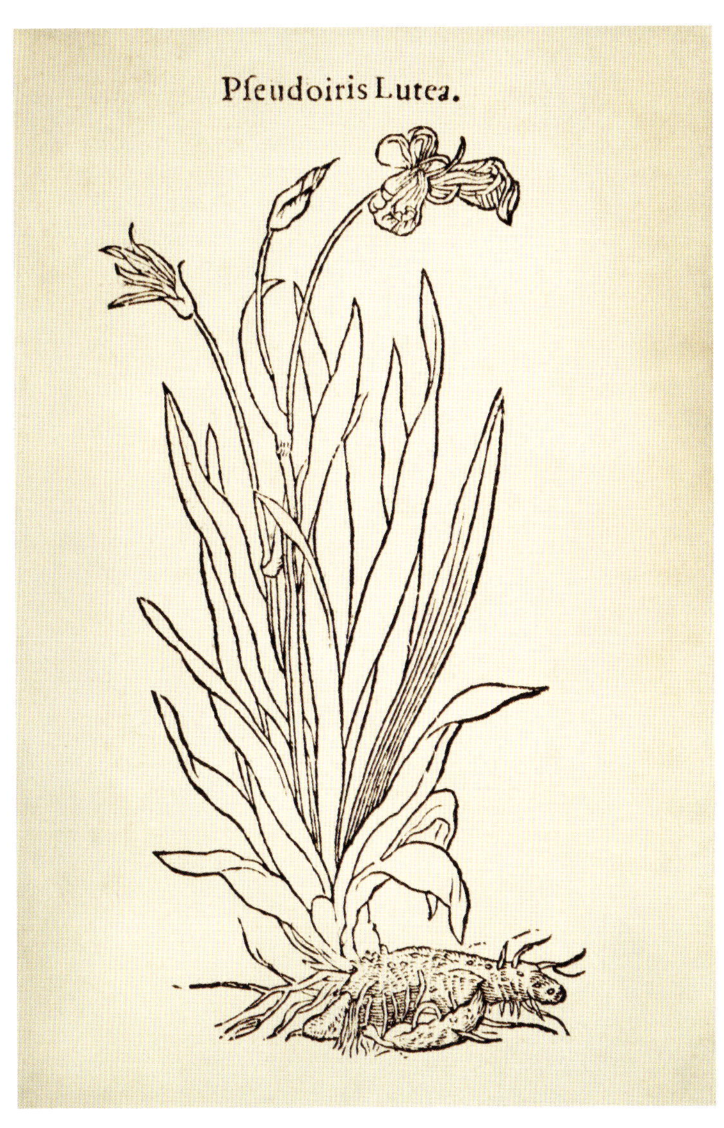

BELOW LEFT: Pseudoiris lutea *from Henry Lyte's* A Nievve Herball.

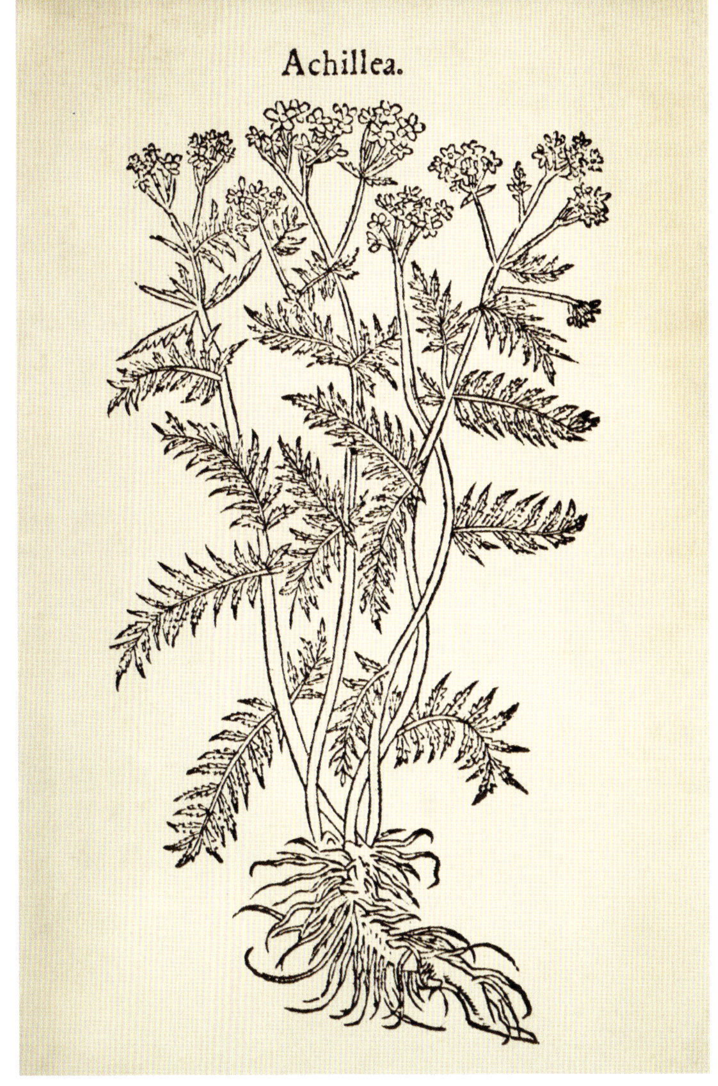

BELOW RIGHT: Another of the woodcut illustrations bought by Plantin and used in A Nievve Herball, *this time of* Achillea.

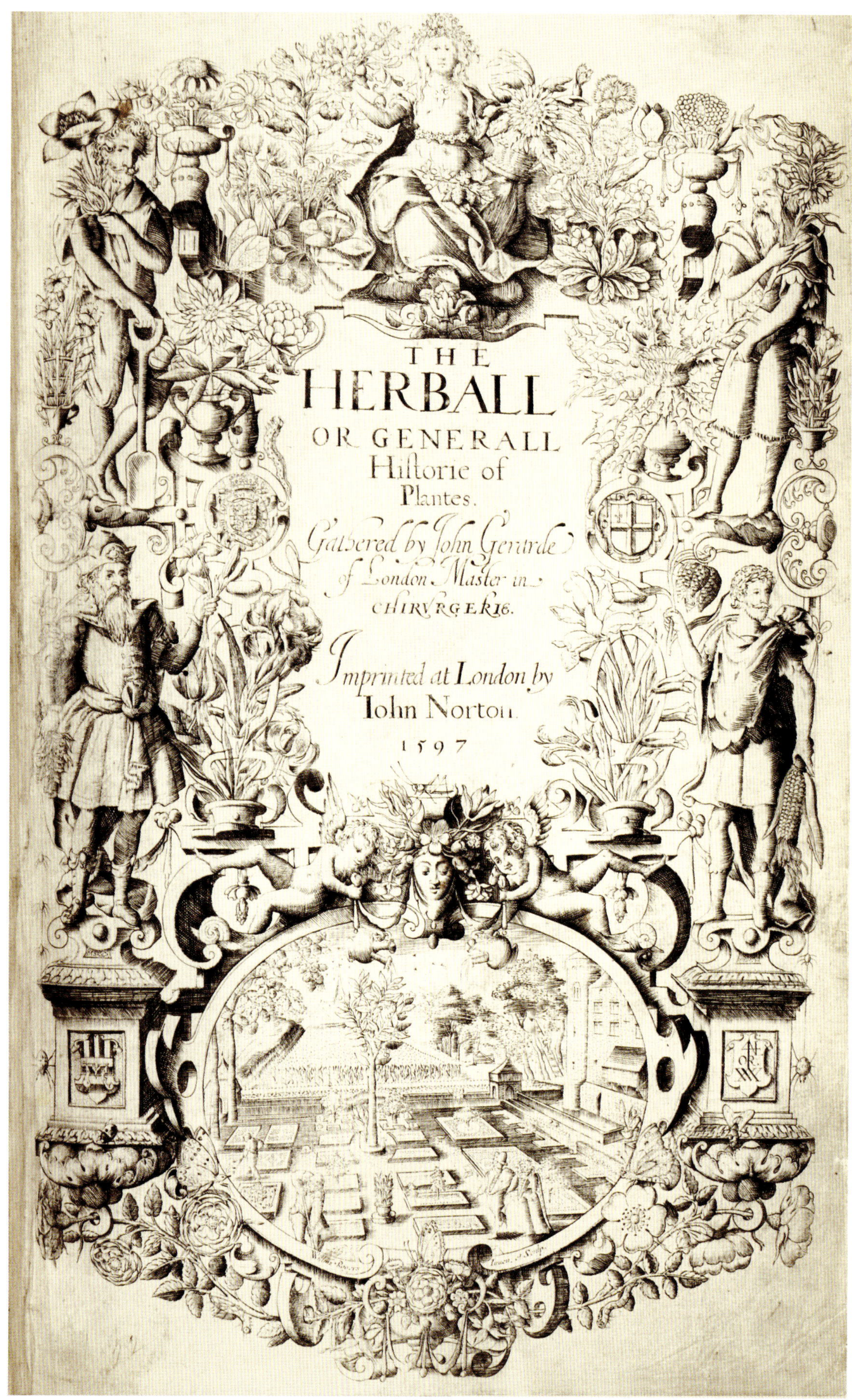
THE
HERBALL
OR GENERALL
Historie of
Plantes.
Gathered by John Gerarde
of London Master in
CHIRVRGERIE.
Imprinted at London by
Iohn Norton.
1597

GERARD'S HERBALL

Of all the English herbals, Gerard's has been the most influential, and has been reprinted many times. The first edition was published by Gerard in 1597, and Thomas Johnson's enlarged edition was published in 1636.

John Gerard (1545–1612) was born at Wisterson, in Cheshire. He was apprenticed to a barber-surgeon in London and, in 1607, was elected Master of the Company of Barber-Surgeons. He was a keen botanist and searched the area round London, looking for new flowers, and went as far afield as Exeter. He may even have travelled to Moscow, as he mentions visiting Poland and other countries en route. Gerard also worked as an advisor on gardens, and was superintendent of William Cecil's (Lord Burghley) gardens in the Strand and at Theobalds in Hertfordshire. In 1586, Gerard was appointed curator of the physic garden of the College of Physicians. His own garden, though, was on the south side of Holborn, and he grew there as many rare plants as he could fit in. In 1596, he published a garden catalogue, and had de l'Obel testify that he had seen the garden and all the plants named. It included over 1000 species, including the first mention in print of the potato. A second edition, published in 1699, is dedicated to Sir Walter Raleigh. De l'Obel's copy of this version, has the attestation struck out and "falsissima" written by it, with de l'Obel's signature!

The first edition of *The Herball or generall Historie of Plantes* was printed by John Norton in London. The title page was engraved in London, and shows botanists and gardeners holding a variety of plants including a crown imperial fritillary, a cob of maize, and a pasque flower, as well as imports from Turkey such as *Iris persica*, *Iris susiana*, a pot of carnations and double anemones. A vignette of a garden, with small raised beds is shown at the base. The illustrations were mostly printed on 5 x 6-inch woodblocks, obtained by the printer from Frankfurt-am-main, where they had already been used for the *Icones Plantarum* of Jakob Theodor, known as *Tabernaemontanus* in 1590. The engravings are made to fill the block, and many are attractive designs; there is a good representation of a potato "Battata Virginiana", showing leaves, flowers, fruit and the roots with tubers.

Gerard's *Herball* was revised by Thomas Johnson, and this edition, known as *Gerardus emaculatus*, published in 1633 and again in 1636. Johnson was a botanist of note and published one of the first local British floras, accounts of excursions to Kent and Hampstead Heath, in 1629. Johnson made many corrections and additions, and included some new illustrations, such as a drawing of a bunch of bananas, but most were the familiar woodcuts from Plantin's collection in Antwerp. The new title page shows Ceres and Pomona, together with Theophrastus and Dioscorides, with two large vases of exotic flowers and a portrait of Gerard himself wearing a large ruff, and holding a flowering stem of potato.

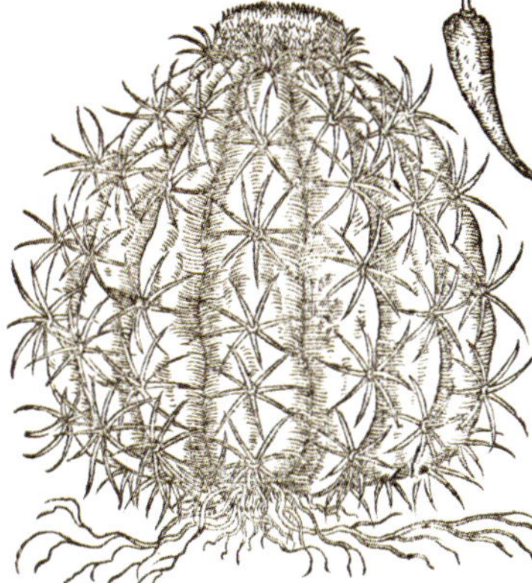

OPPOSITE: The title page of The Herball or generall Historie of Plantes *by John Gerard.*

RIGHT: An illustration from Gerard's Herball, *showing an orchid, probably the early purple orchis (*Orchis mascula*).*

FAR RIGHT: The aptly named Hedgehogge thistle from Gerard's Herball, *an early illustration of a* Melocactus *from the West Indies, and the fruit of a pepper.*

Jacopo Ligozzi

Jacopo (or Giacomo) Ligozzi (1547–1626) was a painter of many subjects, but his flower, animal and fish drawings were better than any before, and in the opinion of Wilfrid Blunt, not surpassed until the eighteenth century. Ligozzi was born in Verona, into a family of painters and artisans, but after a spell in Vienna, spent most of his life in Florence, in the employment of the Medici Dukes. In 1574, he became head of the Accademia e Compagnia delle Arti del Disegno which had been founded by Cosimo I de' Medici. He painted large biblical and classical scenes, designed stage sets for the theatre and produced designs for embroideries and *pietra dura*, which were fashionable at that time.

Ligozzi was also renowned for a series of circa 25 drawings of Turkish costume, sometimes accompanied by animals. Some were said to have been adapted from sketches done by travellers to the east, and certainly Ligozzi's giraffe suggests that he never saw one alive; his elephant is not much better. In the eighteenth century, these drawings were used as designs for porcelain plates by the Ginori di Doccia factory outside Florence, and examples can be seen in the museum there.

Most of his flower paintings and other drawings are now in the Gabinetto Disegni e Stampe in the Ufizzi in Florence; others are in the University of Bologna. The flowers date from circa 1580 and are large, often life size, showing the whole plant with roots, perfectly accurately and beautifully drawn. Many are of Mediterranean origin, and may have reached Ligozzi through the botanist Aldovandri, who founded the Botanic garden in Bologna, and is known to have commissioned flower paintings from Ligozzi. He praised Ligozzi as "a most excellent artist". Many of the plants that he painted, such as *Mirabilis* from Mexico, the mandrake, burning bush (*Dictamnus albus*), black iris (*Iris susiana*), *Anemone coronaria*, the scented *Muscari macrocarpum* and tree spurge (*Euphorbia dendroides*) from Turkey, were quite new and exciting at this time; others such as *Gentiana excisa*, *Dentaria pentaphylla*, *Daphne laureola*, *Digitalis purpurea* and *Pinguicula longifolia* are from the Alps or northern Europe. His plant portraits are painted in opaque watercolour or gouache with brilliant use of shadow to provide modelling on the leaves; the style is very crisp with many similarities to that of Ehret in the eighteenth century.

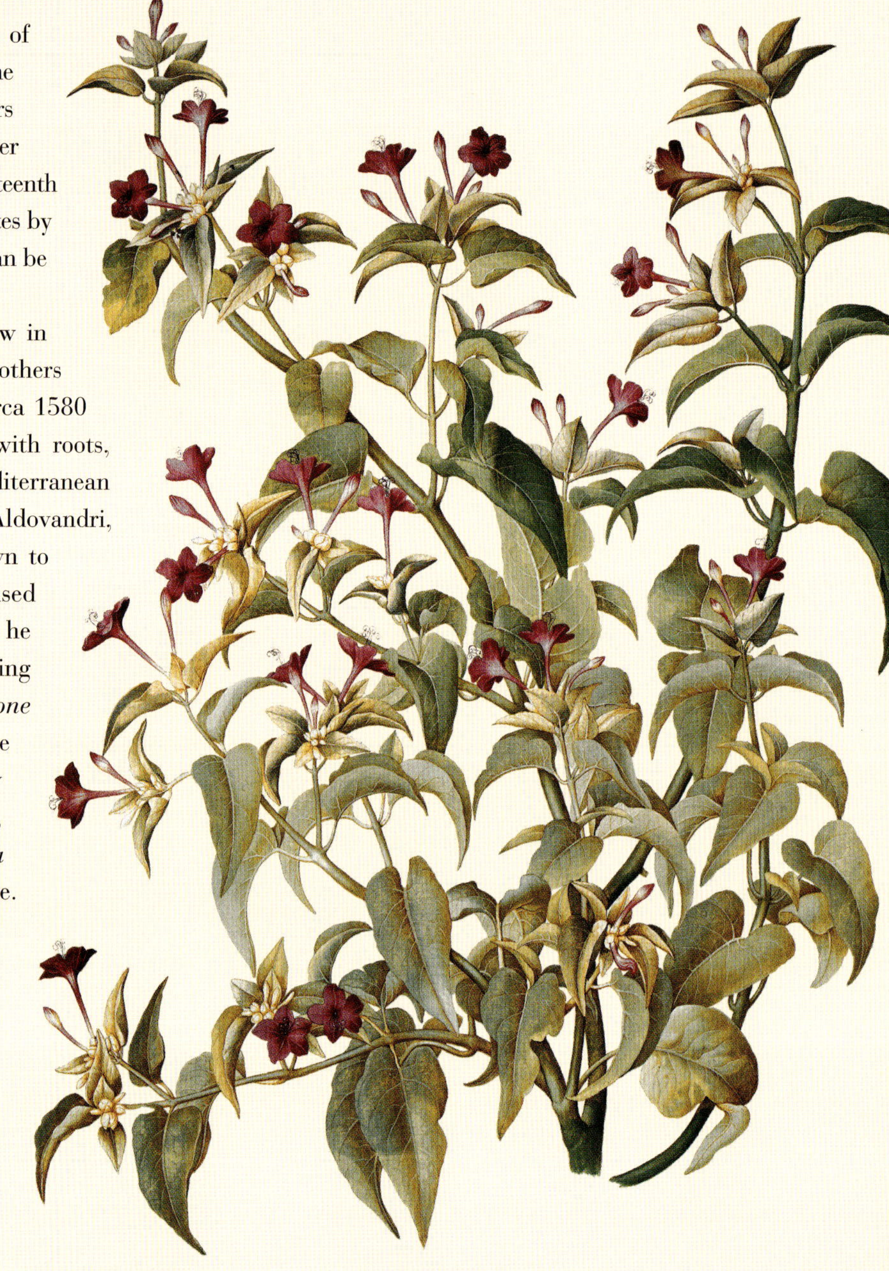

RIGHT: Mirabilis jalapa, *painted by Jacopo Ligozzi c. 1580.*

OPPOSITE: Euphorbia dendroides *with a female Emperor moth and a caterpillar, painted by Jacopo Ligozzi, c. 1580.*

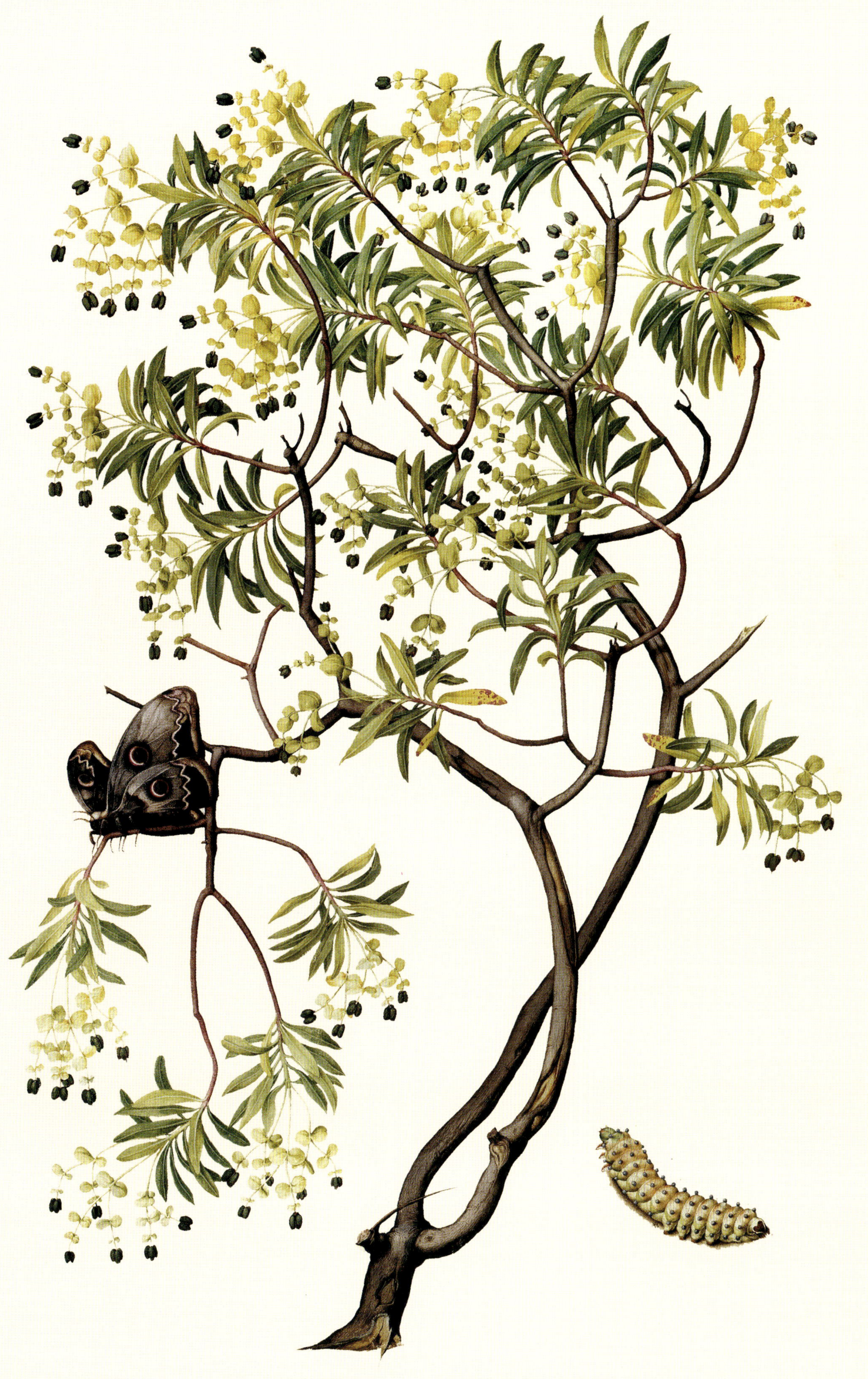

3

SEVENTEENTH-CENTURY FLORILEGIA

NEW ARRIVALS

During the seventeenth century an increasing number of plants were brought into European gardens by travellers and naturalists, who were also often merchants or physicians. Botany initially developed out of the study of medicinal plants and most botanists made a living as medical practitioners or apothecaries.

The earliest printed books describing plants were the Herbals (*see* Chapter 2), whereas volumes known as florilegia (from *florilegium* = the Latin for "a gathering of flowers") were chiefly devoted to plants grown in gardens for their ornamental qualities, rather than for medical or culinary use, and the illustrations tended to dominate the text. Sometimes, as in the case of Besler's *Hortus Eystettensis* (*see* page 37) and Johann Walther (*see* page 44), they also faithfully recorded the contents of a patron's garden, a tradition that continues to the present day in *The Highgrove Florilegium* (*see* page 228) commissioned by HRH Prince Charles to illustrate the flowers in his garden at Highgrove.

Both printed and manuscript florilegia were produced in Europe, particularly France, Germany and the Low Countries, throughout the seventeenth century, reflecting the increasing interest in science and philosophy prevalent at that time. The first collection of illustrations actually described under the title of *Florilegium*, was probably that published by the Flemish artist Aedriaen Collaert, in Antwerp in 1600.

A few years after this, in 1608, another book of the type that we would now describe as a florilegium, was produced by the Frenchman Pierre Vallet (*c.* 1575–1657?) who was born at Orléans, and moved to Paris to become the first botanical painter to the court of Henri IV. *Le Jardin du très Chrétien Henri IV, Roi de France et de Navarre...* (1608) was dedicated to Henri's wife Marie de' Medici, who had brought to the Court the Italian (and familial) tradition of patronage for natural history and science. The garden of the title was that of the Louvre, which had been established by the great Parisian gardener, Jean Robin (1550–1629), and contained a collection of exotic plants "brought back from Guinea and Spain..." some of which had been collected by Robin's son, Vespasien. The book was intended in part as a pattern book for embroidery for the ladies of the court, and in part to document the exotic plants grown in the garden at that time. It contains 75 botanically accurate etched plates of plants, but there is no text. A second edition, with 17 additional plates, was published in 1623.

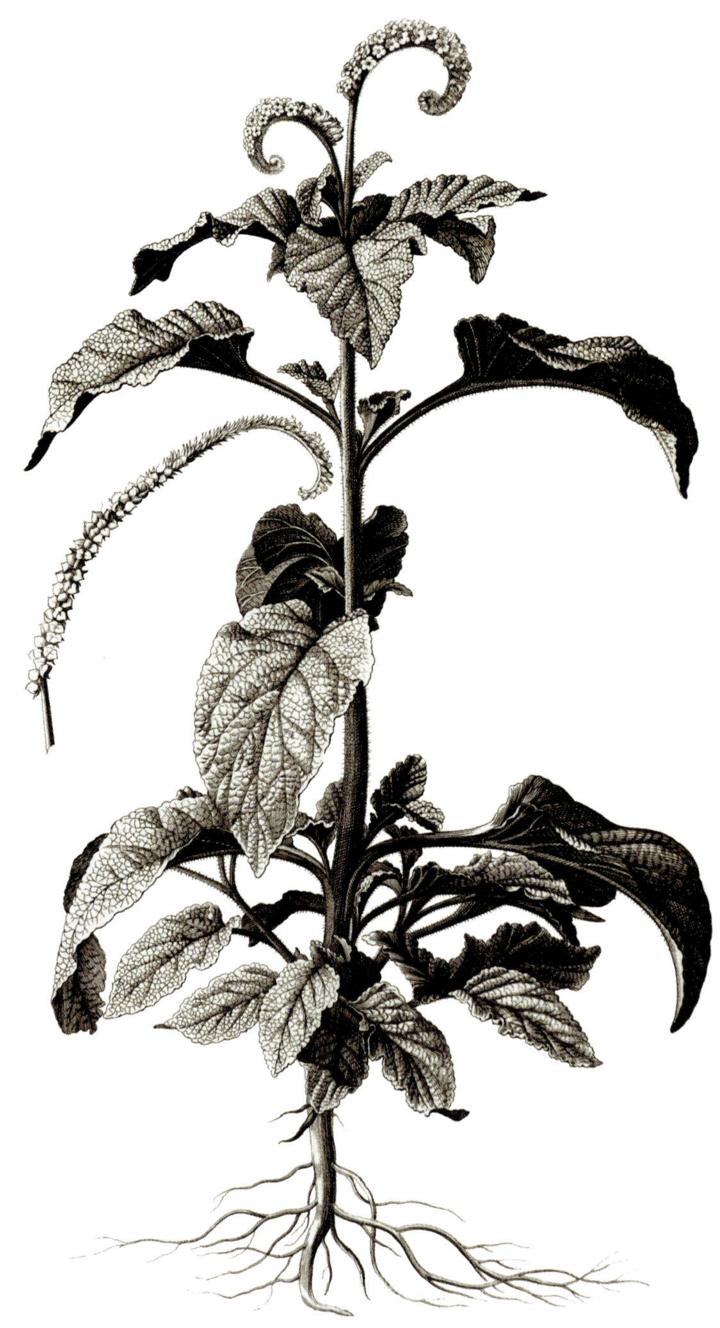

Daniel Rabel (1578–1637), the son of an engraver, was best known to his contemporaries as a designer of stage sets and costumes for the ballets and masques held at the court of Louis XIII, who succeeded Henri IV. Rabel was, however, multi-talented, and painted portraits and landscapes as well as popularizing *cartouche* landscape prints. He succeeded Vallet as court painter and his flower paintings were engraved and published in 1622 under the rather apt title of *Theatrum Florae*. This contained 69 engraved plates of flowers, and was evidently successful, as it was reprinted three times in subsequent years. He also produced detailed flower paintings on vellum (*vélin* in French) for Gaston d'Orléans, younger brother of Louis XIII, and these later formed the basis of the royal collection of *vélins* (*see* pages 86–87).

Work on the *vélins* was continued by the painter and engraver Nicolas Robert (1614–85), who also illustrated a small book of flower etchings entitled *Fiori Diversi*, published in Rome in 1640. His other main contributions to botanical art are found in the *Guirlande de Julie*, a collection of madrigals written by poets of the day for the beautiful Julie d'Angennes (daughter of the Parisian society hostess Mme Rambouillet) illustrated with flower paintings on vellum, and in the rare *Recueil de Plantes*, apparently initially printed in around 1683 for the Académie Royale des Sciences, and later reprinted several times with additional plates. Around 1786, a particularly sumptuous edition was published, with 319 plates and a frontispiece by Sebastien le Clerc, showing Louis XIV visiting a meeting of the Académie Royale des Sciences.

OPPOSITE: Heliotropium americanum *by Nicolas Robert.*

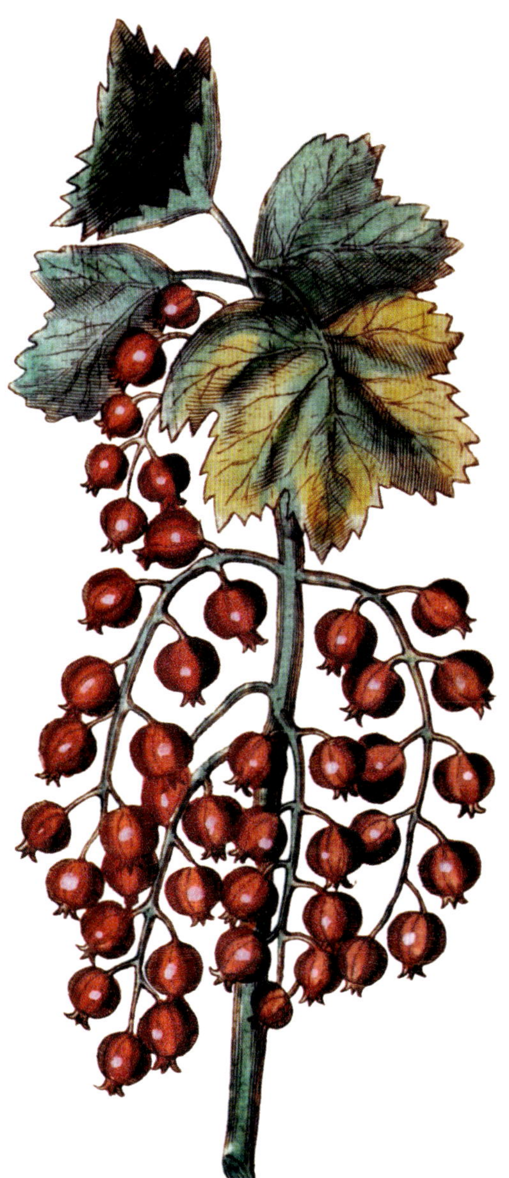

BELOW LEFT: A coloured engraving of red currants from Daniel Rabel's Theatrum Florae.

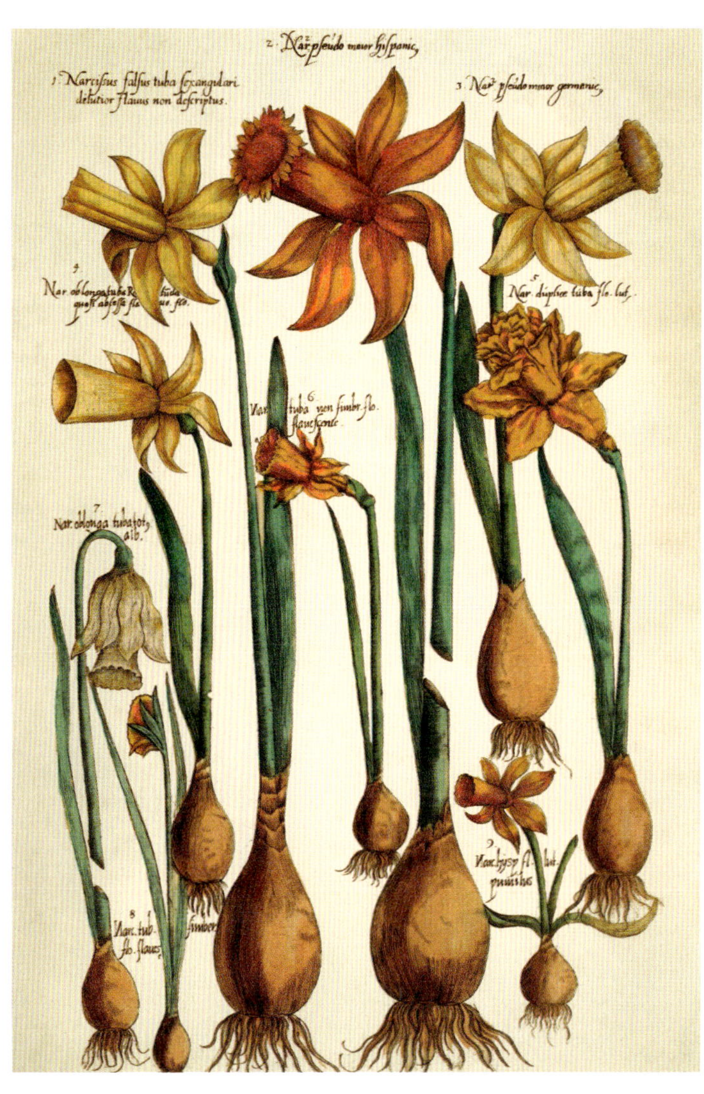

BELOW RIGHT: A page from Daniel Rabel's Theatrum Florae, *this time showing varities of* Narcissus.

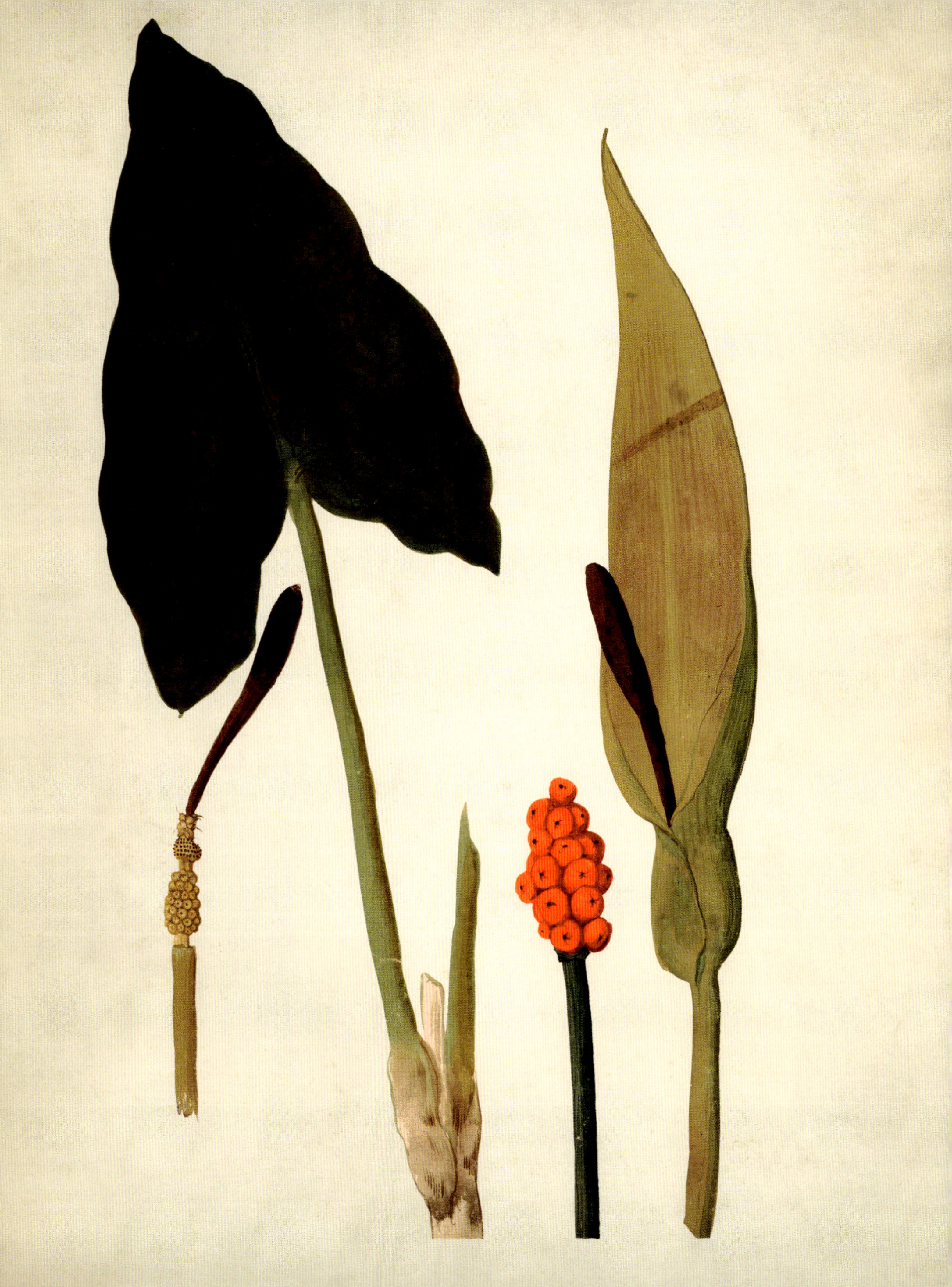

GERMANY

The German Sebastian Schedel (1570–1628) was one of the artists who worked on the great collection of paintings included in Besler's *Hortus Eystettensis*. Schedel's book of watercolours on paper dates from 1610 and is now in the library at Kew. It is known as the *Calendarium*, and is probably a record of plants from various gardens. This collection depicts mostly simple spring flowers such as pinks and pansies, which were the basis for his contributions to Besler's book. The original colour paintings were turned into black-and-white drawings and then engraved on copper, before being hand coloured, using the originals as a guide. Schedel's style is somewhat primitive compared with the more sophisticated painting of the larger, more exotic flowers. The different flowers are shown as if scattered around the page and given depth by the use of shadows.

Hortus Eystettensis is one of the greatest of all the seventeenth-century florilegia, being produced to the highest artistic standards. It provides an extraordinary record of plants grown in northern Germany at that time. The work was commissioned in 1611 by Johann Konrad von Gemmingen, Prince Bishop of Eichstätt from 1595 to1612, and was designed to show the plants in his garden. They were drawn by a team of artists, including Besler (though most are anonymous), and prepared for publication by the finest engravers in Augsburg and Nürnberg.

OPPOSITE: An Arum *from Sebastian Schedel's* Calendarium, *which shows the plant at various stages in its life.*

BELOW: Studies of a martagon lily flower and its various parts of from Sebastian Schedel's Calendarium.

RIGHT: Painted in watercolour like the other images in the Calendarium, *these studies of* Hyacinthus *varieties show the delicate nature of the artworks.*

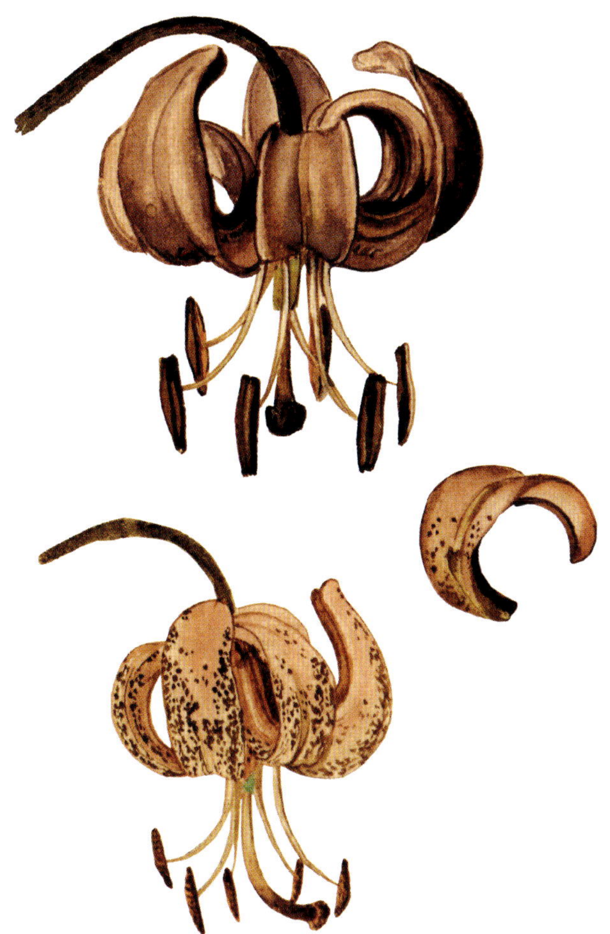

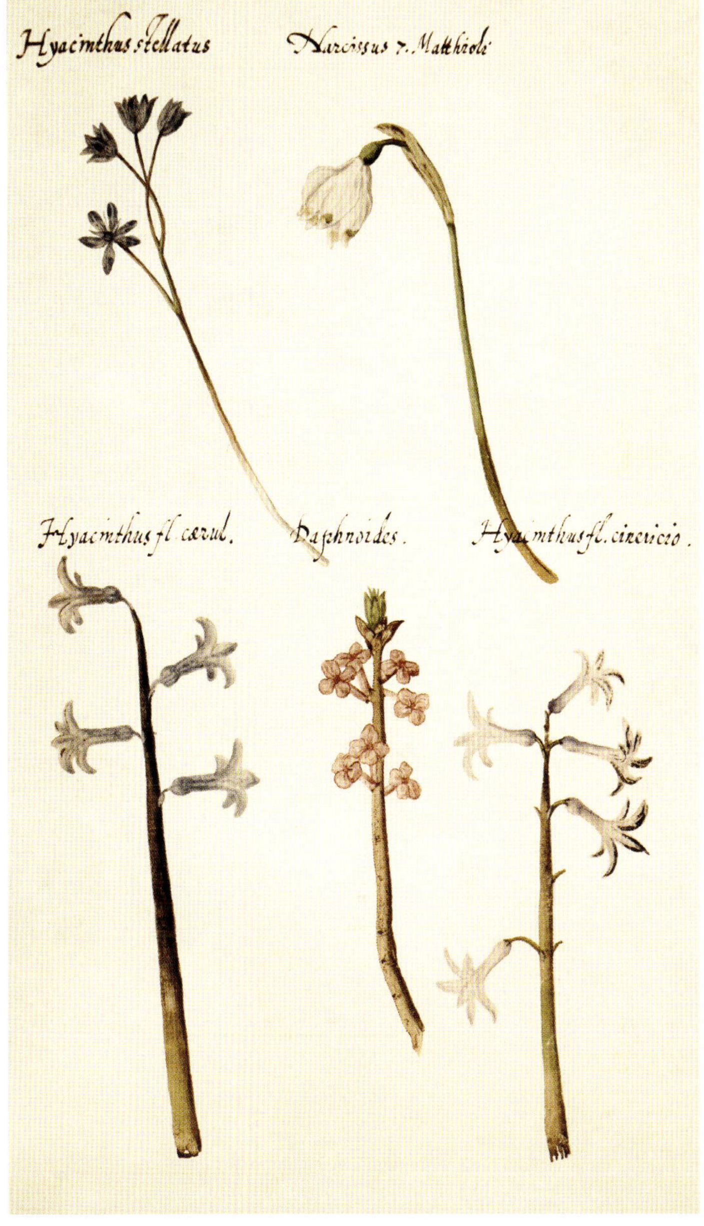

The garden at Eichstätt was in effect a private botanical garden, with terraces surrounding the palace, and it was planted with masses of flowers ordered according to their different countries or origin; bulbs, such as tulips, were a particular favourite. Basilius Besler (1561–1629), an apothecary from Nürnberg, was called in to help manage the garden after the death of the botanist and doctor, Joachim Camerarius, who had started it. Besler encouraged Johann Konrad to produce an illustrated book of his choicest flowers, and the work was set in train; sadly, the Bishop did not live to see the work published, and after his death his successor as Prince Bishop was more grudging of the cost. Despite this, the work was completed, with over a thousand drawings on 367 plates, and the first print run, of 300 copies, came out in 1613, and was followed by several later reprints. The original paintings are now in Erlangen University Library, but the original copper plates were melted down at the Munich mint in 1817!

The book itself was printed on paper of the largest size available (54–55 x 41–47 cm [21¼–21½ x 16–18½ in]), and was produced in two editions, one with Latin text and black-and-white illustrations, designed for use as a reference book, and the other containing high-quality, hand-coloured illustrations only. The coloured copies were extremely expensive, and Besler, who had managed the entire project from the start, engaging the artists and engravers, seems to have made a handsome profit from sales of the book.

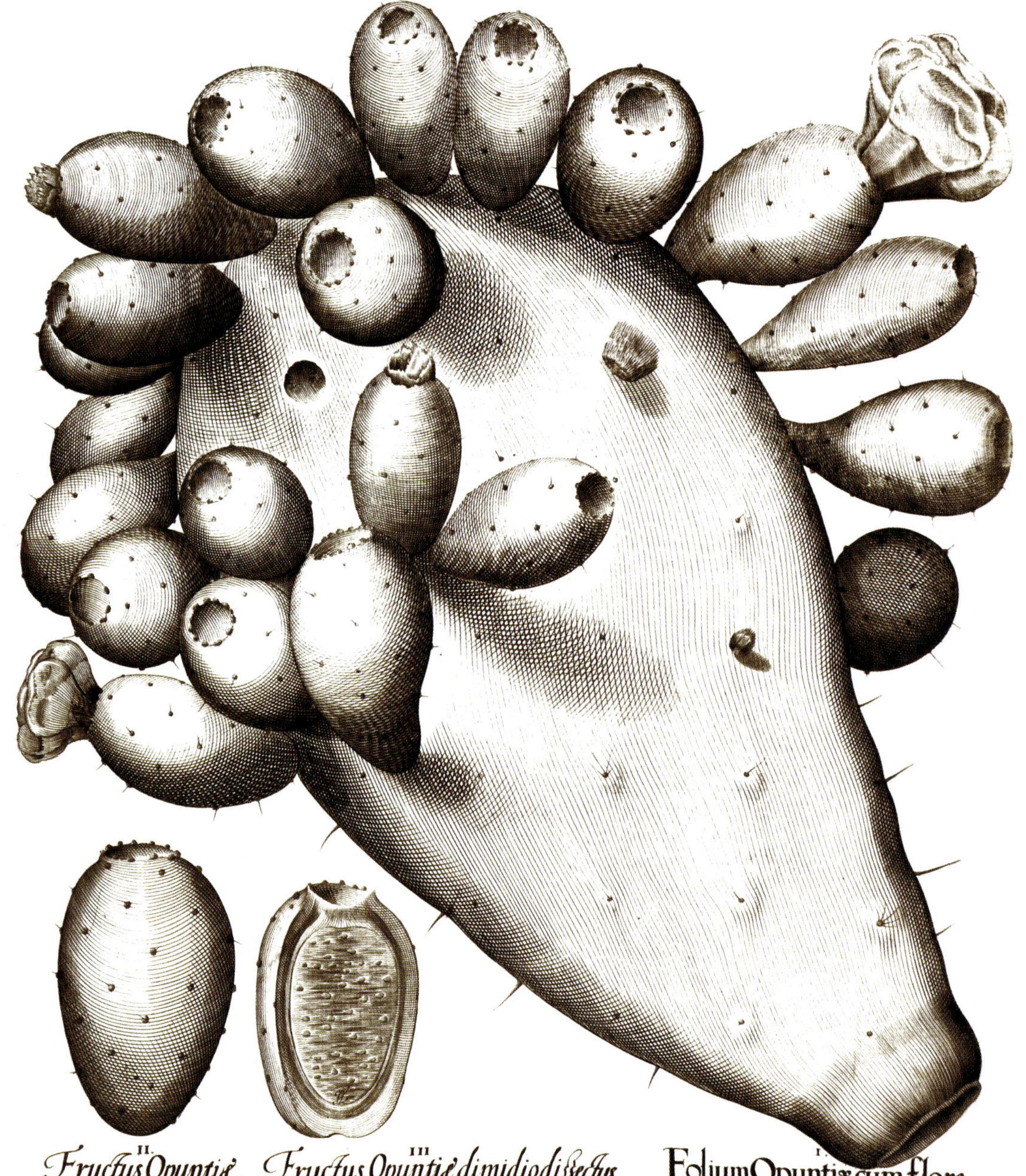

LEFT: These studies of an Opuntia *from the* Hortus Eystettensis *show the plant bearing fruit and flowers as well as the whole fruit and the fruit dissected (far left).*

OPPOSITE: A complete page from the Hortus Eystettensis *with varieties of Viola.*

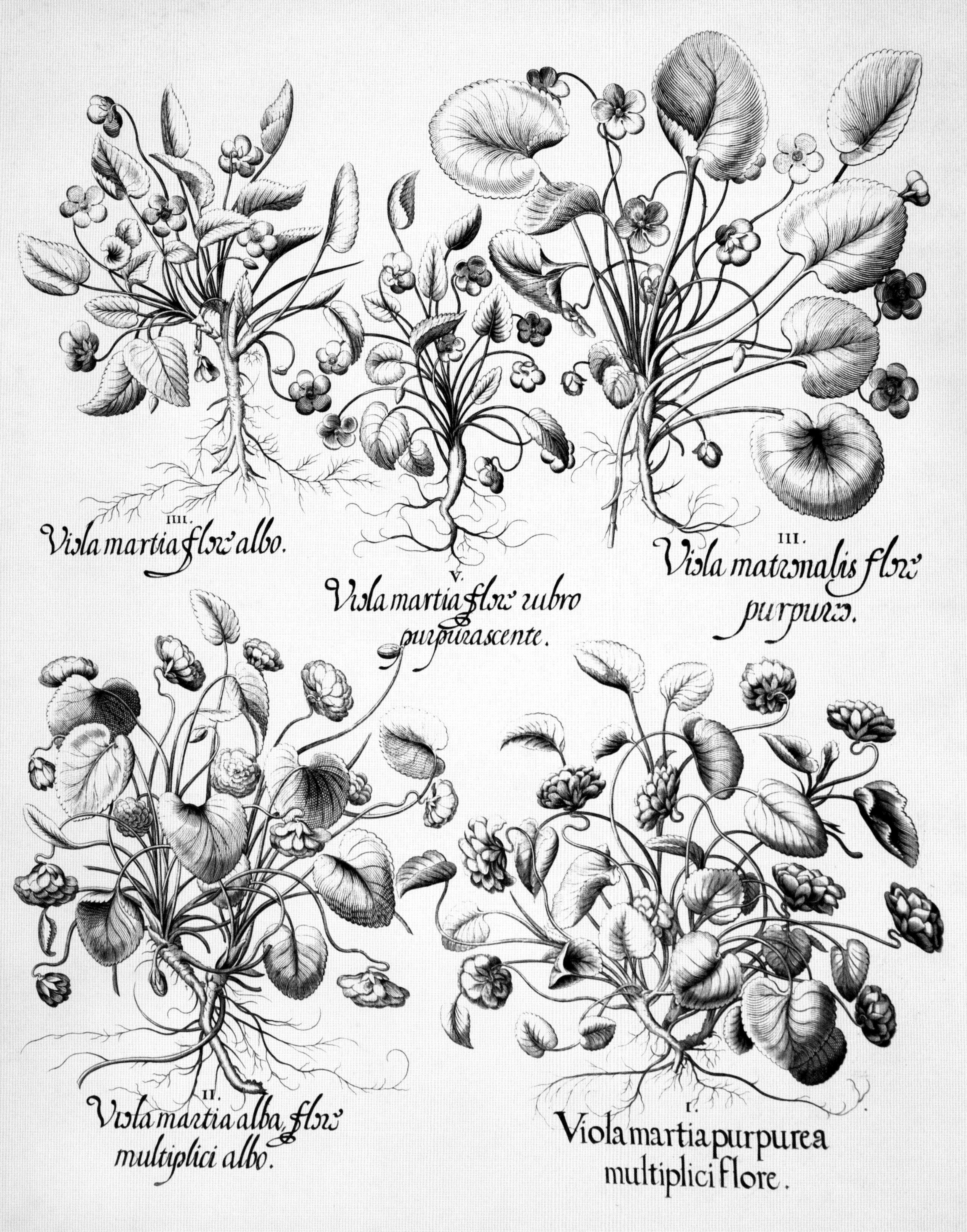
IIII.
Viola martia flore albo.
V.
Viola martia flore rubro purpurascente.
III.
Viola matronalis flore purpureo.
II.
Viola martia alba, flore multiplici albo.
I.
Viola martia purpurea multiplici flore.

THE NETHERLANDS

Virtually nothing is known about the life and work of Pieter van Kouwenhoorn (fl.1630s), although it is thought that he might have worked as a painter on glass in the Netherlands. He painted, in opaque watercolour on paper, an interesting series of garden plants, notably bulbs, and this manuscript florilegium is today in the Lindley Library of the Royal Horticultural Society, London. Van Kouwenhoorn's painting of delicate double flowers and the striped tulips which were then so fashionable is particularly skilful.

Crispin van de Passe the Younger was born in 1589, and came from a well-known family of engravers. His book, entitled *Hortus Floridus*, was published in Utrecht in 1614 and illustrated with charming copper-plate engravings of plants, often accompanied by a bee or butterfly. The following year it was translated from the "Netherlandish Originall" into Latin, French and English, and titled *A Garden of Flowers*; this gives a clue to the fact that this was a book depicting flowers that could be grown by gardeners for ornamental effect, in other words that it was definitely a florilegium rather than a herbal. The book is arranged by season, the first volume "contayninge a very lively and true description of the Flowers of the Springe" – an array of flowers, including many bulbs, such as hyacinths, narcissi and tulips. The summer part contains an eclectic mix "portrayed to the very life" of irises, paeonies, pinks, Canterbury bells and so on; the two remaining seasons dwindle somewhat into autumn crocuses, daphnes, and, strangely, snowdrops.

BELOW: A garden view that was included in Crispin de Passe's Hortus Floridus *and is one of the earliest records of garden design in the late sixteenth and early seventeenth century.*

OPPOSITE: Two pages from Hortus Floridus. *The page above shows varieties of* Melanthium, *while the one below depicts varieties of* Narcissus.

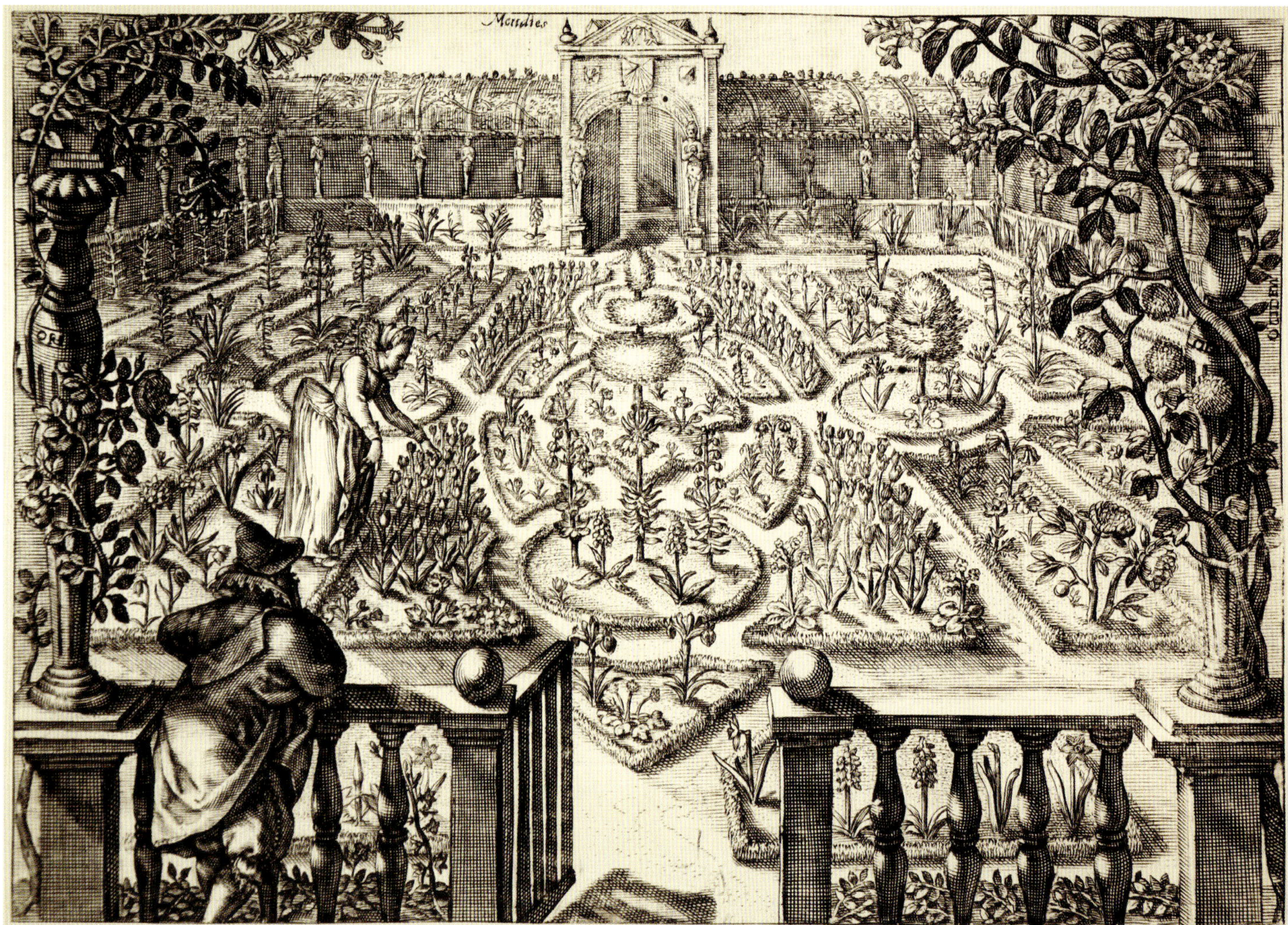

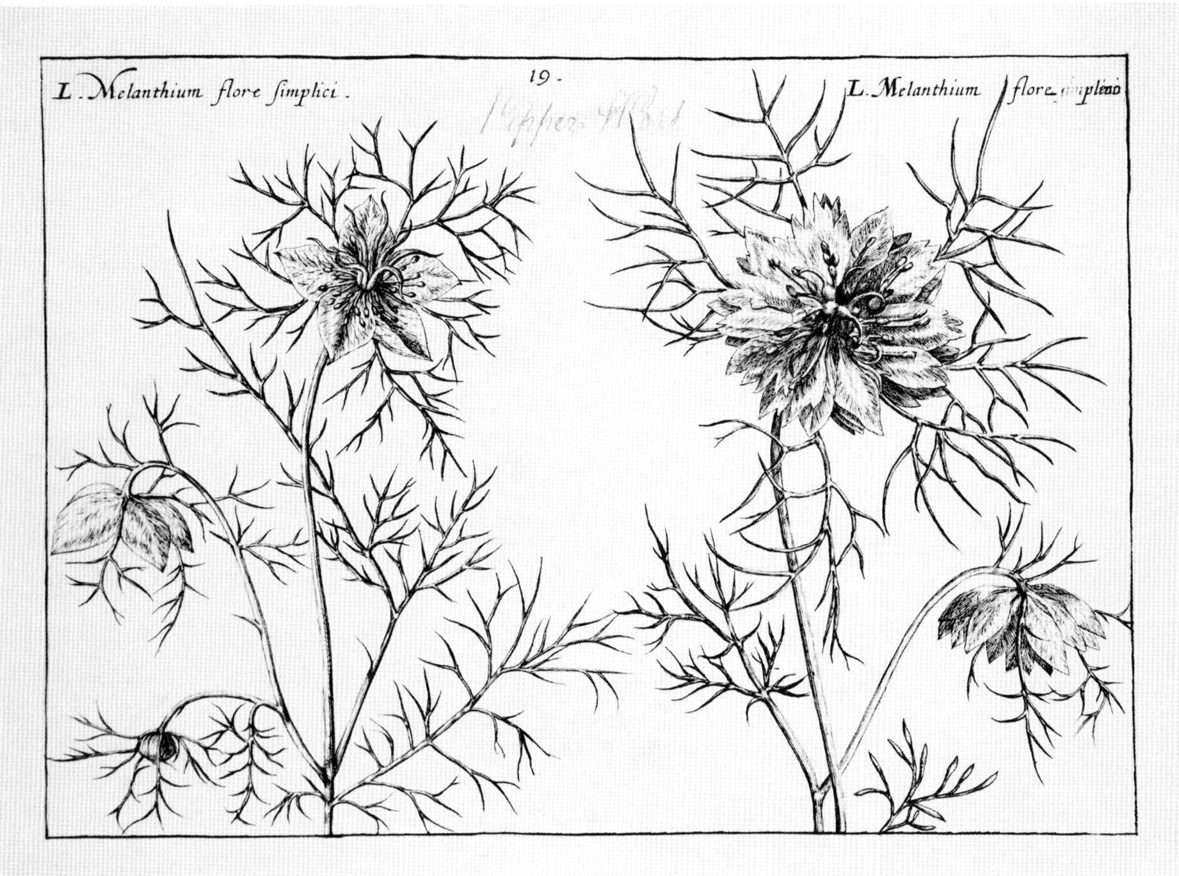
L. Melanthium flore simplici.
19.
L. Melanthium flore pleno

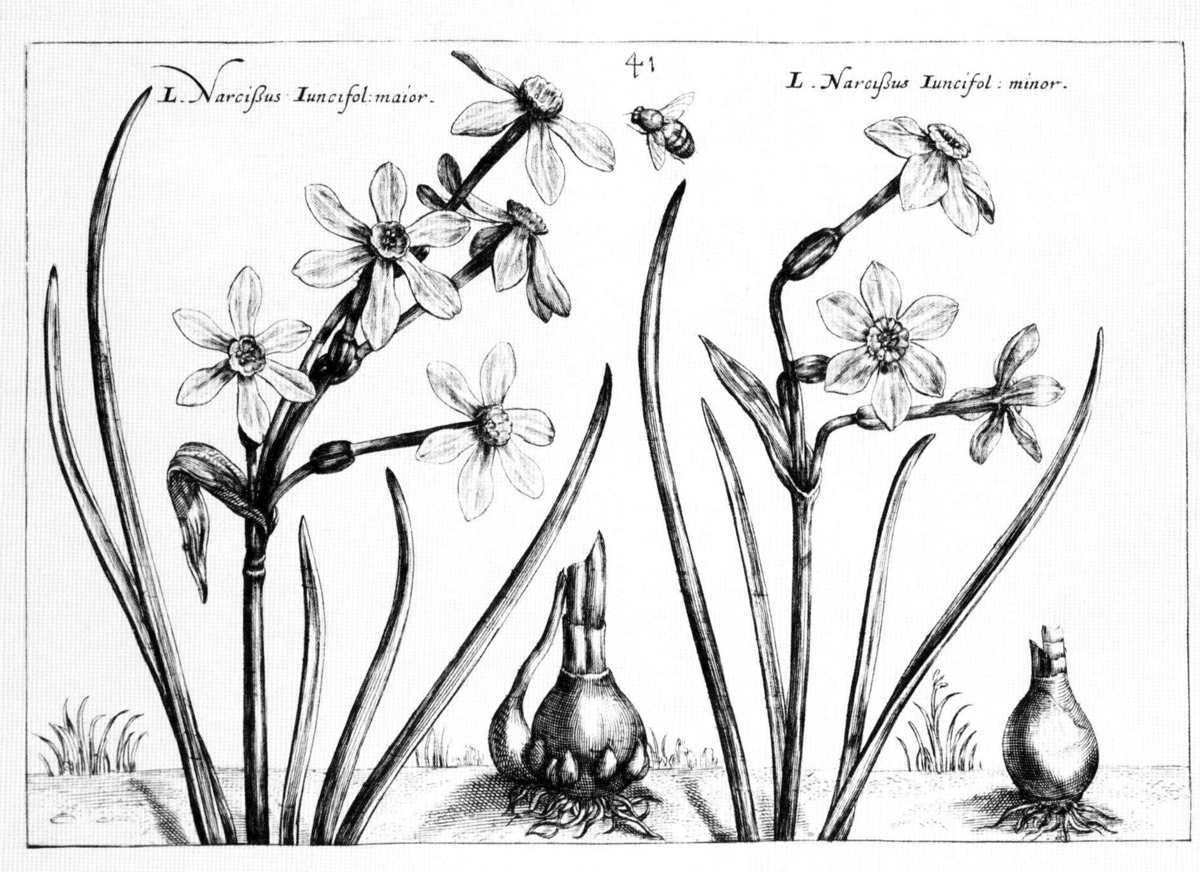
L. Narcißus Iuncifol: maior.
41
L. Narcißus Iuncifol: minor.

LONDON

The London apothecary John Parkinson (1567–1650), was Principal Royal Botanist to King Charles I, and owned a garden in Covent Garden, London. There he grew many "exotic" plants, which formed the basis for his book, entitled *Paradisi in sole paradisus terrestris*, published in 1629. The subtitle promises the reader "A Garden of all sorts of pleasant flowers which our English ayre will permit to be nourished up: with A Kitchen garden of all manner of herbes, roots and fruites…" and it is actually a combination of florilegium, botanical treatise, herbal, and gardening book. The title is a pun on Parkinson's name, (park in sun or earthly paradise), and the title page shows Adam and Eve in a garden, with Adam picking the forbidden fruit from the tree, while Eve finds a strawberry.

In his "epistle to the reader" Parkinson writes that "having perused many Herbals in Latine, I observed that most of them have eyther neglected or not knowne the many diversities of the flower Plants, and rare fruits are known to us at this time, and (except Clusius) have made mention but of a very few… divers Bookes of Flowers also have been set forth, some in our owne Countrey, and more in others, all which are as it were but handfuls snatched from the plantifull Treasury of Nature".

He gives careful instructions on how to avoid pests and protect plants in winter, but the main part of the book is the garden of pleasant flowers, starting with bulbs such as lilies and the Crown Imperial, now "so well known to most persons, being in a manner everywhere common", continuing through a cornucopia of auriculas, violets, marigolds, pinks, roses, passionflowers and cannas, before moving on to the kitchen garden and orchard.

The *Florilegium* of the English Alexander Marshal (*c.*1620–82) consists of 159 pages of fine watercolours depicting over 600 different plants, animals, birds and insects. It is now bound into two volumes, and kept as part of the Royal Collection at Windsor Castle. Few details are known of Marshal's life, but he seems to have been an amateur artist of independent means, a keen gardener and entomologist, who painted the contents of gardens belonging to himself and his friends. Samuel Hartlib, the Puritan intellectual and garden writer described him as a merchant and dealer in "all manner of Roots Plants and seeds from the Indies and else where".

In addition to Hartlib, Marshal was friendly with the keen amateur botanist Bishop Henry Compton, who had a famous garden at Fulham Palace, and with the gardener John Tradescant the Younger. Tradescant, in his catalogue of possessions in 1656, listed "A Booke of Mr. Tradescant's choicest Flowers and Plants, exquisitely limned in vellum, by Mr. Alex Marshall", and as nearly all the plants shown in the Florilegium were grown by Tradescant, it seems that the first work on vellum, which, as far as is known, has not survived, formed the basis of the second.

The *Florilegium* contains 165 plates, depicting 650 individual flowers, and follows the familiar pattern of plants arranged roughly in flowering sequence, beginning with the spring bulbs, particularly numerous varieties of tulip, and continuing through roses, pinks, and so on. It also includes one or two surprises, in the form of probably the first illustrations of two South African plants, which could only recently have been introduced to England – the scarlet-flowered *Sutherlandia frutescens* and the so-called Guernsey Lily, *Nerine sarniensis*.

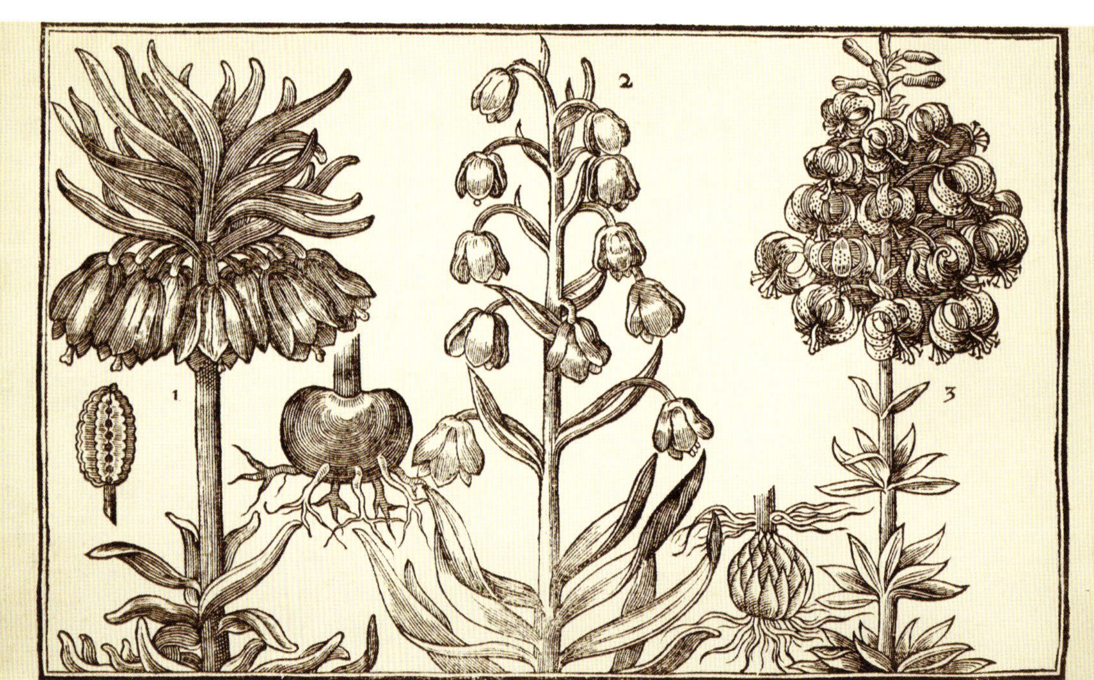

RIGHT: A page from John Parkinson's Paradisi in sole paradisus terrestris. *It shows (left to right)* Corona imperialis *(Crown Imperial fritillary);* Lilium persicum *(Persian lily) and* Martagon imperiale *(Imperial Martagon lily).*

OPPOSITE: A page from the Florilegium *of Alexander Marshal showing studies of* Auricula *made in watercolour* c. *1650.*

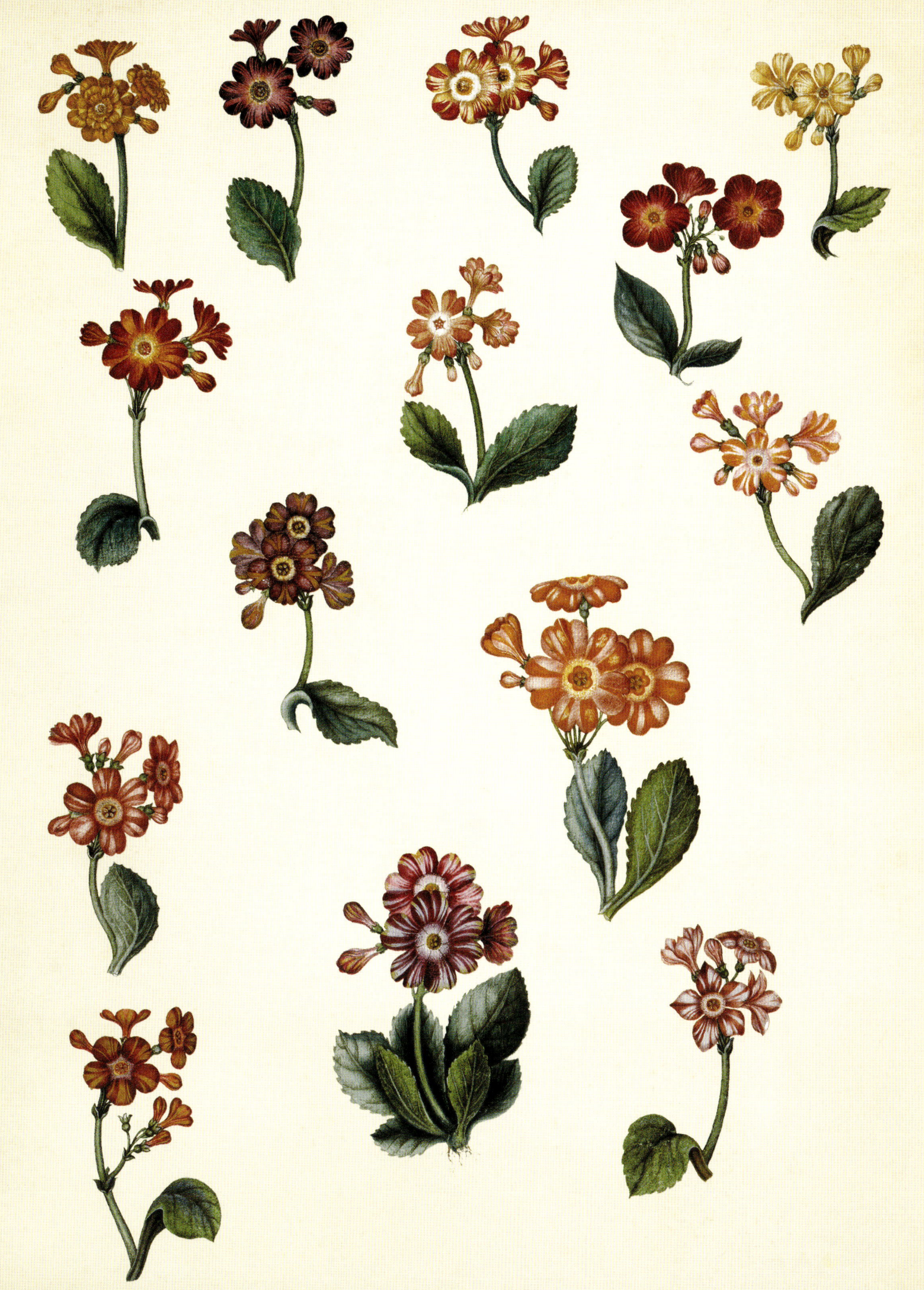

FINE EDITIONS

The two volumes of *Horti Itzteinensis* (c.1654) are reminiscent of *Hortus Eystettensis*, with many of the same flowers arranged in sequence by flowering season. Johann Walther the Elder (?before 1610–c.1676) of Strasbourg was a miniaturist and artist who was commissioned by Count Johann of Nassau to paint all the plants growing in his garden at Idstein, near Frankfurt am Main. The Count, who had been in exile, returned to his castle in 1649 and set about restoring the building and the grounds, laying out a formal garden, which he filled with a splendid collection of plants including a particularly interesting range of bulbs.

Another sumptuous work was produced in Holland towards the end of the seventeenth century. *Horti medici amstelodamensis rariorum plantarum descriptio et icones* (1697–1701), is a record of rare plants growing in the physic garden, or Hortus Medicus, at Amsterdam. This was produced by Jan and Caspar Commelin. Jan, a botanist, and one of the founders of the garden, oversaw work on the first part of the book, published after his death in 1697, and the second part was completed after his death by his nephew, Caspar, and published in 1701. These fine volumes were illustrated with hand-coloured copper engravings of a wide variety of plants, many of them exotics, brought back to the Netherlands by Dutch traders. Most of the paintings were done by Jan Moninckx and his daughter Maria, with the addition of a few extra plates from other artists, including two by Johanna Helena Herolt, daughter of Maria Sibylla Merian (*see* pages 74–75), and the wife of a merchant who traded with South America and the West Indies.

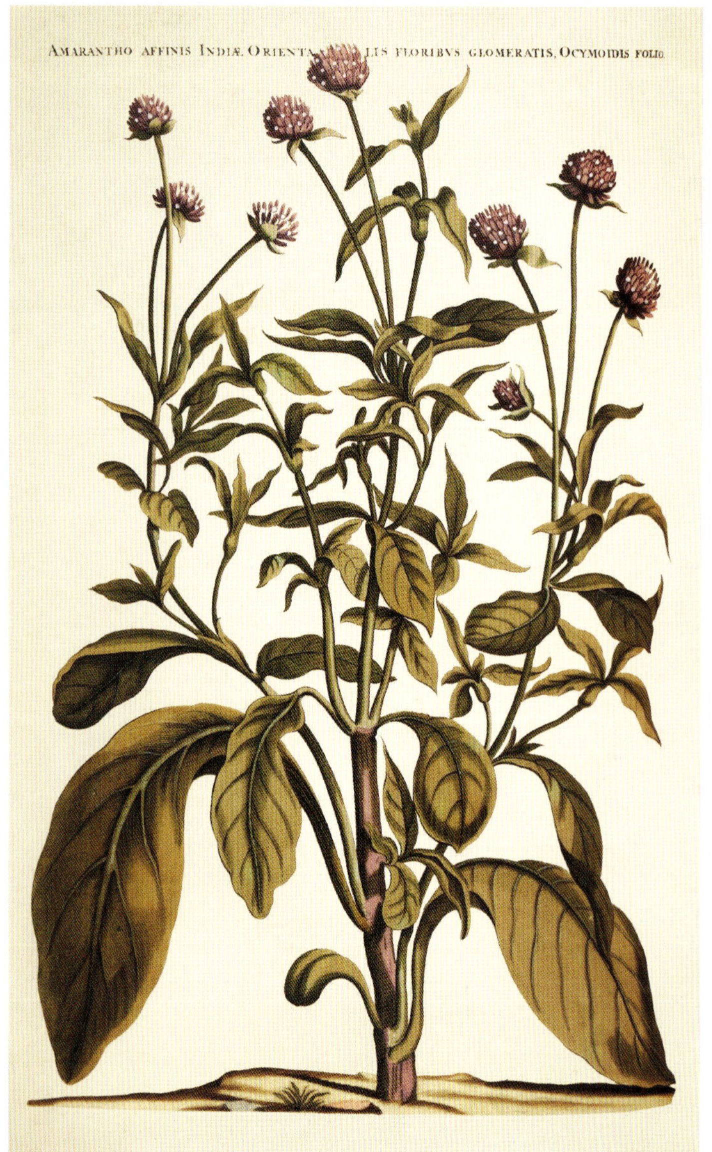

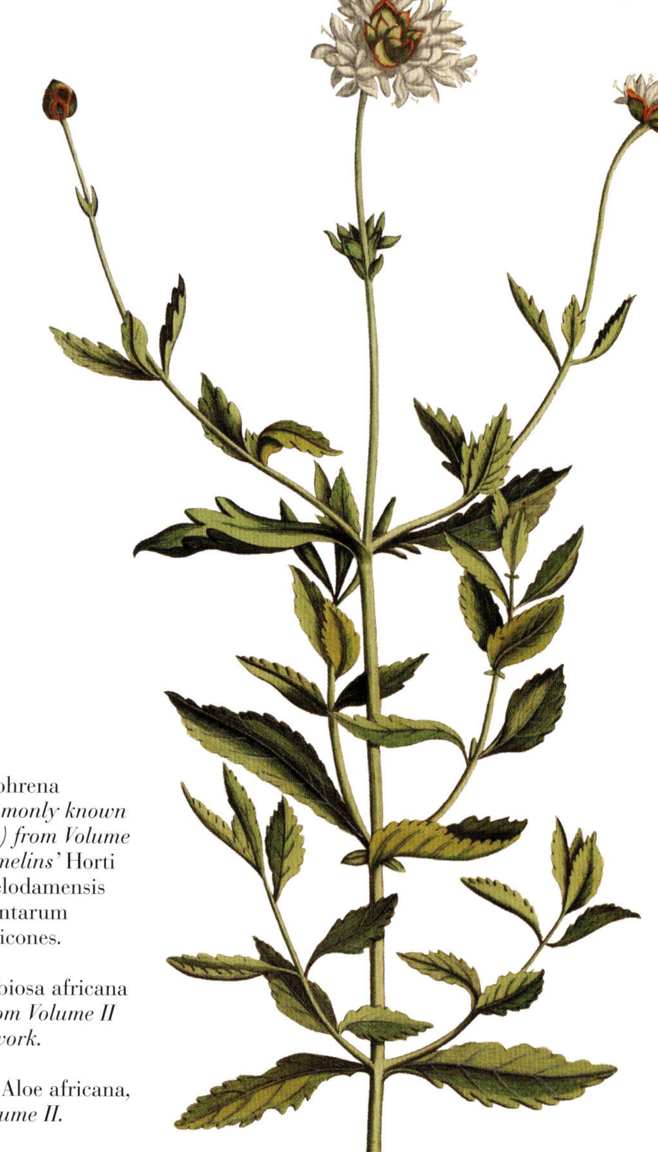

LEFT: Gomphrena globosa *(commonly known as Amaranth) from Volume I of the Commelins'* Horti medici amstelodamensis rariorum plantarum descriptio et icones.

RIGHT: Scabiosa africana frutescens *from Volume II of the same work.*

OPPOSITE: Aloe africana, *also from Volume II.*

ALOE AFRICANA FOLIO GLABRO ET RIGIDISSIMO FLORE SUBVIRIDI.

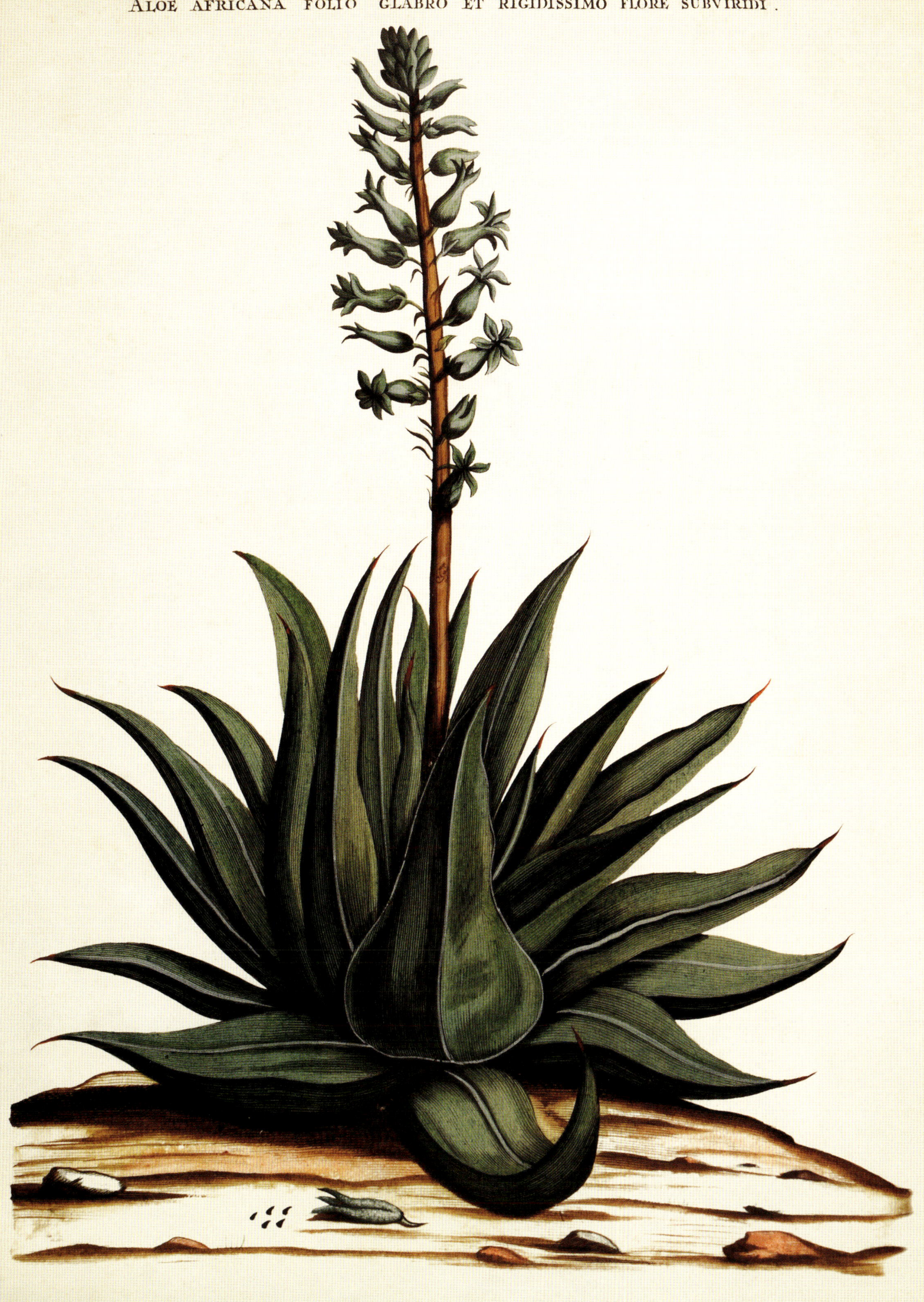

Dutch Flower Paintings

While botanists and gardeners were commissioning paintings of individual flowers, a group of painters in the Netherlands was producing works showing fantastic vases packed with different flowers; it is said that the first was commissioned by a poor lady who could not afford to buy fresh flowers or exotic bulbs.

One of the earliest and most famous exponents of this genre was Jan Brueghel the Elder (1568–1625), and his paintings are interesting both for their beauty and for their wide range of exotic flowers: bulbs then recently introduced from Spain, Italy and Turkey, with newly-bred roses and the then madly fashionable and expensive striped tulips. Most of Brueghel's life was spent in Antwerp, where he often collaborated with Pieter Paul Rubens (1577–1640). In a famous example, *Madonna in a Garland of Flowers*, painted in 1616 and now in the Alte Pinakothek in Munich, the garland is painted by Brueghel, the figures by Rubens. *A wreath of flowers with the Holy Family and Saint Elizabeth with the infant John* shows over a hundred different flowers, and the striking *The Garden of Evil with the Fall of Man* and *The Vision of St Hubert* are other paintings where the two collaborated, the landscapes and objects painted by Breughel, the figures again by Rubens.

Flowers in a Stoneware Vase, now in the Fitzwilliam Museum in Cambridge, is a typical early example of a large flower piece, painted in oils on a wooden panel. In this painting we can identify *Iris susiana*, double *Ranunculus asiaticus* and *Narcissus tazetta*, three favourites of the Ottoman Turks, at least ten tulips, three different dwarf *Narcissus* from Spain, a hyacinth, a fritillary, one Alba rose, probably "Great Maiden's Blush", and two *Rosa centifolia*, as well as numerous other commoner flowers. The flowers are natural and very well painted and the whole effect is rather formal compared with the exuberance of later painters. Brueghel and other Flemish flower painters used the same flower several times in different paintings, and probably had sketchbooks to work from and record rare flowers, but sadly none seem to have survived.

Contemporary with Breughel are Ambrosius Bosschaert (1573–1621), also from Antwerp, and Jan Davidsz. de Heem (1606–1683/4), whose arrangements have a more exuberant style, their vases dominated by large frilled poppies and striped tulips swirling around the canvas. The flower piece continued to be fashionable into the eighteenth and nineteenth centuries. Painters such as Rachel Ruysch (1664–1750), who spent most of her life in Amsterdam, Jan van Huysum (1682–1749) and Jan van Os (1744–1808) painted large vases of flowers, often with the addition of a marble plinth or shelf, strewn with fruit. In the nineteenth century, scientific botanical illustrators such as Gerard van Spaëndonck (1746–1822) and Pierre-Joseph Redouté himself painted vases of flowers, perhaps as a relaxation from their scientific work or as a source of extra income.

BELOW: Peter Paul Rubens's Madonna in a Garland of Flowers *for which the flowers were painted by Jan Brueghel the Elder.*

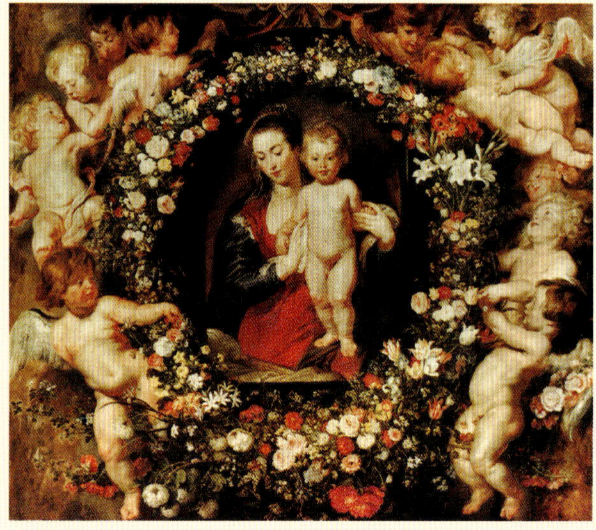

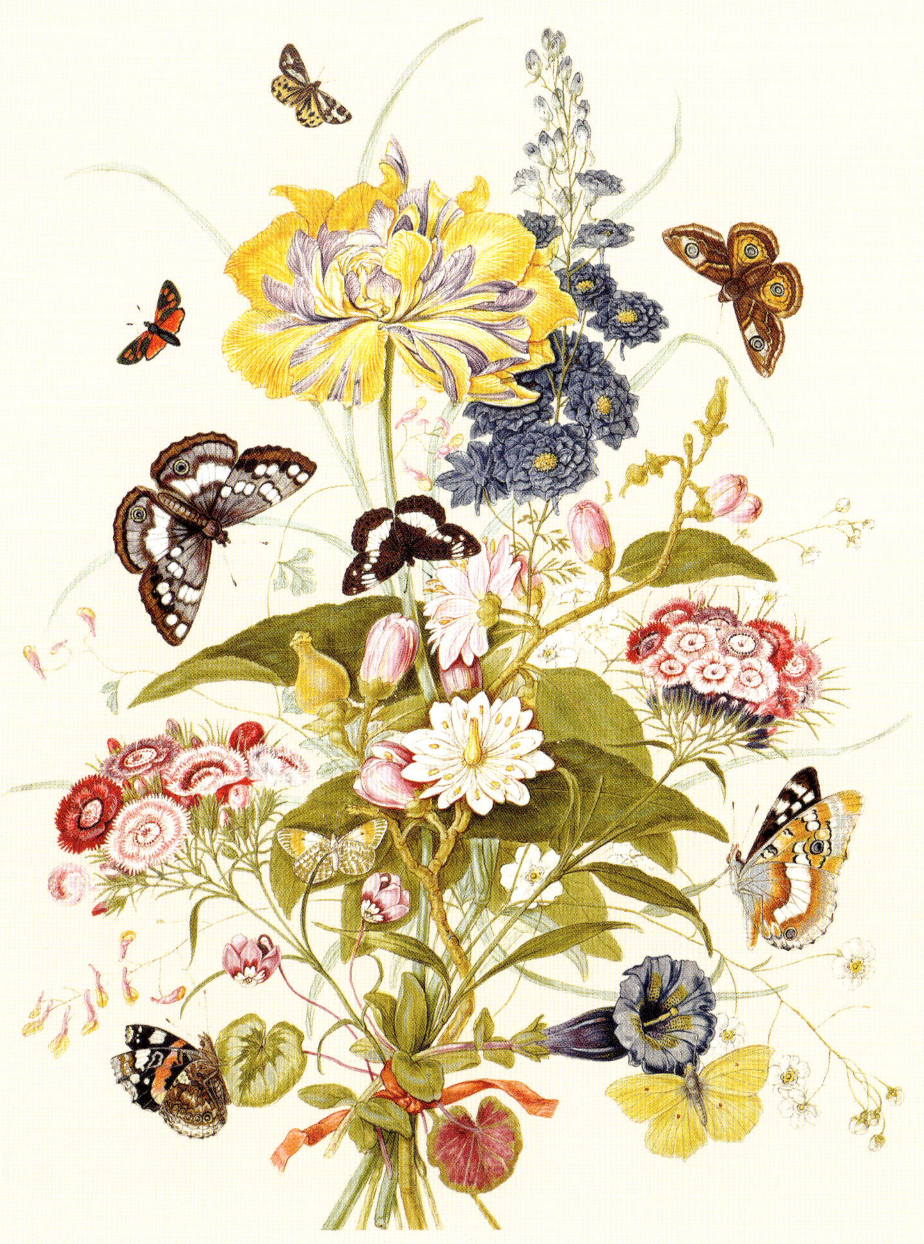

ABOVE: Still Life of Flowers, *which includes a parrot tulip, larkspur, sweet william, gentian and cyclamen, painted in watercolour and gouache by Thomas Robins in 1769.*

The Netherlands and France were the centres of the flower piece, and their painters exported their works to Italy and to England. However, one English painter, Thomas Robins (1716–70), called the elder to distinguish him from his son who also painted flowers, made several paintings of large bunches of flowers, as well as painting floral borders around landscapes, notably a series of the rococo garden at Painswick. The Robins family lived in Bath, then becoming a fashionable resort, and the elder Robins kept a sketchbook, now in the Fitzwilliam Museum, in which he recorded rare flowers; these he used for his paintings and they were probably copied by his son, too, although he specialized in wild flowers. In the elder Robins's work, dating from around 1768, we can see some of the garden flowers that were fashionable in the mid-eighteenth century: double hyacinths, auriculas, passionflowers and the Mexican bulbs *Spreckelia formosissima* and *Polianthes tuberosa*.

An unusual flowerpiece, also in the Fitzwilliam Museum, is a painting on vellum, showing very exotic flowers – Cape heaths, *Phaius tankervilleae*, lobelia, fuchsia and *Pelargonium tricolor* – together with roses and irises. The style is flamboyant, and the flowers are rare and painted with elegance and extreme accuracy. The inscription on the pedestal – "FRANCIS BAUER MDCCXCII" – immediately explains this excellence; Bauer, then working at Kew for Sir Joseph Banks, was perhaps the finest scientific illustrator of the eighteenth century (*see* page 104).

4 NORTH AMERICAN PLANTS

North American plants began to come into gardens in northern Europe from the middle of the sixteenth century; earlier introductions had been mainly from Mexico and the West Indies to Spain, and the plants could not tolerate the English winter. The few that were grown were annuals or tuberous vegetables such as the runner bean (*Phaseolus coccineus*) or the potato, which could be kept away from frost in the winter.

More of these early introductions were grown in southern Spain, from where they were taken to Turkey: Thomas Johnson's herbal of 1636 shows three colour forms of maize, labelled "Turkey Wheat". Few travellers to North America brought back any drawings of flowers, and those American plants that were illustrated came from gardens, and are shown in books such as *Hortus Eystettensis* (1613).

John White went to America as artist and cartographer on the instigation of Sir Walter Raleigh in 1585, and was landed with Richard Greville and around a hundred other men on Roanoke island; they survived the winter, and were rescued the following June. White was put in charge of a second attempt to found a colony in 1587, but had to go to England for further supplies. When he managed to return in 1590, he found the colony deserted, and his daughter and infant granddaughter Virginia Dare, the first colonist to be born in North America, missing. They were never found. While he was in America, White made some paintings of flowers, which are now in the British Museum, as well as more famous illustrations of the local Indians and their customs.

John Tradescant the Younger (1608–62) made three expeditions to Virginia, the first in 1638, "to gather all varieties of flowers, plants and shells". Many of these he grew in his father's garden at Lambeth, and his little book *Museum Tradescantianum* (1656) contains a plant list from the garden as well as a collection of curiosities, which later became the basis of the Ashmolean Museum in Oxford. It had no illustrations, but many of Tradescant's plants are illustrated in Alexander Marshal's florilegium.

John Banister (1654–92) was a graduate of Magdalen College, Oxford, who was sent by Bishop Henry Compton to Virginia, both as a priest and as a botanist, and became one of the founders of the College of William and Mary in Williamsburg. He corresponded widely with collectors in England, and sent back seeds to Compton to grow in the garden at Fulham Palace. He was preparing a natural history of Virginia, but was accidentally shot while on an expedition into the interior before it was published. His library and notes and drawings were dispersed; some of his specimens with some manuscripts are in the Natural History Museum in South Kensington, others are in Oxford. His Virginia plant list was published in John Ray's *Historia Plantarum* (1688).

ABOVE: An engraving of John Tradescant the Younger. His grave can be found in the graveyard of the former church which now houses the Garden Museum in Lambeth, south London.

LEFT: John White's watercolour and graphite study of a pineapple.

OPPOSITE: Sarracenia purpurea, *accompanied by a Leopard frog (perhaps* Rana pipiens*) from Catesby's* The Natural History of Carolina, Georgia, Florida, and the Bahama Islands.

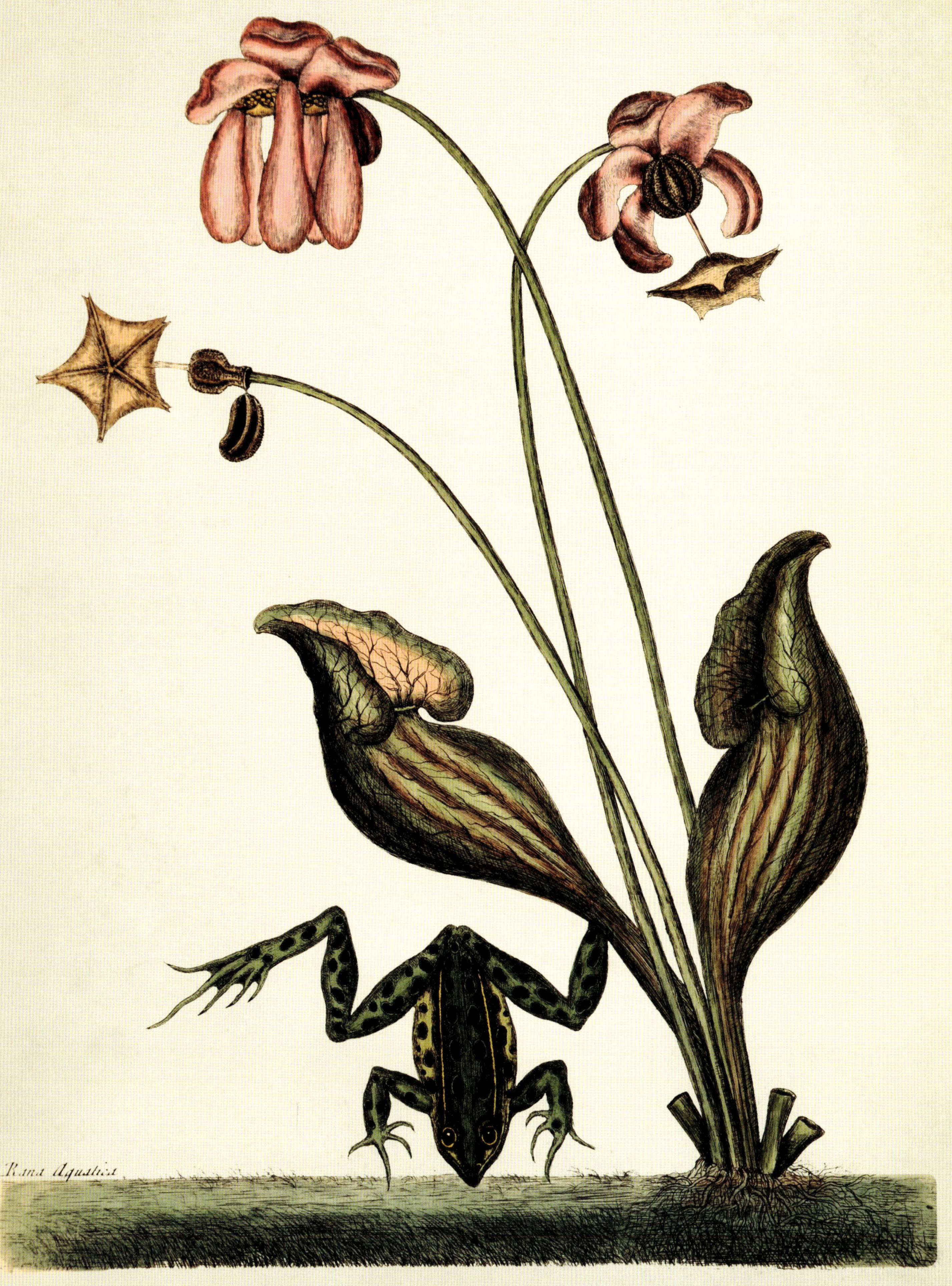
Rana Aquatica

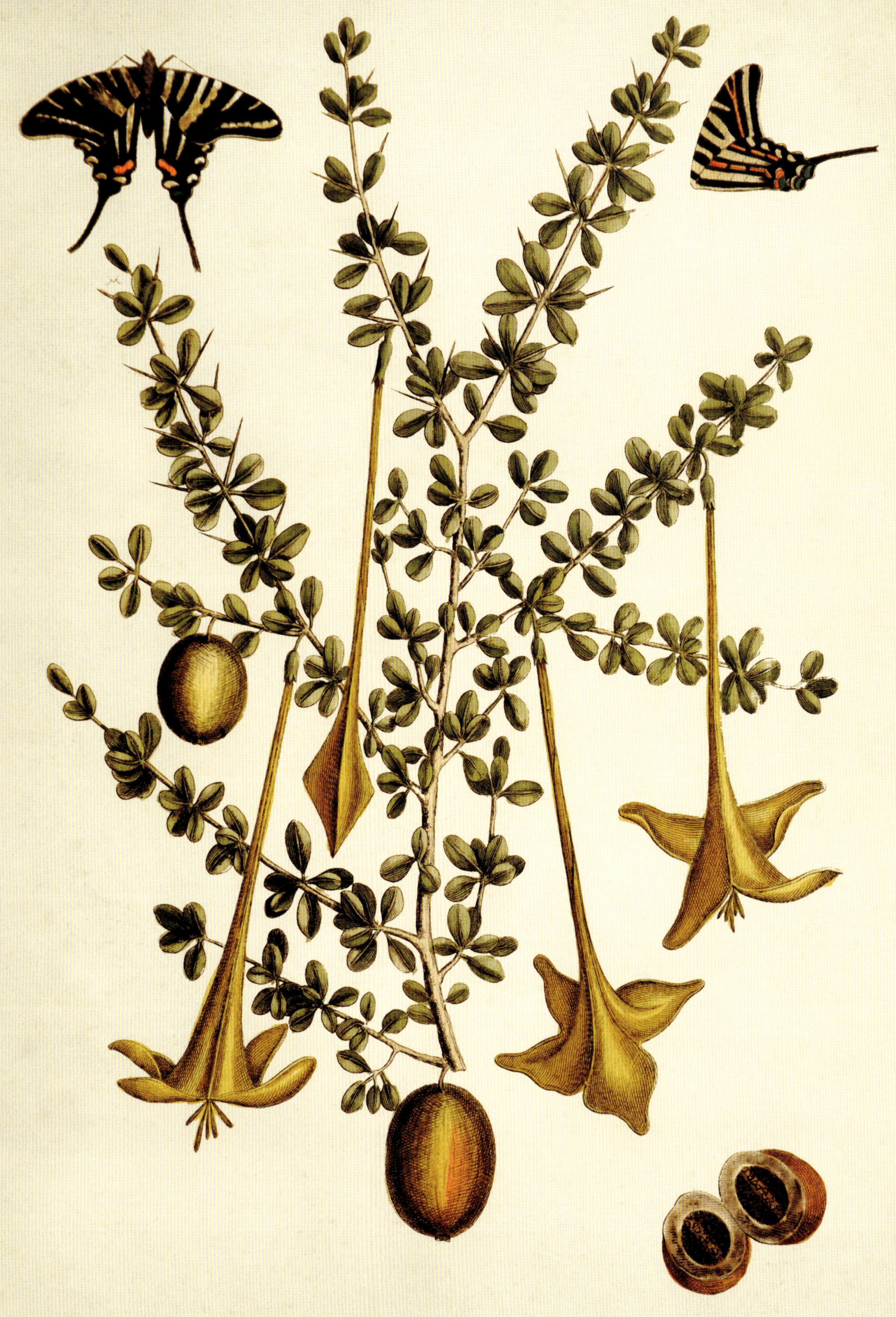

MARK CATESBY

The works of Mark Catesby (1682–1749), 150 years after White, are the first major illustrated books devoted to American plants. *The Natural History of Carolina, Georgia, Florida, and the Bahama Islands* was published between 1730 and 1747. Catesby was born in Essex, probably at Castle Headingham, and brought up in Suffolk. His youthful love of natural history was encouraged by John Ray, a fellow of Trinity College, Cambridge and one of the leading scientists of the day. Catesby's first visit to America was to his sister in Williamsburg, Virginia in 1712, before returning to England with his collections in 1719.

Three years later he made a second expedition, supported by the Royal Society and by a group of subscribers organized by William Sherard, and was based in Charlestown. He travelled into the interior as far as Fort Moore in Georgia, and to the Bahamas in the winter of 1725. He sent back boxes of seeds and plants, and returned with drawings from which he produced his books. He taught himself etching, and produced all 220 plates, which were hand-coloured, and show plants, birds and animals, often together on the same plate. His original paintings are in the Royal Library at Windsor, and have a charming amateurishness, being less stiff than the printed plates. In some of the plates, notably that of *Magnolia grandiflora*, the background is painted black, a very unusual but effective way to show off the white flowers; this tree first flowered in Parson's Green in 1737, and Catesby's plate is based on the same flower that Ehret used for his engraving, published in 1743. Catesby was also one of the first to illustrate the pitcher plant (*Sarracenia purpurea*) which he shows with an attendant frog.

Catesby's second book showed American plants cultivated in England that, in part at least, were grown from seed which he had collected in Virginia. Here again, Catesby drew and etched the plates himself, often putting four different plants on the same page. One plate shows familiar American trees, such as Sassafras, Liquidambar, tulip tree, and snowdrop tree (*Halesia*). Catesby was still working on this when he died, and it was published first as *Hortus Brittano-Americanus* (1763) and then as *Hortus Europae americanus* in 1767.

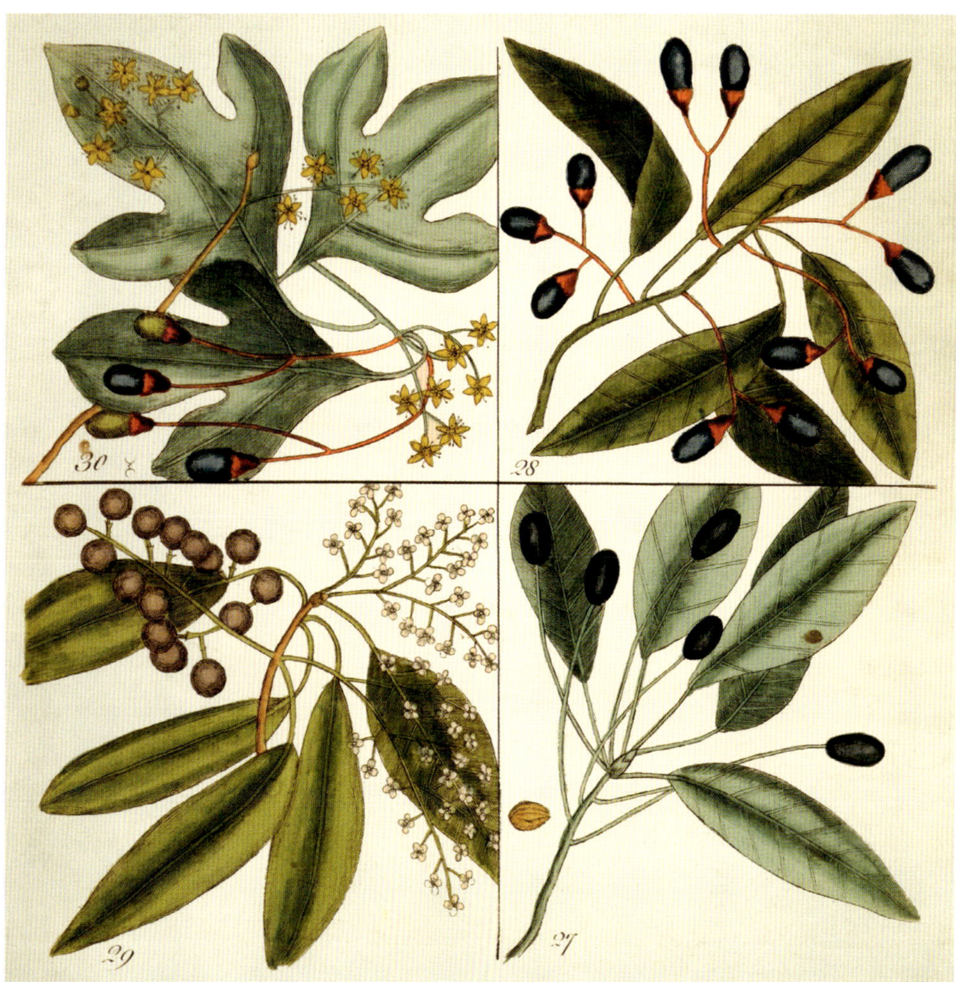

LEFT: Catesboea spinosa with butterflies from The Natural History of Carolina, Georgia, Florida, and the Bahama Islands.

OPPOSITE: A page from Mark Catesby's Hortus Brittano-Americanus, *showing two forms of sassafras, black cherry and tupelo.*

JOHN AND WILLIAM BARTRAM

John Bartram (1699–1777) and his son William (1739–1823) were Quaker farmers who developed an interest in plants and became the first true nurserymen in North America. John set up a botanical garden at his farm on the Schuylkill River at Kingsessing, near Philadelphia, in around 1728: he was named King's Botanist in America by George III in 1765. Both John and William made collecting expeditions into the mountains, and sent back quantities of seed and specimens to Peter Collinson, a fellow Quaker who lived in London (1694 –1768). The Bartrams were excellent botanists and plant collectors, as well as intrepid travellers. William's account of his expeditions, *Travels through North & South Carolina, Georgia, East & West Florida et cetera*, was published in 1791, and became an influential source for the Romantic poets, notably Coleridge who is supposed to have based a passage of *Kubla Khan* on Bartram's description of a spring on the St John's River in Florida, and William Wordsworth used passages from Bartram for his descriptions of America.

Apart from his writing and plant collecting, William Bartram was also a good draughtsman, sometimes producing detailed sketches of a single plant, at others showing a whole habitat, with different plants, insects and birds; his plate of the pitcher plant also shows a snake swallowing a frog. Bartram sent many of these plant and animal illustrations to England, both to Collinson and to John Fothergill (1712–80), and many are now in the Natural History Museum in London (including a fine painting of *Franklinia alatamaha*) while others are thought to be in St Petersburg in Russia with the majority of Fothergill's collection. Few of Bartram's flower paintings or drawings were published in his lifetime, but some were engraved by Philip Miller and published in around 1758, in the second volume of *Figures of the most beautiful, useful and uncommon Plants described in the Gardeners' Dictionary* (1760); most of the plates in both volumes were drawn by Ehret. Others were used by the Philadelphia physician and botanist Benjamin Smith Barton in his *Elements of Botany* (1804), the first botanical textbook for students in America. Among the treasures in the library of the Royal Botanic Gardens, Kew is a small painting of a bunch of dahlias, dating from around 1789. This is interesting as it shows the very first dahlias to be introduced from Mexico, before they had been "improved" by breeders in England. It is by Margaret Meen, who painted flowers at Kew, and exhibited watercolours at the Royal Academy between 1775 and 1785

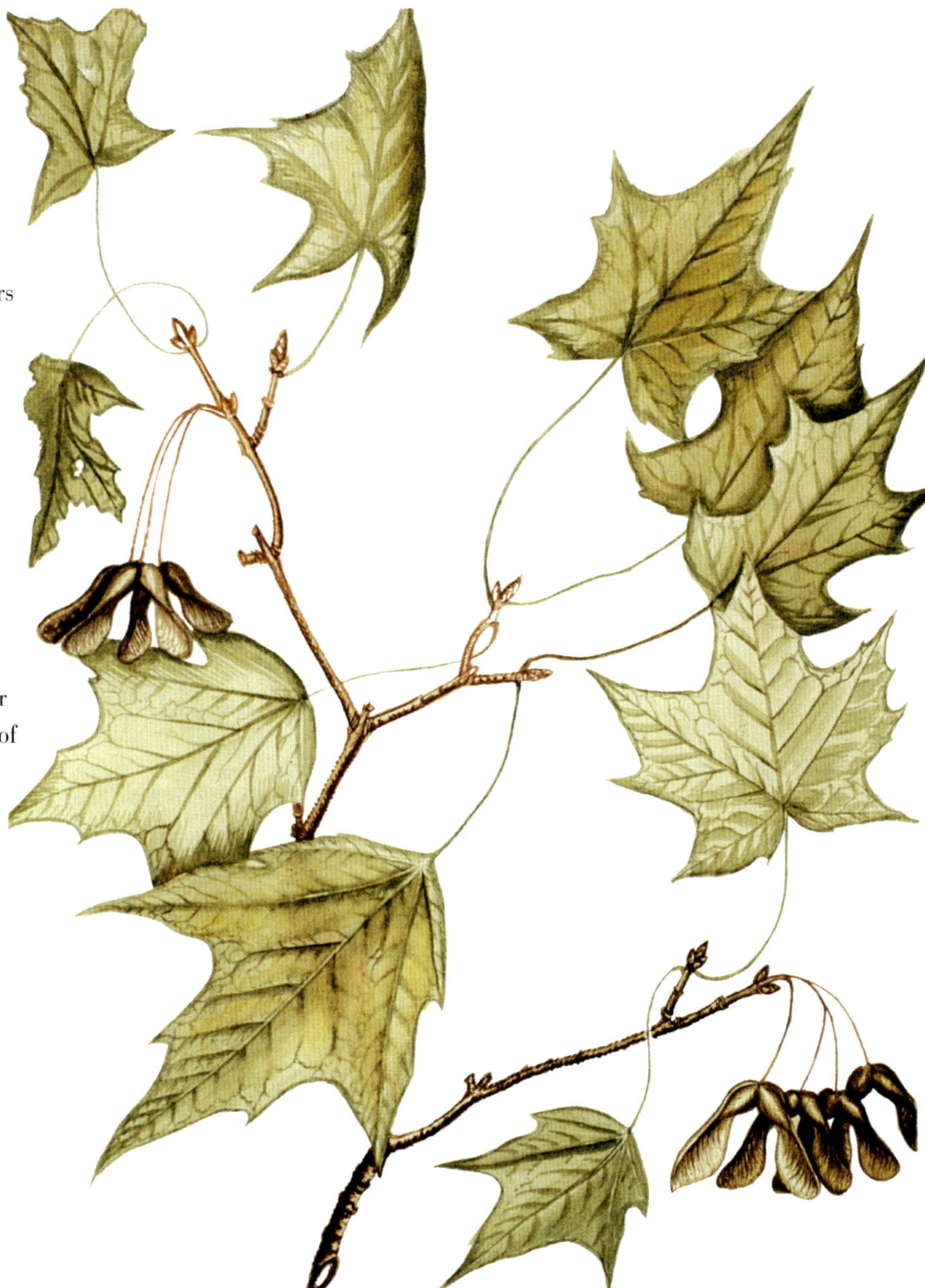

RIGHT: Sugar maple with ripe seeds *painted in watercolour by William Bartram in 1755.*

OPPOSITE: Margaret Meen's painting of c. *1789 of some of the very first dahlias to be introduced from Mexico to Britain.*

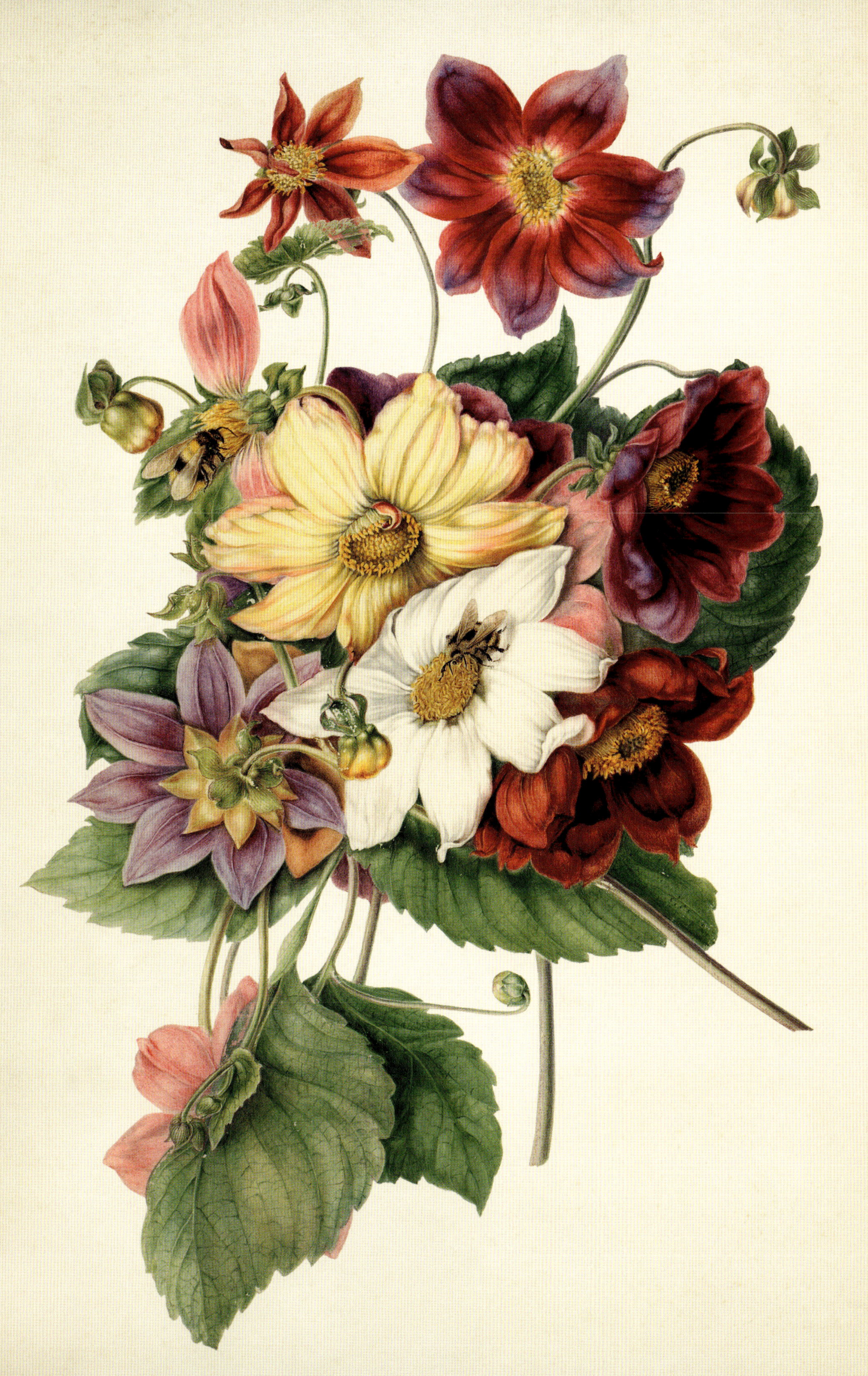

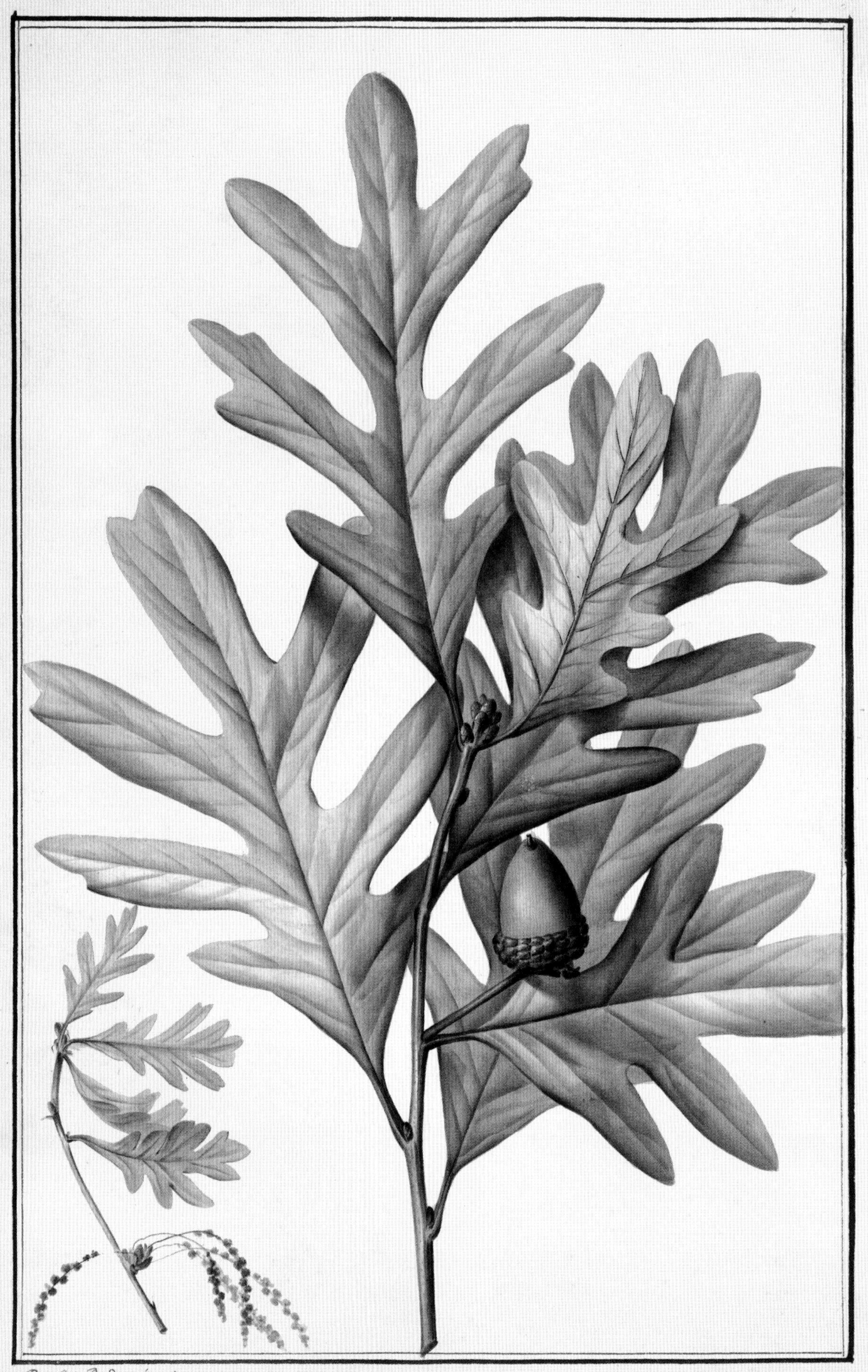
P. J. Redouté pinx.
quercus alba

ANDRÉ AND FRANÇOIS MICHAUX

The American War of Independence, which ended with the Treaty of Paris in 1784, interrupted trade between Britain and North America, but opened up opportunities for France. André Michaux (1746–1802) was a French botanist and traveller. Though born into a simple farming family near Versailles, he became interested in botany and was taught by Bernard de Jussieu, who had been subdemonstrator at the Jardin du Roi and also worked in the garden at Trianon; Michaux's young wife died shortly after the birth of their first child, François André (1770–1855). His first expedition was in 1782 to western Asia with the newly appointed French consul to Persia. Returning to Paris after three years, he was immediately appointed King's Botanist in America and chosen to lead a scientific expedition, encouraged by Thomas Jefferson, to search for new trees to plant in France, where the forests had been depleted by years of war with England.

With two assistants and François André, now aged 15, Michaux sailed for America, arriving in New York in November 1785. Soon, he bought land to set up a Botanic Garden near Hackensack, New Jersey, and the Michaux travelled to Philadelphia to meet Benjamin Franklin and visit the Bartrams with whom they became great friends. The next summer they followed William Bartram's route south, to Virginia and then into the Appalachians and Georgia. Realizing that the climate near New York was much colder than most of France, they bought land and made their headquarters near Charleston. They spent the following years travelling around America, going as far west as the Mississippi, Kentucky and Illinois and as far north as Quebec, collecting plants and sending seeds and specimens to France. With the revolution in France, André's salary ceased, and they were almost destitute: François returned to France in 1789; his father remained until 1796, but on the voyage home lost most of his personal belongings in a shipwreck on the Dutch coast. His herbarium was damaged by saltwater, but saved. His arrears of salary were never paid, and in 1800 he joined Nicolas Baudin's expedition to Australia. However, he left this expedition en route and visited Madagascar, intending to sail from there to India, but in Madagascar he contracted a fever and died in 1802, aged only 56.

André Michaux's only publication in his lifetime was *Histoire des Chênes d'Amérique*, a folio published in 1801, with 36 beautiful line engravings by the Redouté brothers. His *Flora boreali-americana*, an octavo with 51 small engravings was published by François after his father's death.

François Michaux returned to America between 1806 and 1808. On his return to Paris, he published the culmination of his own and his father's work: *Histoire des arbres Forestiers de l'Amérique Septentrionale*, which appeared in three quarto volumes between 1810 and 1813. The 140 elegant, coloured, stipple engravings are by Pancrace Bessa and Pierre-Joseph Redouté. An octavo version was printed in Philadelphia in 1818–19, as *The North American Sylva*. Some of the drawings of oaks, perhaps for this smaller edition, are in the collection at Kew.

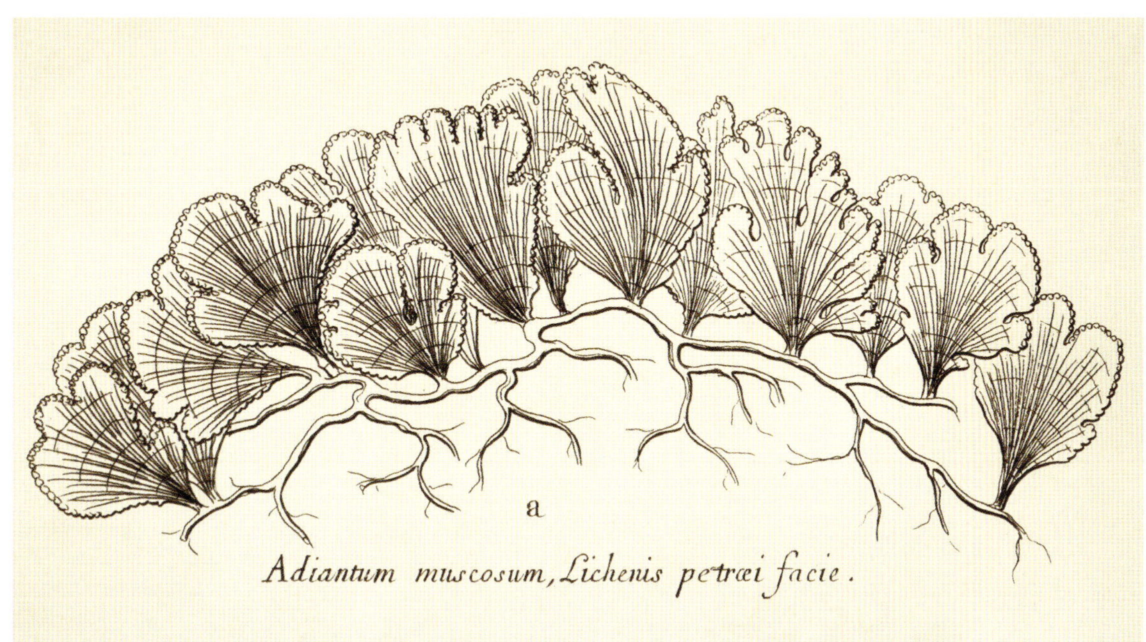

LEFT: Adiantum muscosum *from Charles Plumier's* Description des plantes de l'Amérique *(1693). Plumier's redescription of over 100 genera of American plants was later adopted by Linnaeus for his classification system.*

OPPOSITE: Pierre-Joseph Redouté's drawing of Quercus alba *for André Michaux's* Histoire des Chênes d'Amérique.

GEORG DIONYSIUS EHRET

Georg Dionysius Ehret (1708–70) raised botanical illustration to a new height, both as an art form in itself and as a branch of science. Though he never visited America, he worked at a time when American plants were flooding into Europe from Catesby, the Bartrams or Alexander Garden of Charleston, and painted them when they first flowered in England. He left a brief autobiography, which gives us the details of his early wanderings. He was born in Heidelberg in 1708, in modest circumstances, his father being a smallholder. The father taught his son to draw but died while still young, so the boy was apprenticed to his uncle, a gardener, and managed to continue to draw in his spare time. His mother remarried and her new husband worked for the Elector Palatine of Heidelberg, who put the young Ehret and his brother in charge of one of the gardens; later Ehret worked for the Margrave at Karlsrühe, who was building up a huge collection of rare plants, but again as a journeyman gardener rather than a botanical draughtsman. Ehret's great chance came in 1733, when he met Dr Christoph Trew, a physician in Nürnberg, who had already noticed and admired some of his paintings. Trew was a great scientist, a skilled doctor and student of anatomy, successful and rich, as well as being a keen botanist and collector of natural history objects. He corresponded with botanists from all over Europe, but particularly with a group in England who were becoming increasingly influential: Sir Hans Sloane, Philip Miller, Humphrey Sibthorp and Peter Collinson.

Trew became Ehret's patron and friend and was the major influence in his career, instructing Ehret in botany, and in the importance of

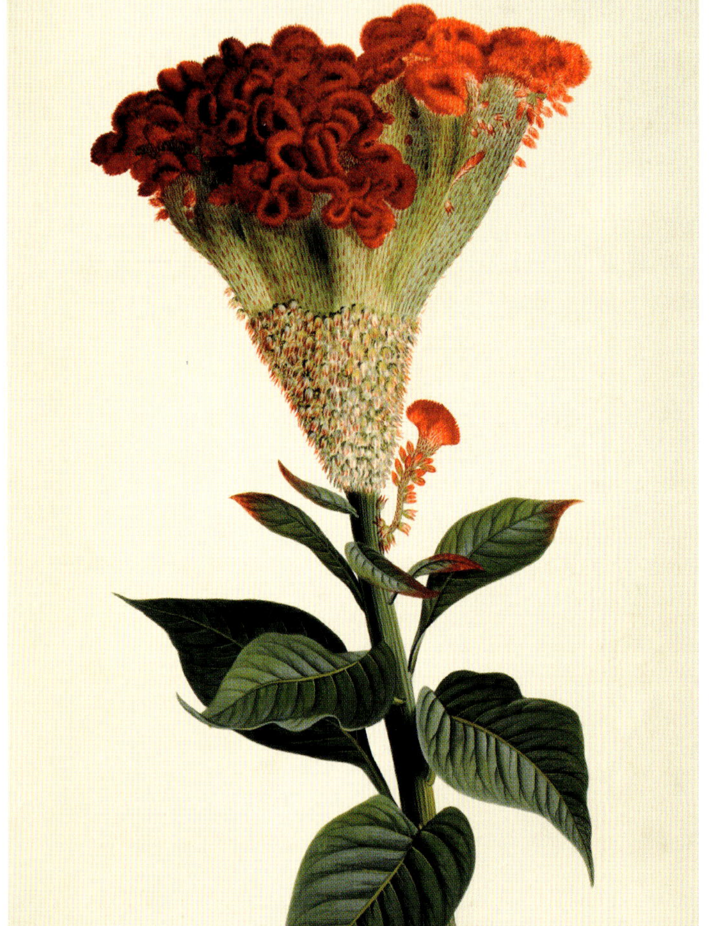

LEFT: Ehret's study of Amaranthus cristatus, *1760.*

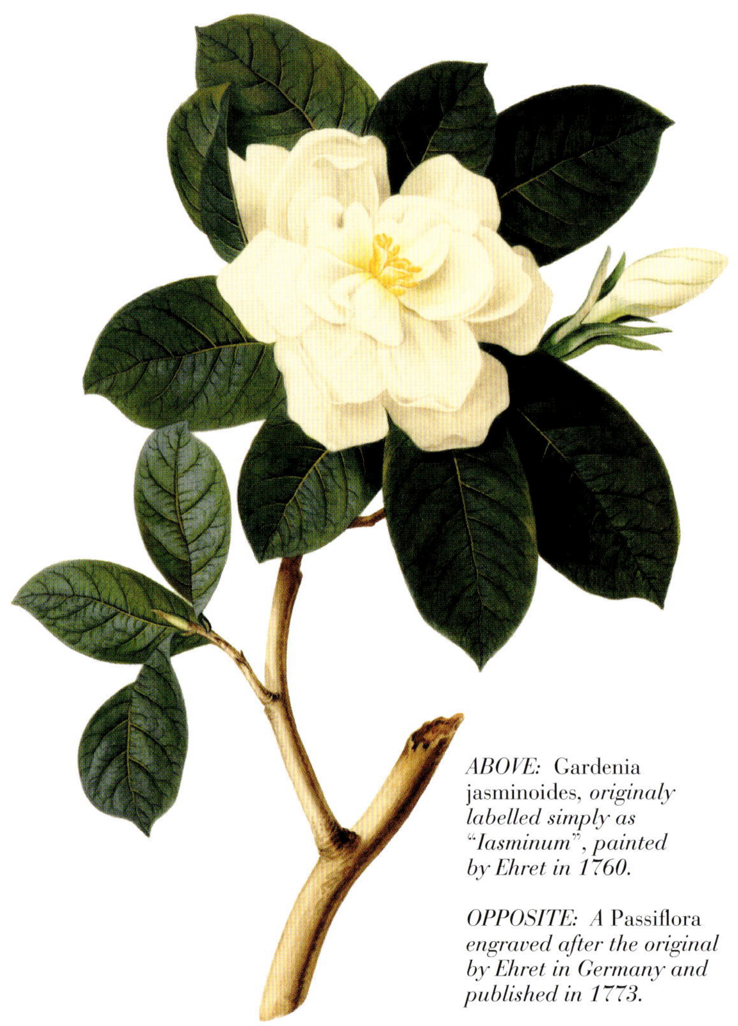

ABOVE: Gardenia jasminoides, *originaly labelled simply as "Iasminum", painted by Ehret in 1760.*

OPPOSITE: A Passiflora *engraved after the original by Ehret in Germany and published in 1773.*

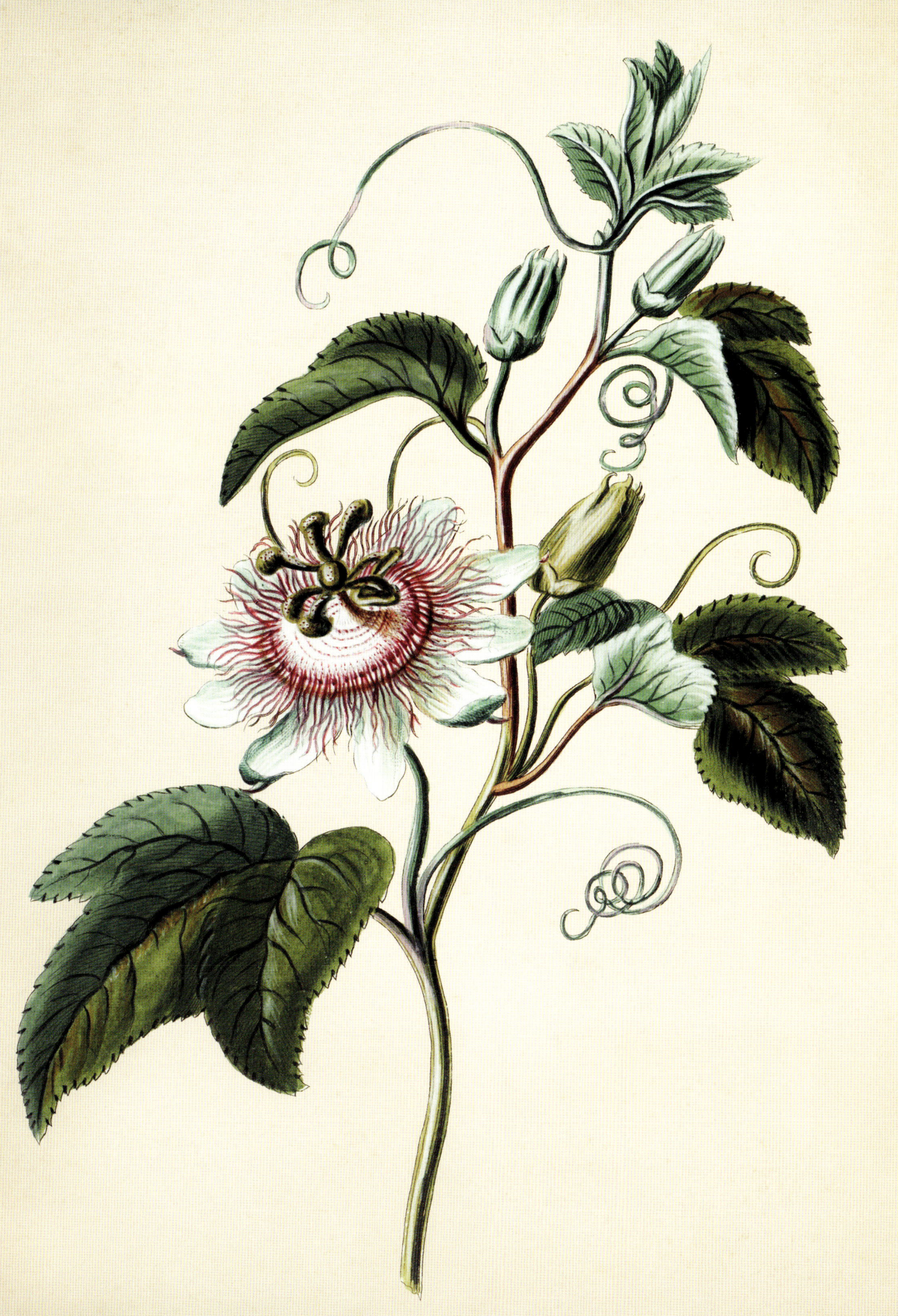

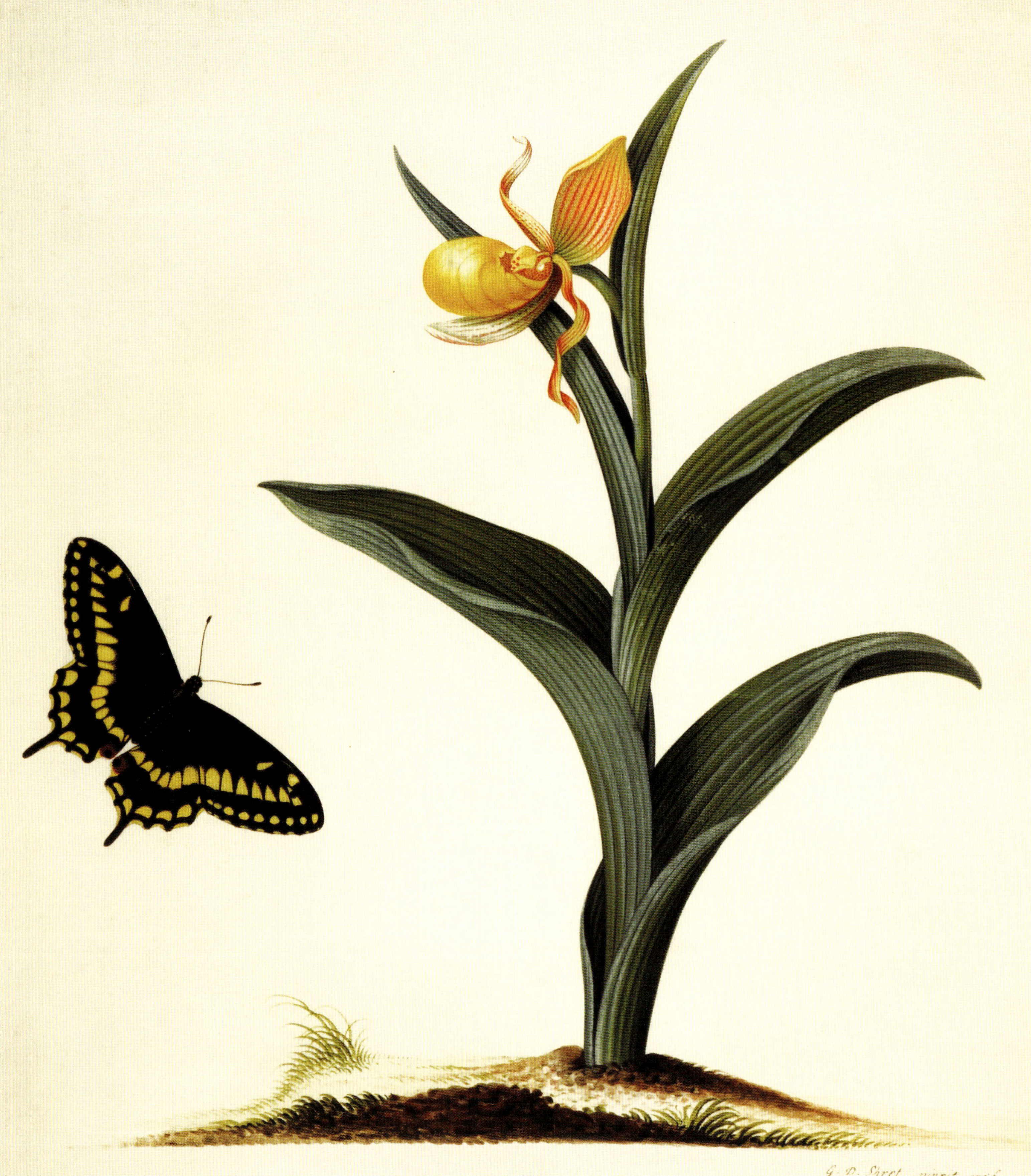

G. D. Ehret. pinxit 1756.

Ladies Slipper from virginia

flower structure in the classification of plants; he also encouraged him to travel and meet botanists in other countries. After visiting Switzerland, Ehret spent the winter of 1734–35 in Paris, where the teacher of botany, Bernard de Jussieu, secured him a lodging at the Jardin du Roi. There, Ehret studied the Royal collection of *vélins*, paintings of flowers on vellum, and met Madeleine Basseporte, then official botanical painter to the King and a pupil of the great Claude Aubriet (*see* page 62).

At de Jussieu's suggestion, Ehret visited England in 1735, meeting Philip Miller and Sir Hans Sloane, both associated with the Chelsea Physic Garden. Here, he continued to paint, sending his work back to Dr Trew in Nürnberg. However, he did not meet with instant financial success, and decided to use an introduction to George Clifford, a banker in Haarlem, which he had been given by the Margrave. The Netherlands were then a centre both for horticulture and botany, and Ehret went first to Leiden, to its famous botanical garden. Here, he heard that Linnaeus was visiting Clifford, so he walked to Haarlem and presented himself, with his letter of introduction and some of his paintings. Ehret produced paintings for Linnaeus' *Hortus Cliffortianus*, but declined his offer of a post in Uppsala and in 1736 decided to return to England, which he considered to have more interesting plants and gardens than Holland. His first position was with Philip Miller, and in 1742 he was addressed as "painter at the Physick Garden". A later appointment, in 1750, as draughtsman in the University Botanic Gardens in Oxford, lasted for only one year, but from then on he managed to make a living selling paintings and engravings, and giving botanical drawing lessons to the daughters of the aristocracy. His artistic leanings were expressed in paintings of garden flowers, such as striped or frilled tulips, auriculas and irises, often accompanied by an exotic butterfly or moth. These are painted in opaque watercolour on vellum, in exuberant style and important patrons for this work included the Duchess of Portland and John Stuart, Earl of Bute. Ehret's scientific paintings showed trees or plants which flowered for the first time in the gardens of his collector friends such as John Fothergill and Peter Collinson, and were often worked up later into engravings for sale; these paintings include details of flower and fruit structure and often have text referring to Linnaeus's new sexual system, as well as a dedication to the discoverer or supplier of the specimen. Some were published in other books, and from there copied onto porcelain by the potteries at Chelsea and Tournai. Other strictly scientific engravings appear in many publications, notably for the Transactions of the Royal Society, for Patrick Browne's *The Civil and Natural History of Jamaica* (1756) and for Alexander Russell's *History of Aleppo* (*see* page 66), as well as illustrations of plants collected by a young Sir Joseph Banks.

Ehret died in London in 1770; the previous autumn he had been demonstrating fungi to the Duchess of Portland at Bulstrode, and Mrs Delany recorded in her diary that though she enjoyed his company, she was sometimes puzzled by his still strong dialect.

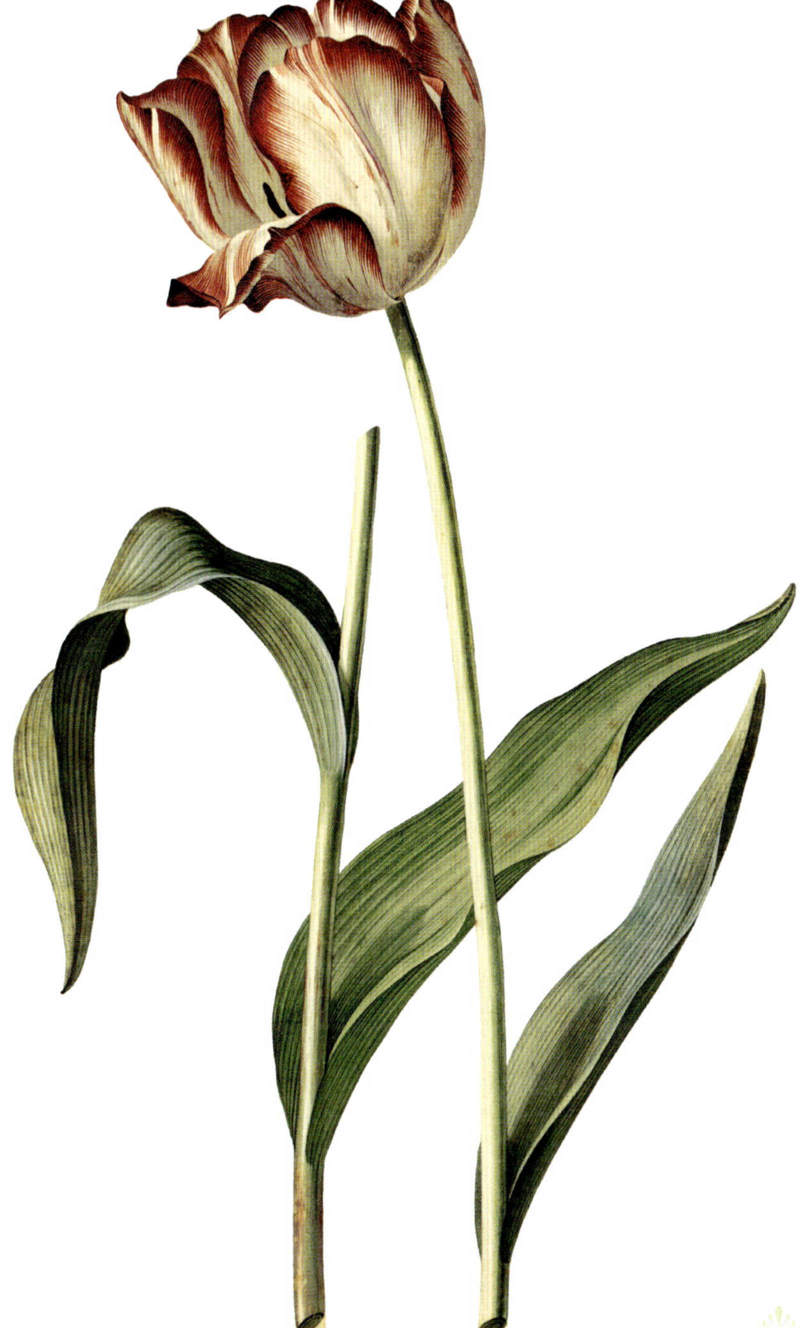

OPPOSITE: Ehret's study from 1786 of Cypripedium pubescens, *a Lady's slipper orchid collected in Virginia, USA.*

RIGHT: An early study of a tulip by Ehret from 1740.

Linnaeus and Plant Classification

The Swedish naturalist Carolus Linnaeus (1707–78), Professor of Botany at Uppsala, is the most famous of all eighteenth-century biologists. His great achievement was to simplify the naming of plants and animals; he wrote huge catalogues of every biological specimen he could find, and gave each two names: a generic name and a species name.

Before Linnaeus there was no definitive plant list, and plant names were many words long, for example Ehret's name for the bull bay, painted in 1743, was *Magnolia altissima Lauro Cerasi folio flore ingenti candido Catesb.* (the tallest Magnolia with a leaf of laurel and a huge white flower of Catesby). Linnaeus, in *Species Plantarum* (1753), put the name *grandiflora* in the margin against its various names; from then onwards it has been known as *Magnolia grandiflora* L. (Animal names tend to use the abbreviation Linn.). Botanical naming begins with Linnaeus's *Species Plantarum*; all earlier names pre-1753 may be ignored, but the earliest name for a plant post-1753 is generally the correct one.

Linnaeus also produced a new classification of plants, using the number of sexual parts in the flower as the basis for his system. This was taken up by many botanists, as it had the advantage of being simple to use, but it was artificial (as Linnaeus himself recognized) and related plants did not always have exactly the same number of floral parts. French botanists, following Bernard de Jussieu, aimed for a more natural system, which relied more on general similarity of the flowers and other characters. While Linnaeus's naming system has survived, his classification has long been superseded.

Few of Linnaeus's many publications were illustrated. As a young man he visited the Amsterdam banker George Clifford, then famous for his large collection of rare plants, and became the manager of his garden. At the same time he started *Hortus Cliffortianus*, a catalogue of the plants in Clifford's garden, and persuaded Clifford to employ Ehret as chief illustrator. This was one of Linnaeus's early publications, leading up to his most important work, *Species Plantarum*. After this meeting, their friendship continued, and indeed Linnaeus offered Ehret the post of botanical draughtsman at the University of Uppsala in 1747. *Hortus Cliffortianus*, a catalogue of Clifford's garden, was illustrated by 20 of Ehret's paintings and published in 1737, when Linnaeus was 30 and Ehret 31. Under Linnaeus's instructions, Ehret began to show floral details alongside the main illustration, a feature he used more extensively in later paintings. The frontispiece shows Clifford's banana plant which flowered in a hothouse in January 1736, and was the subject of a special study by Linnaeus, *Musa Cliffortiana*.

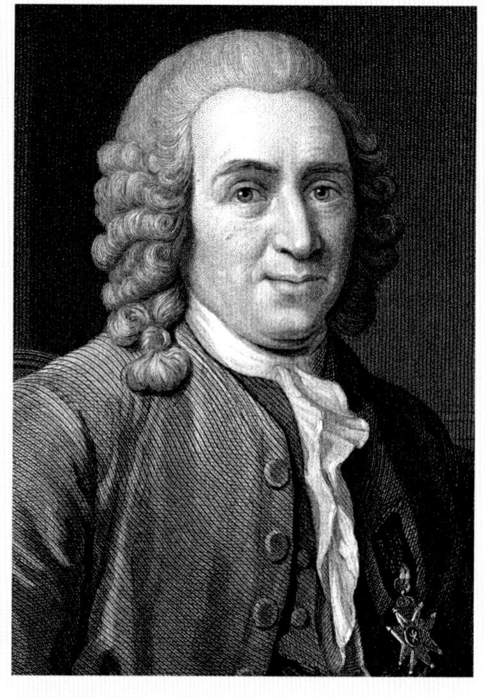

RIGHT, ABOVE: An engraved portrait of Carolus Linnaeus, who was born in Småland in southern Sweden.

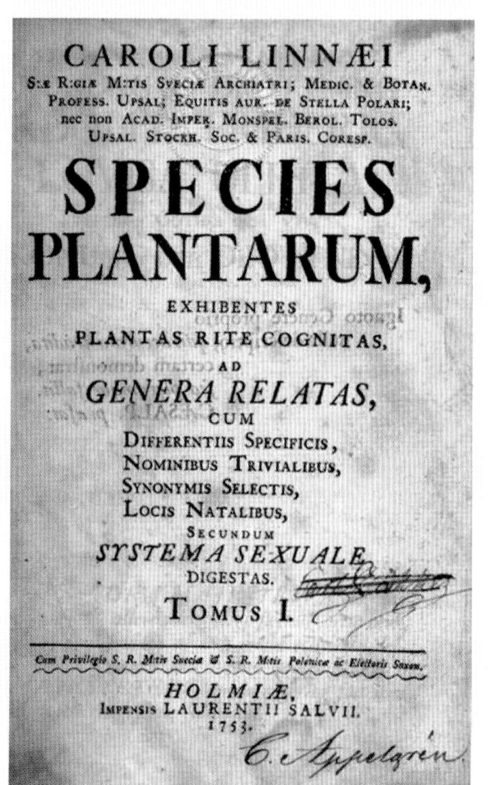

CAROLI LINNÆI
S:æ R:giæ M:tis Sveciæ Archiatri; Medic. & Botan. Profess. Upsal; Equitis Aur. de Stella Polari; nec non Acad. Imper. Monspel. Berol. Tolos. Upsal. Stockh. Soc. & Paris. Coresp.

SPECIES PLANTARUM,

EXHIBENTES
PLANTAS RITE COGNITAS,
AD
GENERA RELATAS,
CUM
Differentiis Specificis,
Nominibus Trivialibus,
Synonymis Selectis,
Locis Natalibus,
Secundum
SYSTEMA SEXUALE
DIGESTAS.

Tomus I.

Cum Privilegio S. R. Mtis Sueciæ & S. R. Mtis Poloniæ ac Electoris Saxon.

HOLMIÆ,
Impensis LAURENTII SALVII.
1753.

RIGHT, BELOW: The original title page of Linnaeus' Species Plantarum *which set the rules for the naming of plants after its publication in 1753.*

OPPOSITE: Aesculus hippocastanum *produced by Ehret as one of the illustrations for* Hortus Cliffortianus.

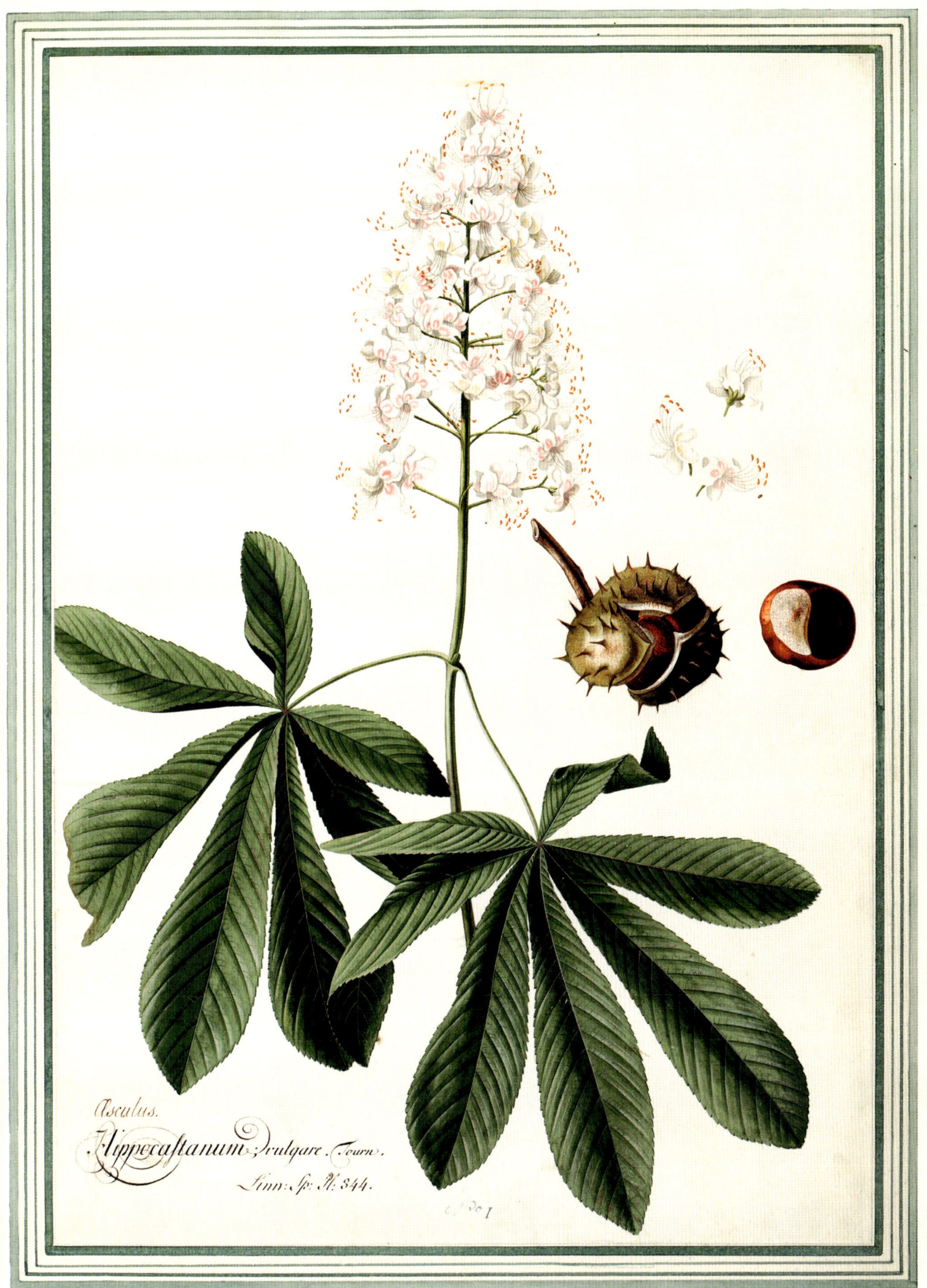
Æsculus.
Hippocastanum, vulgare. Tourn.
Linn: Sp: Pl: 344.

5 TRAVELLERS TO THE LEVANT

JOSEPH PITTON DE TOURNEFORT AND CLAUDE AUBRIET

Europe's fascination with Asia in general, and for the Ottoman Empire in particular, continued throughout the seventeenth and eighteenth centuries. There was always a frisson of terror in the stories of men captured and condemned to the galleys, or women enslaved in the harem. Fear of the Ottomans was never far away, as the siege of Vienna by Merzifonlu Kara Mustafa Pasha in 1683 was only foiled at the last minute by the arrival of a Polish army.

Ambassador Busbecq's letters from Istanbul started the craze, and were followed by those of the instrument builder Thomas Hallam who delivered an ornate clockwork organ, a gift from Queen Elizabeth, to the Sultan Mehmet III in 1599. An even greater sensation were the letters of Lady Mary Wortley Montagu, who followed her ambassador husband to Istanbul in 1716; these described the life of the sultan and the ladies of Sofia and Istanbul and her description of them gathered in the hammam, published in 1782, inspired paintings such as Ingres's *Le bain Turc*.

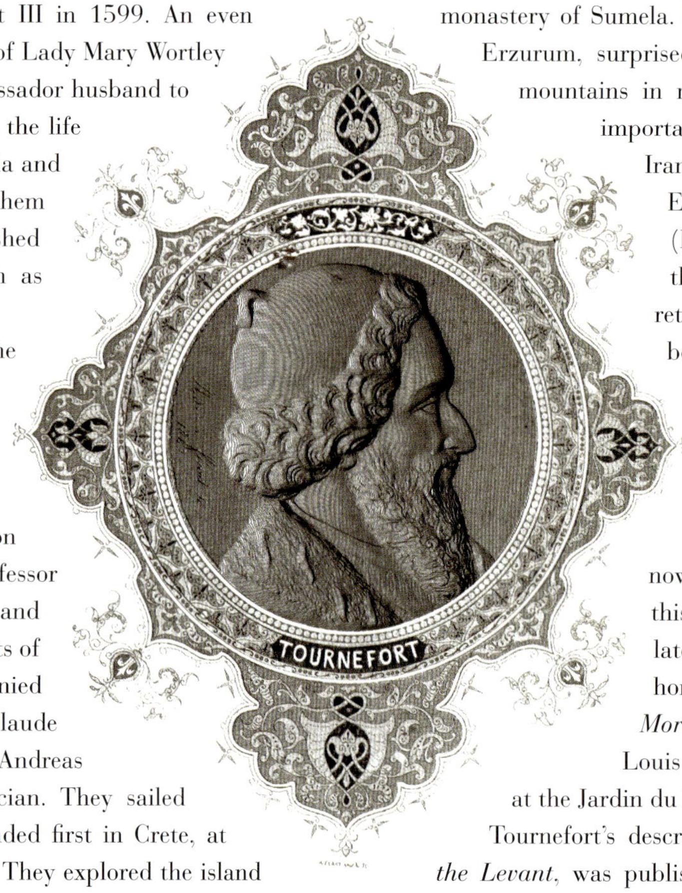

During a period in which the French King had established friendly relations with the Sultan, a small botanical expedition went from Paris to Turkey. It was led by Joseph Pitton de Tournefort (1656–1708), Professor of Botany at the Jardin du Roi, and one of the most influential botanists of the seventeenth century, accompanied by the botanical painter Claude Aubriet (1665–1742) and Dr Andreas Gundelsheimer, a German physician. They sailed from Marseilles in 1700, and landed first in Crete, at that time under Ottoman control. They explored the island for three months, before visiting other Aegean islands and spending the winter on Mykonos. In spring they found a Turkish ship to take them to Lesbos and from there to Istanbul, where they made preparations to visit the east. For protection they joined the caravan of an important pasha and sailed along the coast of the Black Sea to Trebizond, where they visited the still-thriving cliff monastery of Sumela. From here they crossed the pass to Erzurum, surprised to see snow on the surrounding mountains in mid-June. From Erzurum, then an important market town on the trade route to Iran, they went north to Tiflis, passing Erevan and the Three Churches (Etchmiadzin) where they kissed the Armenian patriarch's ring, and returned to Erzurum via Mount Ararat, before crossing Anatolia via Tokat and Ankara to Smyrna. They reached France again early in 1702.

TOURNEFORT

They were thrilled by the unfamiliar plants they collected along the route, which included the now familiar rock plant *Aubrieta* and a thistle-like plant, *Gundelia tournefortii*, later named by Linnaeus in their honour, as well as the very prickly *Morina persica*, named in honour of Louis Morin, the botanical demonstrator at the Jardin du Roi.

Tournefort's description of their travels, *Voyage into the Levant*, was published in 1717 in French and 1718 in English, and became an instant success. It described and illustrated many new plants and contained engravings of scenes and places they visited.

RIGHT: Voyage into the Levant *was also illustrated with views of the countries that were visited. This scene shows the churches of Etchmiadzin, with the two peaks of Mount Ararat in the background.*

BELOW, LEFT: Studies of the flower and seedheads of Papaver orientalis *from Volume II of* Voyage into the Levant.

BELOW, RIGHT: Another study from Volume II of Voyage into the Levant. *This plant was initially given the name* Morina officinalis *(now* Morina persica*).*

OPPOSITE: An engraving of a portrait of Joseph Pitton de Tournefort.

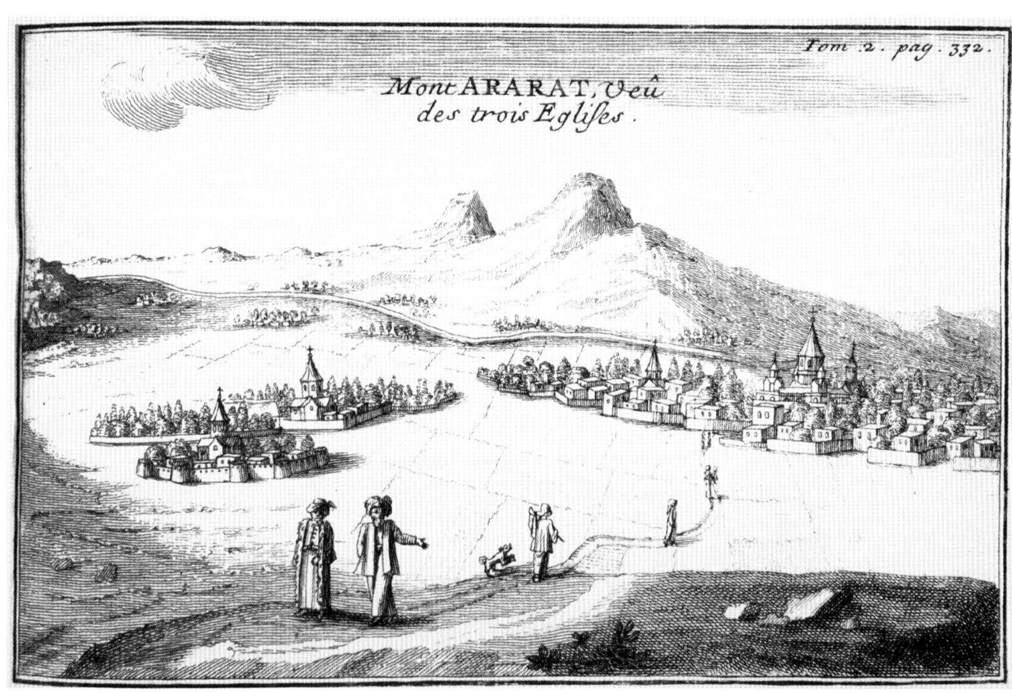

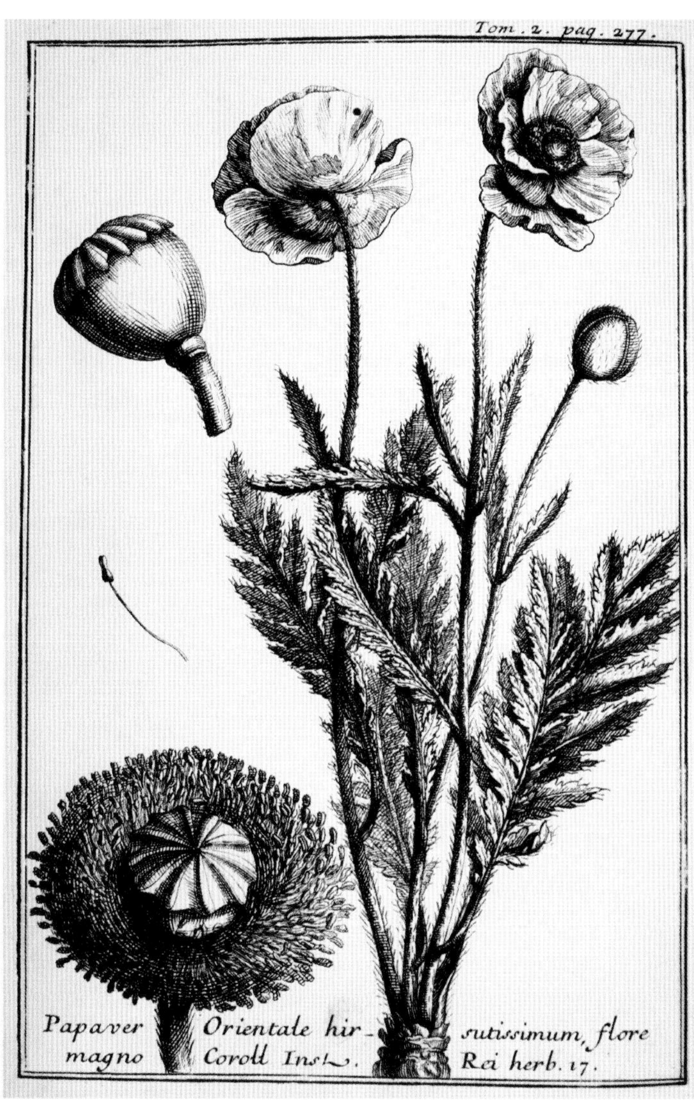

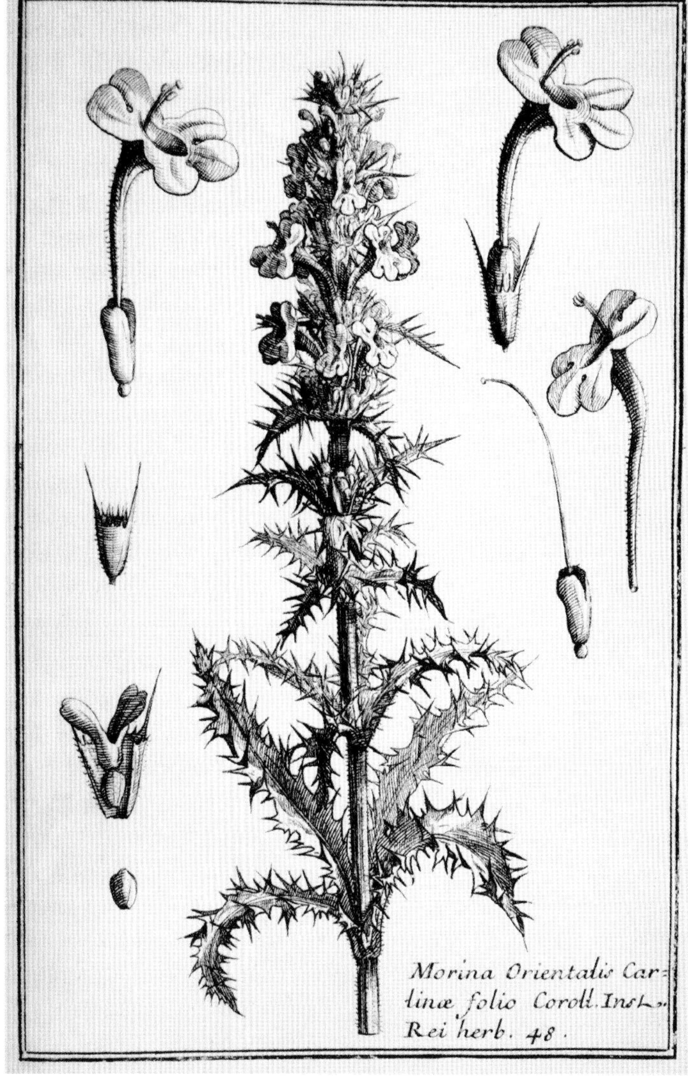

TULIPS FROM THE OTTOMAN EMPIRE

Sultan Ahmet III, who reigned from 1703 to 1730, was renowned for his love of pleasure and of flowers, so much so that his reign came to be known as the Tulip period, *Lale Devri*. A series of spring fêtes culminated in a festival in the Topkapi, with shelves to hold vases of tulips and cages of canaries and nightingales hung in the trees. The tulips of the period are shown in an album dating from around 1725, which was sold at Christies in May 1998. Their flowers are slender, with long, narrow pointed petals, very different from the blowsy striped and flamed tulips popular in the Dutch tulipomania a hundred years before. Over a thousand gold Turkish lira are reported to have been given for one bulb. The tulips in the album have delightful names such as "Dil-süz" (The breaker of hearts), "Naz-dar" (coquettish) and "Cücemoru" (the dwarf's purple, flamed with white), and some of the paintings are edged in gold leaf and mounted on beautifully marbled paper.

Sadly this hedonism came to an abrupt end when the populace revolted, enraged by high taxes to fund the Sultan's luxury, and, led by an Albanian ex-janissary, opened the prisons, freed the galley slaves and destroyed gardens around the palace. The janissaries themselves took over the streets and forced Ahmet to abdicate in favour of his nephew, Mahmut I (1730–54). A period of unrest and relative austerity followed. Having tricked their leaders into an ambush, Mahmut had the janissaries slaughtered and thrown into the Bosphorus, and other minor revolts were put down equally harshly; political discussion was banned and all the coffee houses in the city were closed. A period of peace and prosperity followed a treaty with Iran in 1746, and there was a revival of the tulip fêtes in the Topkapi, and a return to the cheerful excesses of court life. The elegant tulips almost died out, but a few must have survived in Europe, as they can now be purchased as "Acuminata" in shades of orange and yellow.

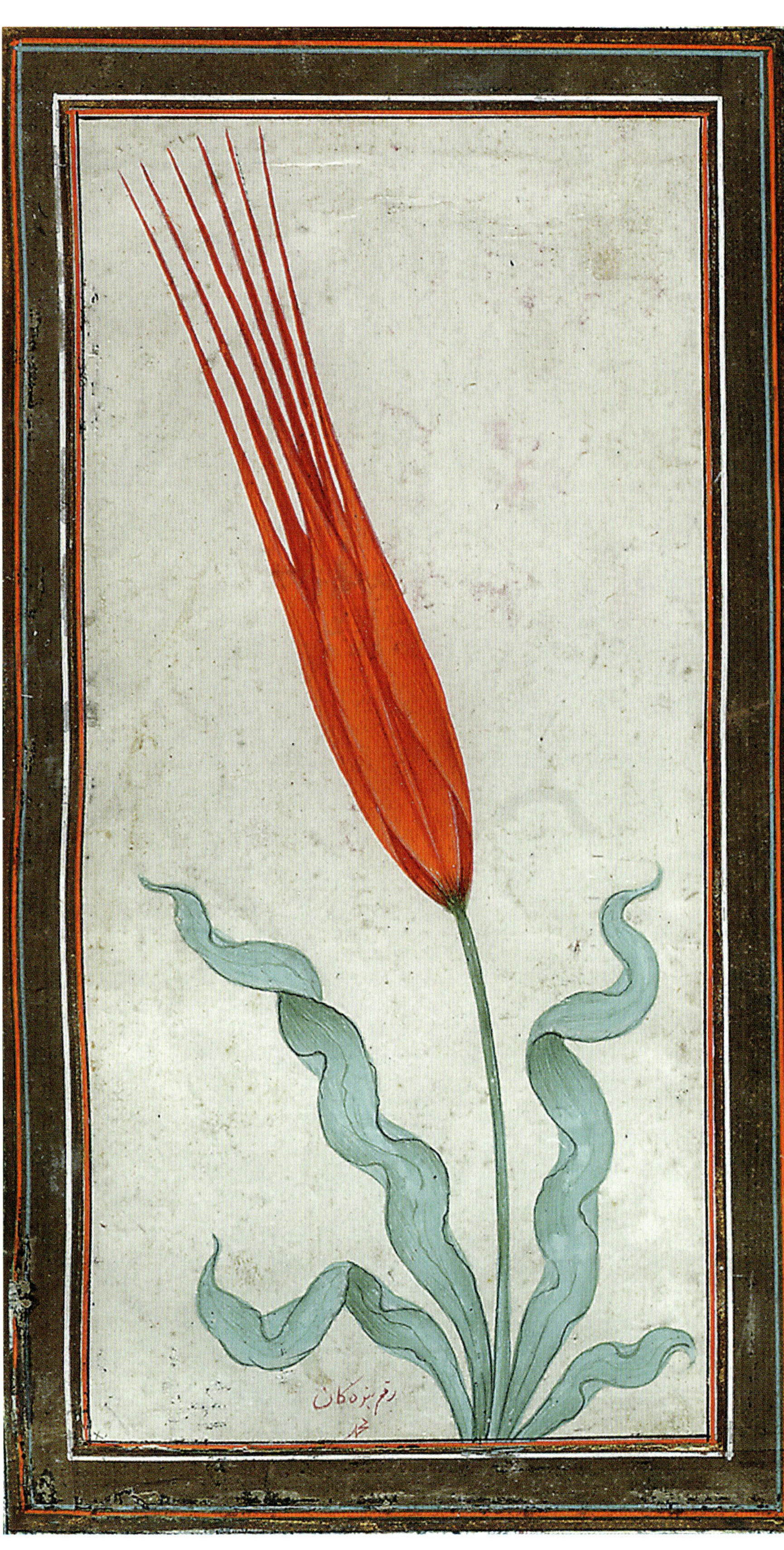

RIGHT: A sample of the illustrations in an album, dating from 1725, during the reign of Sultan Ahmet III.

OPPOSITE: Two beautiful late seventeenth-century studies of tulips by Simon Verelst (1644–1710), a Dutch painter from The Hague.

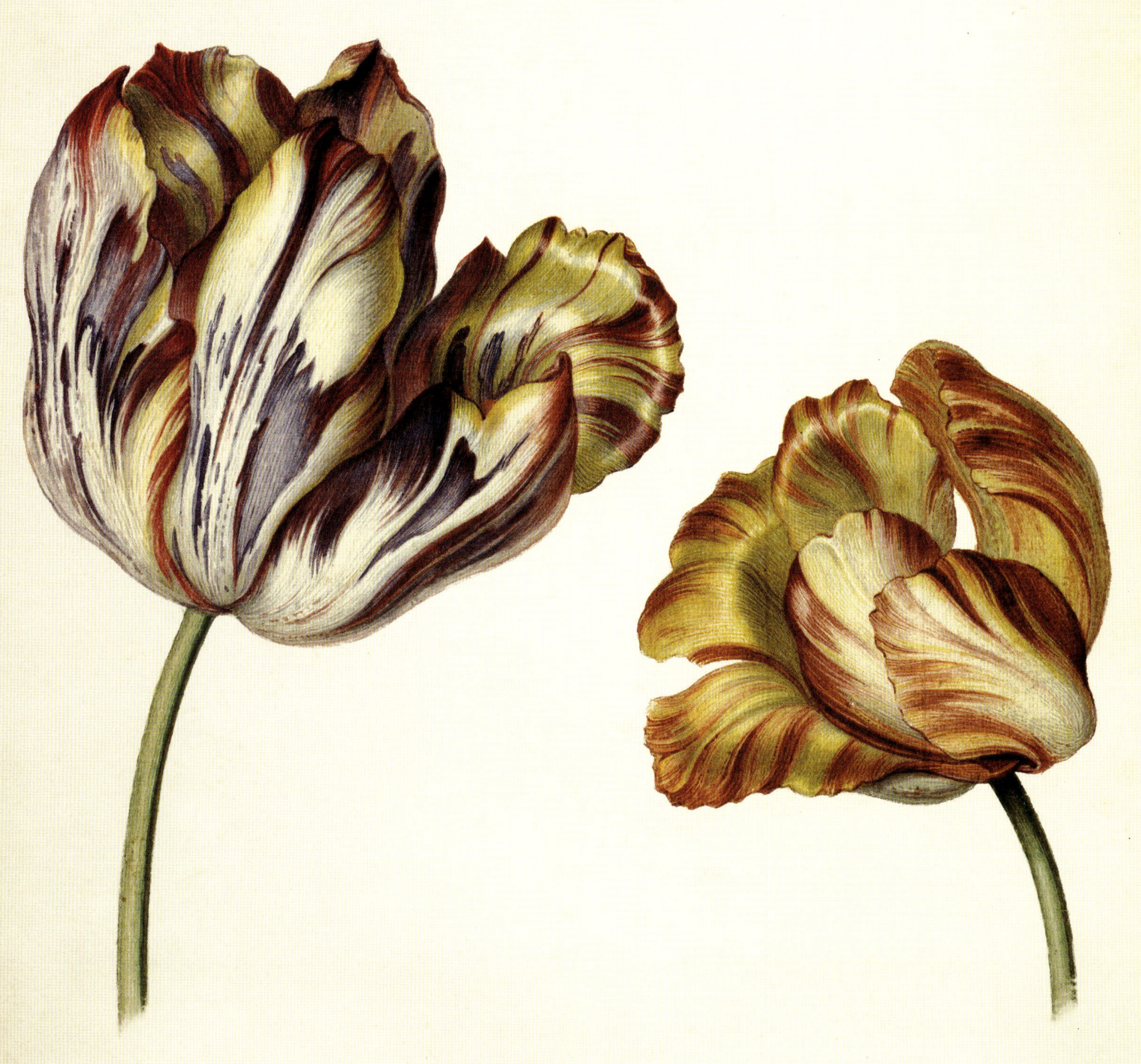

ENGLISHMEN ABROAD

Two English accounts of the east were published in the mid-eighteenth century, illustrated by engravings by Ehret. Richard Pococke, who became Bishop of Ossory and later of Meath, was born in Southampton in 1704 and died at Charleville, County Offaly in 1765. He travelled to western Asia from 1737 to 1740, spending the first six months or so in a detailed exploration of Egypt and up the Nile as far as the first cataract, before visiting the Holy Land where he bathed in the Dead Sea to test Pliny's reports of the buoyancy of the water. After Jerusalem, he explored Baalbeck, and the mountains of Lebanon, crossing the Euphrates at Urfa, before returning to the coast and visiting Cyprus. From here, he returned to Egypt and followed the route of the Israelites to visit Mount Sinai, which he had avoided before because of local fighting. On his return, he visited Crete and climbed Mount Ida, before landing in western Asia Minor where he explored the Troad and tried to locate the site of Troy.

The plants that Pococke collected were named by Philip Miller, and Ehret produced 12 engravings of flowers for his first two volumes of travels, *Descriptions of the East*, published in 1743. In it, Pococke quotes Diodorus Siculus when he finds the huge temple at Luxor, "I am Ozymandias, King of Kings. If anyone is desirous to know how great I am, and where I lie, let him surpass any of my works".

The Natural History of Aleppo, published in 1756, was written by Alexander Russell, MD, who was physician to the British factory owned by the Levant Company circa 1740–53. Russell's original purpose was to write an account of the plague, which raged for three years while he ministered to the inhabitants as well as the merchants, and therefore had an unusually close insight into the customs of the different races and religions in the city. The book contains a detailed account of the city, its buildings, gardens, crops etc., and an account of the wild plants growing around the city. Alexander was assisted by his brother in the compilation of this list, which uses the names of Bauhin and Tournefort. A few of the plants are illustrated by engravings by Ehret, perhaps drawn from herbarium specimens as well as from plants that flowered in England. Russell had sent back seeds to Collinson at Mill Hill near London, and probably also to Philip Miller at Chelsea. Other illustrations in *The Natural History of Aleppo* are of birds, including a small bittern and fish, particularly different species of catfish and loach. Some of the interior scenes are charming and include a Turkish lady, reclining on a divan, smoking a long pipe. Russell was interested to find the Arab physicians using translations of Hippocrates, Galen and Dioscorides, as well as Ebensina, as he calls Avicenna, though he considered them ignorant, particularly of internal anatomy; unlike the medical students in Glasgow where Russell had trained, they did not believe in dissecting cadavers.

Plants from western Asia, and particularly from Turkey continued to be imported to northern Europe during the eighteenth century.

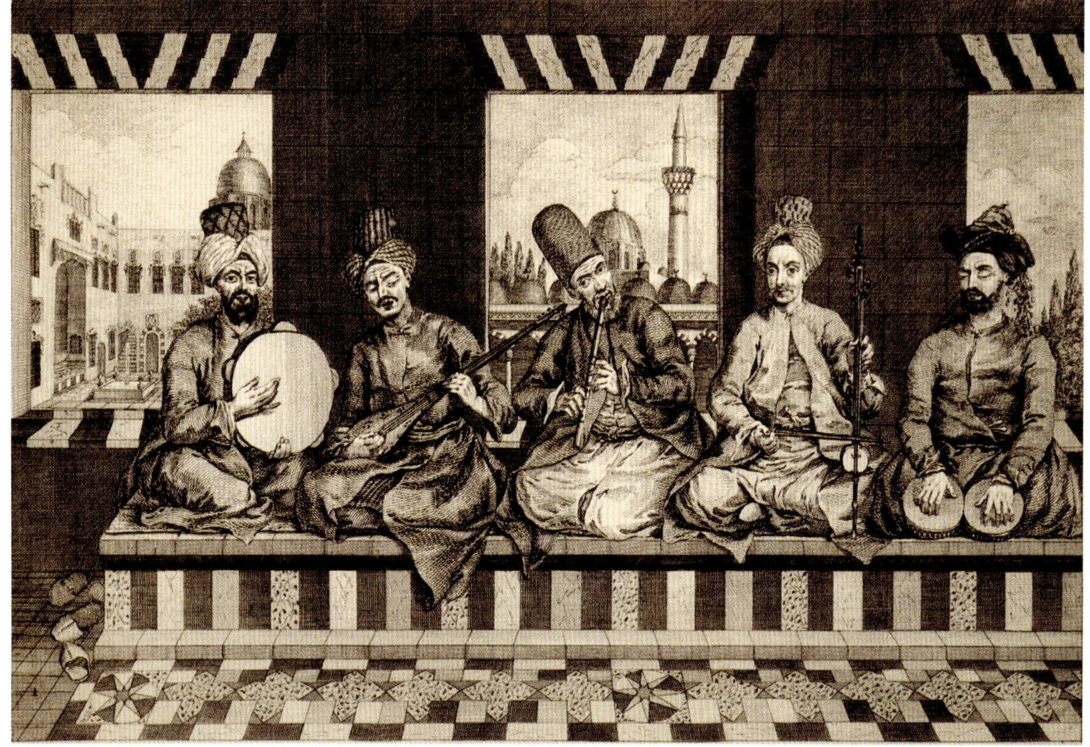

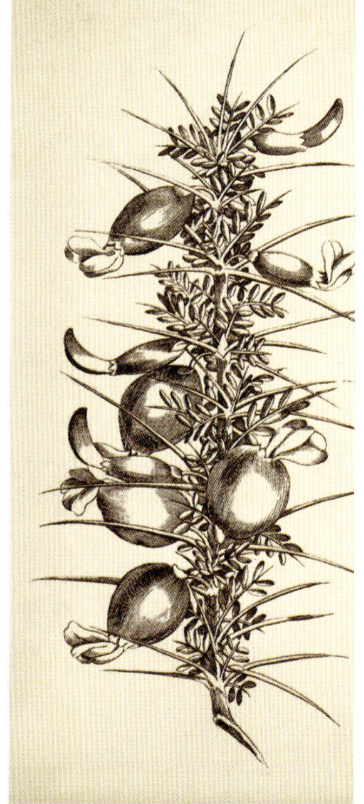

From 1787 onwards many new introductions were illustrated in *Curtis's Botanical Magazine*. The first plant to be illustrated was *Iris "persica"*, a Turkish plant which must have been imported regularly, as it is very difficult to keep in cultivation for more than a year or two. Other plants from the east were beautifully illustrated as individual paintings; Ehret's Horse Chestnut (*Aesculus hippocastanum*) and Margaret Meen's Sweet Chestnut (*Castanea sativa*), painted in 1783, are now in the library at Kew.

Two French works on the southern and eastern Mediterranean appeared in the late eighteenth century. J. J. H. de la Billardière had intended to visit southern Turkey, but on landing in Cyprus in 1787 heard that the plague was raging in Antioch (now Antakya) and had completely depopulated some of the villages north of Aleppo, so he changed his plans and visited the Holy Land, Lebanon and Damascus, finally going as far north as Mons Cassius near the present-day Turkish-Syrian border. His collections were published in *Icones plantarum Syriae rariorum*, between 1791 and 1812, illustrated with beautiful engravings by Pierre-Antoine Poiteau, Pierre Jean François Turpin and the Redouté brothers. Another eminent French botanist, René-Louiche Desfontaines, spent two years in Algeria and other parts of North Africa, and published his finds in *Flora Atlantica* in 1791. It also has elegant and delicate engravings by the Redouté brothers and other artists. Beautiful as these engravings were, they were never coloured, and were totally eclipsed by the *Flora Graeca*, which was in preparation in England at the same time.

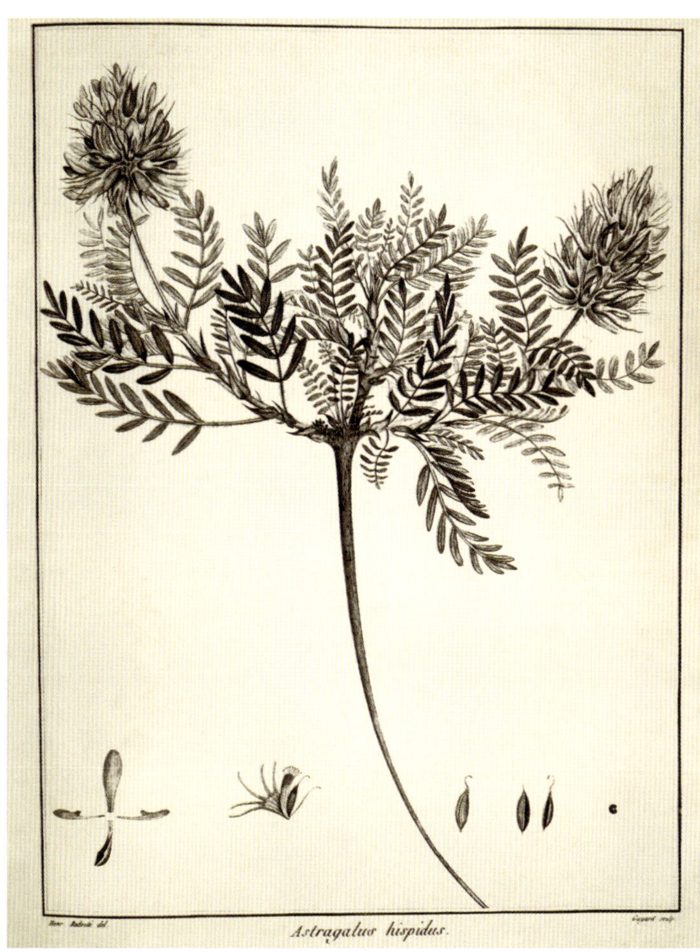

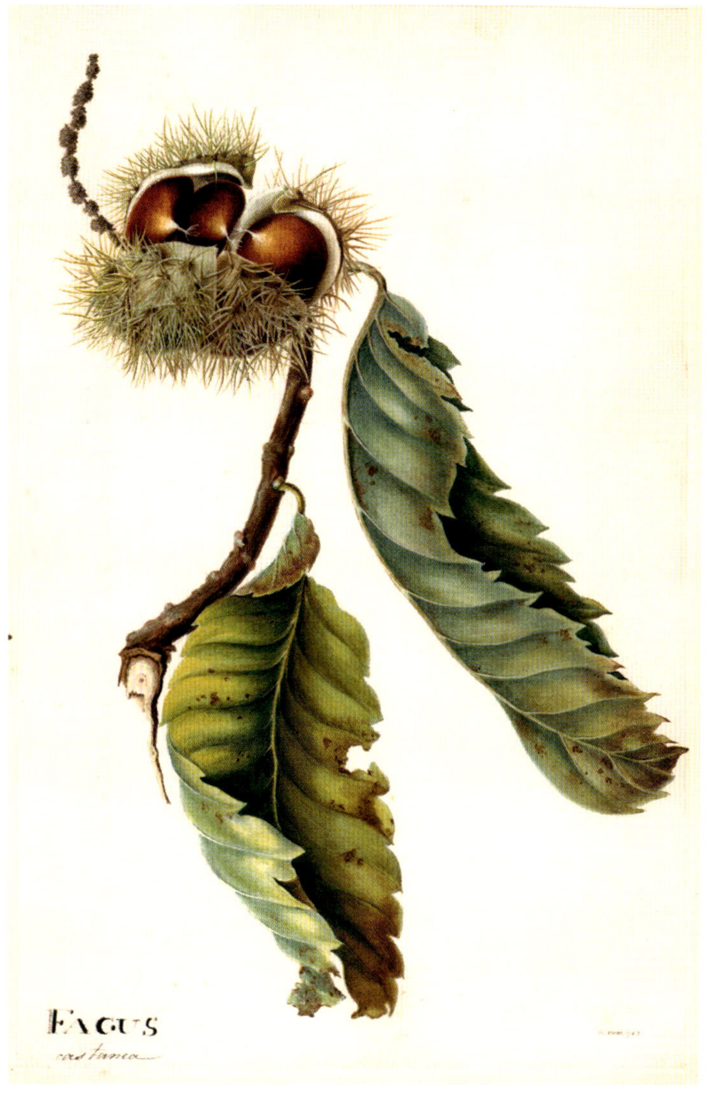

LEFT: Astragalus hispidus *by Pierre-Joseph Redouté from* Icones plantarum Syriae rariorum.

OPPOSITE, LEFT: As well as plants, The Natural History of Aleppo *also contained charming scenes of local life such as this group of musicians.*

*ABOVE: Margaret Meen's study of sweet chestnut (*Castanea sativa*) from 1783.*

OPPOSITE, RIGHT: Georg Dionysius Ehret's study of Tragacantha orientalis versicaria erecto caule *(now* Astragalus sp.*) from* The Natural History of Aleppo.

JOHN SIBTHORP'S *FLORA GRAECA*

Of all the botanical books of the late eighteenth and early nineteenth centuries, John Sibthorp's *Flora Graeca* is probably the most beautiful. The first volume was published in 1806, the tenth and last in 1840. The text of the first volumes was edited by the eminent botanist Sir J. E. Smith, the seventh finished by Robert Brown and the final three by John Lindley. The illustrations were by Ferdinand Bauer.

John Sibthorp (1758–96) was the son of Humphrey Sibthorp, Sherardian Professor of Botany at Oxford from 1747 to 1783. The elder Sibthorp is said to have given only one lecture during his tenure of the chair, and resigned on inheriting Canwick Hall near Lincoln. John took over the professorship from his father, having studied medicine in Edinburgh and at Montpellier, then as it still is, noted for botany, and received his doctorate at Göttingen. At this period, botany was still taught mainly as a branch of medicine, and the classical writers were revered. Sibthorp decided to work out the modern names of Dioscorides' plants, and in 1785 visited Vienna to study the most ancient known manuscript, the *Anicia Iuliana Codex*. While in Vienna, Sibthorp visited the botanist Nikolaus Joseph von Jacquin, and was introduced to the young Ferdinand Bauer, then working for von Jacquin (*see* page 114) and painting rare plants which flowered in the imperial gardens at Schönbrunn Palace, published in *Icones plantarum rariorum* (1781–93). Sibthorp persuaded Bauer to join him on his planned voyage to Greece and Turkey, and record the flowers that they were expecting to find.

Ferdinand Bauer (1760–1826) and his older brother Francis (1758–1840) became England's most accomplished botanical artists; they were born in Moravia (now in the Czech Republic), where their father was court painter to the prince, and they were taught botany by Father Norbert Boccius, Prior of Feldsburg, who added their paintings to his collection. Many of the Bauer brothers' early paintings are now in the Princely Collections in the Liechtenstein Museum in Vienna. Ferdinand worked first for Sibthorp, and then joined Matthew Flinders' expedition to Australia as draughtsman to the botanist Robert Brown. Francis moved to Kew, where he was employed by Sir Joseph Banks (*see* pages 98–99), and spent his life doing botanical and anatomical drawing and painting.

John Sibthorp and Ferdinand Bauer set out for Greece in 1786, accompanied by Sibthorp's brother-in-law John Hawkins (1761–1841), an accomplished classical scholar, botanist and amateur gardener, and one of the founders of the Horticultural Society. They travelled overland to Naples, and from there sailed first to Crete, and then to several Aegean islands and Athens, before reaching Smyrna, then the major trading port on the Turkish Mediterranean coast, where William Sherard, who had founded Sibthorp's professorship, had been consul from 1703 to 1716. From Smyrna they travelled north to Bursa where they climbed Mount Olympus (Uluda), on which they found the purple *Clematis viticella*. They spent the winter in Istanbul, making short excursions into the

LEFT: The flowers and various dissections of Salvia forskaelei *from Volume I of* Flora Graeca.

OPPOSITE: A purple poppy, Roemeria hybrida, *from Volume V of* Flora Graeca.

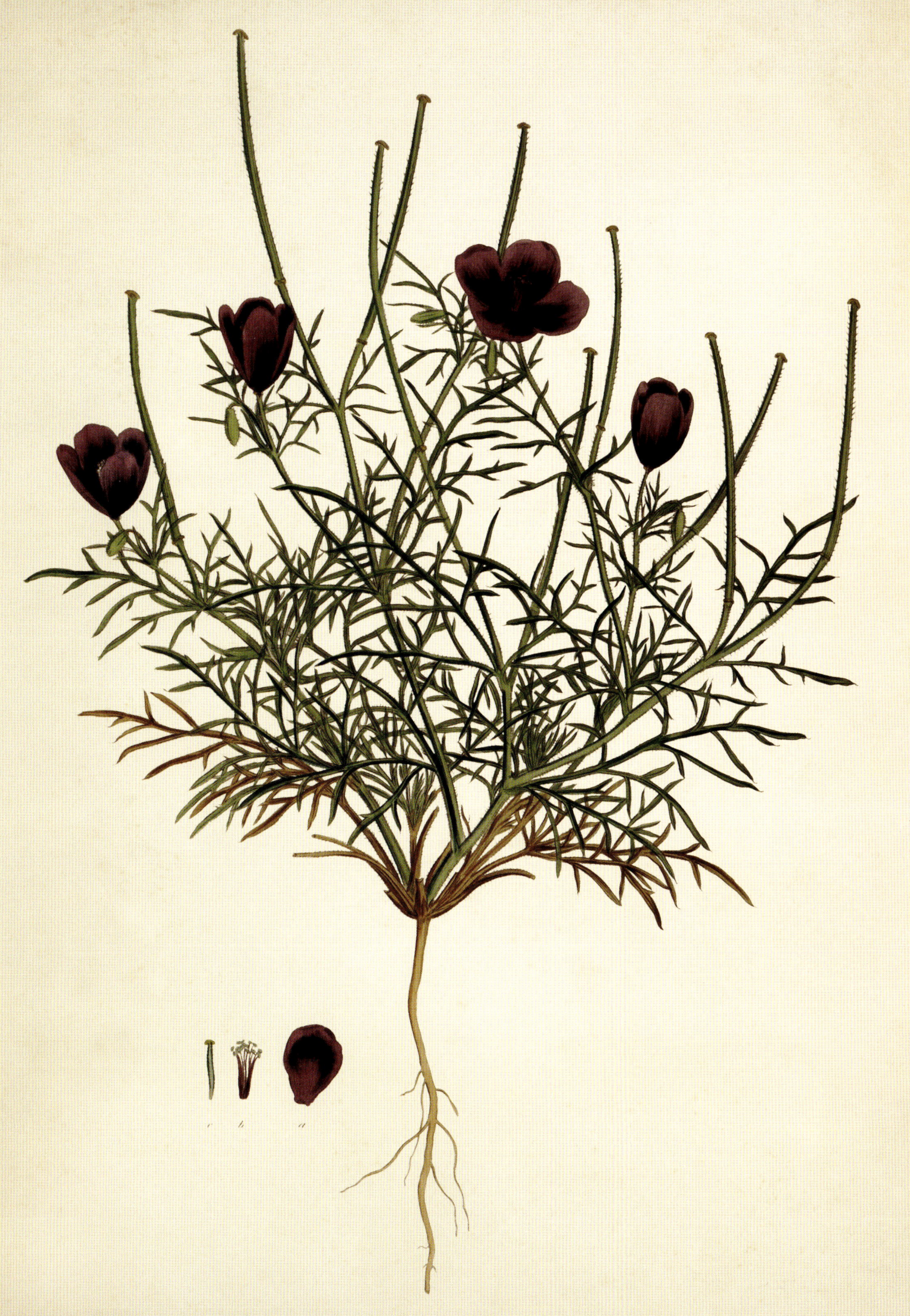

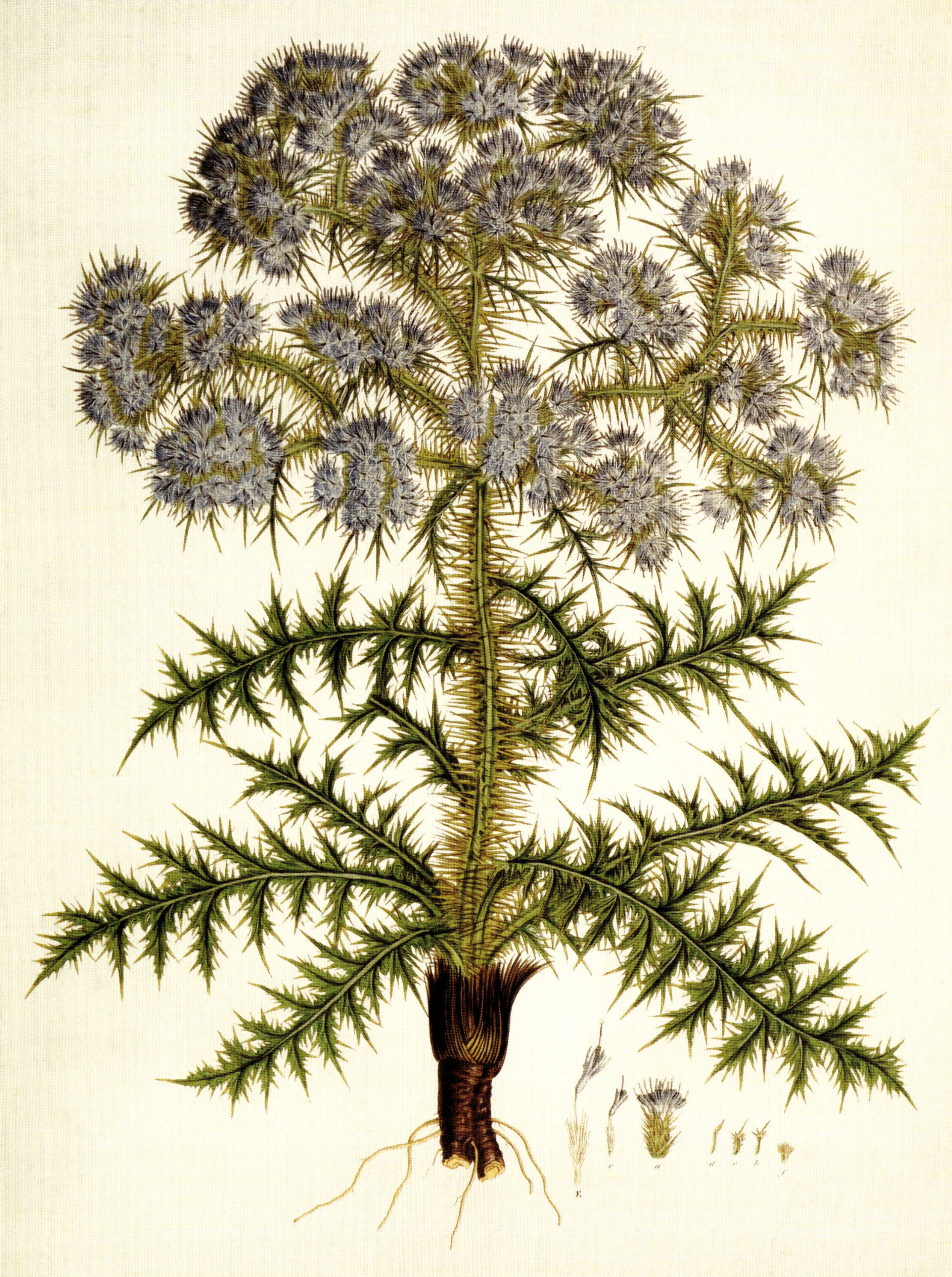

Carthamus corymbosus.

forest and hills near the city, on one of which they found deep pink primroses, now called *Primula vulgaris* var. *sibthorpiana*. They set out again by boat on 14 March 1787, visiting Lesbos, Skiros and Cos, heading for Rhodes, but were forced by southerly winds into Porto Cavalieri, probably in the Bay of Marmaris in southern Turkey. Here they visited the ruins of ancient Cressa, and found a new species of yellow fritillary, which they named *Tulipa sibthorpiana*, a great rarity, not seen again for 200 years. The next day the wind changed, and they were able to cross to Rhodes where they found the town, which had been a stronghold of the Knights of St John, half ruined and still almost deserted 264 years after it had been sacked by the Turks. They sailed then to Cyprus where the ancient city of Famagusta was also in a sad state: here, they found a black speckled arum, which they called *Arum dioscoridis*, and noted that the tubers are edible when cooked. From Cyprus they returned to Athens and climbed Mount Parnassos on 30 June, and were surprised to find snow patches still lying in sheltered hollows. Hawkins conversed with the shepherds and found that they still used many of the same names for plants that had been used by Dioscorides, and that

OPPOSITE: Carthamus corymbosus *(now* Cardopatium corymbosum*) from Volume IX of* Flora Graeca.

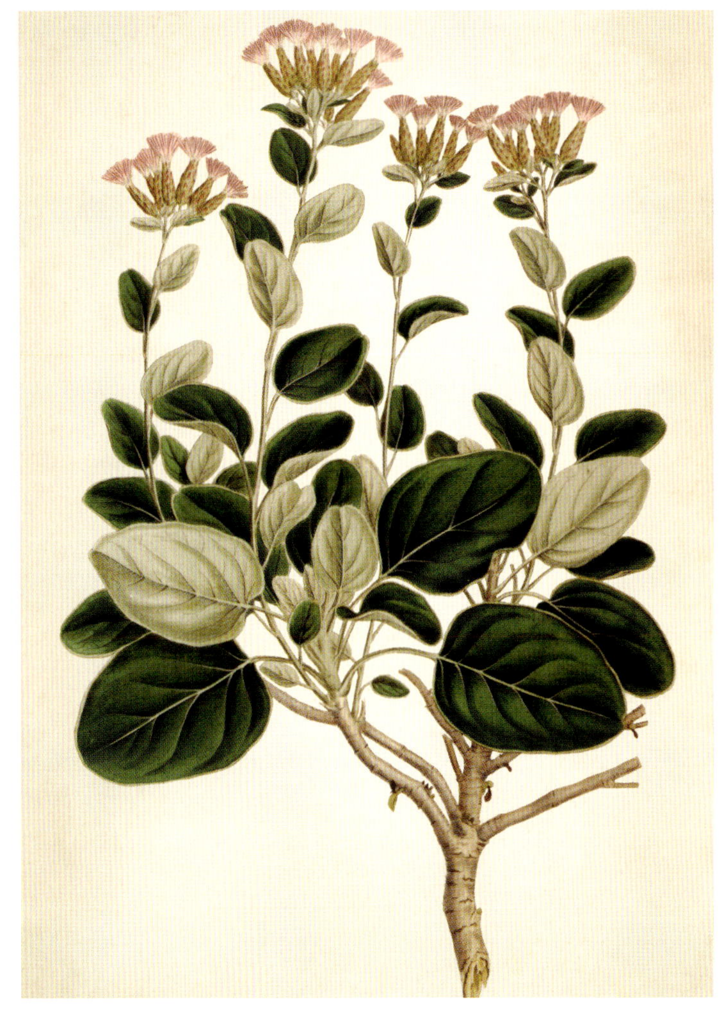

RIGHT: Stachelina arborescens *from Volume IX of* Flora Graeca.

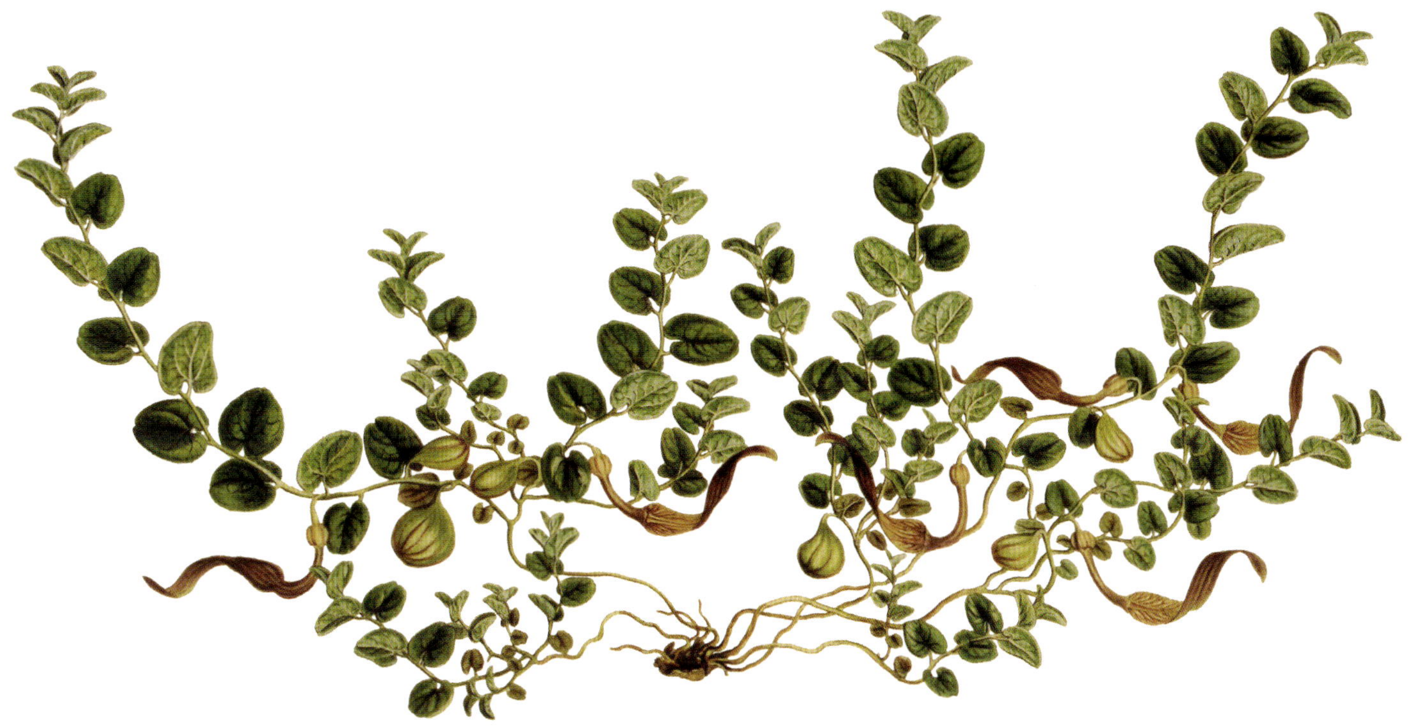

BELOW: Aristolchia *sp. from Volume X of* Flora Graeca.

"their virtues were faithfully handed down in the oral traditions". Passing through Delphi, where few ruins remained on the surface, they found the scented cushions of *Daphne jasminea* growing on the cliffs. By September the party reached Patras, and sailed for home, with drawings and specimens of over 2000 plants.

Sibthorp and Hawkins made a second expedition to Greece in 1794, while Bauer stayed in London to work up his drawings from the first trip. They wintered in Athens, and spent the early spring in the Peloponnese before returning to England. This trip was marred by the death of their young botanical assistant, Francesco Borone, who fell from an upstairs window in Athens while sleepwalking. By the time they reached England, Sibthorp was suffering from "a nasty low fever, with a cough that alarms me, from some affection of the lungs". This was the beginning of his tuberculosis, and in spite of baths in asses' milk and treatment in Brighton, he died in Bath in 1796, at the age of 38. He left "all my freehold estate profits or rents of such estate to be applied in the following manner, first for the publication of a work for which I have collected the materials to be entitled *Flora Graeca*, which is to consist of ten folio volumes, each volume to consist of 100 plates, and also a small octavo, without plates entitled *Prodromus Florae Graecae*." Hawkins was his chief executor, and managed to enlist the excellent and prolific botanist J. E. Smith at a salary of £150 per annum, to complete the writing. Sibthorp's notes, journal and specimens were very disorganized and his handwriting hard to read, because he had expected to rely on his memory, "and dreamed not of dying". Smith, however, persevered and wrote all the descriptions of the plants for the main volumes: he took Sibthorp's lists of Greek plants, to which he had added the citations of Linnaeus and Tournefort, as the basis for the *Prodromus*.

Smith died in 1828 after six volumes had been published, and the seventh was completed by Robert Brown, who had travelled

RIGHT: Each volume of Flora Graeca *was notable for its charming frontispiece. All featured scenes from Greece or Turkey, such as this view of Mount Parnassus.*

BELOW: Cytinus hypocistus, *a parasite on the roots of* Cistus *bushes, from Volume X of* Flora Graeca.

OPPOSITE: Roemeria refracta *from Volume V of* Flora Graeca.

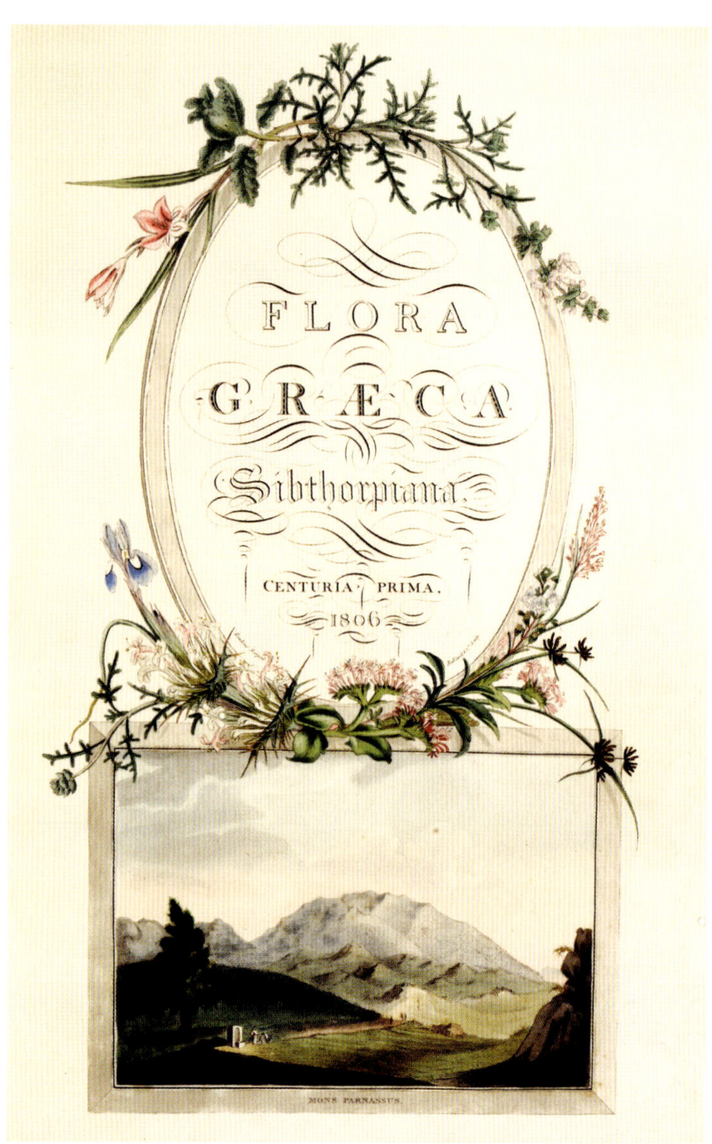

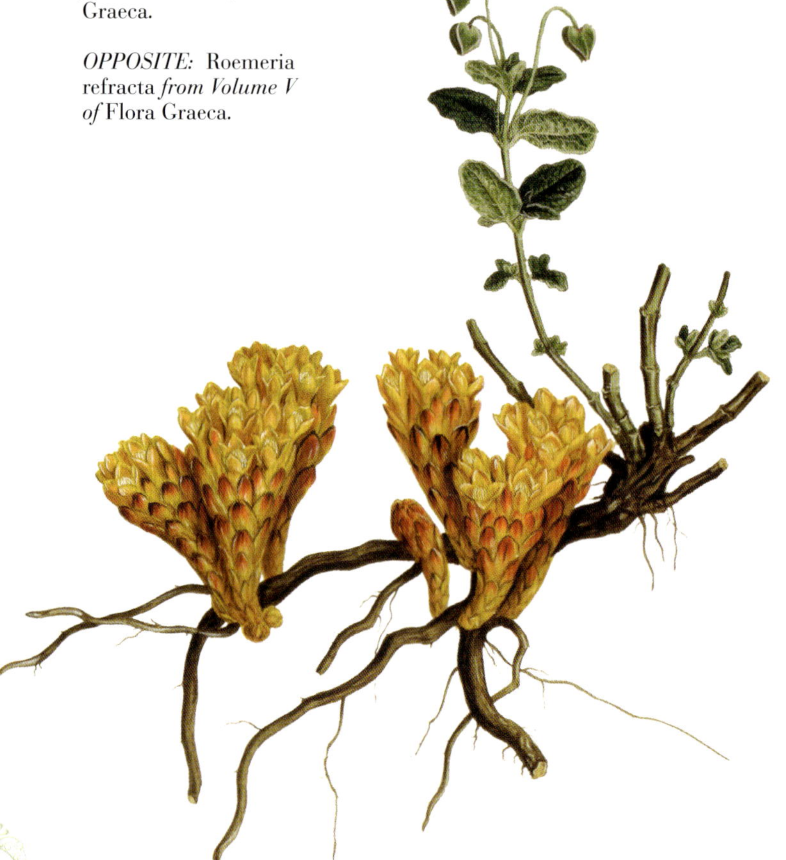

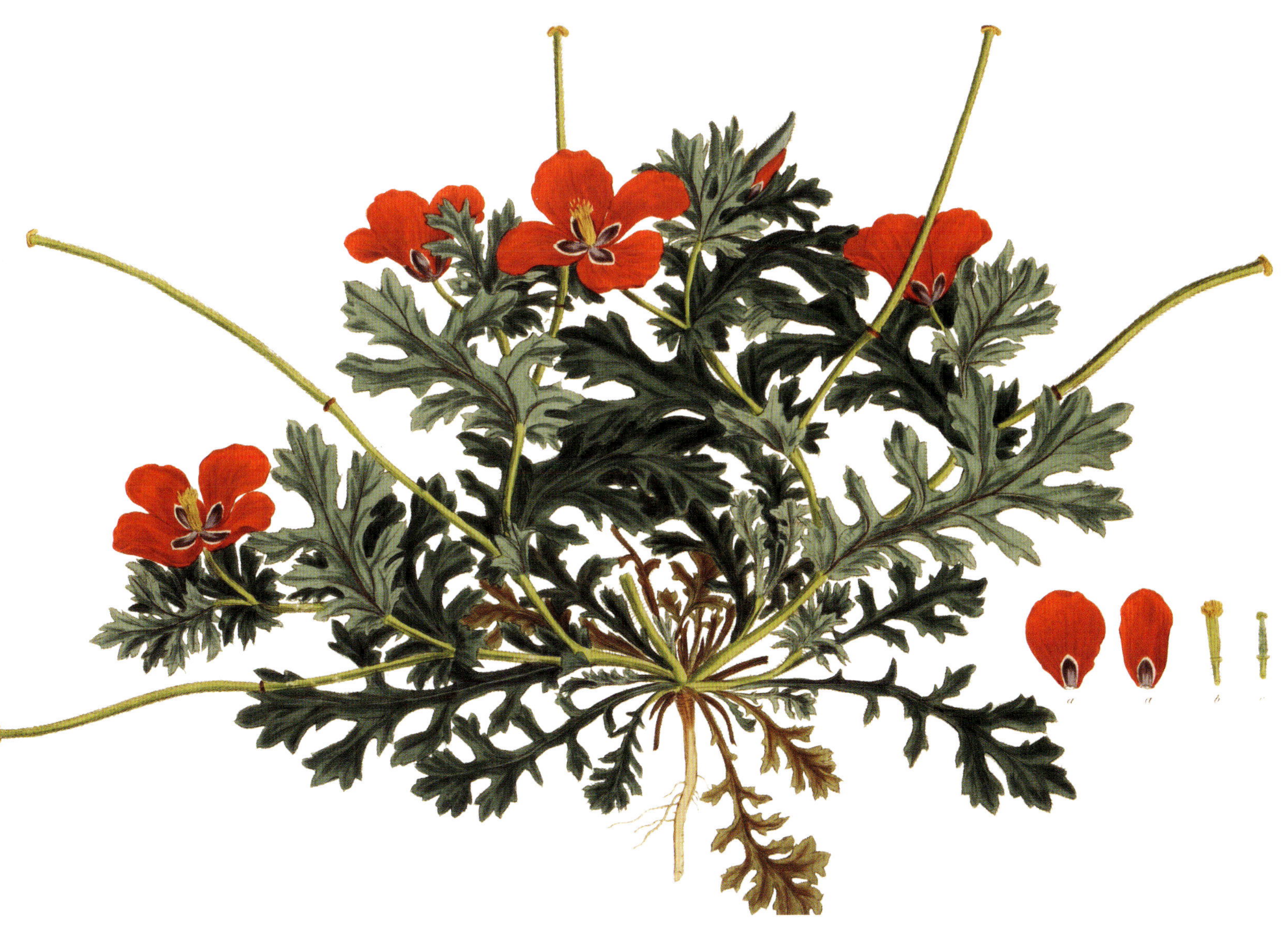

with Bauer on the Flinders expedition. The remaining three volumes were written by a friend of Hawkins, John Lindley, who was working closely with Francis Bauer (*qv*) on the study of orchids. In the last volumes Lindley added a list of the species according to the natural classification of Augustin de Candolle, which by 1840 had begun to supersede the artificial system of Linnaeus; Lindley also added a list of the Greek names in Dioscorides and the vernacular names in use in the Greek countryside.

Flora Graeca is pre-eminent among illustrated botanical books, for several reasons: it is large in size, a folio, with a single plant beautifully illustrated by hand-coloured engraving on each page. The engravings were made by members of the Sowerby family: each volume has a beautiful title page, with a garland of flowers and a vignette of some classical scene or stopping place on their journey: the text is scientifically accurate, and many of the species illustrated were new to science.

By the time the last volume was published, only 25 of the original subscribers out of the original 50 were left. Production had cost over £15,000, or around £620 per set. A reprint of 40 copies was produced in Germany between 1845 and 1856, and in 2009 Koeltz Scientific Books began publishing a smaller-format facsimile.

Maria Sybilla Merian

Maria Sybilla Merian was born in Frankfurt in 1647, the daughter of a well-known publisher and engraver, Matthäus Merian the Elder, who died when Maria was only three, leaving the family in straitened circumstances. Maria's mother remarried, and it was from her stepfather, Jacob Marrel, an artist who specialized in floral still-lifes, that Maria learnt to paint. In addition to her artistic work, Maria was interested in natural history, and particularly fascinated by insects, from an early age.

In 1665 she married Johann Andreas Graff, a publisher, painter and engraver, and in 1670 the couple moved to his home town of Nürnberg, a city of great cultural importance at the time. Here, Maria taught embroidery, drawing and painting to other women. She began to write, illustrate and publish her own work, the first being *The Book of Flowers*, originally designed as a pattern book for her pupils. This was published by Merian in Nürnberg in three slim volumes, each with 12 plates, between 1675 and 1680 (it was reissued posthumously in 1730) and was such a success that she reprinted it and republished it as the *Neues Blumenbuch* (*New Book of Flowers*) in 1680.

In this book, Maria Sybilla brought together both beauty and scientific accuracy; her work is delicate and finely drawn, showing great artistic sensibility. She specialized in painting exuberant garlands, arrangements and bouquets of flowers, somewhat in the style of the Dutch flower painters, rather than the more traditional single botanical specimens, and she often included insects such as butterflies or dragonflies, or occasionally a bird, in the composition. She took enormous care with the hand colouring of her copper engravings, bringing a light touch to the page. A charming painting of various primroses, cowslips and their hybrids is now in the Kew collection.

In addition to her flower book, Maria Sybilla also indulged her interest in insects, working on the first volume of a book on caterpillars, published in 1679. Meanwhile, she had become increasingly disenchanted with her marriage, and she and her husband divorced. She continued to live with, and support, her two daughters, who in turn assisted her with some of her work. In 1685 she, her elderly mother and daughters all joined a religious community in West Friesland, where she continued her work. Following her mother's death in 1691, the family moved to Amsterdam, where, the following year, her elder daughter married a merchant who traded with the Dutch colony of Surinam in South America. In 1698 Maria Sybilla took the opportunity to travel there, with her younger daughter. They spent

LEFT: A hand-coloured copper engraving from De metamorphosibus Insectorum Surinamensium *which features various beetles and a wasp at different stages of their life cycles.*

OPPOSITE: Maria Sybilla Merian's studies of polyanthus and primroses.

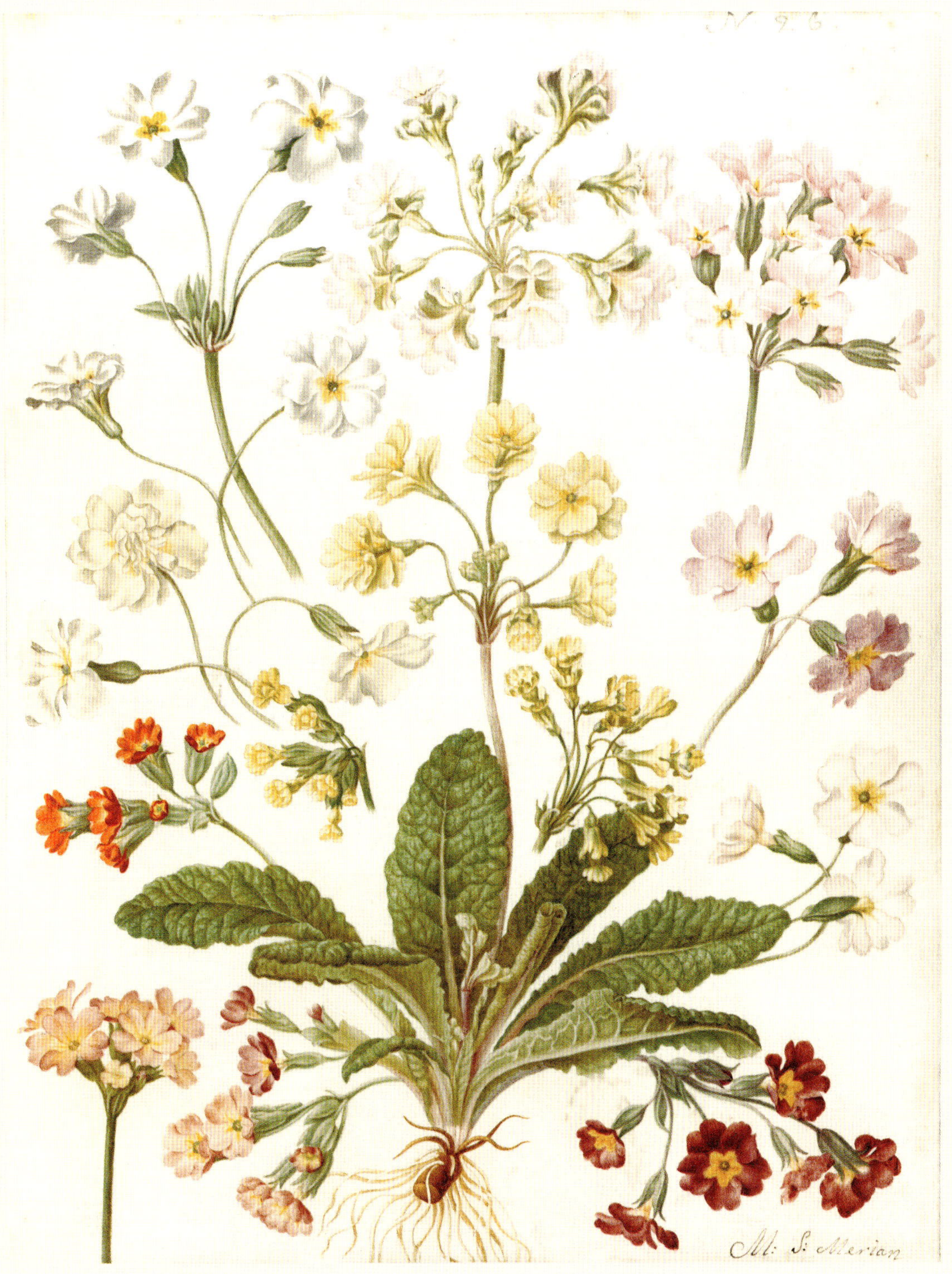

autumn of 1701 returned to Amsterdam, armed with specimens, notes and drawings.

Maria Sybilla's Surinam adventures bore fruit in the form of a large book, produced in 1705 in two editions (one with text in Latin and the other in Dutch), and in an enlarged edition, entitled *Dissertatio de generatione et metamorphosibus insectorum* published in Amsterdam in 1719. The 72 plates are drawn in a bold and confident style, and depict the metamorphosis of various butterflies, reptiles, worms and other insects, as well as the fruits and flowers on which these depend for food. This book made her reputation as perhaps Germany's first truly great female artist; she continued to live in Amsterdam and died there in January 1717.

6

The Exploration of Russia and Japan

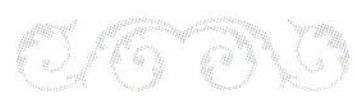

RUSSIA

While English and French botanists and botanical artists were exploring the furthest parts of the Southern Hemisphere by sea, or clarifying the works of classical writers in Greece and Turkey, Russian scientists were making equally exciting journeys and, in the wake of expanding Russian power, exploring the easternmost parts of Siberia, Central Asia and the Caucasus.

In 1720, Peter the Great sent a German scientist, Dr Daniel Messerschmidt, across Siberia; his expedition lasted seven years, and he discovered the first frozen remains of the woolly mammoth and collected thousands of plants, but little of his collection seems to have survived, and his journals have only recently been published.

The first Russian botanical expedition to have lasting results was led by Johann Georg Gmelin (1709–55), whose party left St Petersburg in 1733, reaching Lake Baikal in 1736 and the Sea of Okhotsk, on the northern Pacific, in 1736. Here they found Vitus Bering and his naval companions (who had crossed Siberia more quickly) already camped, but they were unwilling to share their supplies with Gmelin who had lost much of his own in a fire. Gmelin's party was therefore forced to go back to Irkutsk, where they waited for permission to return to St Petersburg. Meanwhile Georg W. Steller (1709–46) had been dispatched to join them, and he arrived in Siberia in 1739. Steller was not so easily daunted and together they pushed on to Kamchatka. There Steller forced his way onto Bering's expedition, who had built themselves boats and were planning to sail to America. Because of bad weather they made only one landing in Alaska, but nevertheless Steller managed to obtain a good collection of plants.

Gmelin's *Flora Sibirica*, published in four volumes between1747 and 1769, described the plants of their expedition, and was illustrated with 302 black-and-white engravings by two artists who had accompanied the expedition. Steller is commemorated by the *Daphne*-like herbaceous *Stellera chamaejasme* L. as well as a sea cow, a sea lion, a huge sea eagle and a jay.

The most famous explorer of northern Asia, Peter Simon Pallas (1741–1811), was brought up in Berlin, obtained his doctorate at Leiden when he was 19 and was offered the post of Professor of Natural Sciences at St Petersburg at the very young age of 23. The Russian Academy of Sciences, under the auspices of Catherine the Great, was planning a series of expeditions to observe the 1769 Transit of Venus at different longitudes, and Pallas decided to join an expedition across Siberia. (This was the same transit that was observed by Captain Cook on Tahiti.)

LEFT: Dryas octopetala *from Pallas's* Flora Rossica.

OPPOSITE: Rheum nutans *(now* Rheum compactum*), a rhubarb from* Flora Rossica.

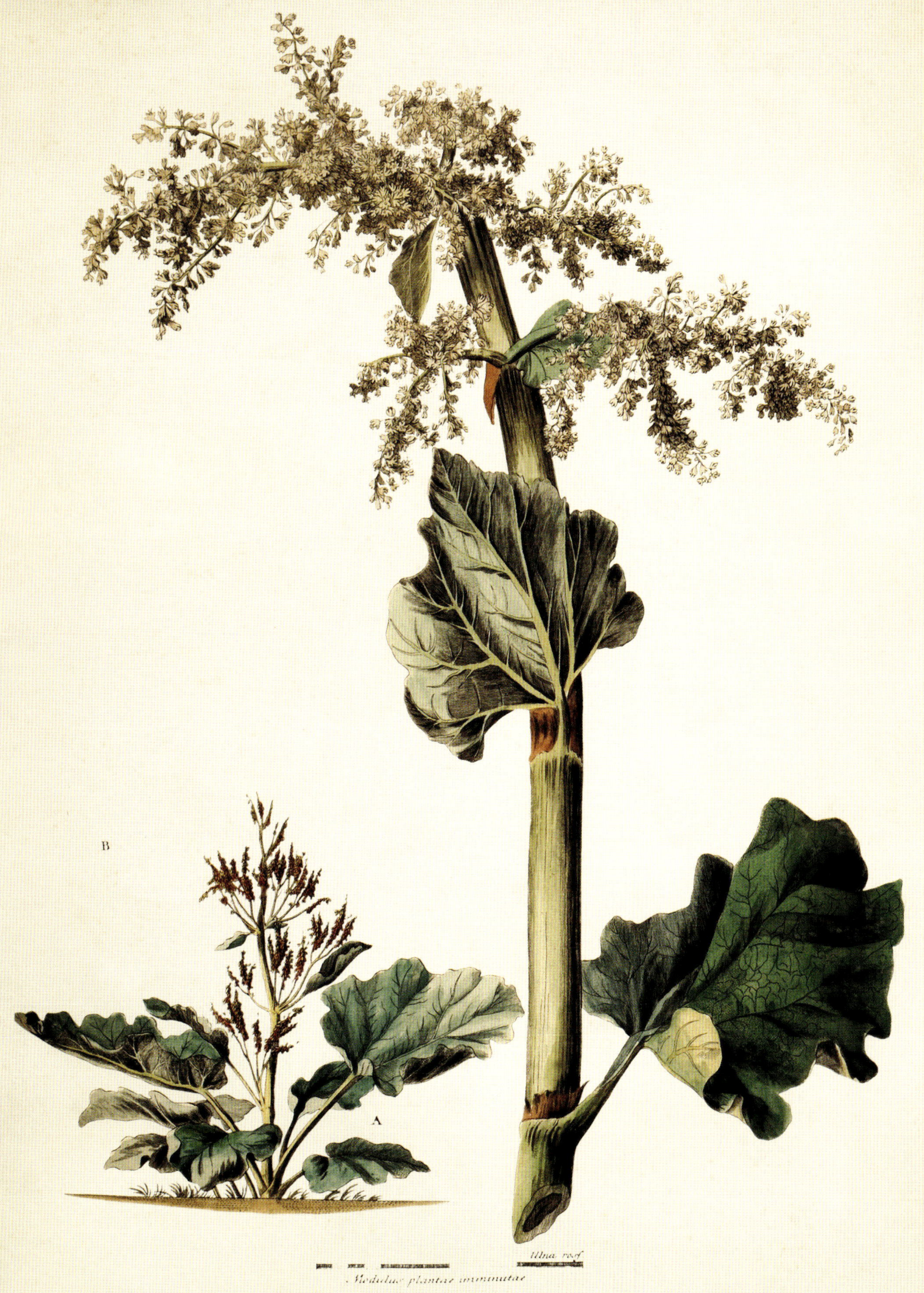

RHEUM *nutans* РЕВЕНЬ *висящій*

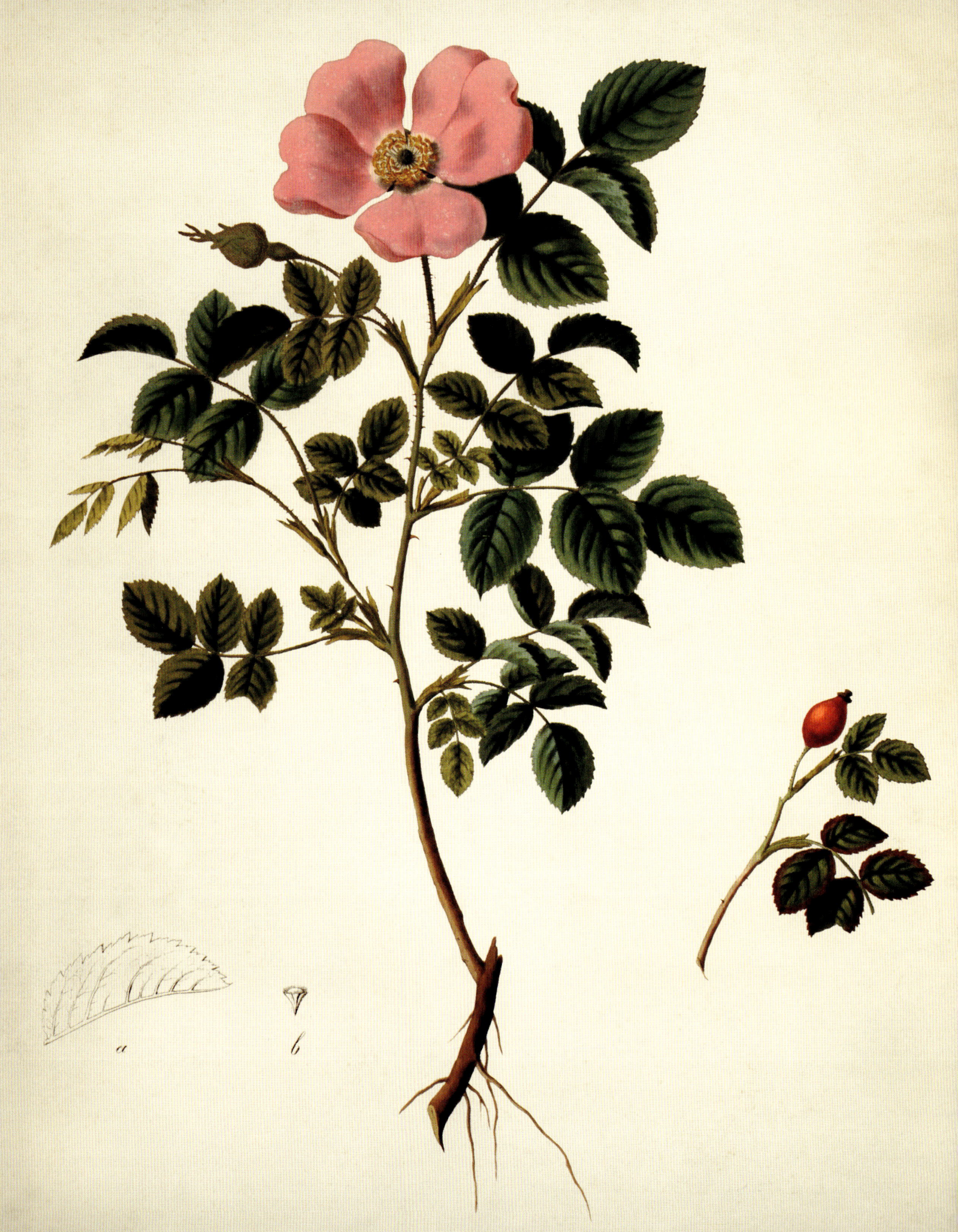
a
b

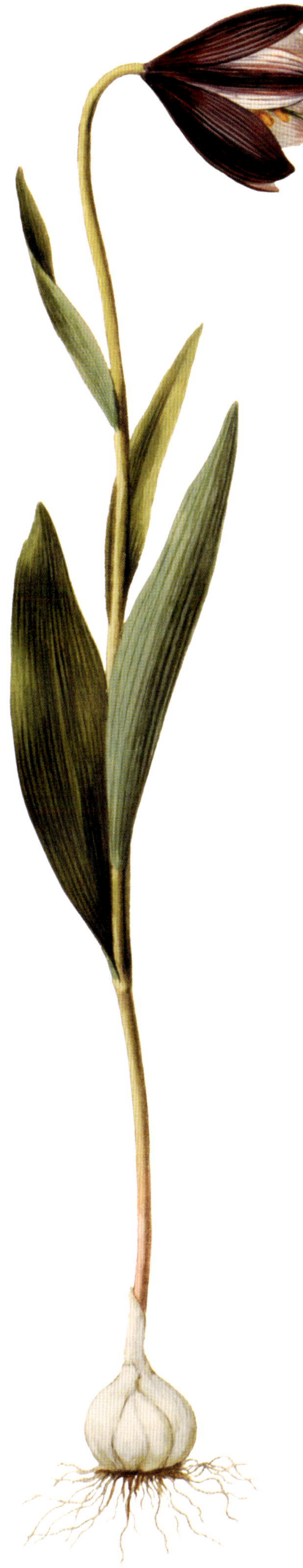

The expedition was away from St Petersburg for six years, and afterwards Pallas stayed in Russia, becoming a favourite of Catherine and publishing the accounts of his travels. His second journey was to the Crimea and to Astrakan in 1793–94. Catherine granted Pallas an estate in the Crimea on the strength of his descriptions of the beauty of the country, but in reality it proved unhealthy; the countryside had been ravaged by Russian soldiers and, under Catherine's successor Tsar Paul, it proved little better than exile. The first two volumes of Pallas's *Flora Rossica* were published in 1784 and 1788, but the last two were never published because of official disapproval. In the first volumes the elegant engravings by Karl Friedrich Knappe were hand-coloured, and include Arctic plants such as *Andromeda polifolia* and *Dryas octopetala*, as well as plants from Central Asia like the white-flowered *Atragene sibirica* and the dwarf almond, *Amygdalus nana*.

The Caucasus has long had a strong hold on the Russian psyche, because so many aristocratic young soldiers were exiled there, fighting the Muslim Chechens, duelling with each other and flirting with the ladies who came to take the waters in Pyatigorsk; their adventures inspired the poetry of Pushkin and Lermontov and many of the stories of Tolstoy. Baron Friedrich August Marschall von Bieberstein (1768–1826) was a German who served in the Russian army, at a time when the Russians were fighting the Turks and the Persians in the Transcaucasus, as well as the tribes in the mountains. In 1795, he joined a military expedition under Count Valerian Zubov which aimed to conquer Azerbaijan, but did little more than capture the city of Derbent on the Caspian coast from the Persians. However, Bieberstein was able to explore much of the western Caucasus and Georgia, discovering many new plants; his *Flora Taurico-Caucasica* was published between 1808 and 1819, and the beautifully illustrated *Centauria plantarum rariorum Rossiae meridionalis* appeared in parts in 1810, 1842 and 1843, and includes familiar garden plants such as *Campanula lactiflora* and *Lilium monadelphum*. The paintings were by Login Andrewitsch Choris, who became a landscape artist and accompanied the poet and botanist Adelbert Chamisso (better known as the author of the romantic novel *Peter Schlemihls wundersame Geschichte*) on his round-the-world voyage in 1815.

Karl Frederick von Ledebour (1785–1851) was another German, and Professor of Botany at the University of Dorpat (now Tartu) in Estonia, then part of the Russian empire. In 1826, he took a university expedition across northern steppes of Central Asia, as far

LEFT: Fritillaria caucasica *from Bieberstein's* Centauria plantarum rariorum Rossiae meridionalis.

OPPOSITE: Rosa oxyodon *from Bieberstein's* Centauria plantarum rariorum Rossiae meridionalis.

as the Altai mountains and Mongolia. Their journeys and botanical finds were published in *Flora Altaica* (1829–34) and magnificently illustrated in *Icones Plantarum novarum vel imperfecte cognitarum floram Rossicam, imprimis Altaicam illustrantes* (published in parts in 1829–34). There were 500 illustrations by various artists, including E. Bommer (fl. 1830), and they were lithographed in Germany. Bommer was also the artist for Karl Eduard von Eichwald (1795–1876), who was primarily a geologist but collected plants around the Caspian. They were illustrated in *Plantarum novarum vel minus cognitarum quas in itinere Caspio-Caucasico observavit*, published between 1831 and 1833.

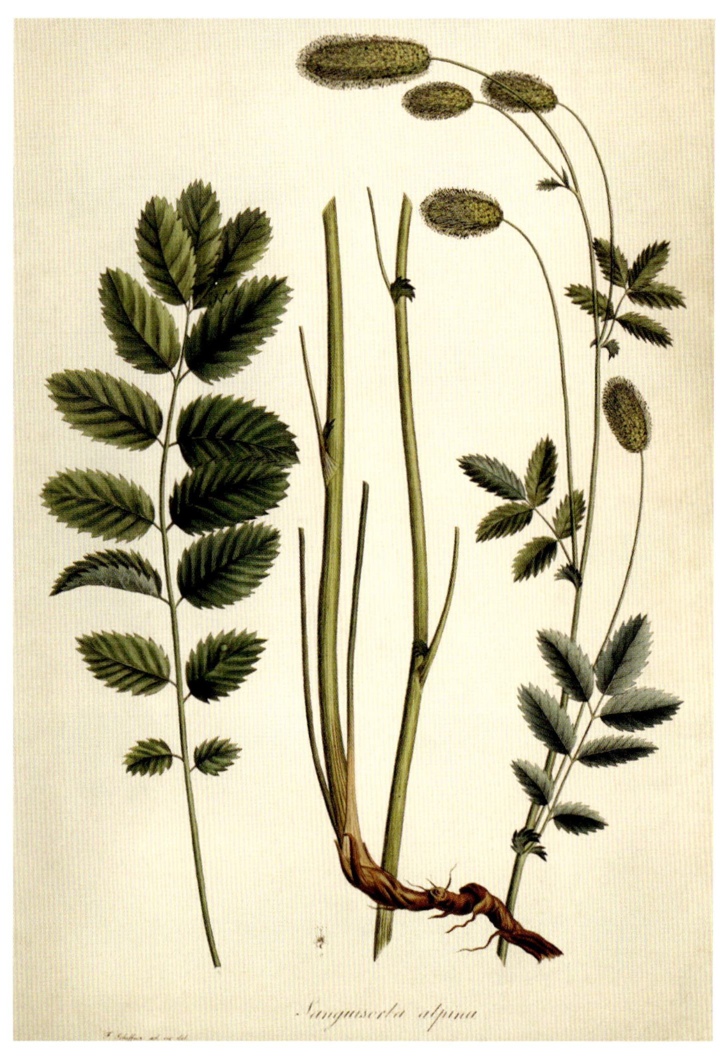

Many of the plants collected by these botanical explorers were sent to the Botanical Gardens in St Petersburg, founded by Peter the Great as a physic garden in 1714 and site of the Komarov Botanical Institute, the most important in Russia. The director in the early nineteenth century was F. E. L. von Fischer (1782–1854), who corresponded with staff at other gardens around the world, and particularly with the Chelsea Physic Garden in London. Fischer had been private botanist to Count Alexei Kirillovich Razumovsky at his garden near Moscow, and on Razumovsky's death in 1822, his botanical library, the rarest plants from his garden and Fischer moved to St Petersburg. In 1846, Fischer started the journal *Sertum Petropolitanum* or *Jardin de St Pétersbourg*, giving illustrations and descriptions of plants which flowered in the garden. *Epimedium pinnatum* from the Caspian forests of Iran, *Almeidea rubra* from Brazil and *Nemophila liniflora* from the Pacific coast of North America are some of the plants illustrated here for the first time. It continued to be published until 1869, under the editorship of Eduard Regel (*see* page 154).

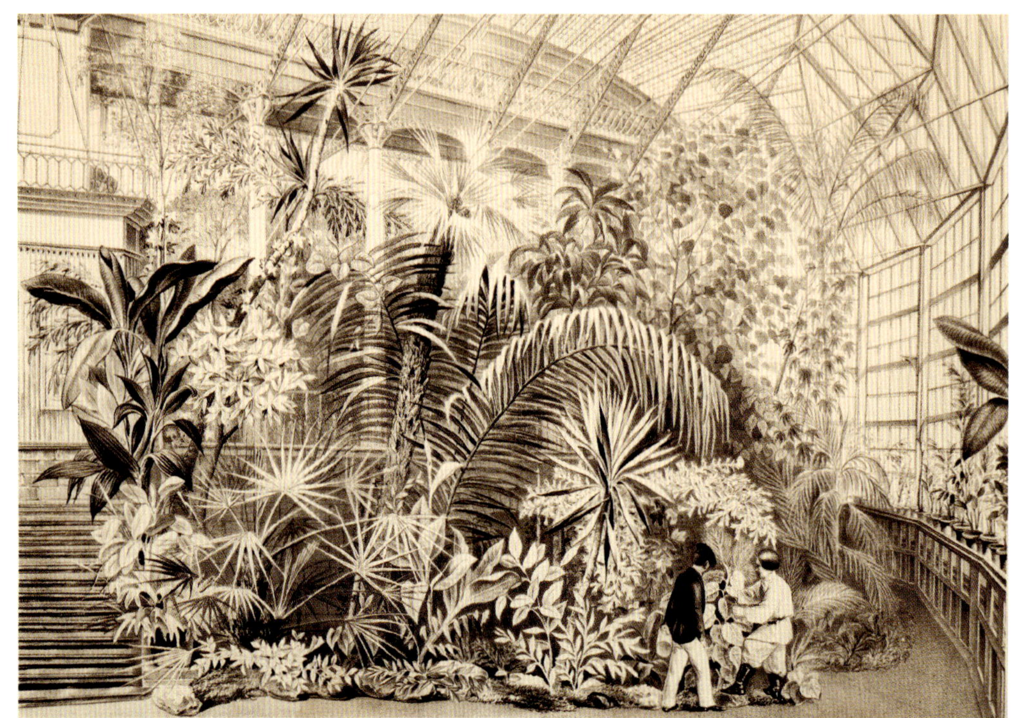

ABOVE RIGHT: Sanguisorba alpina *from von Ledebour's* Icones Plantarum novarum vel imperfecte cognitarum floram Rossicam, imprimis Altaicam illustrantes.

RIGHT: An illustration of one of the large glasshouses in St Petersburg from Fischer's Sertum Petropolitanum.

OPPOSITE: Almeidea rubra, *a greenhouse shrub from Brazil, from* Sertum Petropolitanum.

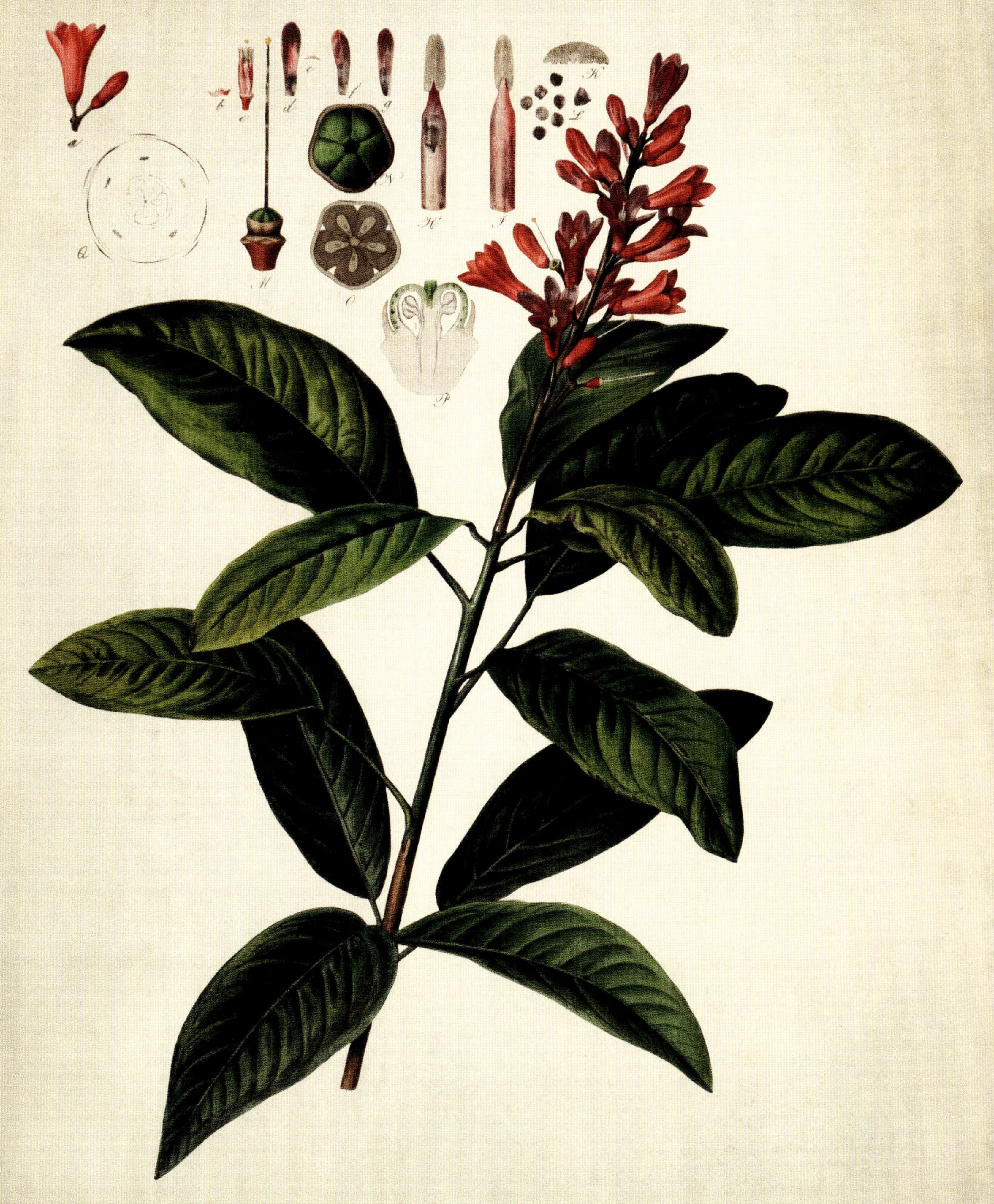
a
b
c
d
f
g
H
I
K
L
M
N
O
P
Q

JAPAN

During the expansion of European scientific exploration in the seventeenth and eighteenth centuries, Japan remained a *terra incognita*, a country that resisted any foreign influence, be it politically, scientifically or even commercially. Only the Dutch and the Chinese were allowed to trade, and they were subject to the most stringent restrictions. The Dutch trading post was on Deshima Island, in Nagasaki harbour, only 32 acres in extent and surrounded by a high palisade. It had been built to house Portuguese traders, but they were expelled in 1638. By 1641 they had been replaced by the Dutch, who were allowed to keep a few staff on the island, to supervise the trade of two ships a year, which were all the Japanese allowed to visit. Medicines from the Dutch East Indies were the favoured imports; the Japanese were also keen to learn more about western medicine, and Japanese medical students were allowed to go to the island to study.

In seventeenth-century Europe botany was considered a branch of medicine, so it is not surprising that some of the doctors who lived on Deshima were also keen botanists. Three of these doctors have names which are familiar to botanists and gardeners, and are associated with Japanese plants: Kaempfer, Thunberg and Siebold.

Englebert Kaempfer (1651–1716) was born in northern Germany and studied medicine at Crakow and Königsburg (now Kaliningrad, an enclave of Russia on the Baltic). In 1683, he was appointed secretary to the Swedish ambassador to Russia, and accompanied him to Moscow; later they visited Astrakhan and Isfahan, where Kaempfer remained behind, travelling around Persia and practising as a physician. In 1688, he joined the Dutch merchant fleet, then in the Persian Gulf, bound for the island of Java (in today's Indonesia) and attained the position of chief surgeon. They visited Ceylon (now Sri Lanka) and Bengal before reaching Batavia (now Jakarta), Java's main city, in 1689. From here, Kaempfer was appointed to the post in Nagasaki and he remained in Japan from 1690 to 1692. Through the use of his medical skill, his knowledge of mathematics and astronomy and by plying the locals with liquor, Kaempfer was able to gather information and get specimens of a number of Japanese cultivated or medicinal plants, including *Camellia japonica*, *Ginkgo biloba*, *Lilium speciosum* and drawings of the tea plant, *Camellia sinensis*. Many were brought to him by his servants, medical students and by the prostitutes who were allowed to visit the Europeans on the island. There were also some opportunities to collect along the road on the embassy (or diplomatic visit) which took place every spring, from Nagasaki to Kyoto and Tokyo.

Kaempfer's travels were published during his lifetime, called *Amoenitates Exoticae*. After his death his papers were bought by Sir Hans Sloane, and are now in the British Museum; his plant specimens are in the Natural History Museum in London. In 1791, some of the Kaempfer's own drawings were engraved and published by Sir Joseph Banks, entitled *Icones Selectae Plantarum quas in Japoniae collegit et delineavit Englebertus Kaempfer*. They are 59 simple but elegant drawings, and interesting as the first Japanese plants to be known in Europe. Lilies, hostas and

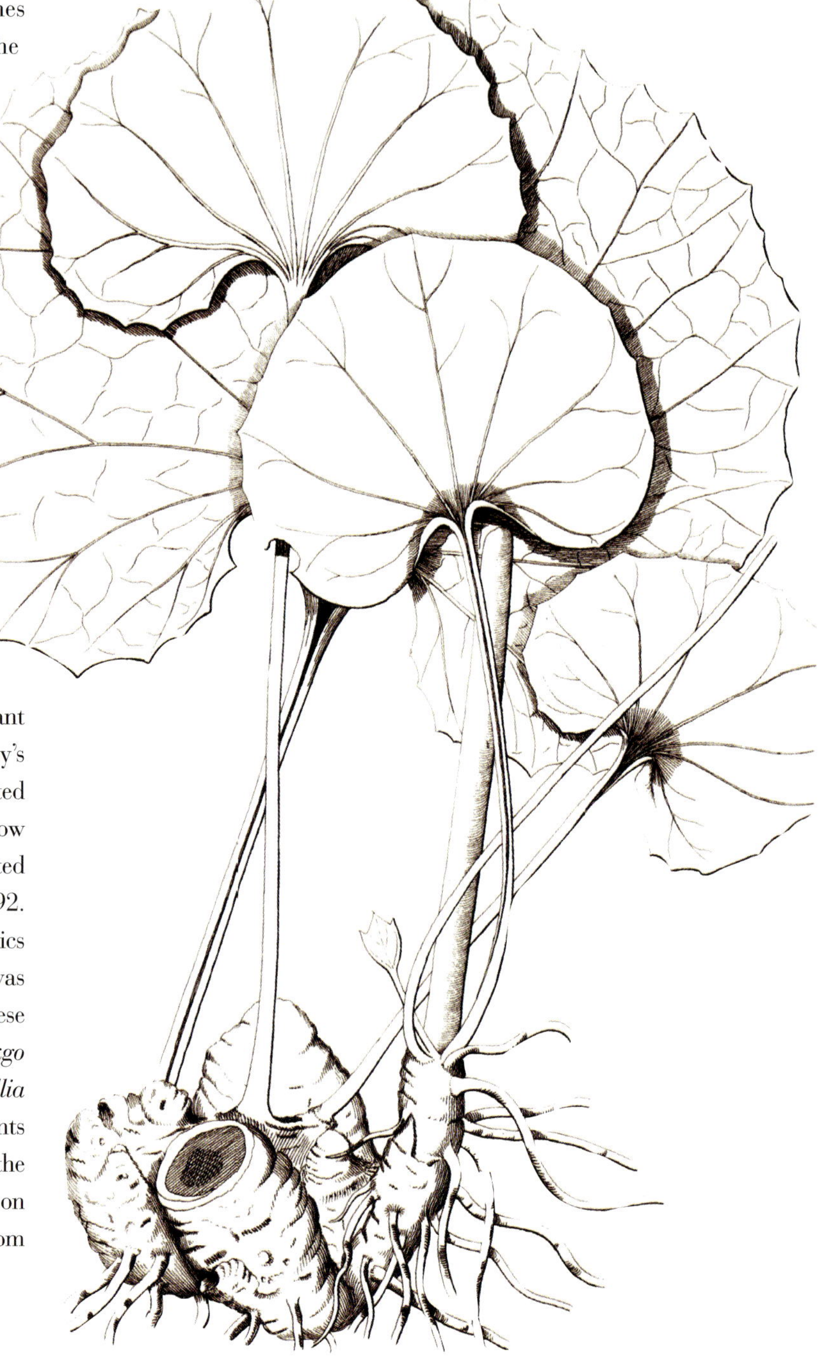

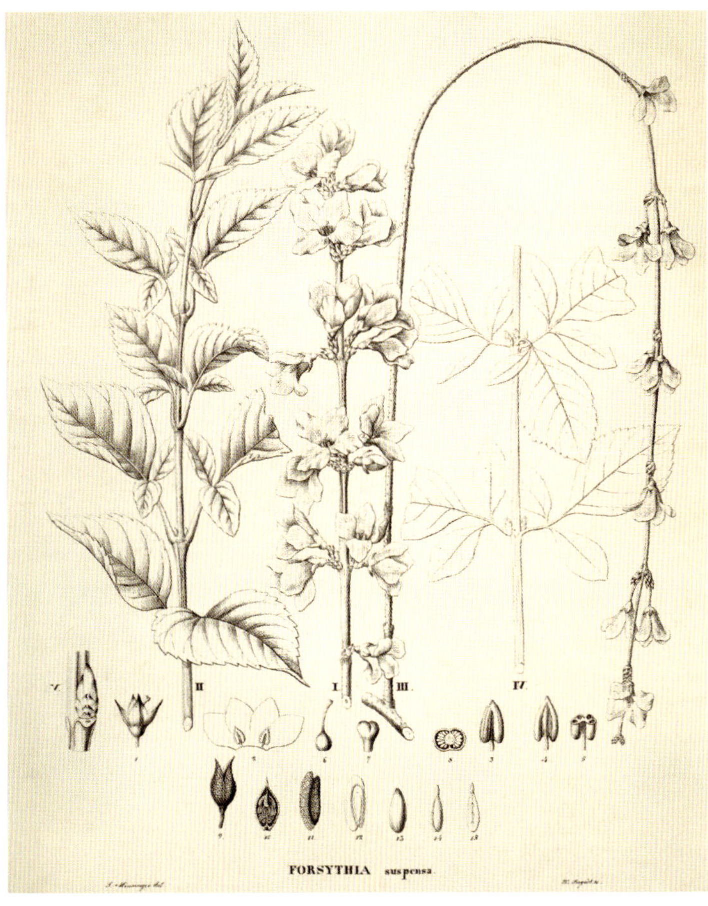

three Magnolia species are among the now familiar garden plants illustrated.

The next botanist to stay on Deshima Island was Carl Peter Thunberg (1743–1828), a Swede and a pupil of Linnaeus. Thunberg was 27 when he left Uppsala, and first visited Professor Burmann in Amsterdam. Burmann persuaded a group of wealthy plant-lovers to sponsor Thunberg, and he joined the Dutch East India Company. First, he landed in the Cape, then a Dutch colony, and spent two years travelling round, collecting plants, on one journey in the company of Francis Masson who had been sent out as a collector from Kew.

Thunberg finally reached Japan in August 1775. Conditions for the Dutch traders had changed little since Kaempfer's visit, and to begin with, Thunberg had to be content with what he was brought and the plants he could find in the hay brought to feed the cows, which they were allowed to keep for milk and meat. Thunberg also travelled as far north as Tokyo on the yearly embassy, and managed to collect some plants along the way. These all formed the basis of his *Icones plantarum Japonicarum* (1794–1805), with 50 large engravings of many familiar plants: *Fritillaria thunbergii*, grown as a medicinal plant in China and brought over to Japan, was called *Uvularia cirrhosa*. Thunberg only saw the top of the plant, so can be forgiven for the misidentification, but he tended to try to squeeze the plants he saw into Linnaeus's names. Thunberg returned to Holland in 1778, where he published *Flora Japonica* in 1784, and worked on his large book of illustrations.

The third great European botanist and collector in Japan was Philipp Franz von Siebold (1796–1866). He was a German, from Bavaria, and came from a large family of eminent surgeons and doctors. He studied medicine and natural history at the University of Würzburg, and qualified in medicine, surgery and midwifery in 1820. Rather than practise in Germany, he decided to be adventurous and joined the Dutch East India Company as a doctor: he reached Batavia in 1823, and was sent on to Japan, where he was nearly refused entry because of his German accent which aroused the suspicion of the Japanese interpreters; however, his description as a *Hochdeutscher* was interpreted as mountain Hollander, and his uncouth accent was accepted.

OPPOSITE: Petasites japonicus *by Engelbert Kaempfer from* Icones Selectarum Plantarum quas in Japoniae collegit et delineavit Englebertus Kaempfer.

LEFT: Forsythia suspensa *from Siebold and Zuccarini's* Flora Japonica.

BELOW: Hydrangea petiolaris *from Siebold and Zuccarini's* Flora Japonica.

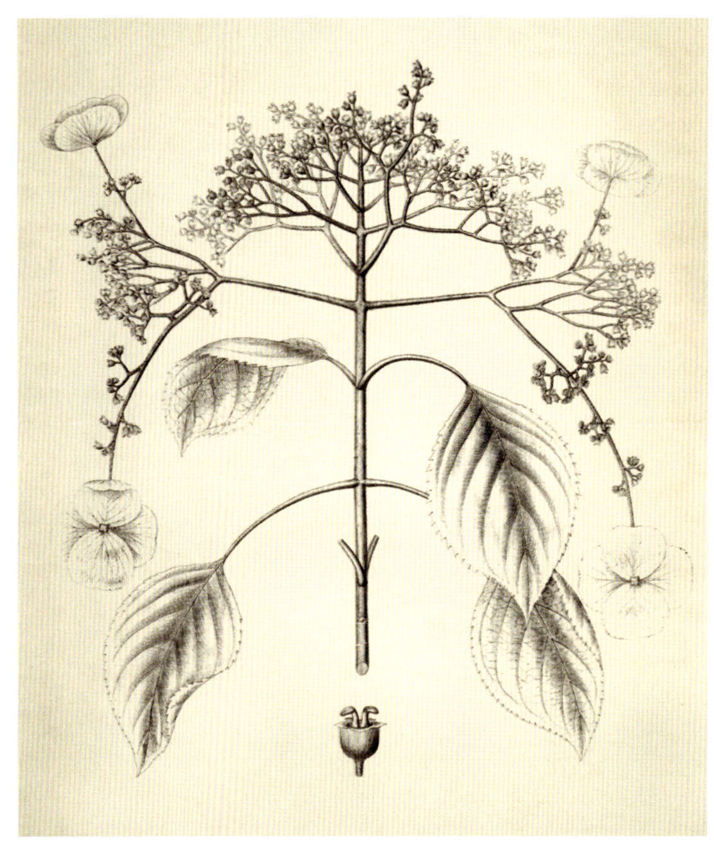

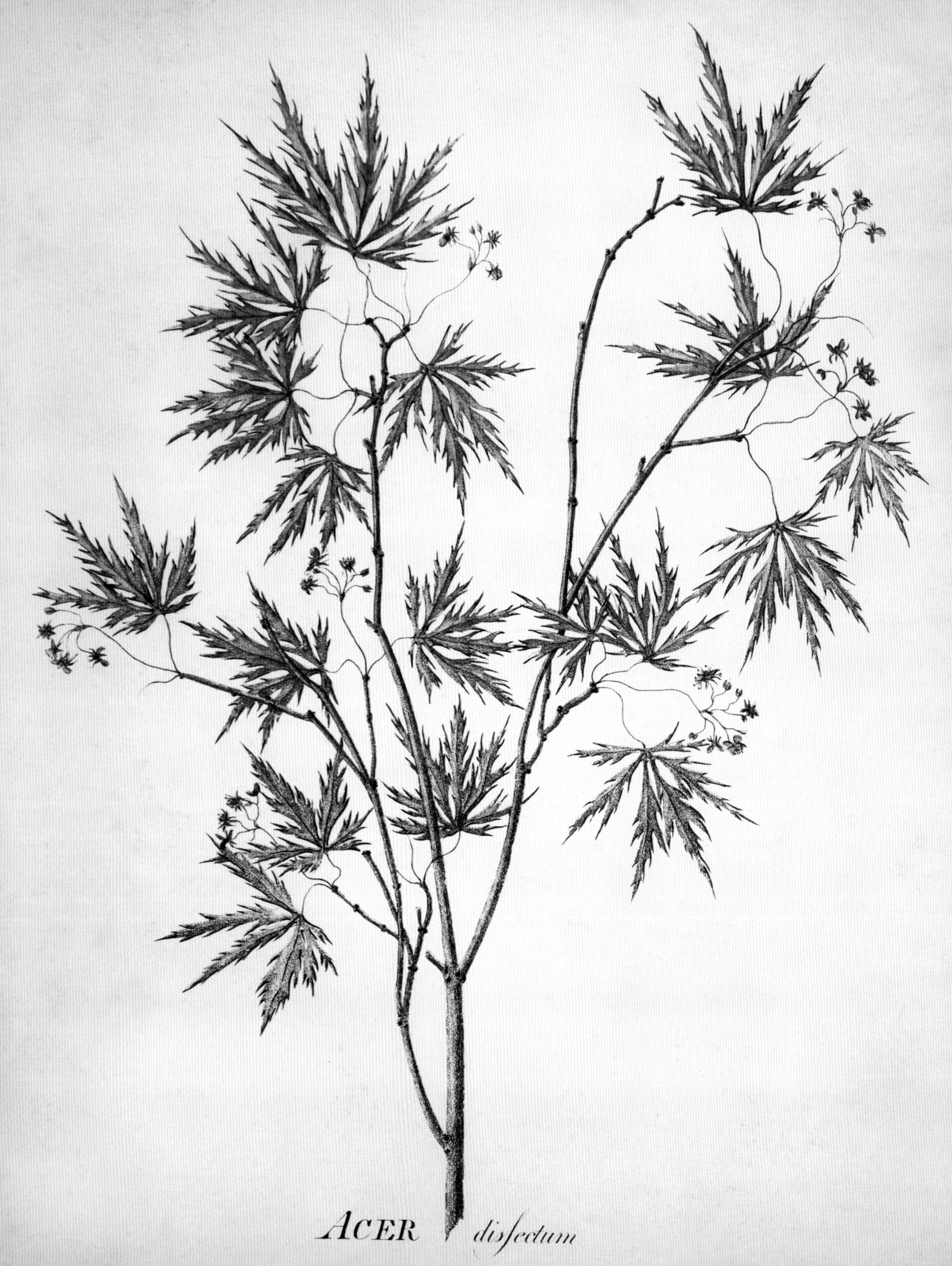

ACER dissectum

OPPOSITE: Acer palmatum *var.* dissectum *from Thunberg's* Icones plantarum Japonicarum.

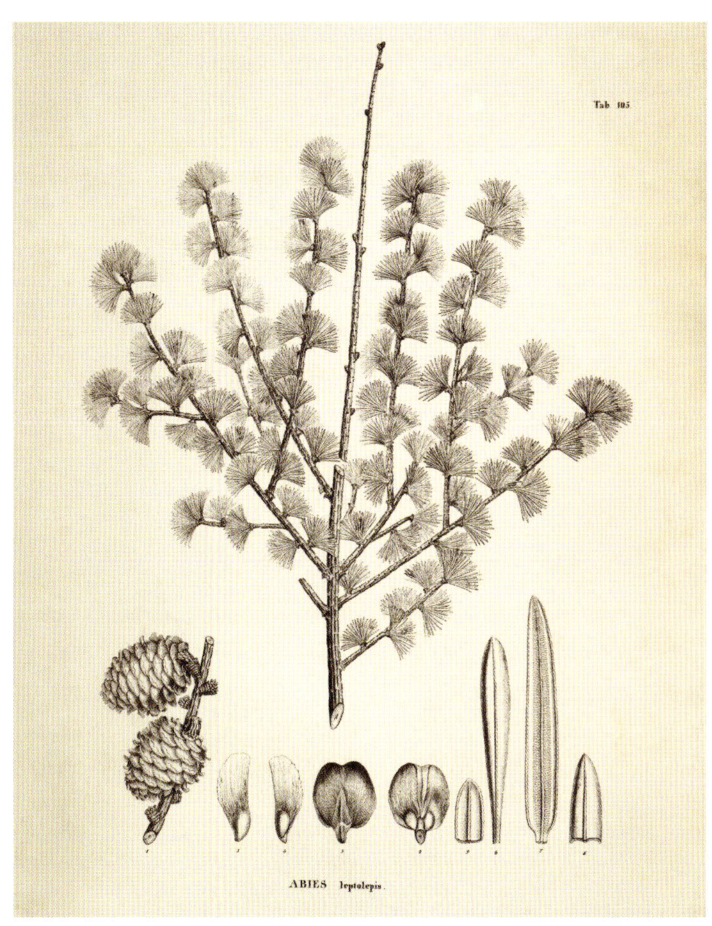

RIGHT: Abies leptolepia *(now* Larix kaempferi*), Japanese larch, from Siebold and Zuccarini's* Flora Japonica.

Siebold was a much more skilful doctor than any who had been on Deshima before, and he was able to have more contact with the local population. He learnt Japanese, and taught the latest techniques of German medicine, notably the operation for removing cataracts, new obstetric procedures and the use of drugs such as digitalis, derived from various species of foxglove (*Digitalis*). His teaching was so popular that he was allowed to open a medical school and clinic on the mainland, where his patients brought him scrolls, screens, pottery and lacquer. He encouraged his students to write on different geographical and ethnographical topics, so he could learn more about the country. While at the house of a patient in Nagasaki, Siebold met and fell in love with a beautiful 18-year-old Japanese girl, whom he called O-Taki-San. As a respectable Japanese, she was not allowed to marry him, but had to register as a prostitute to be able to live with him as his concubine on Deshima Island; their daughter, O-Ine, born in 1827, later became a leading midwife in Nagasaki.

Siebold also trained a young Japanese artist, Kawahara Keiga (1786–1862), to make drawings and woodcuts of Japanese plants and people. At the same time, he was building up a collection of dried plant and animal specimens, and growing Japanese plants on the island in the medicinal garden, which had been abandoned under his predecessors. Siebold was able to learn about Japanese horticulture and realized what a high standard of plant selection and cultivation had been achieved. Many of the ancient forms of Japanese cherry, Mume, chrysanthemum, *Dianthus* and camellia are still preserved in Japan.

The yearly embassy to Tokyo, which had given Kaempfer and Thunberg their glimpse of the interior of Japan, had been changed to every four years, and it was not until 1826 that Siebold managed to do the journey. He took his artist Kawahara, an artist from Batavia and his pharmacist, all three disguised as servants. In Tokyo, he gave demonstrations of surgery, and was shown works of art including a secret map of the Japanese coast, kept by the court astronomer who promised to provide him with a copy. He also collected over a thousand more plants, and was given their local Japanese and Chinese names. Siebold was due to leave Japan in 1828, and had all his collections loaded onto a ship for transport to Europe – bulbs, plants, specimens, works of art, as well as forbidden books and the copy of the secret map. Shortly before the ship was due to leave, a devastating typhoon struck Japan, and the loaded ship was stranded in Nagasaki harbour for the next three months. Meanwhile, the court astronomer had been arrested for providing Siebold with the secret map, and he was suspected of being a spy, working for the Russians as were so many of his fellow German scientists. Siebold himself was arrested, along with many of his Japanese acquaintances. So the ship had sailed without him, and reached Amsterdam, with most of the living plants now dead. Siebold awaited trial, was interrogated and in October 1829, was expelled from Japan and forbidden to return.

Many of Siebold's plant introductions thrived in Europe, and *Hosta sieboldiana*, *Magnolia sieboldii* and *Primula sieboldii* are some of the many that have familiarized his name. Some were illustrated in journals at the time, and 150 appeared in Siebold and Zuccarini's *Flora Japonica* (1835–41), but the majority were only published in 1995, after 981 of the drawings made for Siebold were rediscovered in St Petersburg, having been purchased by Maximowicz from his widow in 1869. Siebold's *Florilegium of Japanese Plants*, published by the Komarov Botanical Institute in St Petersburg, includes 263 of Keiga's beautiful drawings and is a fitting tribute to Siebold's remarkable years in Japan.

Les Vélins du Muséum

The history of the collection of paintings known as *Les Vélins du Muséum*, today one of the highlights of the library of the Muséum national d'Histoire naturelle in Paris, begins in the seventeenth century with Gaston d'Orléans, third son of King Henri IV of France and younger brother of Louis XIII.

Gaston was a keen amateur botanist and naturalist, who established a botanic garden and private zoo in the Loire Valley at the Château de Blois, which had been given to him by his brother as a wedding present. He commissioned the artist and engraver Nicolas Robert (1614–85) to record the most interesting and rare animals, birds and flowers in his collection. These paintings, made in gouache on vellum, together with some slightly earlier ones made by another flower painter, Daniel Rabel (1578–1637), were to form the nucleus of a magnificent collection constantly added to by successive court painters over the years. On Gaston's death his collection passed to his nephew, Louis XIV, who, in 1664 appointed Robert "peintre ordinaire de Sa Majesté pour la miniature". Robert continued to work on the *vélins*, which were at that time housed in the Louvre, and also taught other artists who worked alongside him.

Robert was succeeded as the King's flower painter by Jean Joubert, who employed Claude Aubriet (1665–1742) to assist him. The pupil outshone the master, and was fortunate to be taken up by the great botanist Joseph Pitton de Tournefort, with whom he travelled to the Levant, and from whom he learnt much about plant structure. On Joubert's death, Aubriet succeeded him as royal artist, in which role he continued until 1735, when he handed over the job to his pupil, Madeleine Françoise Basseporte (1701–80). She was succeeded by the Dutch artist Gerard van Spaëndonck (1746–1822), who was appointed professeur de peinture florale at the Jardin du Roi in 1780.

Meanwhile, the young Pierre-Joseph Redouté had also appeared at the Jardin du Roi, having been appointed, with unfortunate timing, draughtsman to Marie-Antoinette. In 1792, the monarchy was abolished, and Marie-Antoinette and her husband Louis XVI were executed the following year. During the Revolution the ever-increasing collection of *vélins*, by now declared the property of the state, was transferred to the renamed Jardin des Plantes (previously the Jardin du Roi), the chief botanic garden in Paris, and a department of the Muséum national d'Histoire naturelle; van Spaëndonck was appointed as professeur d'iconographie naturelle at the museum. Van Spaëndonck initially used gouache for his paintings on vellum, but started to experiment with pure watercolour, and from 1784 onwards the *vélins* were done exclusively in this medium. Fortunately for Redouté, van Spaendonck arranged for him to be taken on to the staff at the Jardin des Plantes, and as well as doing scientific drawings for botanical publications, he continued to contribute to the collection of *vélins*.

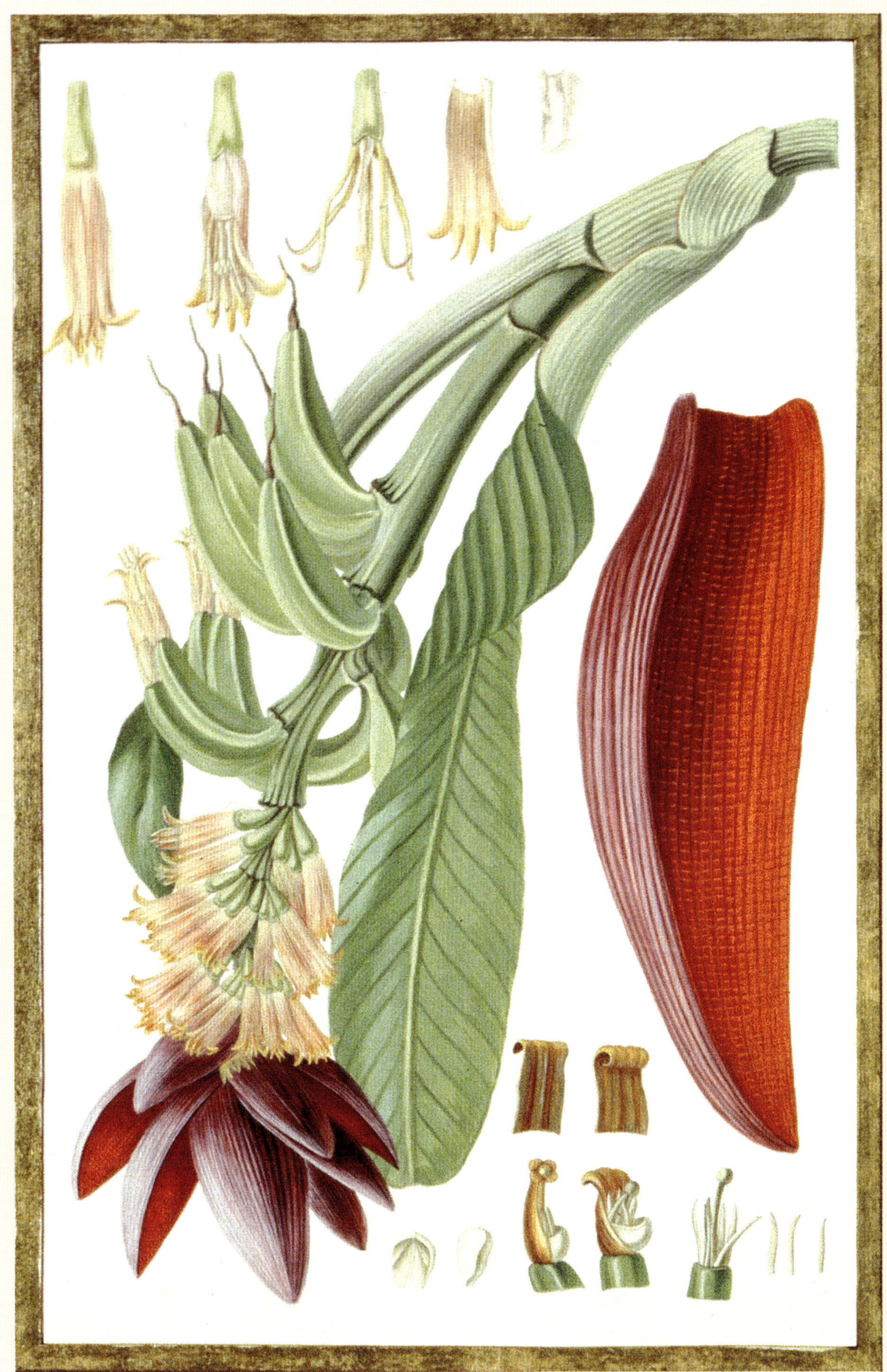

OPPOSITE: An example of the engraving work of Nicolas Robert: Sanicula, siue Cortusa indica *(now* Mitella diphylla*).*

RIGHT: Madeleine Basseporte's vélin *of* Musa × paradisiaca *l, the banana.*

7 BOTANY BAY AND BEYOND

The first Europeans to see Australia were probably Dutch or Portuguese traders who strayed off course on voyages to the East Indies, and the first definite sighting was in 1606 when a Dutch ship sailed down the western side of Cape York, having missed the Torres Strait. Then, in 1616, the Dutchman Dirk Hartog landed on what is now Dirk Hartog Island, near Carnarvon in Western Australia, and in 1688, William Dampier (1651–1715) (dubbed "Australia's first natural historian"), landed on the same coast; on his second visit in 1699 he made a detailed survey of the western coastline.

William Dampier was a tenant farmer's son, born in East Coker in Somerset. He was orphaned by the age of 14, and went to sea in a Weymouth trader at the age of 18, having received an academic education at a "Latin school". He traded around the Caribbean, then imported timber and spent some years privateering against Spanish ships along the Pacific coast of South America, where, in 1685, he transferred to the *Cygnet*, captained by Charles Swan. They crossed the Pacific to Guam, then sailed south, passed the western end of Timor, and on 14 January 1688 landed north of Broome, and grounded the ship in a sandy bay on a spring tide, to repair the sails and repair the hull. Dampier noted the plants and birds there and they saw the footprints of dingos. He eventually reached England again in 1691, and wrote *A New Voyage Round the World*, which was published in 1697. It was very popular and was soon translated into Dutch, French and German. Two years later, he wrote *A discourse on Trade Winds, Breezes, Storms, Seasons of the year etc.*, a manual for sailors in the tropics, which was issued as part of a second volume, together with a revised and corrected edition of volume 1.

Dampier was now a sailor of repute, and persuaded the Admiralty to provide a ship to explore New Holland, and investigate the prospects for trade. He was given command of the *Roebuck*, a three-masted, square-rigged barque with a crew of 50. They fitted out at Deptford on the Thames, and left England in January 1699, calling at Bahia in Brazil, before taking the westerlies south of the Cape, towards Australia. By early August Dampier reckoned he was nearing the coast (there was no accurate way of calculating longitude until the time of Captain Cook's second voyage in 1772), and the crew began to see seaweed and other signs of coastal life floating in the sea. They approached the coast with great caution, sailing north and south until they spotted land on 10 August.

As the weather was very squally they sailed northwards along the coast, before anchoring off Dirk Hartog Island, where Dampier landed, studied the trees and heath-like vegetation, and particularly noted an abundance of blue flowers: "There were also some plants, herbs and tall flowers, some very small flowers, growing on the ground, that were sweet and beautiful, for the most part unlike any I had seen elsewhere." This shows unexpected sensitivity and appreciation of beauty in a pirate. One of the blue flowers Dampier mentioned, which still grows where he saw it, is *Dampiera incana*, a creeping shrub, named in his honour by Robert Brown in 1810.

Dampier's most exciting plant discovery was Sturt's Desert Pea, which has had more Latin names than English. For many years it was called by Lindley's name *Clianthus dampieri*, but has recently been put into its own genus, *Willdampia* A. S. George. The stems creep along the sand, and the flowers stand up like a group of red parrots, with a purple-black boss in the middle. Dampier's plant specimens are in Oxford, and some simple engravings drawn from his herbarium specimens were published in his *A Voyage to New Holland, &c. in the Year 1699*, published in 1703; as well as plants, this also contained the first drawings to be published of Australian fish, birds and other wild life. Dampier's collections never received the attention they deserved: indeed, they were ignored by Linnaeus, even though he visited Oxford before publishing his *Species Plantarum* in 1753.

The most famous of all the early voyages to Australia is without doubt that of Captain James Cook. Cook left Plymouth on the *Endeavour* in 1768, commissioned by the Royal Society to view the transit of Venus from Tahiti, due to take place on 3 June 1769, enabling them to gather information necessary for calculating the distance between the earth and the sun, vital for navigation. Cook also carried sealed orders from the Admiralty for the second

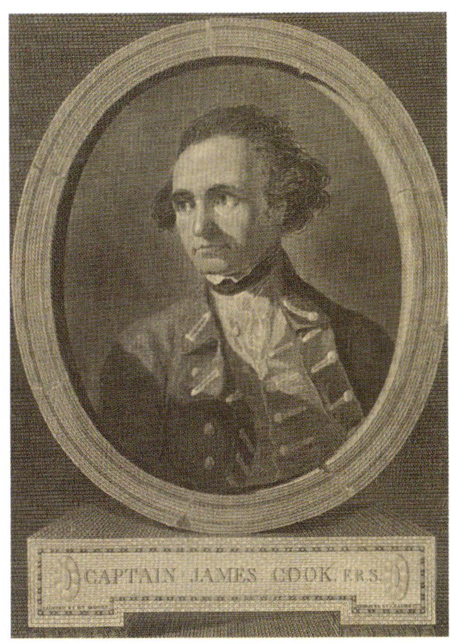

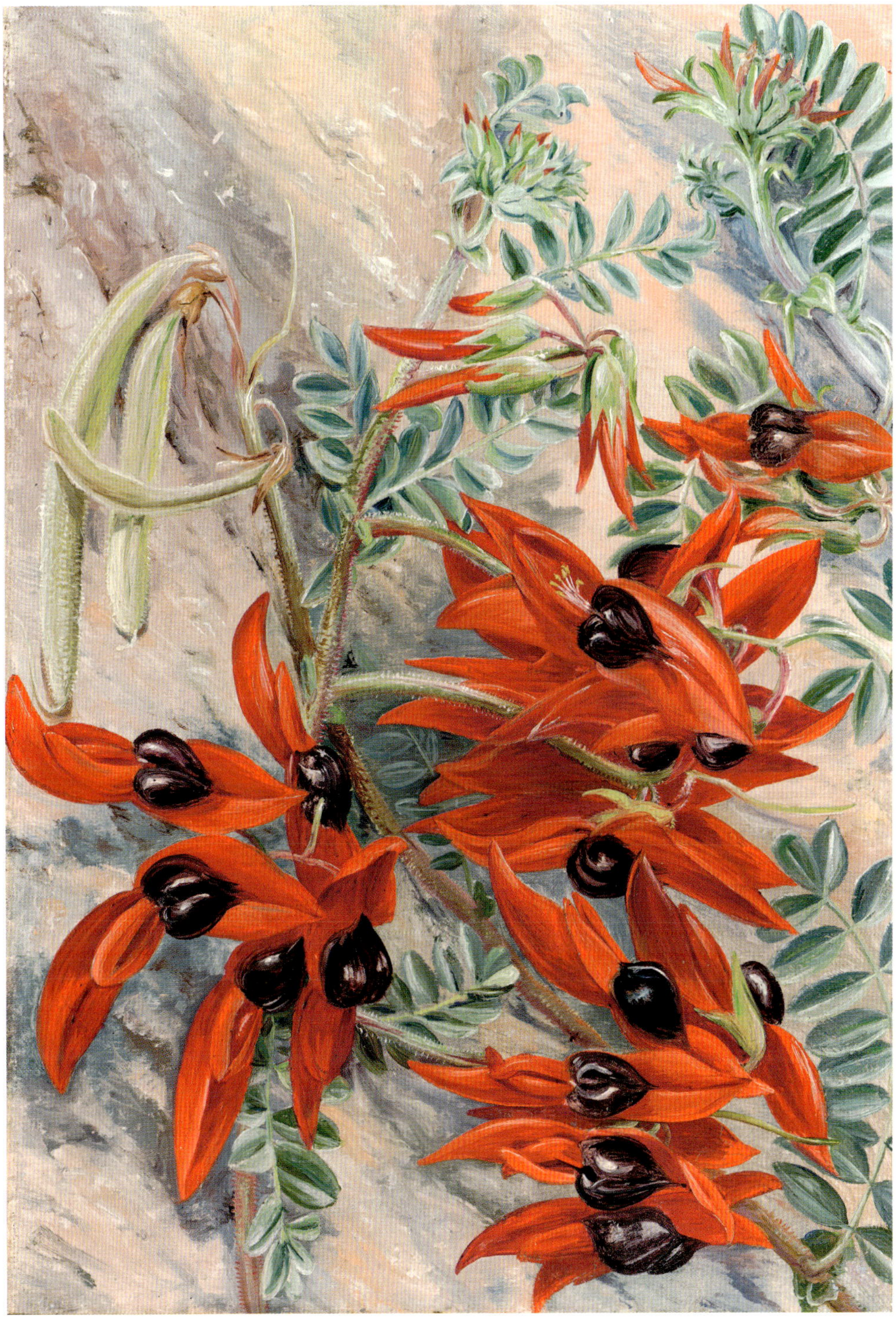

ABOVE: Captain James Cook, who named Botany Bay in Australia after the "great number of plants discovered there by Mr Banks and Dr Solander".

OPPOSITE: Captain William Dampier after whom a number of plants were named.

RIGHT: Sturt's Desert Pea, a member of the genus Willdampia *A. S. George, which has very bright red flowers.*

part of the trip, instructing him to search for *Terra Australis*, the southern continent, and to note the geography, flora and fauna that were encountered en route. On board were Sir Joseph Banks (1743–1820), then aged 25 and already a well-travelled scientist and fellow of the Royal Society (*see* pages 98–99), his assistant, the botanist Daniel Solander (1733–82) who had been a pupil of Linnaeus, and Sydney Parkinson (1745–71), as natural history artist, together with landscape artist Alexander Buchan, Finnish physician and artist Herman Spöring and four field assistants. After landing in Brazil, they stopped in Tierra del Fuego to make an expedition inland, where the weather was much colder than they expected; they ran out of food and two of the field assistants died from exposure. After rounding Cape Horn, they reached Tahiti in April 1769, and recorded breadfruit and sweet potatoes, both new to science, and made calculations of the transit of Venus across the sun on 3 June. Here, Alexander Buchan died of epilepsy. After opening the sealed orders, they sailed south to search for *Terra Australis*, charting and circumnavigating both islands of New Zealand, which had been seen but not explored by Abel Tasman in 1642.

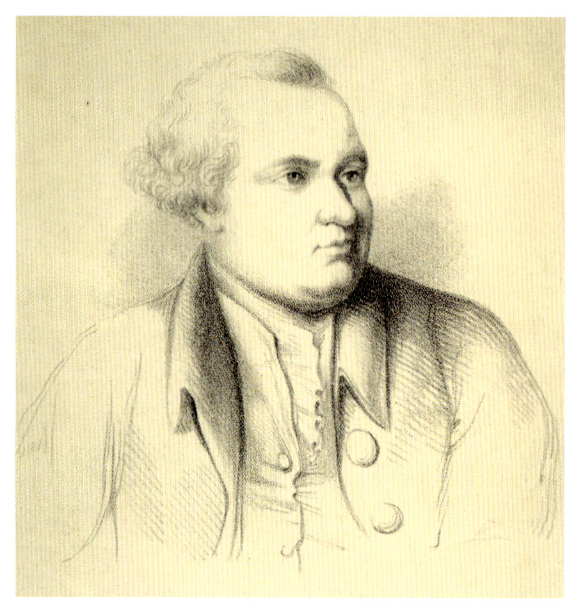

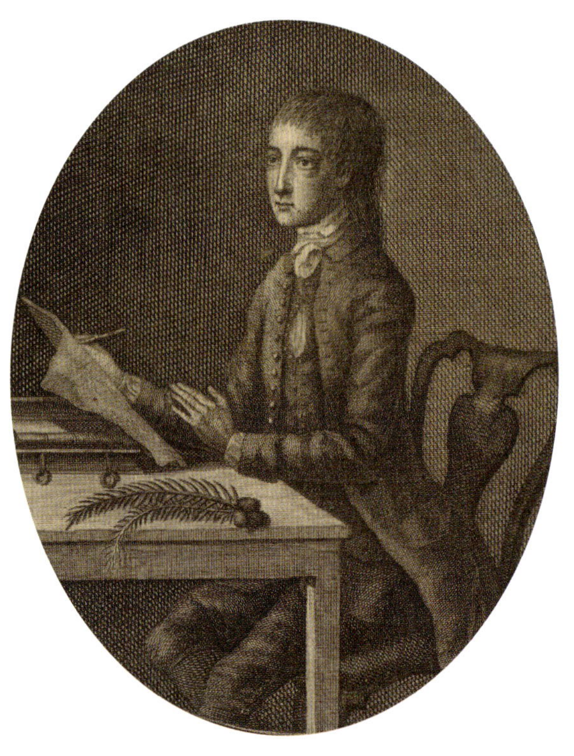

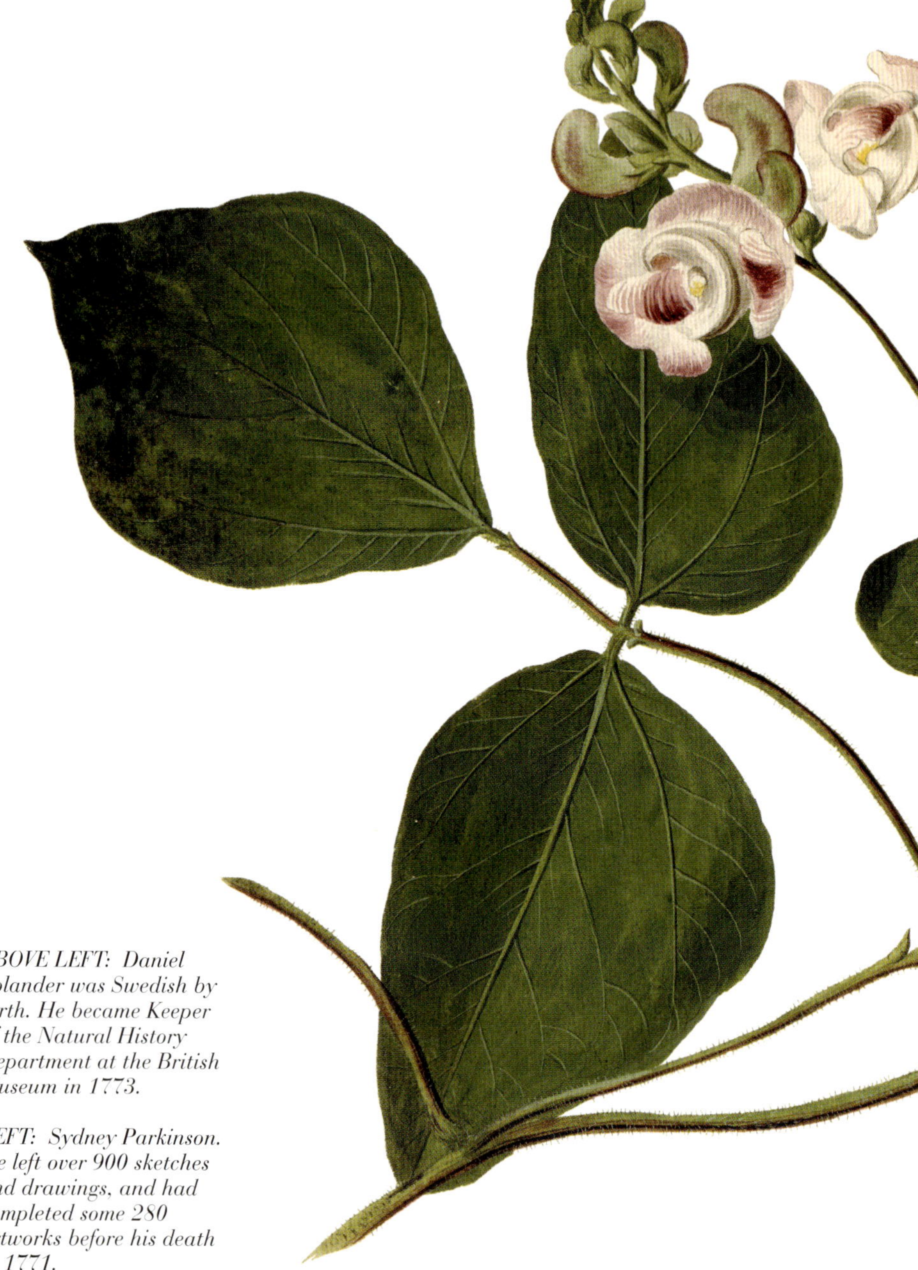

ABOVE LEFT: Daniel Solander was Swedish by birth. He became Keeper of the Natural History Department at the British Museum in 1773.

LEFT: Sydney Parkinson. He left over 900 sketches and drawings, and had completed some 280 artworks before his death in 1771.

Here, they collected over 400 new plants and were fascinated by the similarities of many of the plants with those of South America. They then sailed northwest, sighting the Australian mainland in April 1770.

The bay where they landed, near present-day Sydney, was so full of new flowers that they named it Botany Bay. Sailing north along the coast, they tried to reach the open ocean north of Cairns, but ran aground on the Great Barrier Reef; their ship was badly damaged but they managed to nurse her into the mouth of the Endeavour River, where she was beached for repairs to her hull. This gave the botanists another six weeks for further collecting.

On the return journey in late 1770, they visited Java. Until then, Cook's crew had been healthy, but in Java they picked up both malaria and dysentery, and in January Spöring died of fever, followed shortly afterwards by Parkinson, Tupara, a Tahitian chief who had helped them communicate with the Maories, and seven of the crew. Almost 20 more sailors died before they reached Cape Town, where the sick were put ashore. The ship and surviving crew and scientists finally reached London after calling at St Helena, in July 1771.

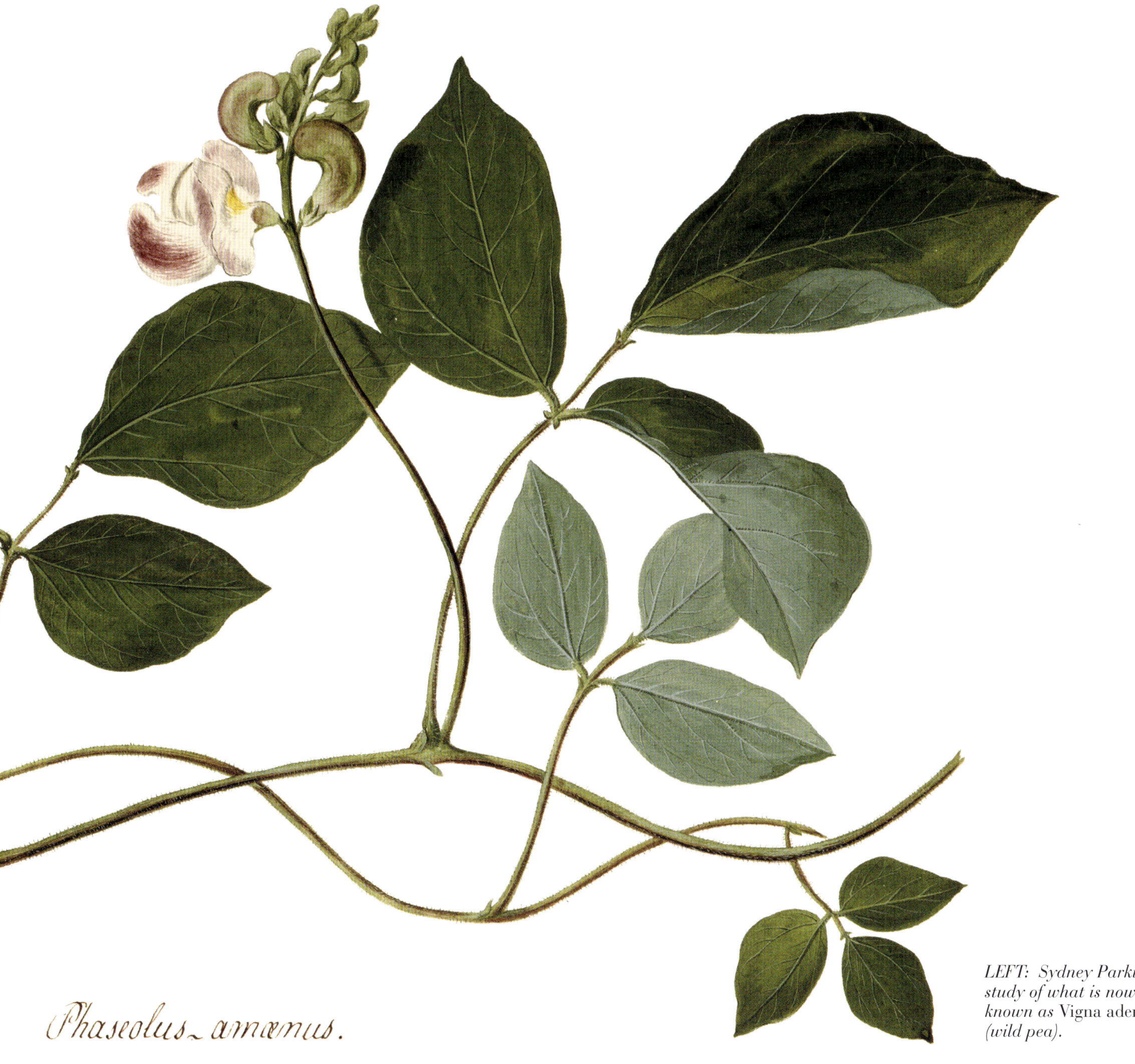

LEFT: Sydney Parkinson's study of what is now known as Vigna adenantha *(wild pea).*

Banks and his men had collected over 30,000 plant specimens, which included 1400 species new to science. Parkinson had done 674 outline drawings and 269 finished paintings. Back in London, Banks supervised the completion of many of Parkinson's paintings, which consisted of a pencil drawing, in which important parts had been coloured in, the rest to be filled in later. Banks employed 18 engravers for over ten years preparing engravings for this project, intending that they should be printed in colour; a total of 738 copper sheets were engraved. They were not printed, however, and the death of Solander in 1782, and Banks' other interests, intervened.

Finally, between 1900 and 1905, 315 of the plant illustrations were published as lithographs by the Natural History Museum as *Illustrations of the botany of Captain Cook's voyage round the world in H.M.S. Endeavour in 1768–71, by Sir Joseph Banks and Daniel Solander; with determinations by James Britten: 1. Australian Plants* and in 1973 more were published as *Captain Cook's florilegium*. Finally, from 1980 onwards, most of the surviving copper plates were restored, and hand-printed in colour by the Royal College of Art: 100 sets were produced under the title of *Banks' Florilegium*, with an account of the voyage by Wilfrid Blunt and of the botanical exploration by William Stearn.

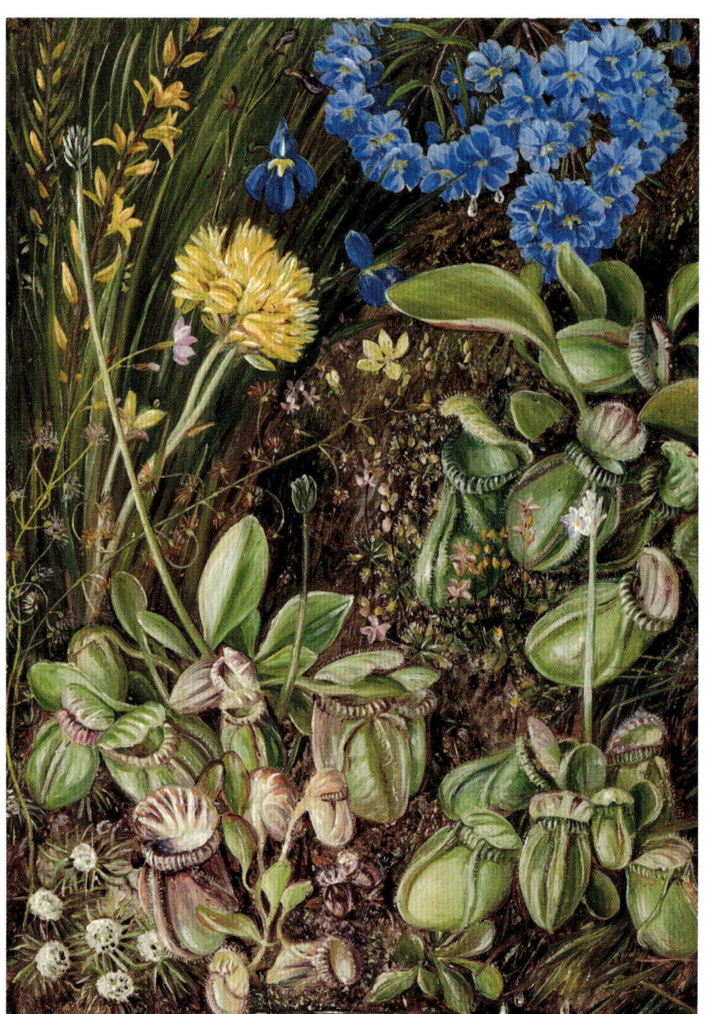

In 1791, a French expedition under Admiral Bruni D'Entrecasteaux sailed to Australia, with the botanist J. J. H. de Labillardière (who had previously collected plants in Syria) and a gardener from Paris, Félix de Lahaie (or Delahaye). Their main aim was to search for the explorer Admiral Lapérouse and his ships, which had vanished after leaving Sydney in 1788. D'Entrecasteaux landed briefly on the coast of Western Australia, where de Labillardière made some important collections, before returning via Tasmania and Java where their specimens were confiscated by the Dutch, now at war with France. The Dutch sent the collection to England, but Sir Joseph Banks returned them unopened to de Labillardière, who published his discoveries in *Novae Hollandiae plantarum specimen* between 1804 and 1806. It contains elegant uncoloured engravings after Redouté, Turpin, Poiteau and other French artists, drawn from herbarium specimens, so they do not have the fresh feeling of those drawn from living specimens.

After several years in Oxford, working up his paintings for Sibthorp's *Flora Graeca* (*see* pages 68–73), Ferdinand Bauer was invited to go as botanical artist on another major voyage of scientific exploration to Australia, organized by Sir Joseph Banks. In the 30 years since Bank's own voyage with Captain Cook, the penal colony had been set up at Botany Bay, and the London nurserymen Lee and Kennedy had sent a seed collector to Australia. This was a period of great popularity for plants from the Cape, which required dry heat in winter under glass, and Australian plants thrived in the same conditions.

Bauer sailed from Spithead in 1801 aboard HMS *Investigator*, with Robert Brown as botanist, assisted by Peter Good, a foreman gardener from Kew, under the command of Mathew Flinders. The sloop *Investigator* was an old and leaky vessel, but Flinders was a skilled and experienced navigator, having sailed to Tahiti with Captain Bligh's second expedition to Tahiti to collect breadfruit plants, and visited Australia in 1795 and again in 1798 when he made the first circumnavigation of Tasmania. Flinders was to undertake a hydrographic survey, while the botanists were to collect living plants and seeds for Kew and dried specimens for Sir Joseph Banks' herbarium; Bauer was to make drawings of animals and plants.

LEFT: Cephalotus follicularis *(the pitcher-plant), one of the specialities of Western Australia can be seen in the foreground of this painting.*

OPPOSITE: This watercolour of Phormium tenax *(New Zealand flax) was commissioned by Sir Joseph Banks from Frederick Nodder who completed the sketch begun by Sydney Parkinson.*

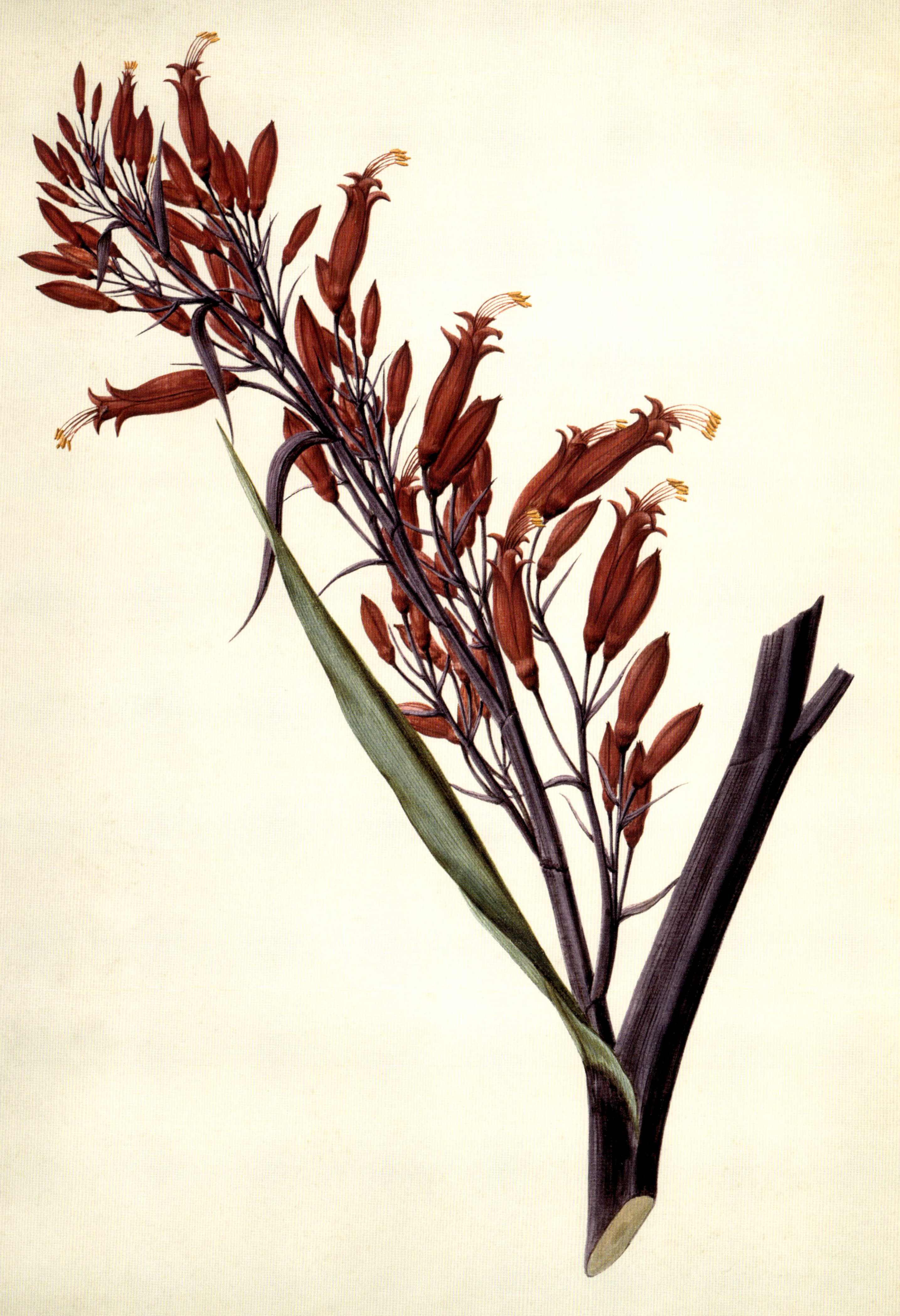

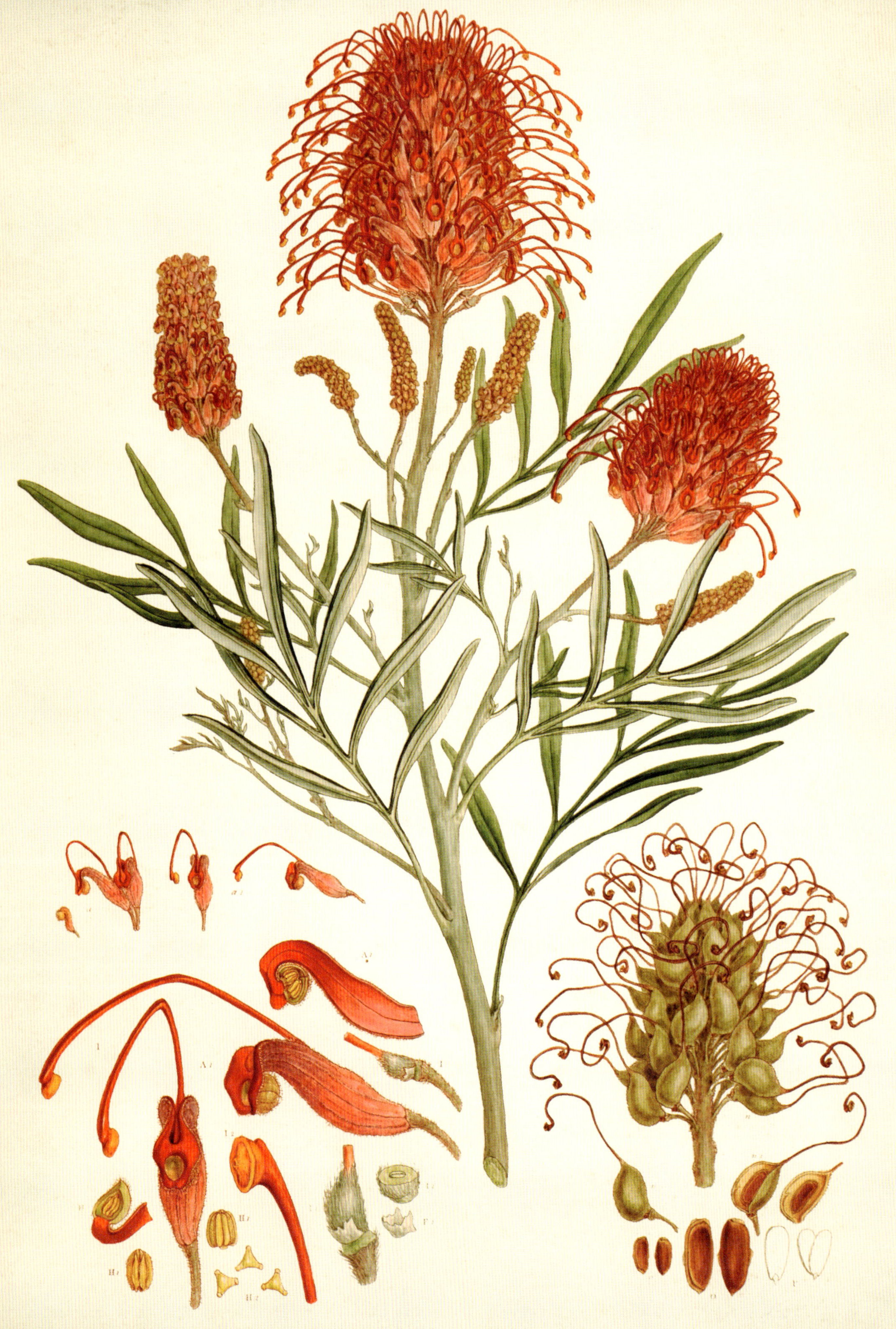

OPPOSITE: Grevillea banksii *painted by Ferdinand Bauer for* Illustrationes florae Novae Hollandiae.

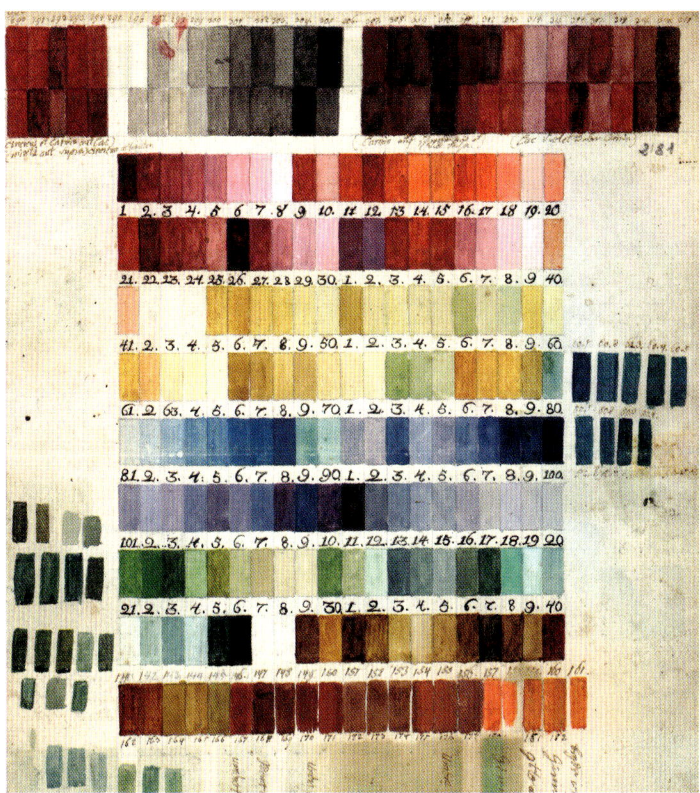

RIGHT: A section of Ferdinand Bauer's colour chart, which shows the variation in shades and the numbering he used for each one.

The *Investigator* sailed first to the Cape and then eastwards, more or less following Dampier's route, and in December anchored inside King George's Sound, off present-day Albany in Western Australia. Here they spent a month exploring, and later, in a letter to Sir Joseph Banks, Brown commented on the exceptional richness of the flora of the south of Western Australia, since confirmed as one of the world's hotspots of biological diversity; here they collected, and Bauer drew one of its specialities, the pitcher-plant, *Cephalotus follicularis.* Sailing eastwards, they reached Sydney, where they overhauled and patched the ship, reaching the Gulf of Carpenteria in November. Here, they found *Grevillea banksii*, another of the plants Bauer illustrated, named in honour of Sir Joseph and Scottish botanist, Robert K. Greville. They again beached the ship and found her in such bad condition that they decided to return to Sydney, calling first at Timor to pick up supplies. Like Cook and Banks before them, they contracted dysentery or some other disease, and by the time they had reached Sydney in June 1803, some of the crew had died, and others were unfit to continue.

Bauer and Brown stayed on in Australia until 1805, Bauer travelling to Norfolk Island, while Brown visited Tasmania (Van Dieman's Land). Bauer's method of working in the field has been investigated; he drew the specimen in pencil, and noted the colour of each leaf or flower with reference to a numbered colour chart which he had created. On returning to England, he used these notes to work up the finished painting. Some of his field drawings, with colour reference numbers are now in Vienna. Bauer's early colour chart, with 150 shades, has survived: his Greek colour chart, with 250, and his Australian chart with 992 shades do not seem to have survived; there were 200 tints of green and 100 of pinks, reds and purples etc, which shows the exceptional care he took to reproduce the colours of the living plant as perfectly as possible.

After returning to England in 1805, Brown began to work on the Australian collections, while Bauer worked up his sketches into paintings. Flinders returned separately, and after shipwreck on the Great Barrier Reef, and imprisonment by the French in Mauritius, reached England in 1810. Brown's publication of the botany of this and earlier voyages, *Prodromus florae Novae Hollandiae et Insulae Van-Diemen*, was a financial failure but a scientific triumph as the first full account of the Australian Flora. He paid for the first copies to be printed, but sold very few.

Flinders prepared the topographical journal, maps and other material as *A Voyage to Terra Australis*, which included ten of Bauer's botanical illustrations, but he died, at the early age of 40, before it was published. Bauer himself engraved and coloured the most beautiful plates for his own work, *Illustrationes florae Novae Hollandiae*, but sold few copies, and he ceased publication after only 15 plates had been finished. They are, however, perhaps the finest of all scientific botanical plates, beautifully drawn with dissections and enlargements of floral details, delicately engraved and carefully hand-coloured. Over 200 of Bauer's Australian flower and animal paintings remained hidden in the Natural History Museum library, which had acquired them from the Admiralty, until 1975 when 25 were published by the Basilisk Press. These show the unrivalled delicacy and detail of Bauer's finished watercolours.

Plants collected by these expeditions soon began to flower in gardens in England and France, and were illustrated in various journals, such as *Curtis's Botanical Magazine*. One of the most spectacular plants was the giant gymea lily, *Doryanthes excelsa*, which flowered in the garden at Woodhall, Lanarkshire in 1826; the painting of this specimen, by E. Weddell is now in the collection at Kew.

RIGHT: Brunonia sericea *painted by Ferdinand Bauer for* Illustrationes florae Novae Hollandiae.

OPPOSITE: Doryanthes excelsa *(the giant gymea lily) painted by E. Weddell.*

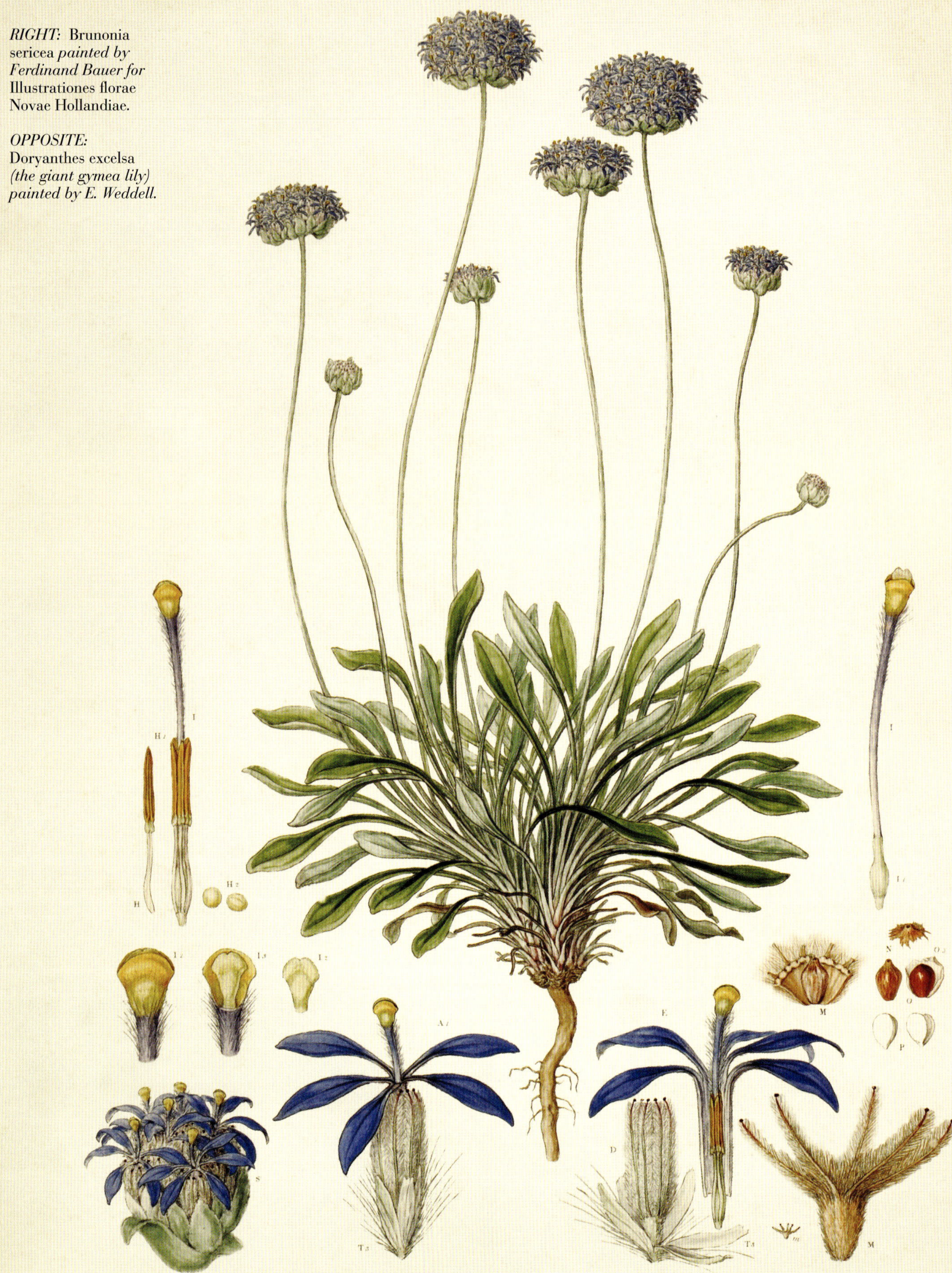

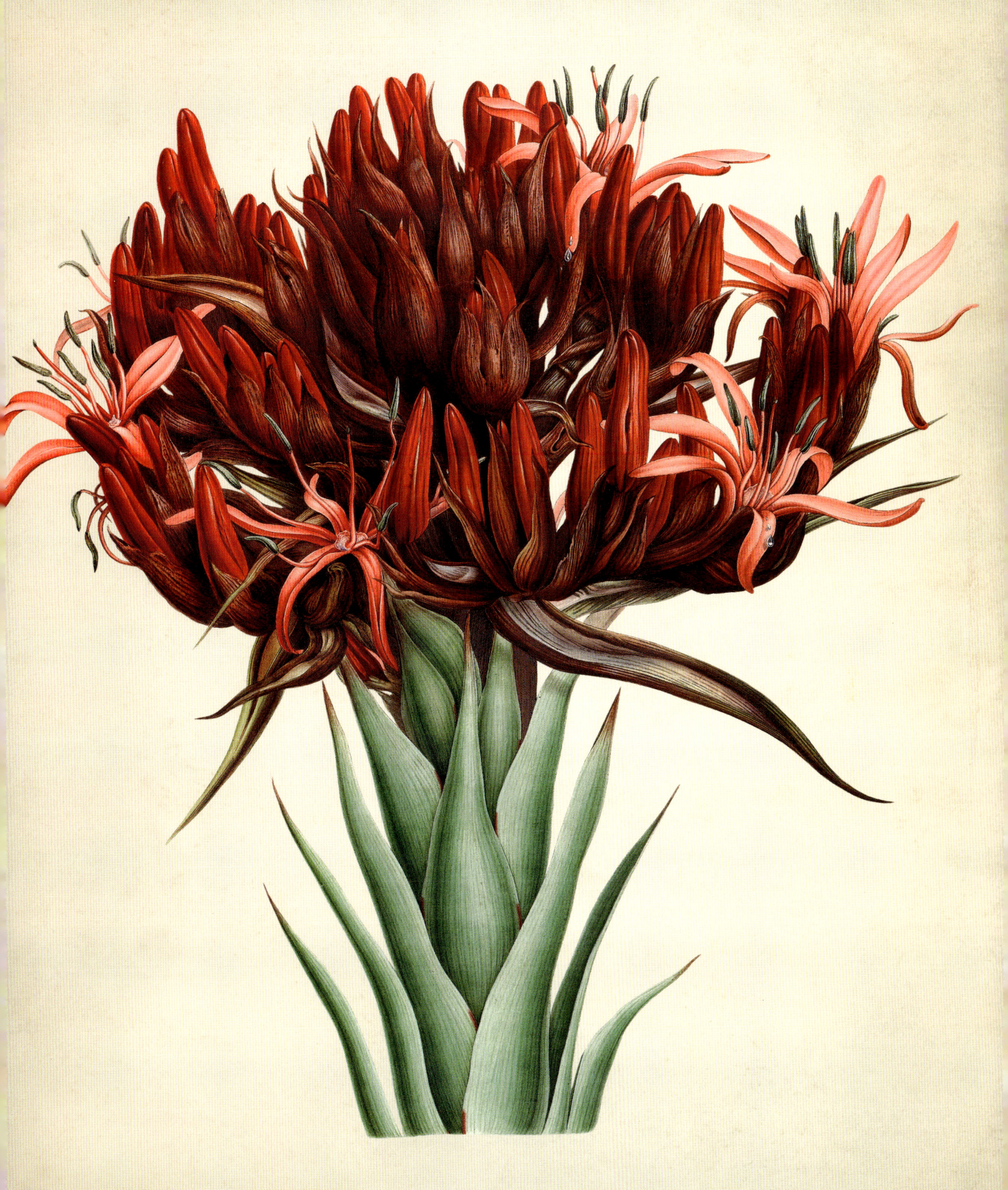

Sir Joseph Banks

Sir Joseph Banks (1743–1820) and his voyage to Australia with Cook and Parkinson have been mentioned, but his influence in England on science, and particularly on the promotion of botany, cannot be overestimated. He was involved with many scientific institutions, serving as President of the Royal Society and Director of the Royal Gardens at Kew. He was also a founder member of both the Linnean Society and the Horticultural Society (now known as the Royal Horticultural Society).

Banks was born into a well-to-do family, and received a gentleman's education at Eton and Christ Church, Oxford. There, he became increasingly fascinated by natural history, and botany in particular. He became a wealthy young man at the age of 21, having inherited his late father's estates in Lincolnshire, and was able to travel and follow his own interests.

In 1767, he undertook his first expedition, to Labrador and Newfoundland, and later in the same year became a Fellow of both the Royal Society and the Royal Society of Antiquaries. The following year, he joined Lieutenant James Cook's first great expedition to Tahiti, on board the *Endeavour*, a voyage that was to last until the summer of 1771. Banks played an important role in financing and directing the natural history observations, and was meticulous in keeping his own journal.

Banks planned to accompany Cook on his second expedition to the Pacific in 1772–75 but eventually withdrew, going instead to Iceland. Despite this, he was much involved in

BELOW: Stapelia reticulata *from Francis Masson's* Stapeliae novae.

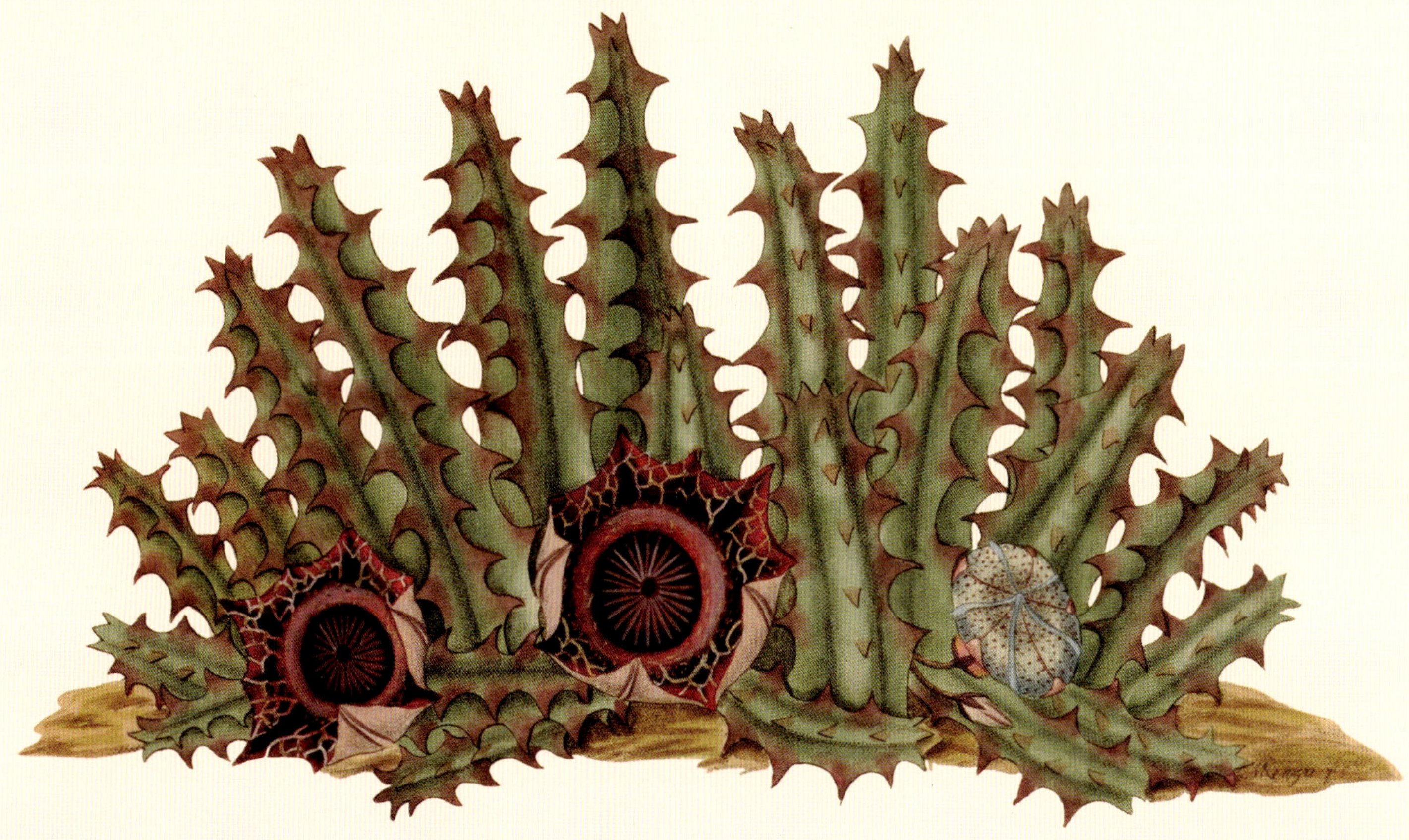

the later publication of all three of Cook's Pacific voyages in three volumes, illustrated with maps and engravings, and entitled *A Voyage to the Pacific....*

Meanwhile, he had set up house in Soho Square, London, and married Dorothea Hugessen three years later, in 1779, but had no children. In addition to containing his library and his natural history collections, his home became a meeting place for scientists of all nationalities and a centre for intellectual and scientific discussions. He was also instrumental in promoting horticulture, and it was he who, with John Wedgwood, of the famous pottery family, organized the inaugural meeting in 1804 of what was eventually to become the Royal Horticultural Society.

Banks, who was knighted in 1781, was much involved with various aspects of colonial life, such as the settlement of Botany Bay, and between 1787 and 1795 he was instrumental in supporting the project (culminating in the mutiny on the *Bounty*) to transport the breadfruit tree and other plants from the South Seas to feed the slaves on West Indian plantations. In 1801, he also helped to organize Matthew Flinders's voyage around Australia on the *Investigator*, and frequently underwrote the expenses of botanists, such as Francis Masson (*see* page 83) to South Africa and Archibald Menzies, who travelled to western America, searching for new plants. These specimens augmented Banks's collection, while the living plants went to Kew, where Banks had acted as unofficial Director since 1773, becoming official Director in 1797. He died on 19 June 1820.

RIGHT: Stapelia grandiflora *from Francis Masson's* Stapeliae novae.

BELOW: Sir Joseph Banks.

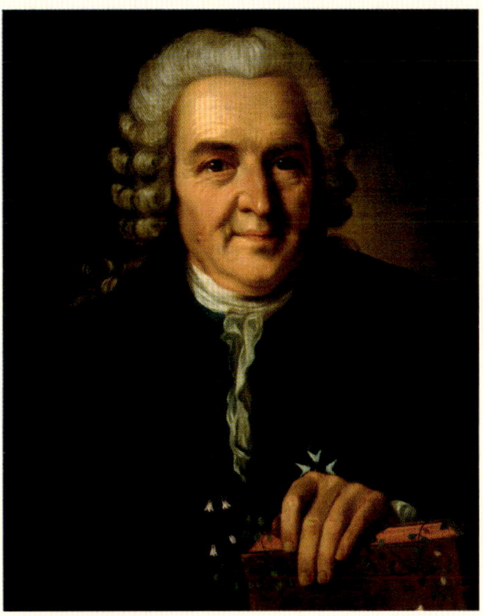

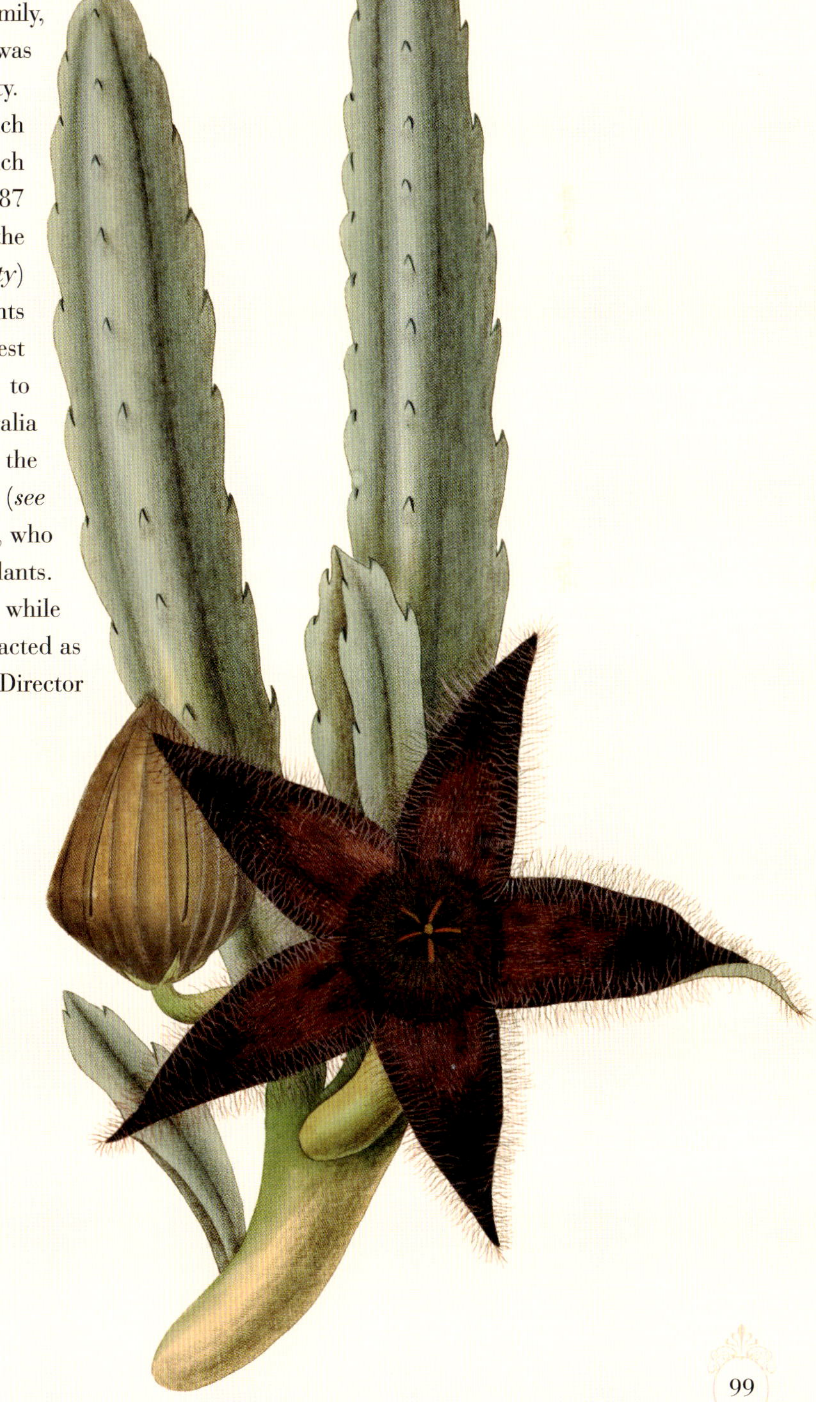

8 THE GOLDEN AGE IN ENGLAND

THE ROYAL GARDENS AT KEW: BANKS, BAUER AND MASSON

The Royal Botanic Gardens, Kew had their origins in the garden around Kew Palace; they were first developed by Princess Augusta, daughter-in-law of George II, with the help of John Stuart, 3rd Earl of Bute (1713–92). In 1752, a year after the early death of her husband, Frederick, Prince of Wales, Augusta instructed her head gardener John Dillman to "compleat all that part of the Garden at Kew that is not yet finished in the manner proposed by the Plan", and Bute meanwhile expressed his hope that the garden would "…contain all the plants known on Earth".

Bute also introduced the architect William Chambers to the Princess, and Chambers embarked on a series of decorative buildings in the garden, the chief of which was the Pagoda. By 1759 Bute and Augusta had established the first botanic garden at Kew, with William Aiton (1731–93) as chief gardener; Aiton remained at Kew after Augusta's death, and in 1789 produced his influential, three-volume *Hortus Kewensis*, a catalogue of plants at Kew, still used as a reference for the dates of early introductions of plants to gardens. He was succeeded by his son, William Townsend Aiton (1766–1849), who published a second edition in 1810–13.

Lord Bute was a politician, bibliophile, patron of the arts and keen amateur botanist, who had a garden of his own designed by Capability Brown at Luton Hoo, Bedfordshire, and he commissioned the artist Simon Taylor (1742–c.1796), to draw and paint many of the plants not only there, but also at the royal gardens at Kew. From around 1760 Taylor assisted Ehret (*see* pages 56–59) at Kew, becoming, after Ehret's death in 1770, the chief artist. Taylor was also one of a number of artists employed to record his collection of rare and exotic plants by the Quaker botanist and physician Dr John Fothergill (1712–80) at his botanical garden at Upton,

OPPOSITE: Robinia chamlagu *(now* Caragana sinica*) from L'Héritier's* Stirpes novae.

RIGHT: Delphinium pelegrinum *by Simon Taylor, from the collection of watercolours commissioned by Lord Bute.*

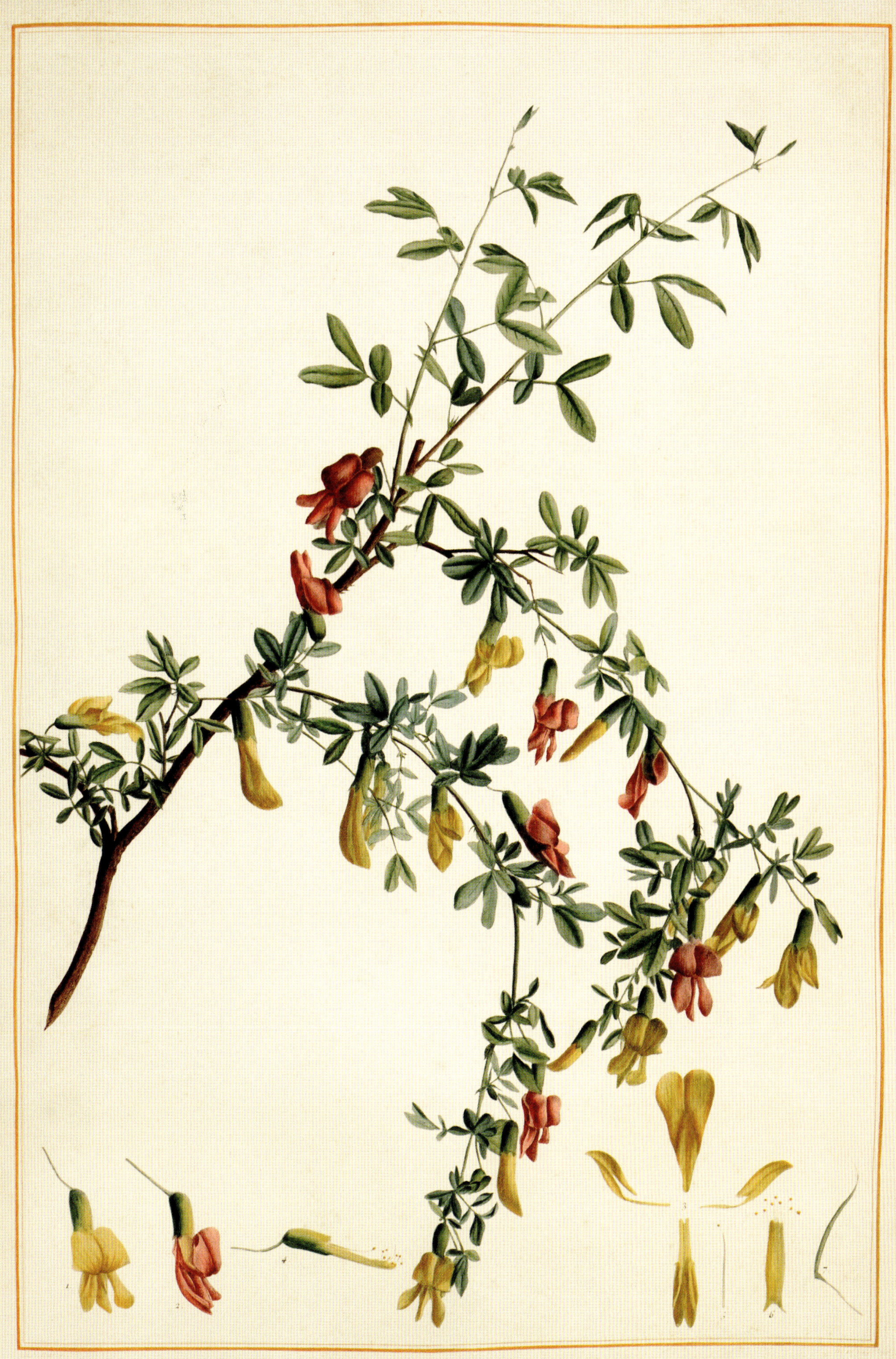

Robinia Chamlagu

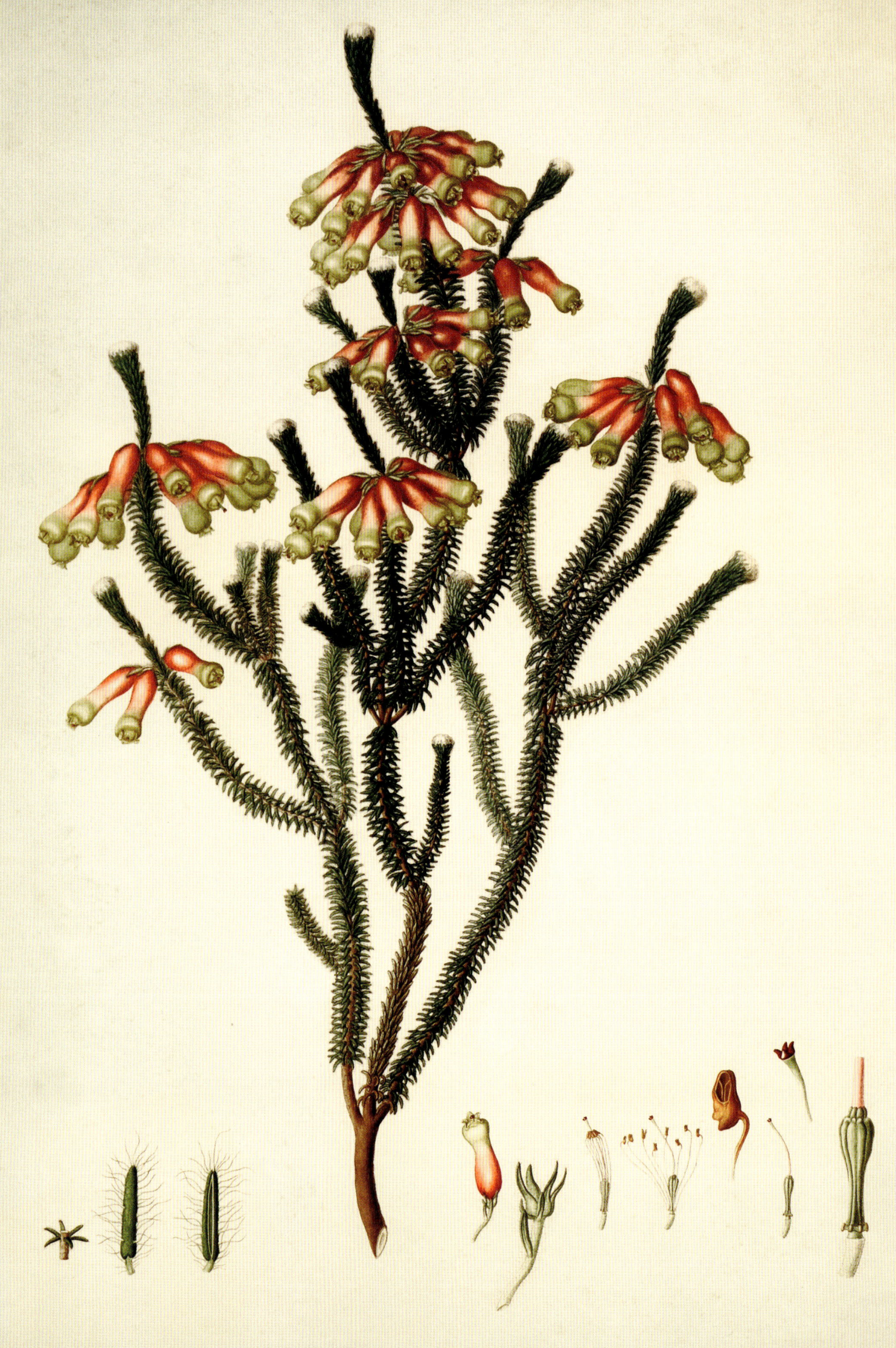

near Stratford, then a village near London. Taylor's output was considerable, as can be seen from the fact that 15 volumes of his work, containing 684 watercolours on vellum, were broken up into smaller lots and sold separately at an auction of part of Bute's personal library in 1794. One of these volumes, containing 48 unsigned, coloured drawings, is in the collection at Kew.

Sir Joseph Banks was a key figure in the botanical world during the late eighteenth and early nineteenth centuries as well as being the unpaid director of the Royal Gardens at Kew from 1773. He welcomed botanists from all countries to work in his herbarium, and during 1786 Charles-Louis L'Héritier de Brutelle was a regular visitor at Soho Square. The following year Pierre-Joseph Redouté arrived from Paris, and by the end of 1788 L'Héritier, with Redouté and the English artist, James Sowerby, had commenced publication of a work describing the new plants being grown in and around London and Kew, entitled *Sertum Anglicum*, and produced as a thank-you for the hospitality they had received in London. This had 35 plates, and was published in four instalments, of which the first contained text only.

By employing the most skilled artists to illustrate plants seen on the various expeditions with which he was involved, Banks has left us with an insight into the exciting new discoveries made at that time, and these illustrations are backed up by a large body of correspondence (now mostly at Kew and the Natural History Museum) between the indefatigable Banks and his collectors, such as the Scotsman, Francis Masson (1741–1805) and other protégés. Banks requested a place for a Kew gardener to travel as far as the Cape of Good Hope, on James Cook's second expedition of 1772, and it was Masson who was chosen. He undertook a series of field trips, and met up with the Swedish botanist Carl Thunberg (*see* page 83), with whom he travelled into the interior. He appears to have spent the winter sorting his collections for dispatch to Kew, before undertaking another expedition with Thunberg the following year. Masson returned to England in 1775 with a large collection of plants, and by the next year he was travelling once more, to Madeira, the Canaries, the Azores and the West Indies, where he collected about 200 species of plants. Again, Masson returned to England, but continued to travel for many years; his only book was entitled *Stapeliae novae: or a collection of several new species of that genus discovered in the interior parts of Africa* (1796–97). The 41 illustrations were (with one exception) engraved from sketches made by Masson himself and, as he explains in his preface, "The figures were drawn in their native climate, and though they have little boast in point of art, they possibly exhibit the natural appearance of the plants they represent,

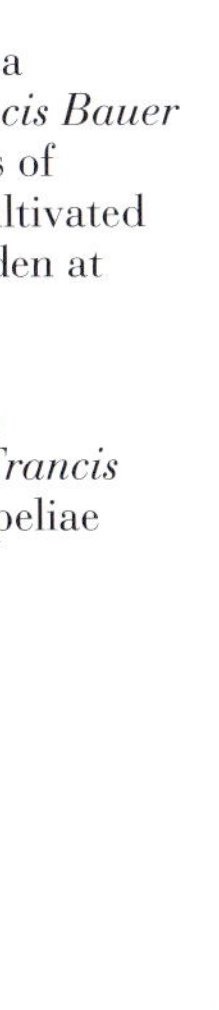

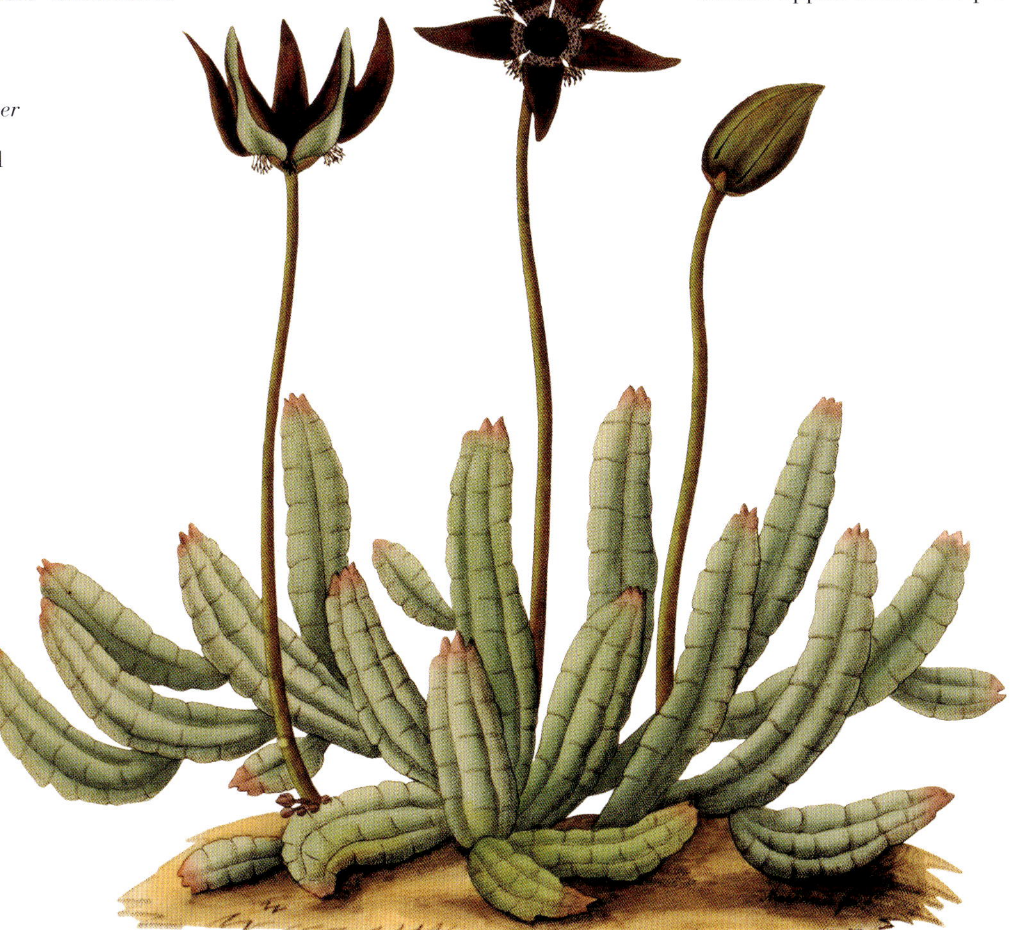

OPPOSITE: Erica massonii *by Francis Bauer from* Delineations of Exotick plants cultivated in the Royal Garden at Kew.

RIGHT: Stapelia pedunculata *by Francis Masson from* Stapeliae novae.

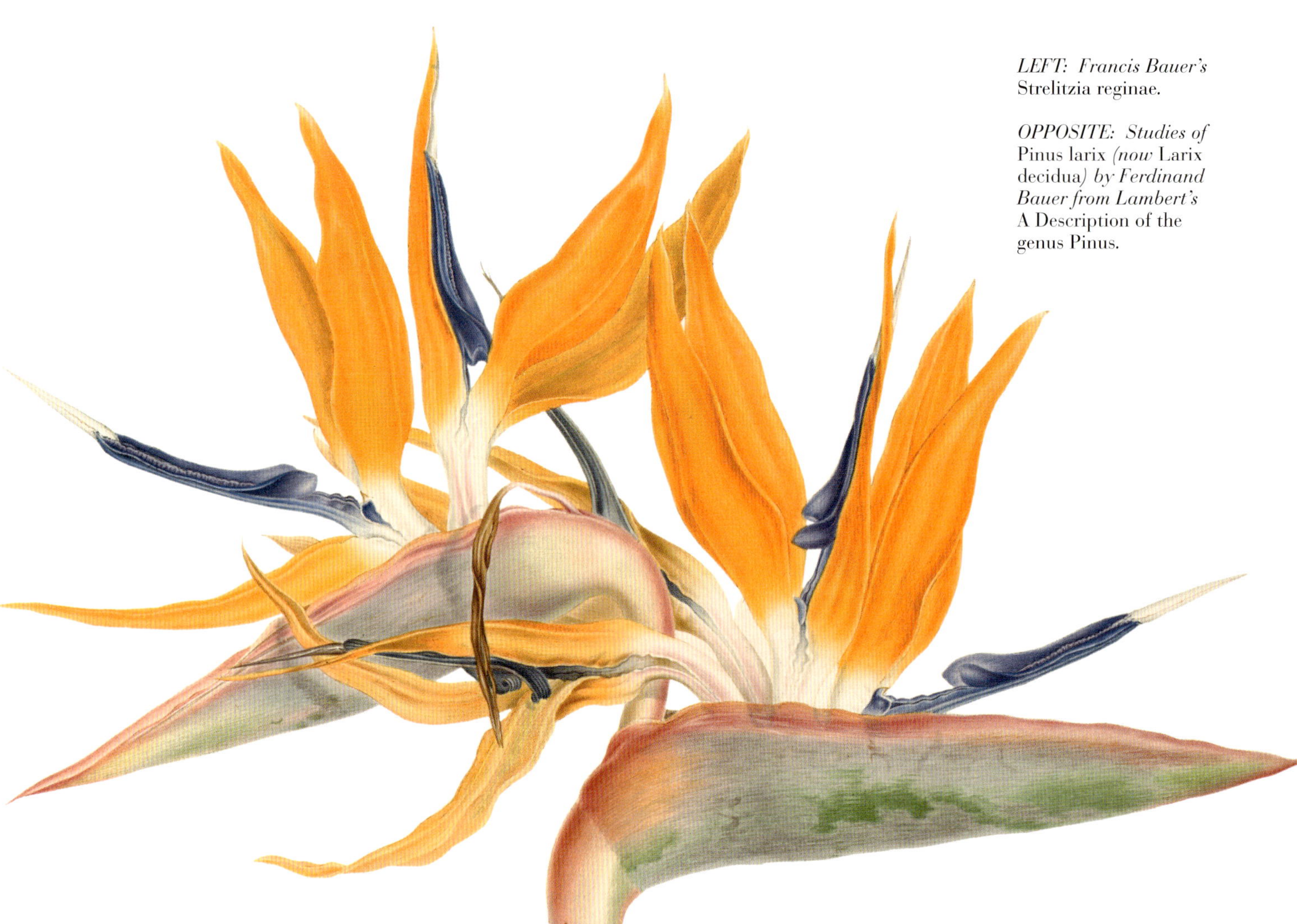

LEFT: Francis Bauer's Strelitzia reginae.

OPPOSITE: Studies of Pinus larix *(now* Larix decidua*) by Ferdinand Bauer from Lambert's* A Description of the genus Pinus.

better than figures made from subjects growing in exotic houses can do". Masson, who seems by all accounts to have been a thoroughly charming and modest man, died in Montreal in 1805.

In his role as Director of the Royal Gardens at Kew, Banks wielded possibly ever more influence in the sphere of botanical illustration. Princess Augusta had died in 1772, leaving the estate at Kew to her son, King George III, and both the King and Banks were keen to develop economic uses for exotic and native plants. Banks employed, at his own expense, the exceptionally talented artist Francis (Franz) Bauer (1758–1840), as "Botanick Painter to his Majesty". Bauer, originally from Feldsberg, then in Austria, had spent some time illustrating plants for Nikolaus Joseph von Jacquin, Professor of Botany and Director of the Botanic Garden of the University of Vienna, before coming to England with his brother Ferdinand in 1788. By 1790 Francis was busy at Kew, drawing and painting the new plants flowering in the garden, and he remained at Kew for the rest of his life. Some of his finest paintings are of the Cape heaths, introduced by Francis Masson, which are included in *Delineations of Exotick plants cultivated in the Royal Garden at Kew*, published in three parts between 1796 and 1803. Only three parts were completed before publication ceased; the original paintings and some which were not published are in the collection at Kew. *Strelitzia depicta: or coloured figures of the known species of the genus Strelitzia from the drawings in the Banksian Library*, published in 1818, shows detailed paintings of the known species of the South African genus *Strelitzia*, named in honour of Queen Charlotte, whose family were from Mecklenburg-Strelitz.

In 1797, Ferdinand Bauer supplied the illustrations for *A description of the genus Cinchona* for Aylmer Bourke Lambert, who was particularly interested in South America and the properties of the shrub *Cinchona*, the source of quinine. The two Bauer brothers also drew the majority (Ehret supplied one) of the 100 plates for Lambert's fascinating and beautiful book, *A Description of the genus Pinus*, published in two volumes from 1803 to 1824; as well as pines, it includes larches, cedars and other members of the pine family. Lambert received many specimens for illustration in his book from his friend Sir Joseph Banks, to whom the book is dedicated, and notes that a number of the trees illustrated were well established in the collection at Kew. Banks also helped by letting Lambert have some unpublished

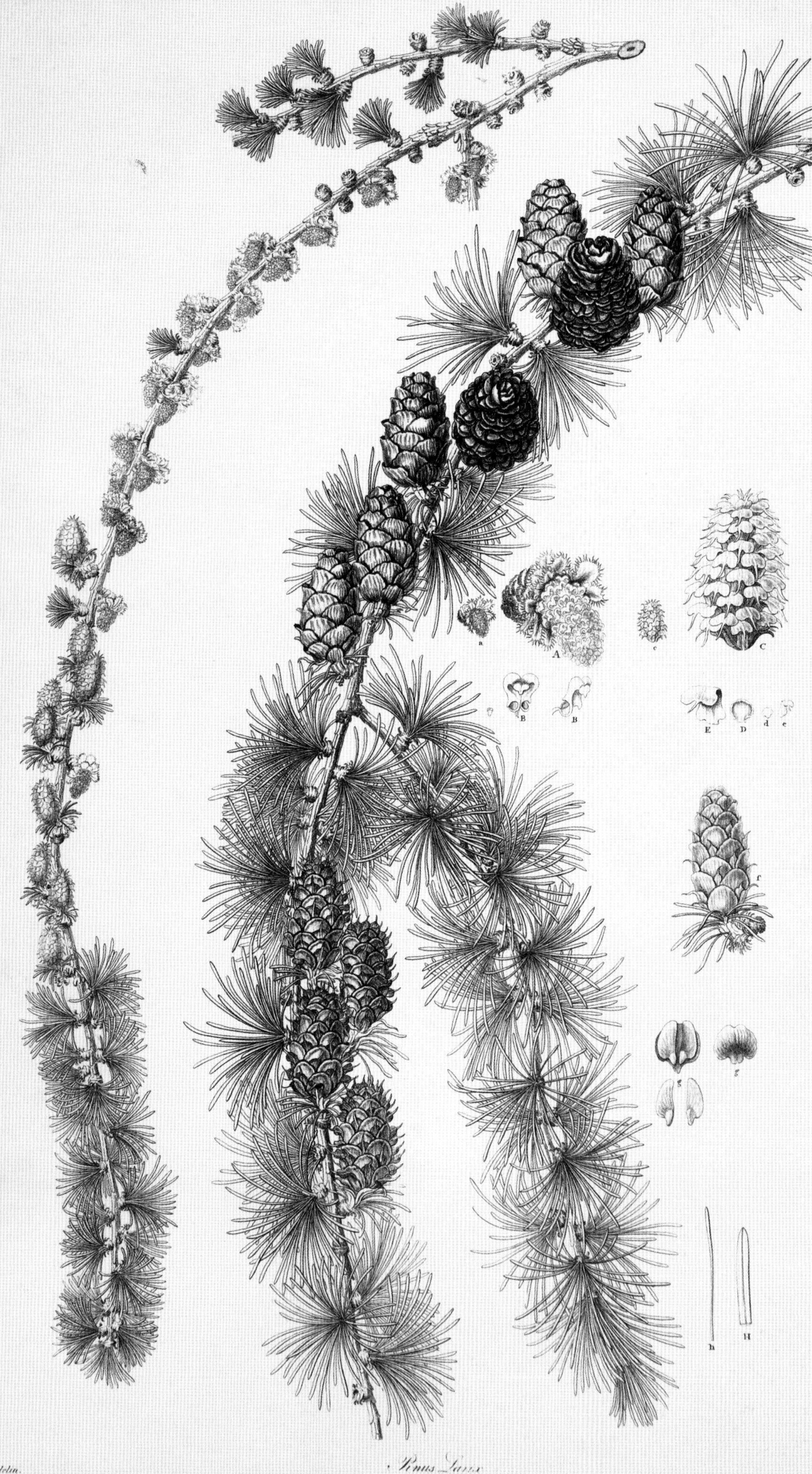

F. Bauer delin.

Pinus Larix

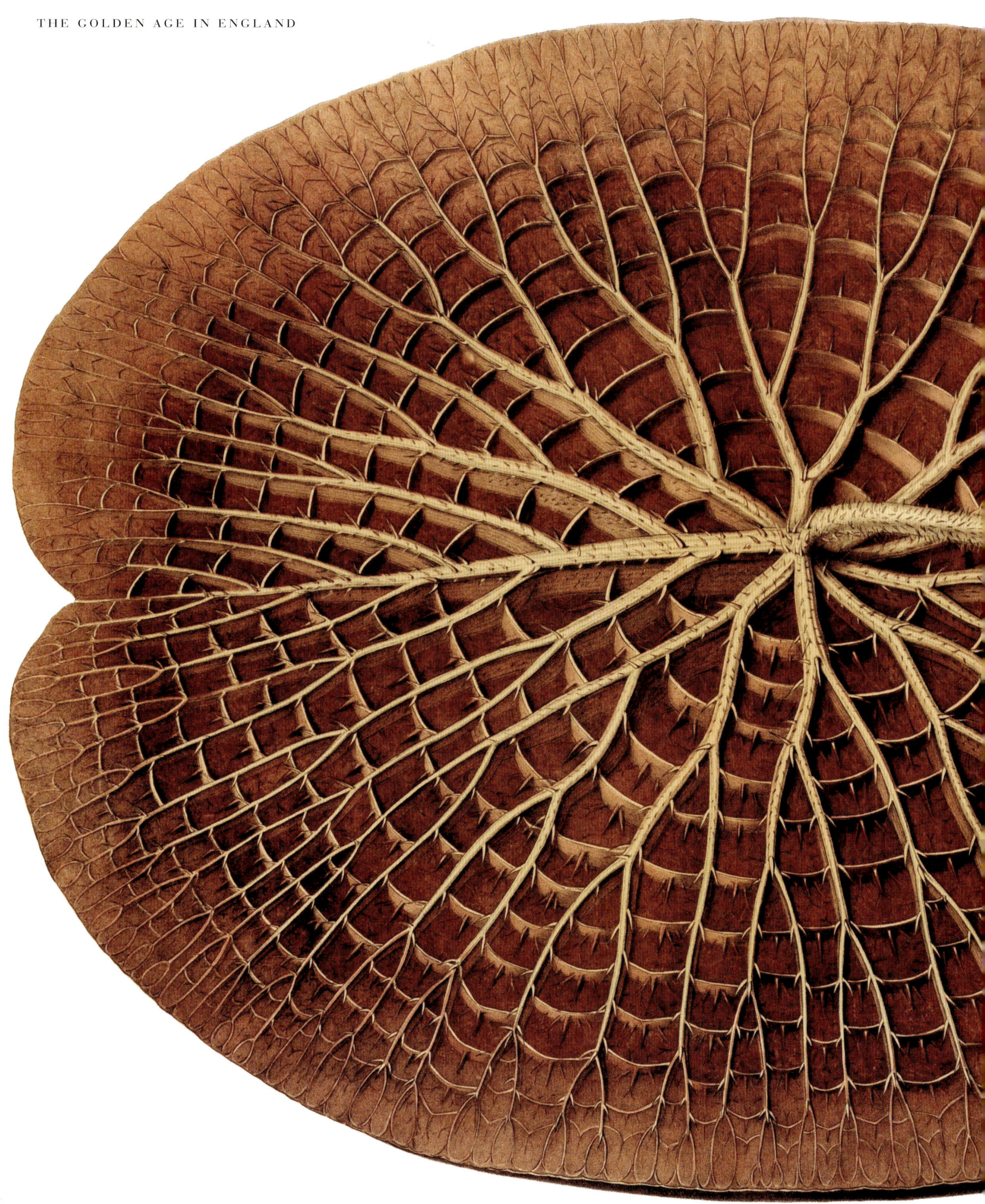

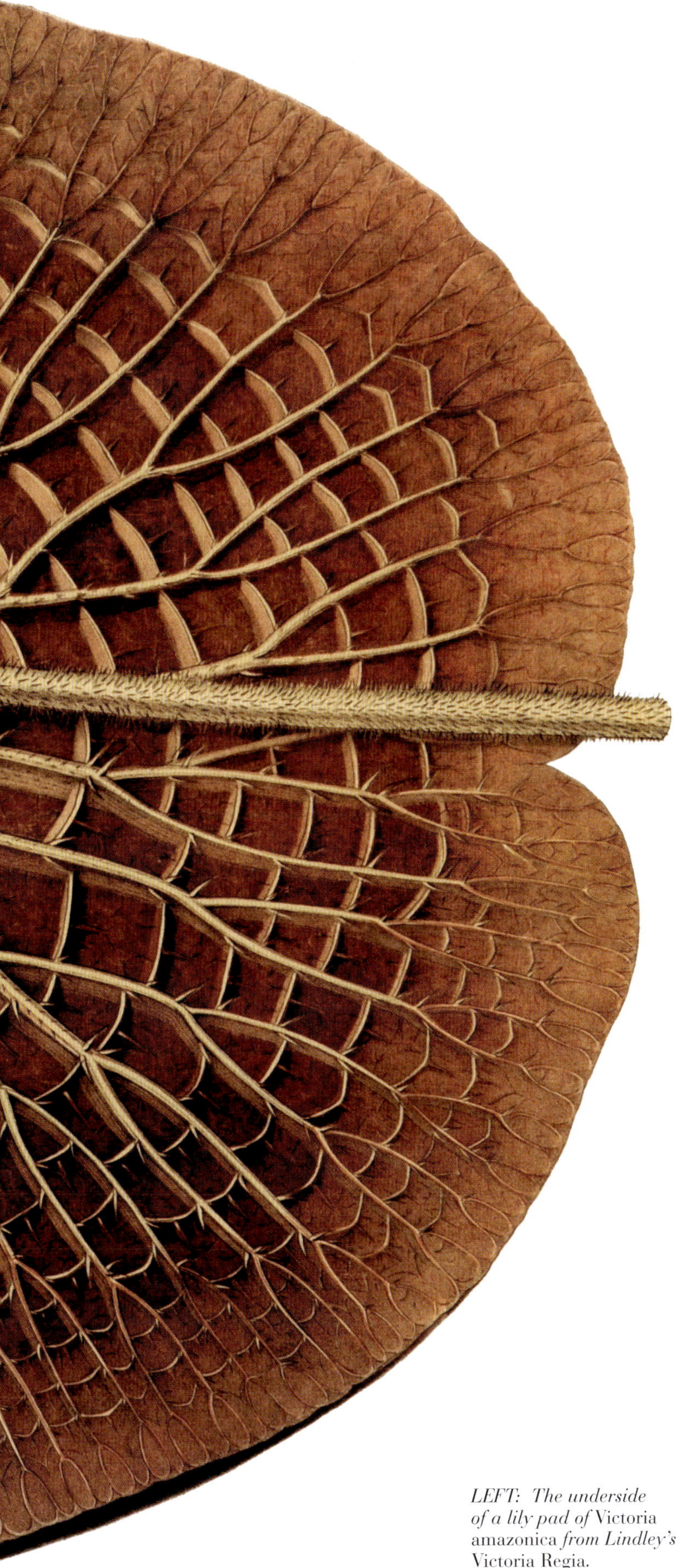

drawings of the New Zealand genus *Dacrydium*, done by Sidney Parkinson during his voyage with Banks and Captain James Cook many years earlier. Lambert was, like Banks, a keen botanist of independent means, and had met John Sibthorp while at Oxford. He formed a private herbarium, containing around 30,000 specimens, which he housed at his home, Boyton House, near Heytesbury in Wiltshire, and to which he regularly welcomed members of the scientific establishment.

Francis Bauer worked with another influential figure, John Lindley (1799–1865), a keen gardener, who had a particular passion for orchids. An eminent and industrious man, Lindley had been brought up among plants, his father being a nurseryman and pomologist in Norfolk. Lindley was an artist, as well as being a prolific author, university professor, editor of *The Botanical Register*, and, for many years, administrator of the Horticultural Society of London. Lindley, alongside his contemporaries Darwin, Paxton and William Jackson Hooker, was one of the great Victorian biologists, and started his working life alongside the botanist Robert Brown at Sir Joseph Banks's herbarium and library in Soho Square. Here Lindley specialized in the study of roses and in 1820 published *Rosarum Monographia; or a botanical history of roses*, with descriptions of 76 species and 19 engraved plates, of which 18 were drawn by himself, and one by John Curtis. When Banks died in the same year, Lindley was employed by a merchant with many useful business contacts, named William Cattley. Cattley was a keen grower of orchids and employed Lindley to study these amazing plants; Lindley reciprocated by naming a genus of orchids *Cattleya*.

In 1822, Lindley was appointed assistant secretary at the Horticultural Society's garden at Chiswick, where he remained for the rest of his working life. Lindley was responsible for the numerous new plants and seeds flooding in from the plant collectors sent abroad by Banks and others, and also held the post of Professor of Botany at London University. He died in 1865, and is commemorated by the Royal Horticultural Society's Lindley medal, and the RHS Lindley Library, the nucleus of which was Lindley's own library. Among the publications associated with Lindley are the *Illustrations of orchidaceous plants, by Francis Bauer... With notes and prefatory remarks by John Lindley*, containing 35 lithographed folio plates, and published in two parts in 1830 and 1838; *The Genera and Species of Orchidaceous Plants* (1830–40) and *Collecteana Botanica*, published in 1821 and containing "figures and botanical illustrations of rare and curious exotic plants, chiefly cultivated in the gardens of Great Britain". The plates for this were contributed by Ferdinand Bauer, and others, including William Hooker and William Jackson Hooker. These two were not related; the first a skilful painter, particularly of fruit, who made many paintings for the Horticultural Society, is remembered in the pigment

LEFT: The underside of a lily pad of Victoria amazonica *from Lindley's* Victoria Regia.

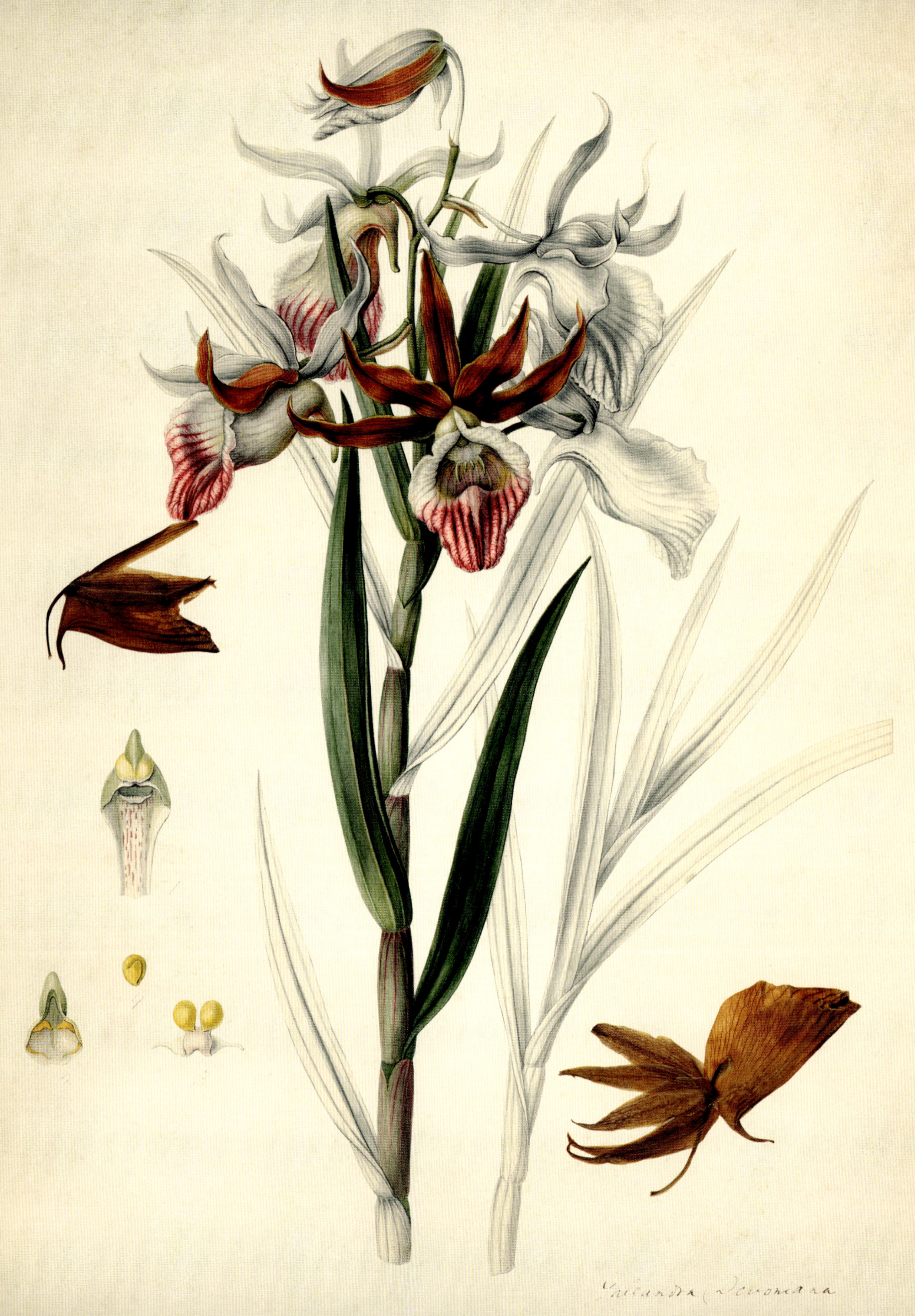

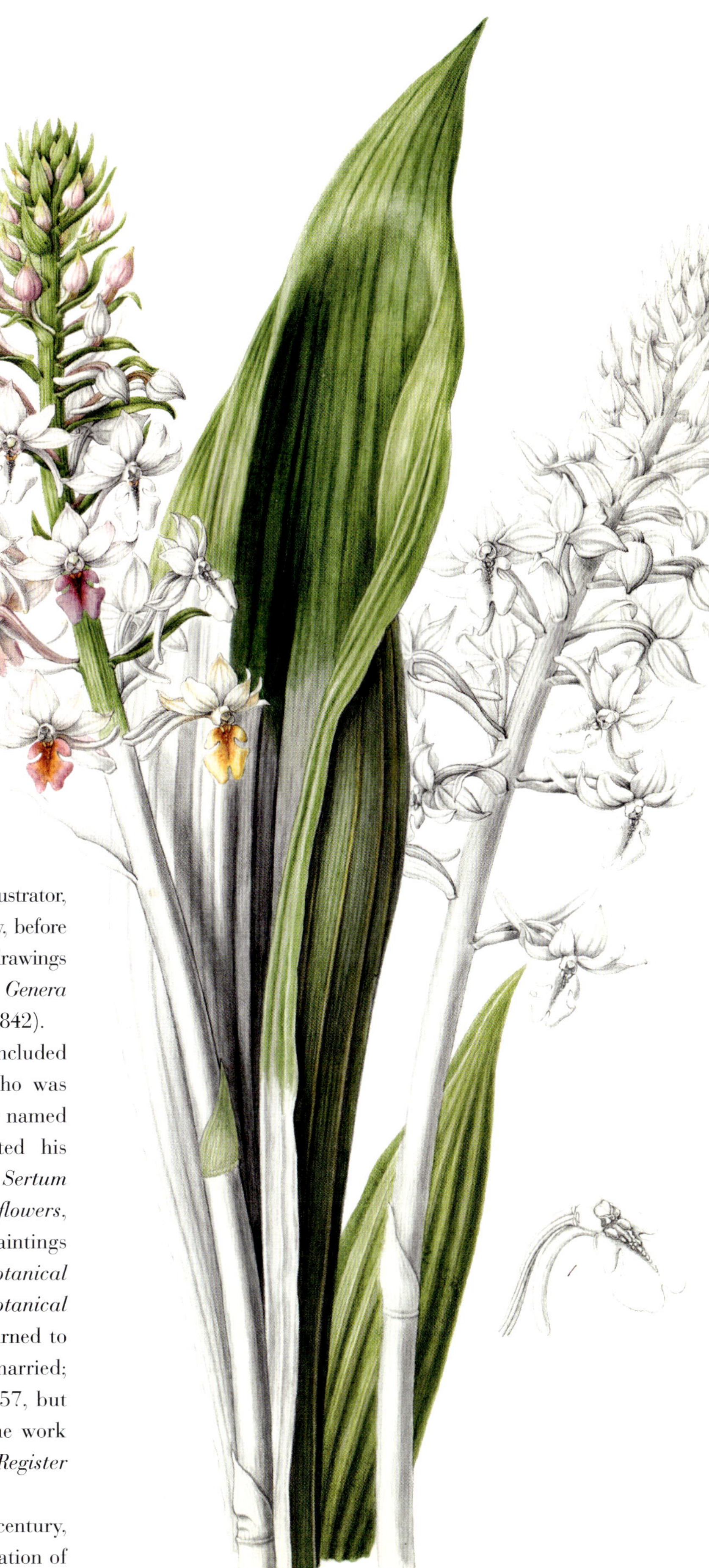

OPPOSITE: Galeandra devoniana, *an orchid from Brazil, Venezuela and Guyana, by Miss Drake.*

RIGHT: Calanthe versicolor *by Miss Drake, annotated "From Sion Gardens, Aug. 31".*

Hooker's Green. The second, W. J. Hooker, also a good plant illustrator, was a botanist and Professor of Botany at Glasgow University, before becoming Banks's successor as Director of Kew. He also used drawings by Francis Bauer, publishing these, with a descriptive text in *Genera filicum; or Illustrations of the Ferns, and other allied Genera* (1842).

Another artist who contributed to works by Lindley included a family friend, Miss Sarah Anne Drake (1803–57), who was his chief artist for several years, and after whom Lindley named a genus of Australian orchids, *Drakea*. She illustrated his *Ladies' Botany* (1834–37) and the great folio entitled *Sertum orchidaceum, a wreath of the most beautiful orchidaceous flowers*, published in 1837–41. She also found time to contribute paintings to *The Botanical Register*, a short-lived rival to *Curtis's Botanical Magazine*, of which Lindley was editor for a time. *The Botanical Register* ceased publication in 1847, and Miss Drake returned to her home county of Norfolk, where she subsequently married; sadly, she survived only five years after this, dying in 1857, but her work lives on, and as recently as 1991 a two-volume work containing 325 of her orchid paintings from *The Botanical Register* was reissued by Birkhauser Verlag of Basle.

Mosses and ferns were popular during the nineteenth century, and became increasingly sought after following the publication of

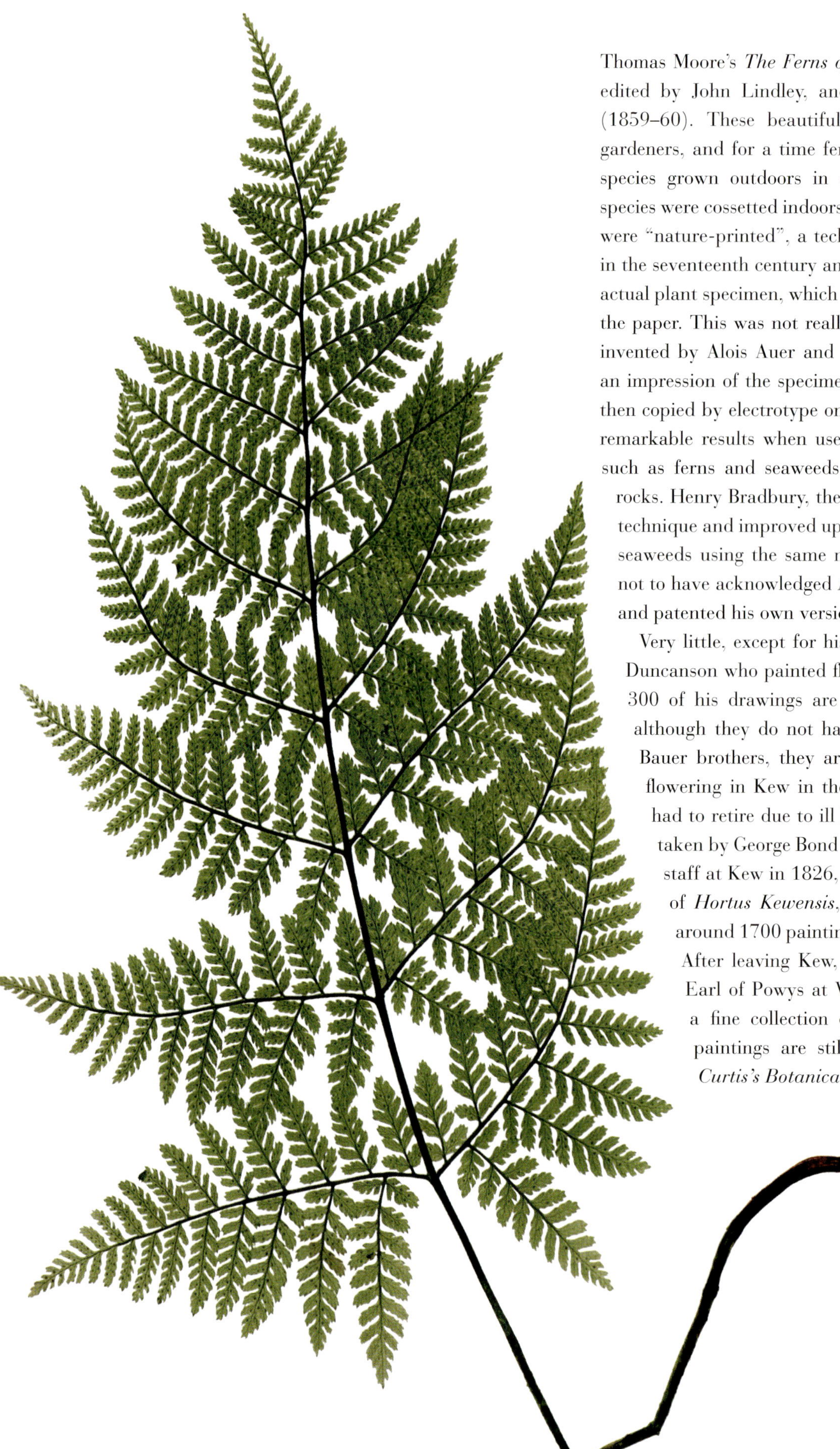

Thomas Moore's *The Ferns of Great Britain and Ireland* (1855), edited by John Lindley, and *The nature-printed British Ferns* (1859–60). These beautiful books caught the imagination of gardeners, and for a time ferneries were all the rage, with hardy species grown outdoors in dripping grottoes, while the tender species were cossetted indoors in humid glasshouses. Moore's books were "nature-printed", a technique that had first been attempted in the seventeenth century and which involved applying ink to the actual plant specimen, which was then passed through a press with the paper. This was not really successful, but a modified process, invented by Alois Auer and Andreas Worring in 1852, in which an impression of the specimen was captured on a lead plate, and then copied by electrotype onto a copper printing plate, produced remarkable results when used for rather two-dimensional plants such as ferns and seaweeds, and other natural objects such as rocks. Henry Bradbury, the printer of Moore's book, took Auer's technique and improved upon it, and published a book of British seaweeds using the same method. Sadly, Bradbury, who seems not to have acknowledged Auer's part in developing the process, and patented his own version, committed suicide in 1860.

Very little, except for his paintings, is known about Thomas Duncanson who painted flowers for W. T. Aiton at Kew. About 300 of his drawings are now in the collection at Kew, and although they do not have the brilliance of the work of the Bauer brothers, they are an important record of what was flowering in Kew in the 1820s. It appears that Duncanson had to retire due to ill health in 1826, and his position was taken by George Bond (*c.*1806–92). Bond joined the garden staff at Kew in 1826, and drew plants for a major revision of *Hortus Kewensis*, planned by W. T. Aiton. He made around 1700 paintings before the project was abandoned. After leaving Kew, Bond became head gardener to the Earl of Powys at Walcot in Shropshire, which still has a fine collection of large trees. Most of his original paintings are still at Kew, and some were used in *Curtis's Botanical Magazine* in the 1830s.

LEFT: An example of nature printing shows Lastrea dilatata *(now* Dryopteris dilitata*) from Moore's* The Ferns of Great Britain and Ireland.

OPPOSITE: Adiantum capillus-veneris, *also from* The Ferns of Great Britain and Ireland.

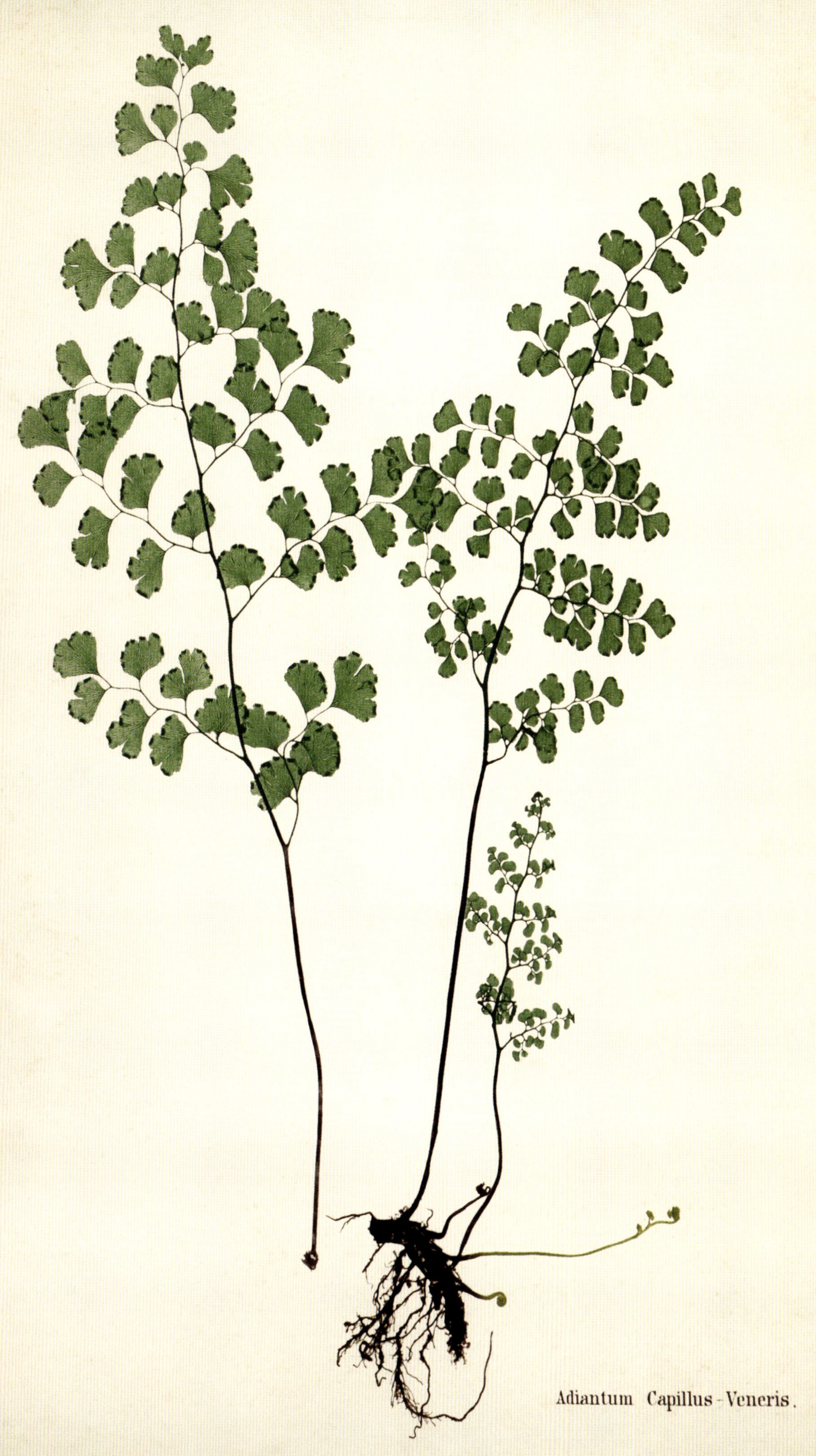

Adiantum Capillus-Veneris.

Mrs Delany and her Paper Mosaicks

The early life of Mrs Delany (1700–88), born Mary Granville, gave little indication of her later success; although from a good family, she was married off at a young age to a considerably older man, an unhappy arrangement which lasted until his death from gout and excessive drinking, in 1724. At this point, she moved to London, where she enjoyed a lively social life, took painting lessons and became friendly with many of the great men of the day, including George Frederick Handel and Jonathan Swift.

Swift introduced the young widow to Patrick Delany, a protestant Irish clergyman, poet and tutor at Trinity College, Dublin, and they married in 1743, settling at Delville, in Glasnevin near Dublin. This marriage seems to have been in complete contrast to the first, with the pair happily engaged in improving and laying out their garden, which extended to 11 acres and included such necessities as a grotto, temple and orangery, the result being satirized by Swift: "But you forsooth, your all must squander/On that poor spot call'd Del-Vill yonder". In addition to gardening, Mary developed her artistic skills, such as shell work, landscape painting and silhouette-cutting.

After her second husband's death in 1768, Mrs Delany spent the summers with her friend Margaret Bentinck, Duchess of Portland, at Bulstrode in Buckinghamshire. The Duchess took a keen interest in natural history and built up a museum of curiosities and an extensive collection of plants. She also employed Georg Dionysius Ehret to teach her daughters to draw plants, and Sir Joseph Banks and the botanists Daniel Solander and John Fothergill were frequent visitors to her house. The garden at Bulstrode provided plenty of subjects for Mrs Delany to paint, and it was from here in 1772 that she wrote to her niece that she had "invented a new way of imitating flowers". This was the technique of flower collage, the "flower mosaicks" that were to make her famous. These mosaics were made by cutting numerous minute pieces of coloured paper and carefully building them up in layers before sticking them on to a black background to represent the different parts of each specimen. Occasionally, these were enhanced with watercolour and, in some cases, by gluing on small seeds or parts of pods, and the finished result was not only attractive but also botanically accurate.

Mrs Delany persisted with this painstaking work for nearly ten years, during which time she produced almost a thousand collages, but eventually, unsurprisingly, her eyesight began to fail and she was forced to give up. The Duchess of Portland had introduced her to King George III and Queen Charlotte, who provided her with a house (marked today with a blue plaque) in St Albans Street, Windsor from 1785 until her death, just before her 88th birthday, three years later. Her collection of collages eventually filled ten albums, known as the *Flora Delanica*, and these were given to the British Museum by her family in 1897.

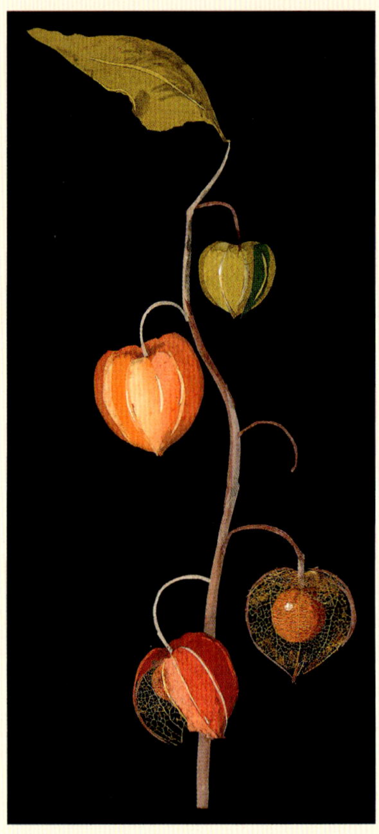

ABOVE: Mrs Delany's collage of Physalis alkekengi, *also known as winter cherry.*

OPPOSITE: A collage of Bombax ceiba *made c. 1780. It used coloured paper and watercolour on a background of black ink.*

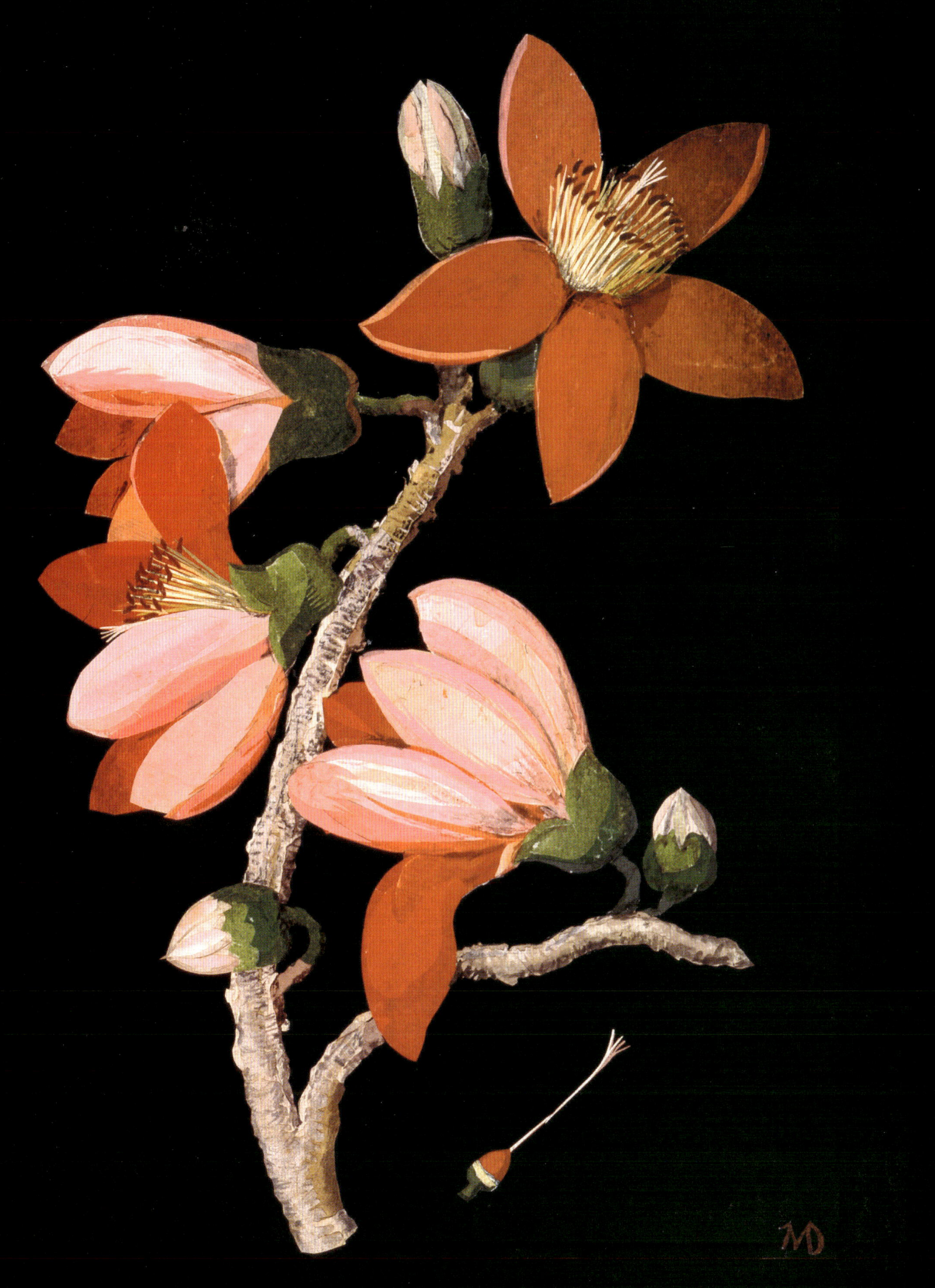
MD

9

South American Adventures

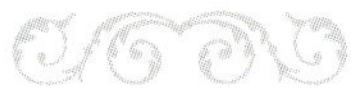

During the seventeenth and eighteenth centuries the Spanish positively discouraged foreigners from visiting their dominions in South America as well as keeping all the trade with their American colonies for themselves. The Spanish had little interest in science and so the wonderfully rich flora of the Andes or even of the coast of Chile was almost unknown.

Vegetables and some ornamental plants that had been cultivated by early cultures in South America were introduced to Spain and spread through Europe and the eastern Mediterranean, as well as being taken by the Portuguese to India. The Caribbean, where the Spanish did not have such complete control, was explored by the English, who were usually at war with Spain, and by the French and Austrians, often their allies.

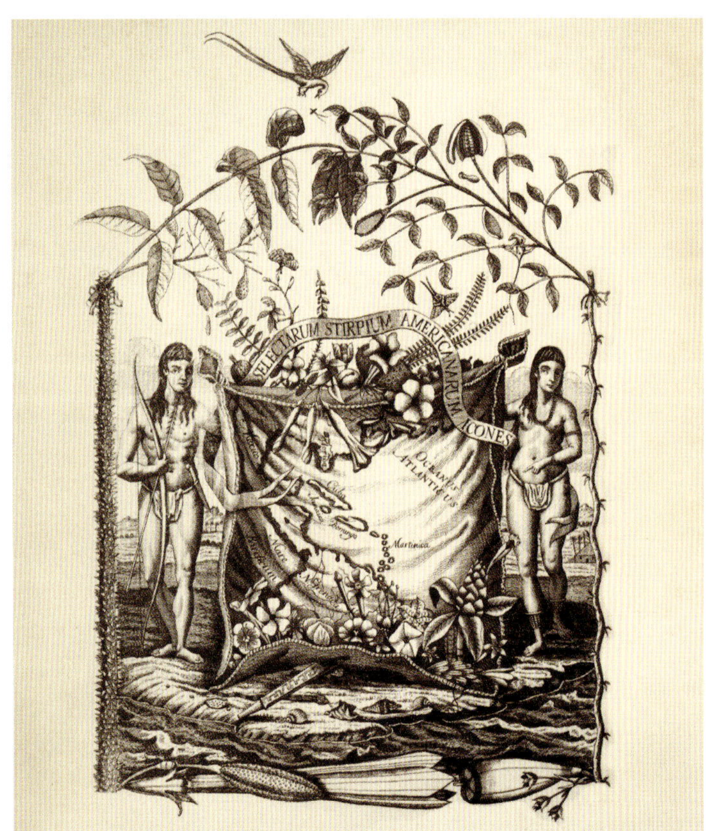

In 1707, when France and Spain were allies, Louis Feuillée (1660–1732), a French priest, managed to get permission to visit Chile and Peru. He reached Concepción in 1709, and having based himself in Lima, spent the next four years travelling and exploring. His main aim was to make astronomical and geological observations, and in 1714 he published his *Journal des observations physiques, mathématiques et botaniques*, which included maps and drawings of cities: his botanical studies were mainly on medicinal plants and the 100 plant illustrations were drawn by Feuillée himself, showing some now familiar garden plants such as *Alstroemeria* and *Salpiglossis*, as well as medicinal herbs.

In 1754, at the age of 27, a botanist born in Leiden, Nikolaus Joseph von Jacquin, made his first expedition to Central America. He was collecting seeds and plants for the Imperial gardens at Schönbrunn in Vienna. He took with him his Dutch head gardener and two Italian zoologists, and initially they concentrated on Grenada, Martinique and Domingo, then under the control of the French. Von Jacquin sent the others home, in succession, laden with plants, but was himself captured by the British and kept prisoner for over a year. On his release, he remained in America, visiting Cuba and Jamaica to collect more plants before returning to Vienna in 1759. His books are among the finest of the period: *Selectarum stirpium Americanarum historia* was first published in 1763. Von Jacquin was unhappy with the engraving of his paintings, and a more lavish version, of only about 15 copies, appeared in 1781. This had

LEFT: The frontispiece from von Jacquin's Selectarum stirpium Americanarum historia.

OPPOSITE: Capparis cynallophora *(now* Capparis cynophallophora*) from* Selectarum stirpium Americanarum historia.

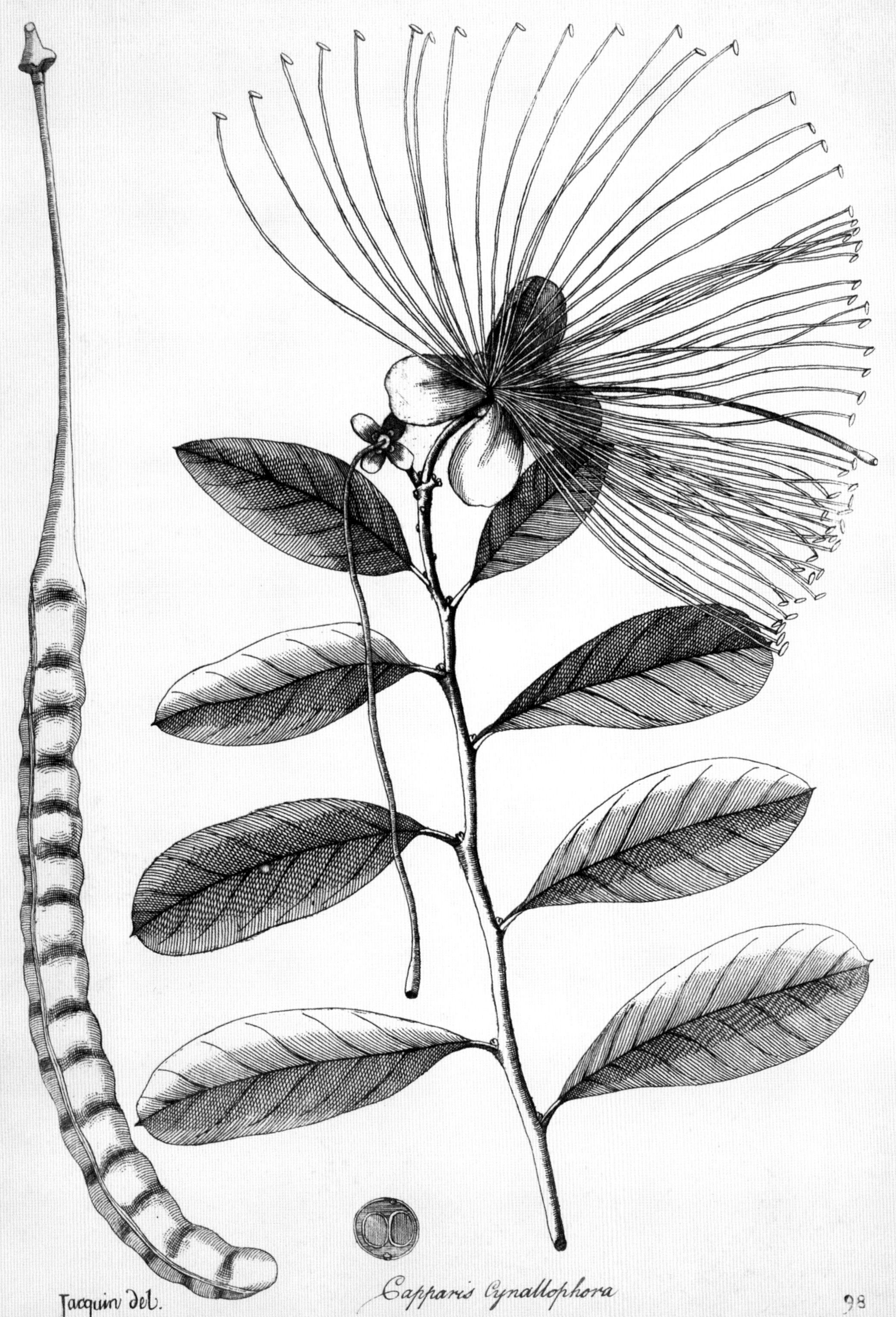

Jacquin del. *Capparis Cynallophora* 98

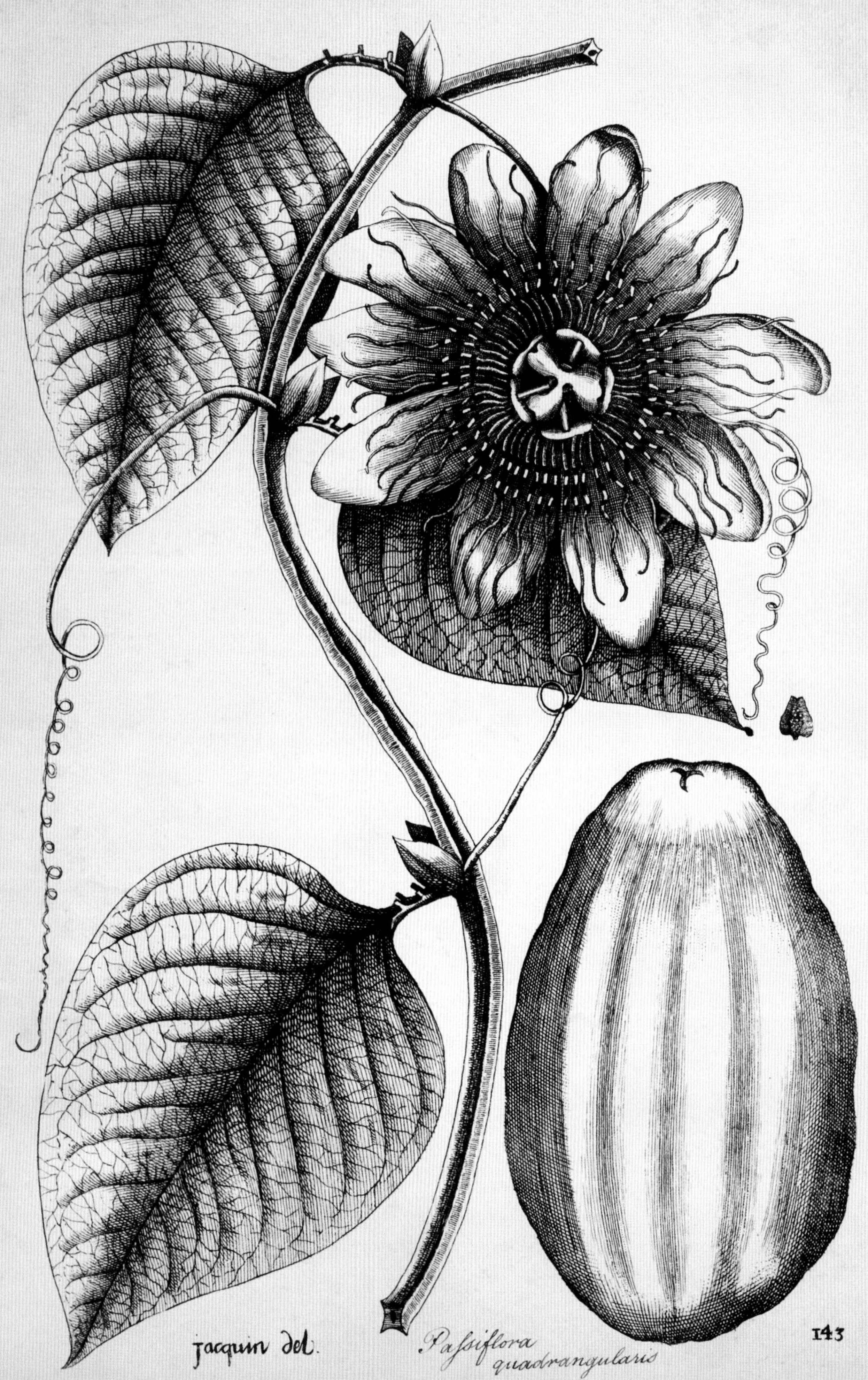
Jacquin del.
Paſsiflora quadrangularis
143

This had 264 watercolours, copied from von Jacquin's originals, and each has an original title page by Francis Bauer, who was, of course, working in Vienna before he came to London. Von Jacquin's *Icones plantarum rariorum* (1781–93) included plates by both Bauer brothers, and later, around 1800, he produced a series of folios with hand-coloured engravings by other artists, *Fragmenta botanica*, which includes some of the plants he had collected. Von Jacquin's finest work was published between 1797 and 1804 and covered plants grown in the Imperial gardens in Vienna: *Plantarum rariorum horti caeseri Schoenbrunnensis Descriptiones et icones* contained 500 hand-coloured engravings, after paintings by Johann Scharf (1765–94) and Martin Sedelmayer (1766–99). Under the patronage and personal interest of the Emperor Francis (1768–1835), the gardens had huge greenhouses built into steep terraces. Francis ruled as the last Holy Roman Emperor, Francis II, from 1792 to 1806, and as Francis I, Emperor of Austria from 1804 to 1835; during much of this time von Jacquin held academic posts around Vienna, as well as being director of the Botanical Gardens. He died in 1817, and some of his botanical work was continued by his son, Joseph Franz von Jacquin (1766–1839), a close friend of Mozart and his family.

At the same time that von Jacquin was writing his *magna opera*, another artist was producing illustrations of flowers in the garden for Francis I. Mathias Schmutzer (1752–1824) painted large watercolours, each with a gold frame on the page, in the style of the *vélins* produced in Paris. Schmutzer had the title of "Imperial Botanical Painter and Court Art Master" and was active between 1794 and 1824. His paintings were probably never intended for publication, and have only recently been published, in 2006, under the title *Florilegium Imperiale*, with an historical introduction, and botanical explanation by Hans Walter Lack.

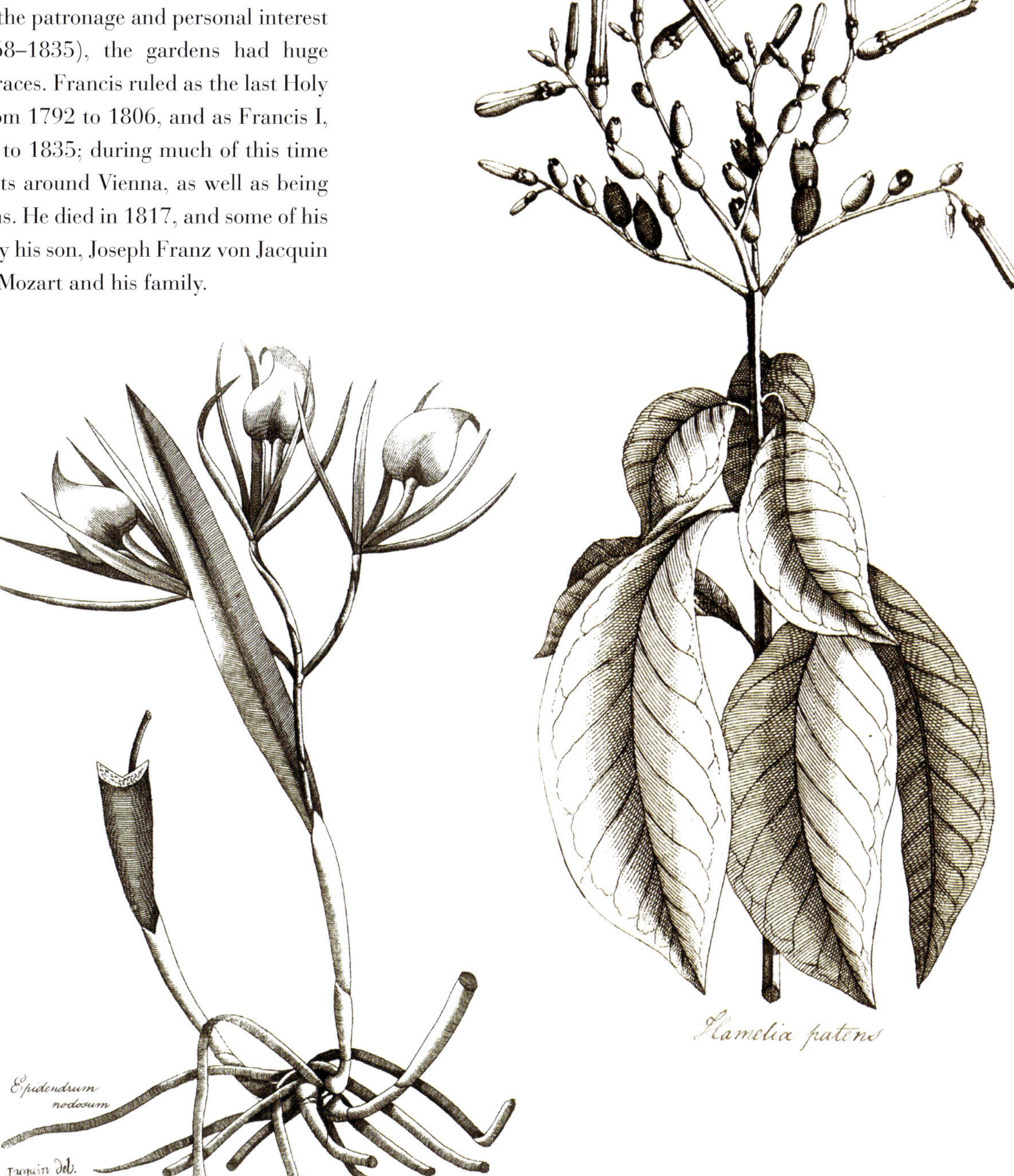

OPPOSITE: Passiflora quadrangularis, *the giant grenadilla, from von Jacquin's* Selectarum stirpium Americanarum historia.

RIGHT: Epidendrum nodosum, *now* Brassavola nodosa, *from* Selectarum stirpium Americanarum historia.

FAR RIGHT: Hamelia patens *from* Selectarum stirpium Americanarum historia.

A series of very thorough expeditions to study the flora of South America was conducted in the late eighteenth century; it was initiated by King Carlos III of Spain, who was told that valuable medicinal and other plants might grow in the Spanish possessions in the New World. Four famous botanists dominate this period: José Celestino Mutis (1732–1808), Don Hipólito Ruiz López (1754–1816), Don José Pavón Jimenez (1754–1840) and Joseph Dombey (1742–94). The first expedition was set up by royal decree in April 1777, and led by two outstanding students from the royal gardens in Madrid, on the recommendation of their professor: Don Hipólito Ruiz and Don José Pavón finally left Cadiz on 4 November. At the special request of the King of France, they were accompanied by Joseph Dombey, a more experienced French botanist, who had studied in Montpellier and then moved to Paris and become botanist at the Jardin du Roi. Two artists and an apprentice botanist and apprentice artist also joined the expedition. This lasted 11 years and mainly covered Peru and Chile, but also visited parts of present-day Ecuador, Bolivia and Colombia. Ruiz and Pavón returned to Spain in 1788.

The fourth botanist, José Mutis, was already established in Bogotá, where he had lived since 1761, practising medicine, which he had studied at the University of Cadiz. He realized the potential of the flora, especially the medicinal properties of some of the plants, and had already studied *Cinchona* species, the source of quinine, as well as teaching mathematics and astronomy and improving silver mines. He sent several petitions to the King in Madrid to set up a botanical institute and fund an expedition, but they were all ignored and it was not until 1783, with the help of a new Archbishop-Viceroy in New Granada, that he managed to get himself an official position as director, botanist and astronomer, based first in Mariquita and then in Bogotá; they set up the Instituto Botanico del Nuevo Reino de Granada with a botanic garden, which became an important centre for the study of the Colombian flora. This remarkable institution operated until 1816, and trained and employed altogether 40 botanical artists (around 15 at any one time), to make illustrations of the Colombian flora. The most famous, the Colombian Francisco Javier Matiz, was described by Humboldt, as "the best painter of flowers in the world", and contributed 326 of the paintings. There are detailed dissections of grasses and other flowers, but most are beautiful folio-size paintings, with elegant, if rather formal, design; the style is somewhat reminiscent of Company School paintings produced in India, where there is a sense of flatness, as if the specimen has been laid out for pressing. It seems likely that the paintings were done in the studio, perhaps from freshly-pressed plants with colour notes. Some of the pictures of climbers such as *Tropaeolum*, *Bomarea*, *Fuchsia* and *Mutisia clematis* (named by Linnaeus's son in honour of Mutis) are particularly successful as designs, but they do not have the natural poise

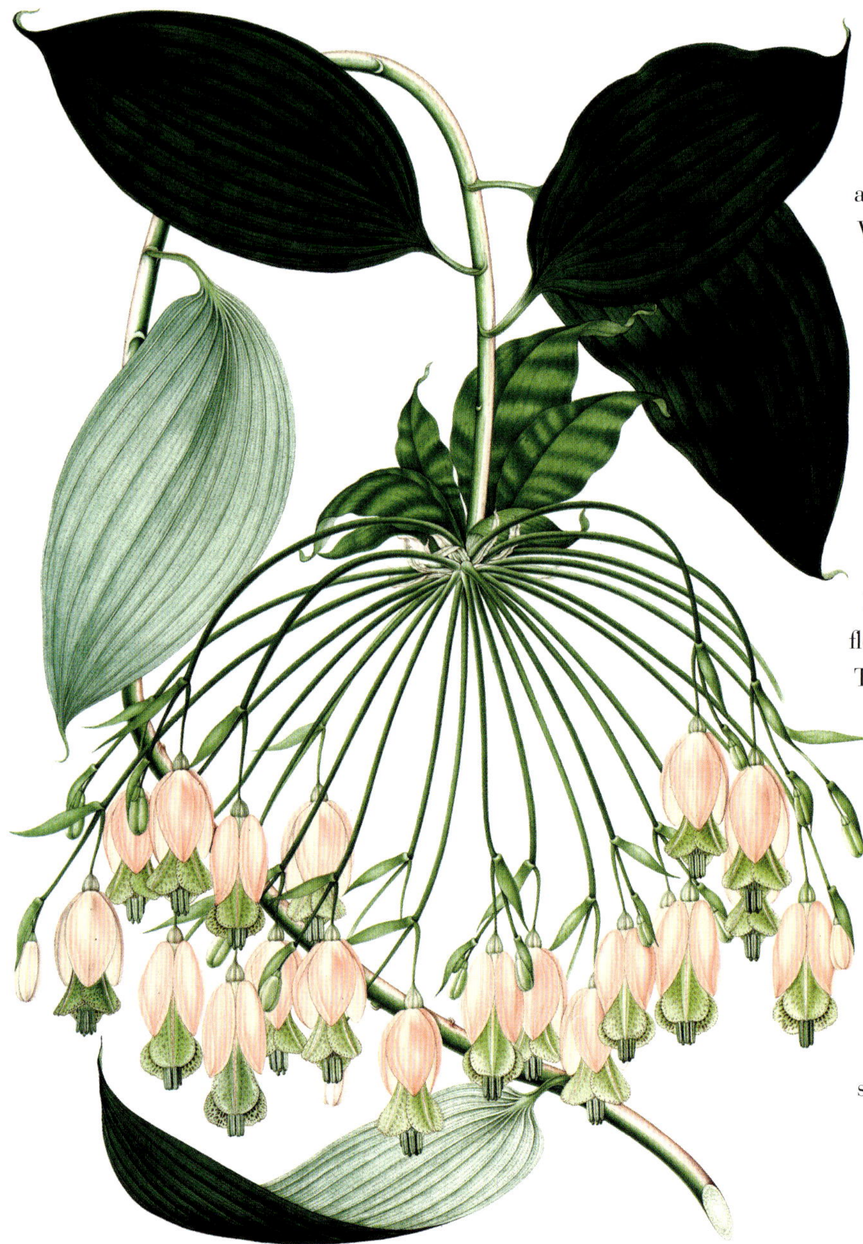

LEFT: Alstroemeria salsilla, *now* Bomarea salsilla, *from the workshop of José Mutis.*

OPPOSITE: Mutisia clematis, *named in honour of Mutis and painted by Salvador Rizo Blanco.*

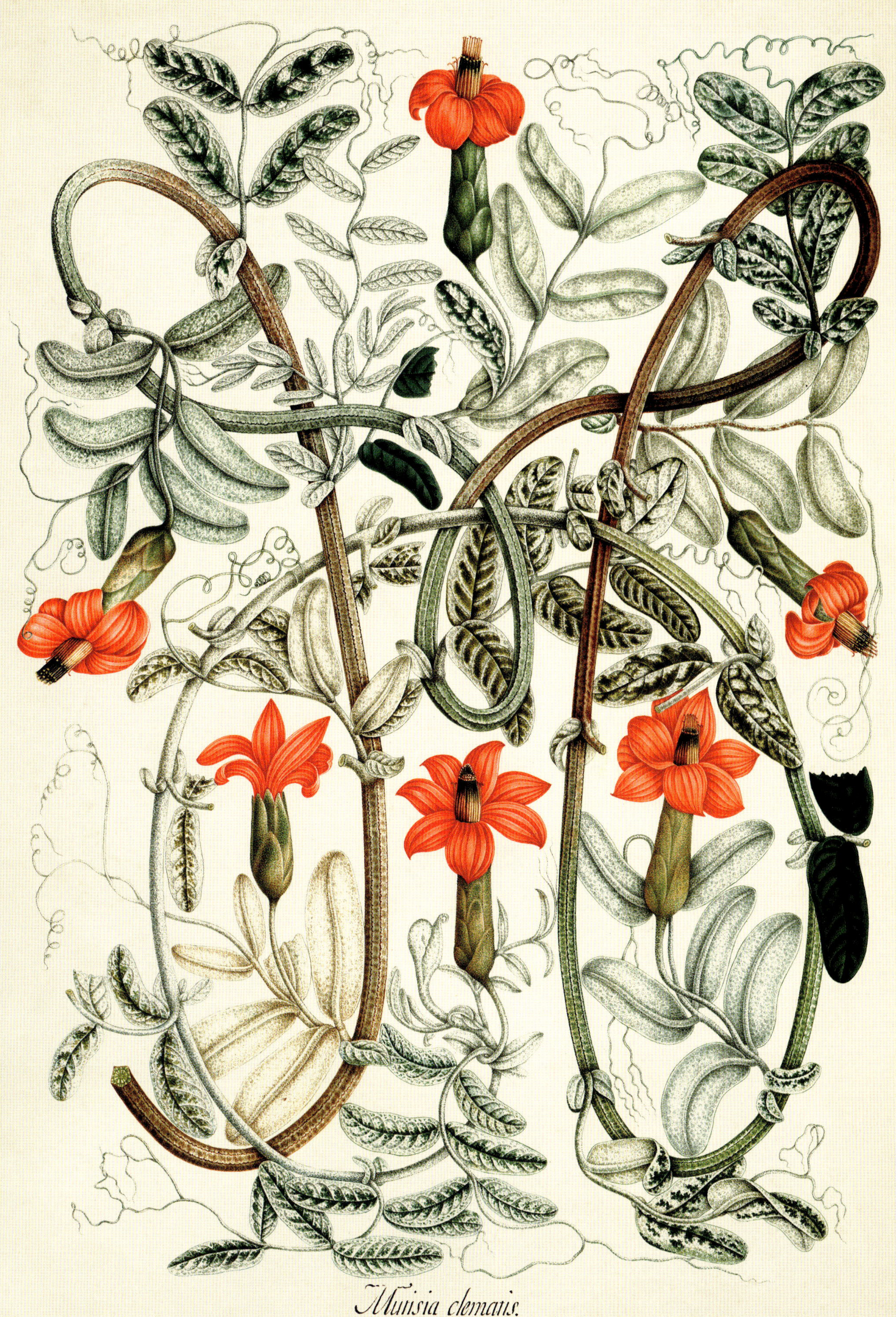

Mutisia clematis.

Tropaeolum

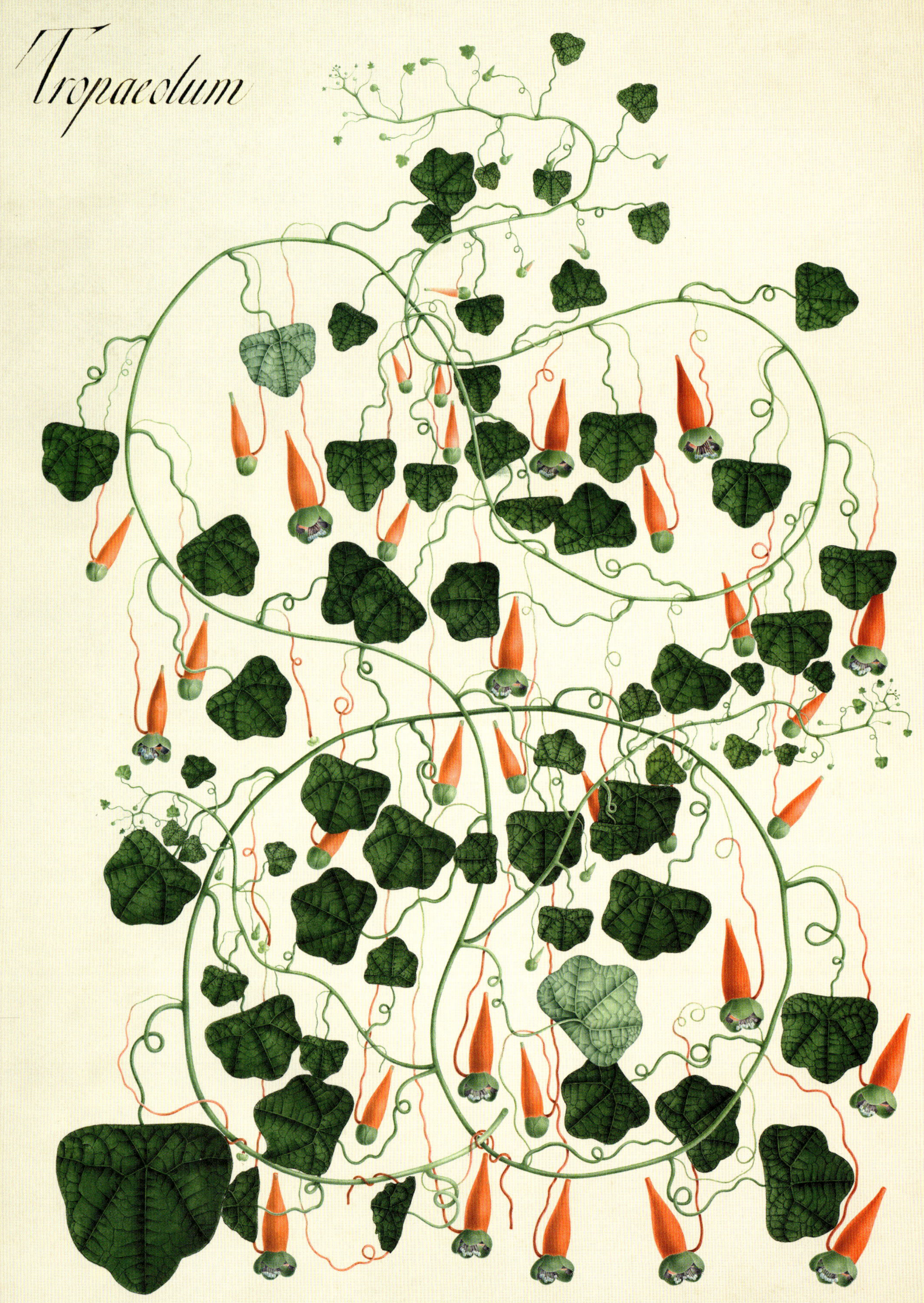

of flower paintings by Ferdinand Bauer or Pierre-Joseph Redouté who were working in Australia and Europe at about the same time.

Soon after 1810, with unrest and revolution in Colombia, the institute was disbanded and its work was confiscated by the government. Mutis had died in 1808 and his post was taken by his nephew; his projected Flora of Bogotá was never published, and his botanical papers were lost. The paintings, however, numbering more than 6,500, which were made under his direction, were taken to Madrid, and remained in the archives of the Real Jardín Botánico.

Meanwhile, Ruiz, Pavón and Dombey were exploring Peru and Chile. We can read the details of their journeys and the plants they found in each place in *The Journals of Hipólito Ruiz*. The manuscript for this was discovered by a Colombian surgeon and diplomat in the Natural History Museum in London during the Second World War, while he was studying the botany and history of quinine. Ruiz's journal probably arrived in London in the early nineteenth century through contacts between Aylmer Bourke Lambert and Pavón. The chapters describe the geography of each area they investigated, the towns, cities and villages, noting the animals, crops and habits of the local Indians. The perils of travel in a country that could be both lawless and difficult to cross are also described and each chapter ends with a detailed list of the plants found on that part of the expedition, noting any with local medicinal uses or even potential as garden ornamentals. For instance, Ruiz describes the common nasturtium, a native of Peru: "*Tropaeolum majus*, *astuerzo* (cress) or *capuchinas* (nasturtium); the natives of Peru frequently use this plant to treat the mouth sores of scurvy, and innkeepers and other interested people pickle the buds, just as is done in Spain with capers. The flowers ... are likewise added to uncooked salads, to which they lend a peppery and appetizing flavour." They also include a detailed description of the cultivation and use of the coca plant, the source of cocaine.

The three botanists usually travelled together, or to neighbouring areas accompanied by the artist Isidor Galvez, at first around Lima and then, in December 1781, sailing south to Chile. They were thrilled by the new species they found near Concepción, studied them carefully and drew them with "the greatest care".

During an outbreak of plague in Concepción, Dombey "temporarily abandoned botany", ministered to the poor and himself paid for medicines and nurses. He is said to have been offered a post as physician at a salary of 10,000 livres a year (twice his original salary), and the hand of a beautiful heiress who had taken a fancy to him, but he refused in order to return home. After spending two years in Chile, the party sailed from Valparaiso north to Callao, the port for Lima, arriving in November 1783. Here, they prepared to send a consignment of 800 coloured drawings, 55 boxes of dried plant and animal specimens, rocks, Indian artefacts and clothing on a ship bound for Spain, together with six glass cases containing 33 potted trees and a man to look after them, but the ship was caught in a storm off Chile and the cases had to be jettisoned overboard.

At this point in the expedition, Dombey returned to Spain by ship, and Ruiz and Pavón remained in Peru, finally sailing from Callao in March 1788. Ruiz kept the potted plants in a cabin next to his own, and put them out on deck on warm, calm days! They sailed well to the south of Cape Horn, reaching 59°S, before turning east and north again. During this time Ruiz continued writing up his plant notes and descriptions, adding some of the bulbous plants that had flowered in their pots on board. They finally reached Cádiz on 12 September.

The results of this expedition were published first as a *Prodromus* in 1794, with 37 engravings of plants after Galvez's drawings, and later, from 1798 to 1802, as *Flora Peruviana, et Chilensis* in three volumes with 325 engravings. This was only part of their proposed publication, and further paintings are in the Natural History Museum in London. The engravings were probably never meant to be coloured and are very different to the work done for Mutis. They are realistically drawn, with detailed dissections of the flowers included on the plate.

Joseph Dombey had returned earlier, but half his collection had been captured by the British and sold at auction in Lisbon, from where they were redeemed by the Spanish. The government in Madrid demanded half the remainder, because many of the Ruiz and Pavón cases had been lost in a shipwreck, and briefly imprisoned him, but a dispirited Dombey eventually reached Paris in October 1785, and was soon caught up in the French Revolution, which he was lucky to survive. He at length returned to the West Indies but, after several adventures, died in Montserrat, disguised as a Spanish sailor, a prisoner of the British, in 1796.

OPPOSITE: Tropaeolum dekerianum, *painted by José Jerónimo Triana under the direction of José Mutis.*

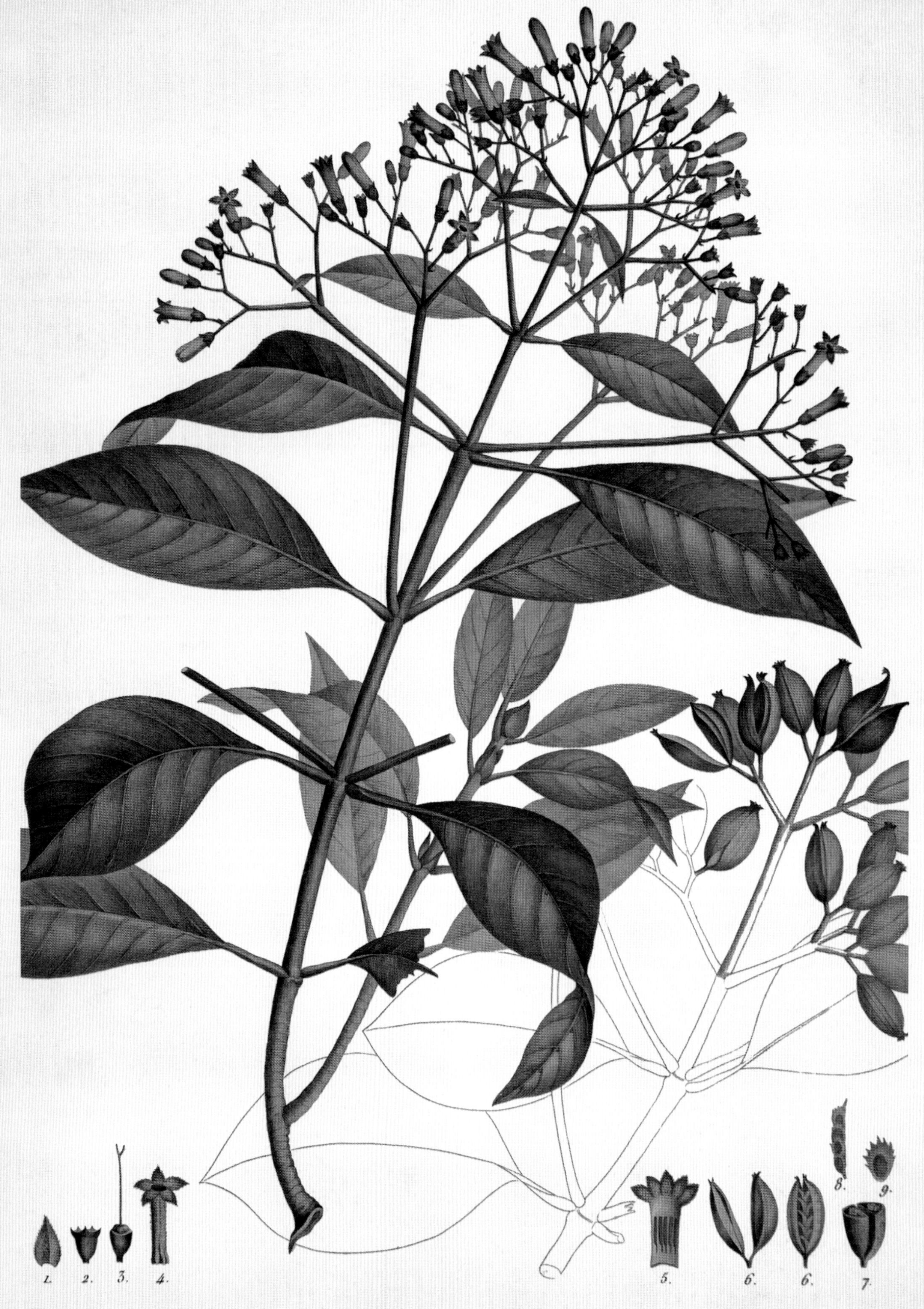
1.
2.
3.
4.
5.
6.
6.
7.
8.
9.

The great scientist Alexander von Humboldt (1769–1859), made a typically thorough and efficient exploration of America from 1799 until 1804, in the company of the French botanist, Aimé Bonpland (1773–1858). It was a totally remarkable journey: they managed to get permission from Spain's King Charles IV through a friend of Humboldt's who was ambassador from Saxony to the Spanish court, and set out with a passport and royal letter of recommendation in June 1799. In 1800, they explored the Orinoco and Rio Negro. In 1801, they visited Mutis in Ecuador, and admired the work of his botanical artists. The following year, they visited Ecuador and climbed to 5878 metres (19,285 feet) on Mount Chimborazo, then the highest recorded climb, and Humboldt made a special study of volcanoes. They sailed down the coast of Peru, where they measured the sea temperature and deduced the existence of the cold Humboldt current, as well as noting the deposits of guano on the coast (the beginning of the important phosphate fertilizer trade). They then spent a year in Mexico, before returning home via North America where they met Thomas Jefferson, arriving in Bordeaux in August 1804. The results of this expedition were written up in the 30-volume *Le Voyage aux Régions équinocxiales du Nouveau Continent*. It was so expensive that Humboldt himself could not afford a copy. Part 6 contained the plants, in 15 volumes, and was published between 1805 and 1829, with the earlier volumes written by Bonpland, and later by C. S. Kunth. The illustrations were by Poiteau and Turpin, and are scientifically excellent, but rather sterile, and almost certainly drawn from dried specimens. Of the plates 896 are printed in colour and finished by hand, 220 are hand-coloured engraved plates, and a further 145 are uncoloured.

Mutis himself is honoured with a portrait in Volume 1. Some of the original paintings for one of the volumes, a monograph of the Melastomataceae, are in the Fitzwilliam Museum in Cambridge.

Even more stupendous than Humboldt's work, was that Carl Friedrich Philipp von Martius who was born in Erlangen, Bavaria, in 1794, where he graduated in medicine. A keen naturalist, he wrote his thesis on the plants in the botanic garden of the university, and later travelled to Brazil as part of the entourage accompanying Archduchess Maria Leopoldina, who had recently married the future Brazilian Emperor, Dom Pedro de Alcantara. From 1817 until 1821 Martius travelled across the country by foot, on horseback or by canoe, accompanied by a zoologist, Johann Baptist von Spix.

On his return to Europe, loaded with several thousand plant specimens, Martius became conservator of the botanic garden at Munich, and was later appointed Professor of Botany at the university there. He published two substantial works; the first, *Historia naturalis palmarum*, with descriptive Latin text and 240 chromolithographs, almost all of which were based on his own

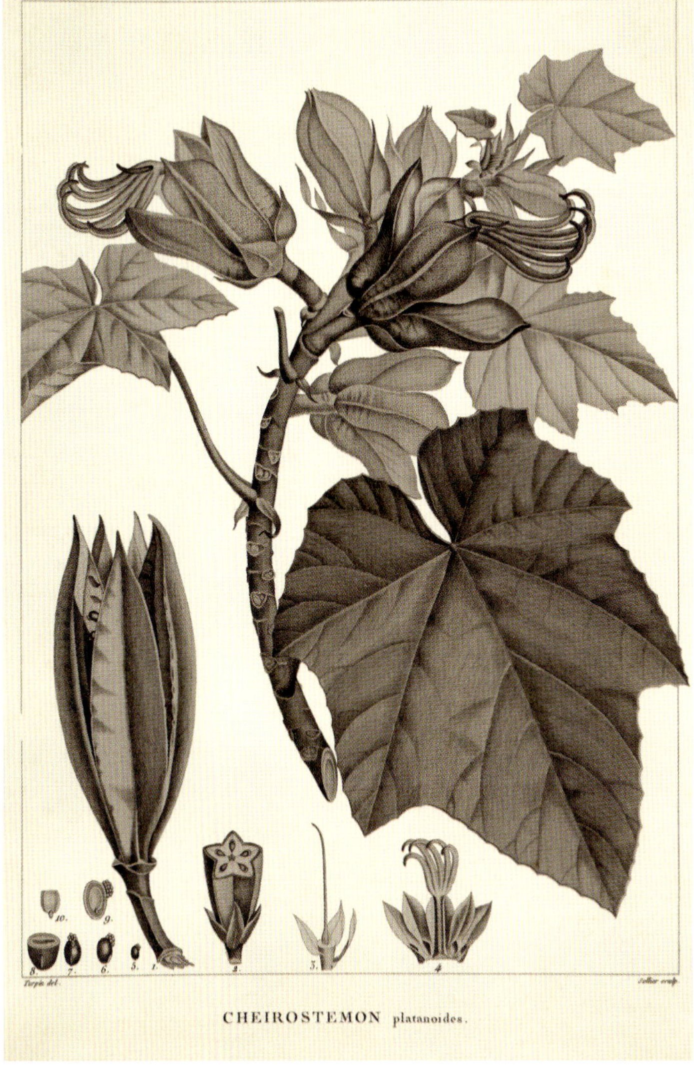

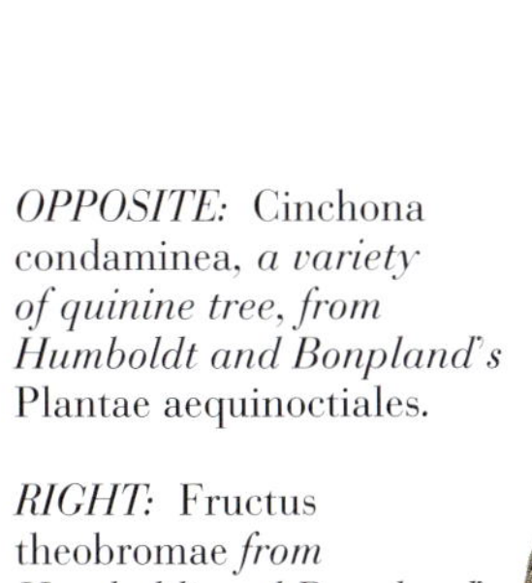

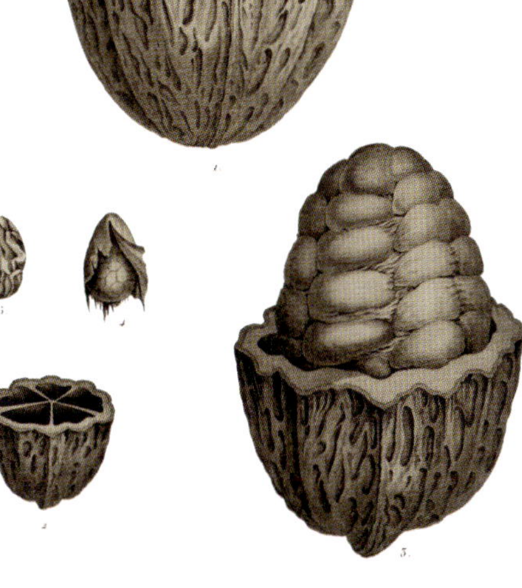

OPPOSITE: Cinchona condaminea, *a variety of quinine tree, from Humboldt and Bonpland's* Plantae aequinoctiales.

RIGHT: Fructus theobromae *from Humboldt and Bonpland's* Plantae aequinoctiales.

FAR RIGHT: Cheirostemon platanoides, *now* Cheirostemon pentadactylon, *from Humboldt and Bonpland's* Plantae aequinoctiales.

drawings, was published in three folio volumes in Leipzig between 1823 and 1850. In 1840, Martius, with the botanist S. L. Endlicher, started work on the immense *Flora brasiliensis*, producing 46 fascicles before his death in 1868. This enormous work, containing nearly 4000 lithographs and some nature prints, was continued by August Eichler, and eventually completed by Ignatius Urban and a team of other contributors in 1906.

The discoveries in Central and South America helped to fuel the craze for orchids that took hold in England and Germany in the second quarter of the nineteenth century. Before this time, the dry heat used to heat greenhouses favoured plants from dry areas such as heathers and pelargoniums from the Cape and Banksias from Australia. When steam heating and heating by hot water pipes were introduced, humidity could be kept high and tropical palms, ferns and particularly orchids became all the rage. This resulted in many splendidly illustrated books, but none finer, or larger than *The Orchidaceae of Mexico and Guatemala* by James Bateman. It is one of the largest of all botanical books, an elephant folio, with hand-coloured lithographs after paintings by Miss Drake and Mrs Augusta Withers, flower painter to Queen Adelaide, and later to Queen Victoria. The paintings are wonderfully accurate, and done from living plants; some of the originals are in the RHS Lindley library, and are little better than the printed version. James Bateman of Knypersley Hall in Staffordshire, and later of Biddulph Grange was a rich landowner and keen grower of orchids: he grew many of the plants himself, and others came from the Duke of Devonshire at Chatsworth or the nurseryman George Loddiges of Hackney. The original collectors of many of the plants were Theodore Hartweg who had been sent out to Mexico by the Horticultural Society in 1836 and George Ure Skinner, who lived in Guatemala, working in the cochineal and indigo trade. Bateman rated Skinner as the most important of all introducers of orchids to cultivation, and his book was an early example of orchidomania which "even extended to Windsor Castle itself".

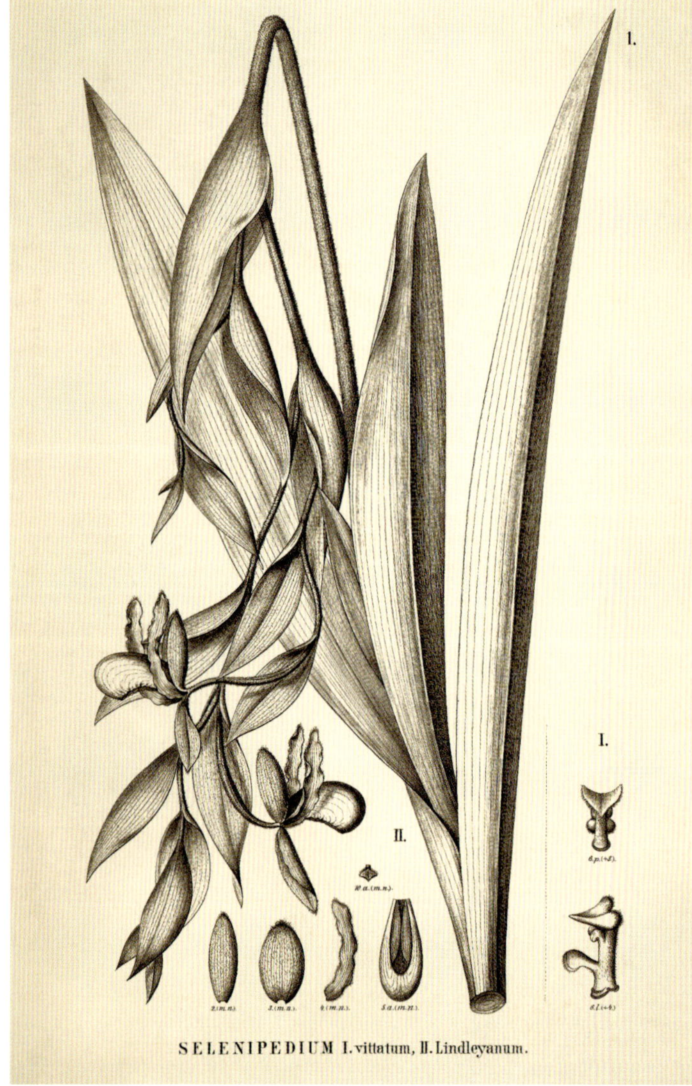

LEFT: Selenipedium vittatum *and* lindleyanum *from* Flora brasiliensis.

BELOW: Psychotria peoppigiana *and* colorata *from* Florabrasiliensis.

OPPOSITE: Cattleya skinneri *from Bateman's* Orchidaceae.

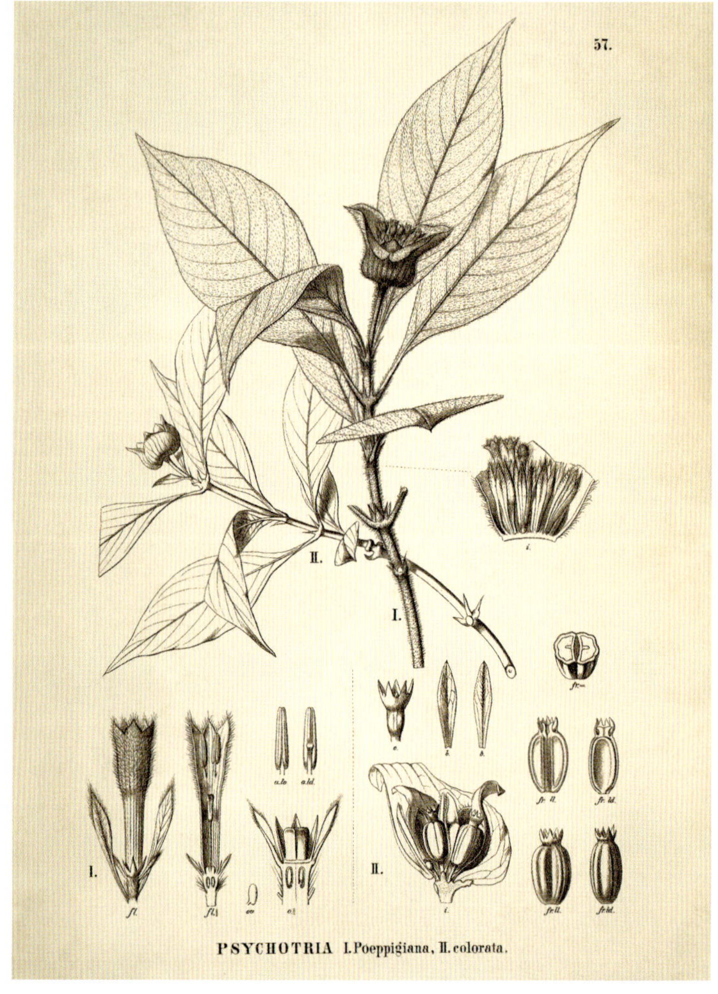

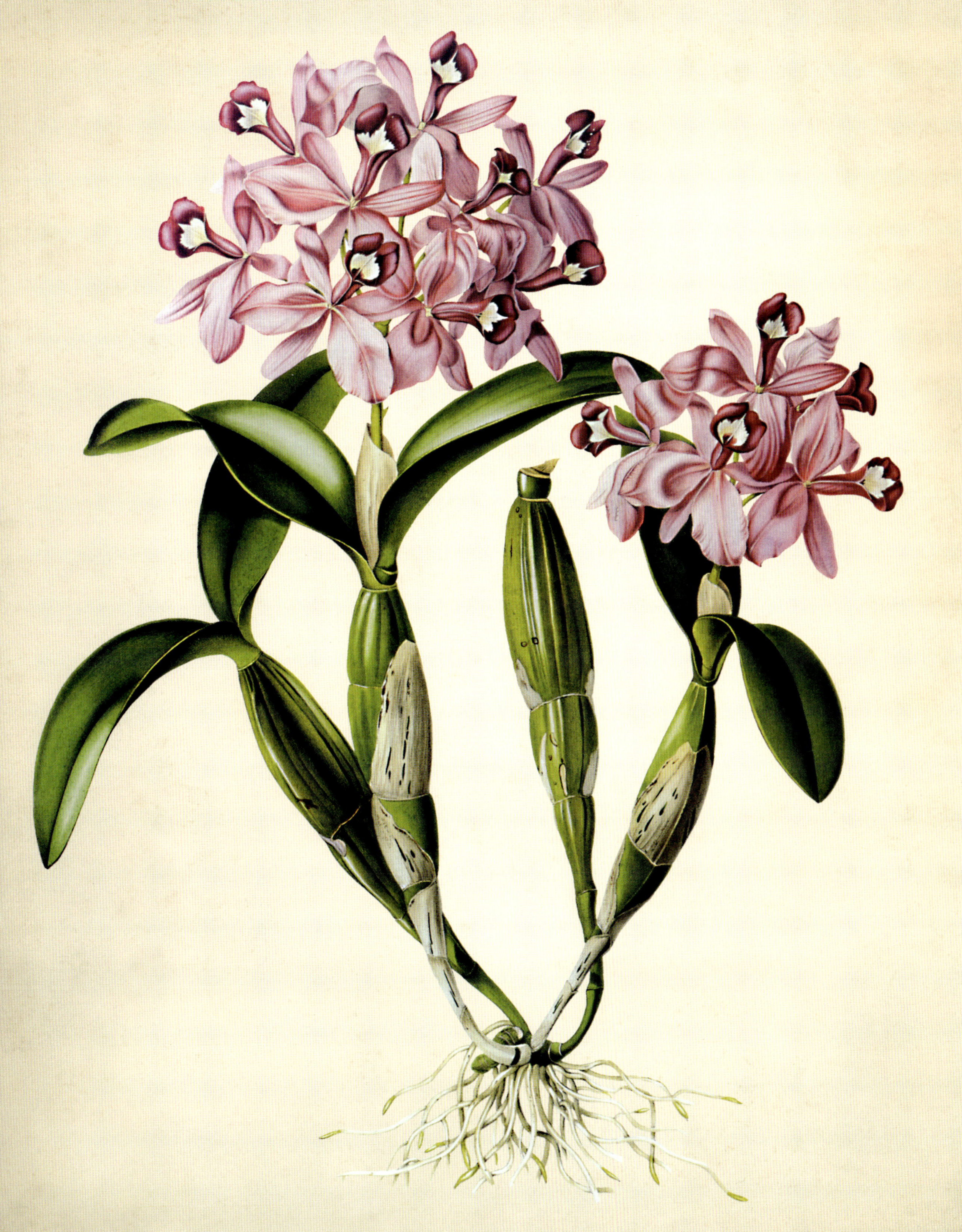

Thornton's The Temple of Flora,

OR GARDEN OF NATURE

Thornton's *The Temple of Flora, or Garden of Nature* contains the most extraordinary engravings of plants set against dramatic backgrounds designed to show either the plant in an appropriately romantic habitat or depicting its habit of growth. For example, the night-blowing *Cereus* is shown with a church, the clock at midnight.

The exact date of Robert John Thornton's birth is unknown, but he was the son of a fashionable doctor, who died in 1768, when Thornton was a young child. As a boy, he was fascinated by all aspects of natural history, but most particularly plants, birds and insects. He proceeded to Cambridge, where he attended lectures on Linnaeus by the Professor of Botany, Thomas Martyn, and thence to Guy's Hospital to study medicine, going abroad before setting up his practice in London in 1797.

On the death of both his mother and elder brother, Thornton came into his inheritance, and decided to produce a grandiose scientific work, of atlas folio size, provisionally entitled the *New Illustration of the sexual system of Carolus von Linnaeus....* This was published in parts from 1799 onwards, and by 1807 was complete. The final part of the work is that commonly known today as *The Temple of Flora*, consisting of "picturesque, botanical, coloured plates, of select plants, illustrative of the same, with descriptions".

Thornton's interest in Linnaeus, and his new system for naming plants, was probably encouraged by the fact that J. E. Smith, whom he had succeeded as lecturer of medical botany at Guy's, had bought Linnaeus's herbarium and library, and was one of the founders of the Linnean Society in London. In addition, at a time when Britain was at war with France, Thornton felt it incumbent on those who could afford it to support the national interest in the great botanical discoveries of the time, and dedicated the work to Queen Charlotte, consort of George III.

BELOW: The Superb Lily, Lilium superbum, *from Thornton's* Temple of Flora, *published in June 1799.*

OPPOSITE: The Dragon Arum, Dracunculus vulgaris, *from Thornton's* Temple of Flora, *published in December 1801.*

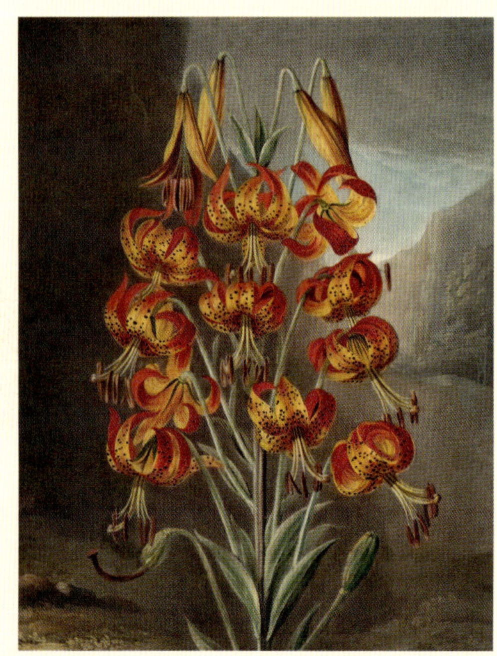

Oil paintings for the work were commissioned from Peter Henderson (a miniaturist who exhibited at the Royal Academy), Philip Reinagle (initially a portrait painter), Abraham Pether (famous for skilful depictions of moonlight) and Sydenham Teast Edwards (who also worked on *Curtis's Botanical Magazine*). In addition, a band of engravers was employed, using various processes – aquatint, line engraving, mezzotint, stipple engraving – to reproduce the plates, which were hand coloured for publication. Later editions were produced, with a varying number of illustrations, but eventually, despite strenuous efforts at recouping his losses by exhibiting the paintings to the paying public and a disastrous attempt at running a lottery with prints as prizes, poor Thornton was reduced to penury and died in 1837.

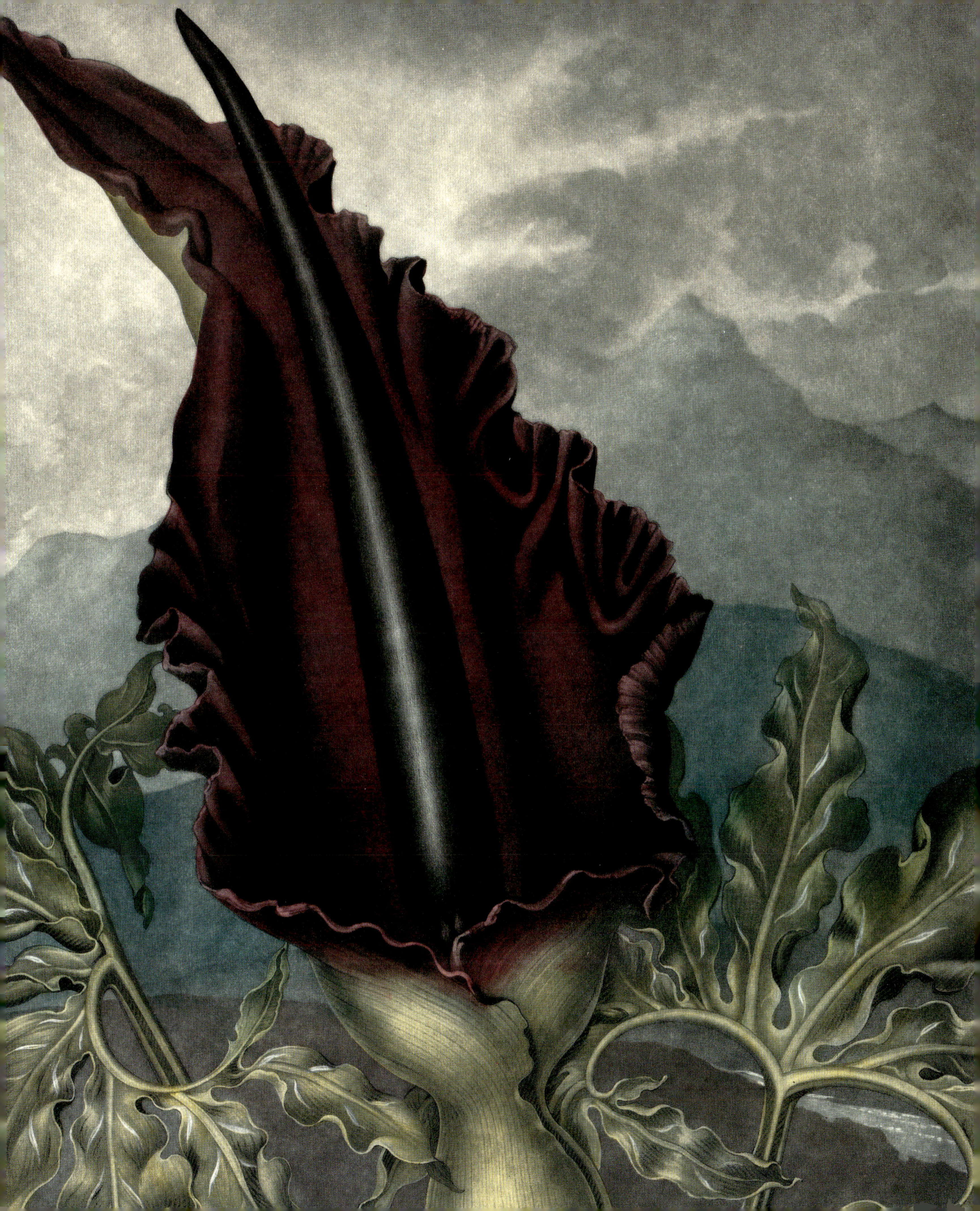

10 THE GOLDEN AGE IN FRANCE

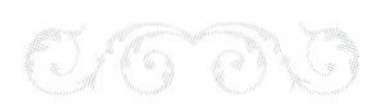

GERARD VAN SPAËNDONCK

One of the most influential artists of eighteenth-century France was, ironically, a Dutchman named Gerard van Spaëndonck (1746–1822) who had arrived in Paris in 1766. Born in Tilburg in southern Holland, both Gerard and his younger brother Cornelis went on to become fine artists, originally producing flower pieces in the Dutch manner of Jan van Huysum.

Both van Spaëndonck brothers studied with the Flemish painter Willem Jacob Herreyns in Antwerp before moving to France, where Cornelis worked at the Sèvres porcelain factory, and Gerard became a fashionable painter of snuff boxes. In 1774, Gerard was appointed miniature painter to Louis XVI, and a few years later succeeded Madeleine Basseporte as Professor of Flower Painting at the Jardin du Roi, and it was here that he found his métier as a teacher of flower painting and botanical drawing. He was one of the first members of the Academie des Beaux Arts, created in 1795, part of the prestigious Institut de France.

Gerard van Spaëndonck himself published only one book of 24 fine drawings, engraved by P. F. Legrand, entitled *Fleurs dessinées d'après nature* (1799–1801), but he contributed over 50 paintings to the royal collection of *vélins*. The author and art historian Wilfrid Blunt, who studied these *vélins* carefully, realized that van Spaëndonck used only gouache for the paintings before 1782, but then changed his medium and by 1784 was using pure watercolour – a technique subsequently popularized by his most famous pupil, Pierre-Joseph Redouté.

In addition to Redouté, and his younger brother Henri-Joseph Redouté, who was both zoological and botanical draughtsman at the Jardin des Plantes, van Spaëndonck numbered among his pupils several really fine artists. Of these, Pancrace Bessa (1772–1846), Pierre Antoine Poiteau (1766–1854), Jean Henri Jaume Saint-Hilaire (1772–1845), and the Danish flower painter Johan Laurentz Jensen (1800–56) are of particular note.

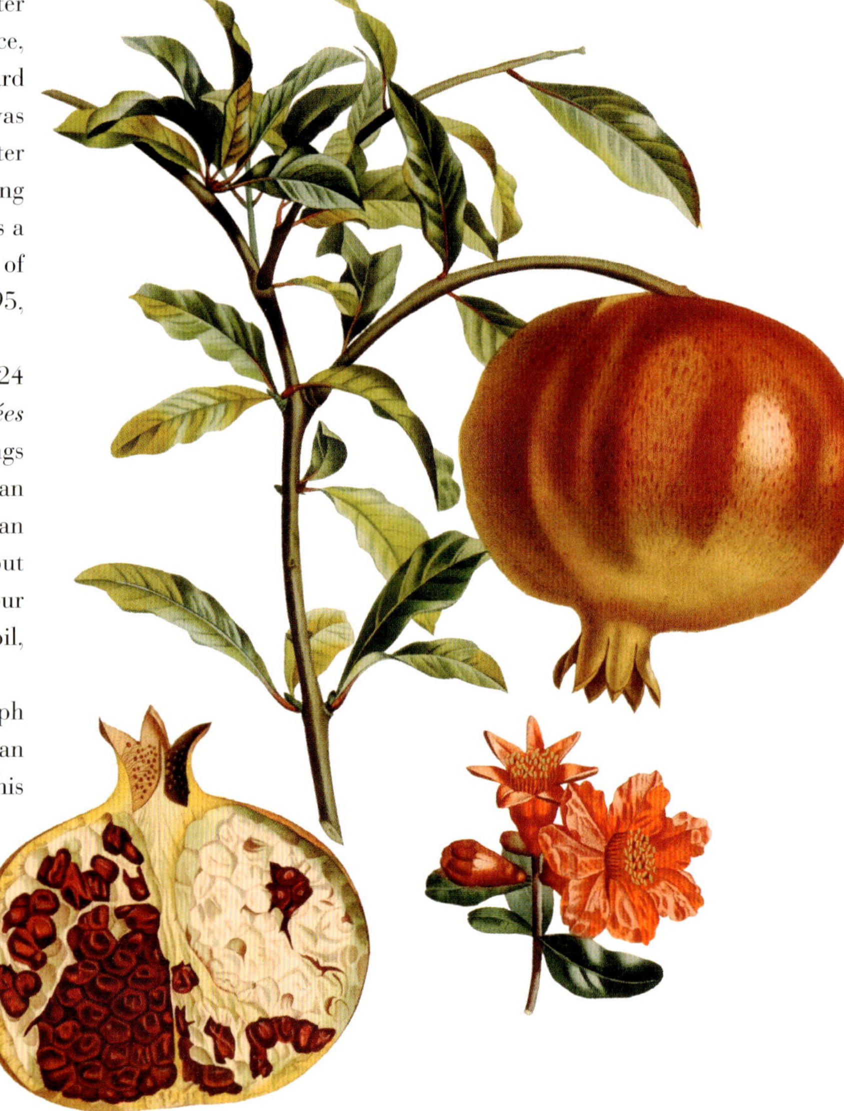

RIGHT: Studies of a pomegranate by Pierre Antoine Poiteau from his La Pomologie française.

OPPOSITE: Gerard von Spaëndonck's study of Hibiscus rosa-sinensis *made in watercolour on paper.*

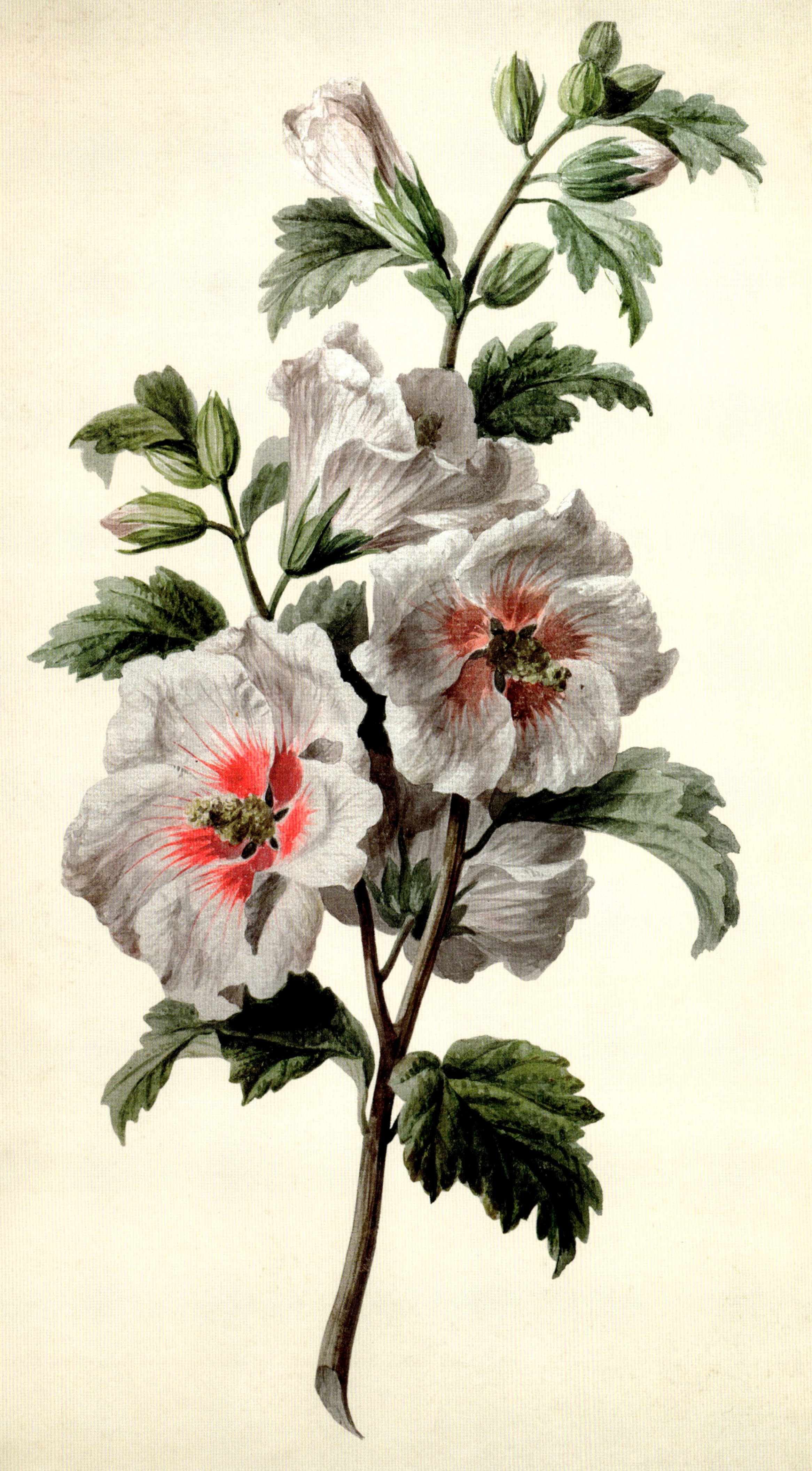

21.

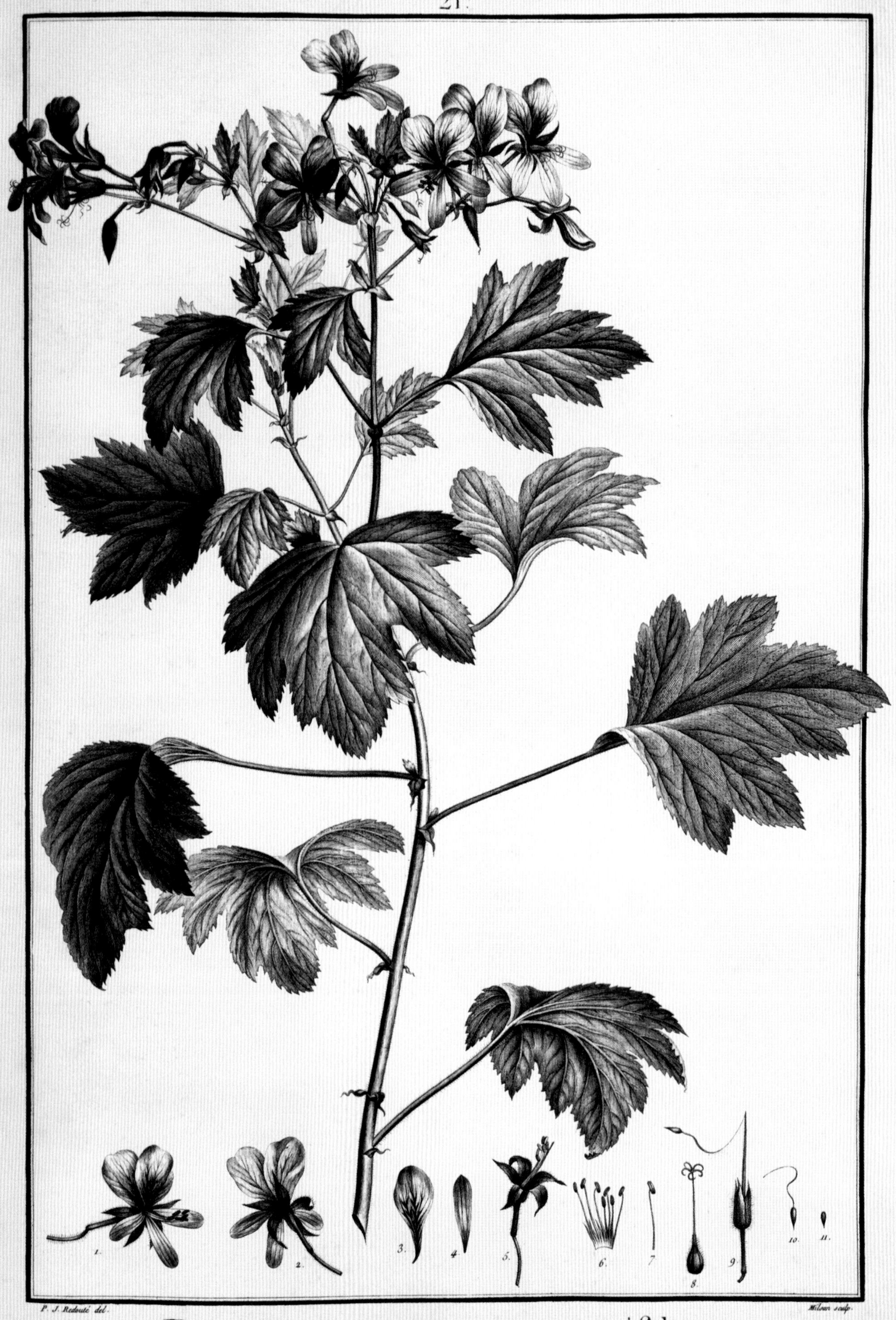

PELARGONIUM acerifolium.

PIERRE-JOSEPH REDOUTÉ

Pierre-Joseph Redouté (1759–1840) is probably the best known flower painter of all time – countless prints of his works (especially of roses) have been reproduced, with various levels of expertise, and are to be found in houses and hotels around the world, while the original prints are rightly regarded as treasures in fine libraries and private collections.

Redouté's fame is due partly to his undoubted talent as an artist, his love of flowers and his sustained hard work, and partly to the fact that he lived and worked in Paris during the age of the Enlightenment, at a time when the city was the scientific, philosophical and cultural centre of Europe. In addition, in the Empress Joséphine (*see* pages 144–45), wife of Napoleon I, he found both a genuine lover of plants and an extravagant, enthusiastic and generous patron. Thanks to her, Redouté enjoyed for many years the freedom to live and work without having to worry about money, even today the scourge of most botanical artists. His original paintings were reproduced for publication using the finest new methods of engraving, and his name became synonymous with the highest artistic achievement of his time.

Pierre-Joseph Redouté was born in the Ardennes (today part of Belgium), the son of a painter and decorator, and left home to study painting at the age of 13. He spent several years wandering through the Low Countries, during which time he came to know the work of the Dutch flower painters, including Rachel Ruysch and Jan van Huysum, both of whom influenced his later work.

Redouté's elder brother, Antoine-Ferdinand, meanwhile, had set himself up in Paris as an interior decorator and theatrical scenery designer, and in 1782, Pierre-Joseph joined him. The younger brother continued to paint flowers, with moderate commercial success, and learnt the techniques of line engraving and colour printing. He began to haunt the Jardin du Roi, the royal botanic garden, where many unusual plants were grown, and where he found inspiration for further paintings. His work, some of which had been engraved and sold by an art dealer, was seen by Gerard van Spaëndonck, then official artist at the garden, and also by an aristocratic bibliophile and keen amateur botanist, Charles-Louis L'Héritier de Brutelle (1746–1800). The encouragement of these two men gave Redouté the chance to develop his talent. L'Héritier (as he is normally known) had been appointed, thanks largely to family connections, to the post of superintendent of water and woods of the Paris region. He immediately set about studying the subject, becoming as a result a keen botanist, while at the same time also working as a magistrate, a job he apparently carried out with great integrity.

In the spring of 1785, L'Héritier began to publish, in instalments, the first of a series of descriptions and illustrations of rare plants, under the title *Stirpes novae*, and the next year, Redouté was drafted in to help, contributing three plates. In 1787, Redouté, by now a good friend of L'Héritier, visited him in London, where he was studying plants growing in and around London and Kew, and the two, along with the English artist, James Sowerby, produced another book, *Sertum Anglicum* in 1788. This time, Redouté produced 22 plates and Sowerby 13, and Redouté also made the acquaintance of the engraver Francesco Bartolozzi, from whom he was to learn about the technique of stipple engraving. In a monograph on Cornus, also published 1n 1788, Redouté was still using line engraving, but in coloured copies, the coloured inks were applied onto the copper plate.

On his return to France, Redouté continued to work for L'Héritier, and was also commissioned by van Spaëndonck to contribute some coloured drawings to the great collection of *vélins* at the botanic garden; he eventually produced over 500. These were difficult and dangerous times, however, with the revolution in France, and although Redouté was appointed official artist to Marie-Antoinette, this career was cut short by her execution. Meanwhile, poor L'Héritier, who had supported some of the early revolutionary reforms, but was disenchanted by the later violence of the Terror, lost most of his money, was left a widower and was finally murdered outside his house in 1800, a

ABOVE: An engraved portrait of Pierre-Joseph Redouté.

LEFT: Lachenalia aloides *by Pierre-Joseph Redouté from* Les Liliacées.

OPPOSITE: Pelargonium acerifolium *(now* Pelargonium cullatum *subsp.* strigifolium*) by Pierre-Joseph Redouté from L'Héritier's* Geranologia.

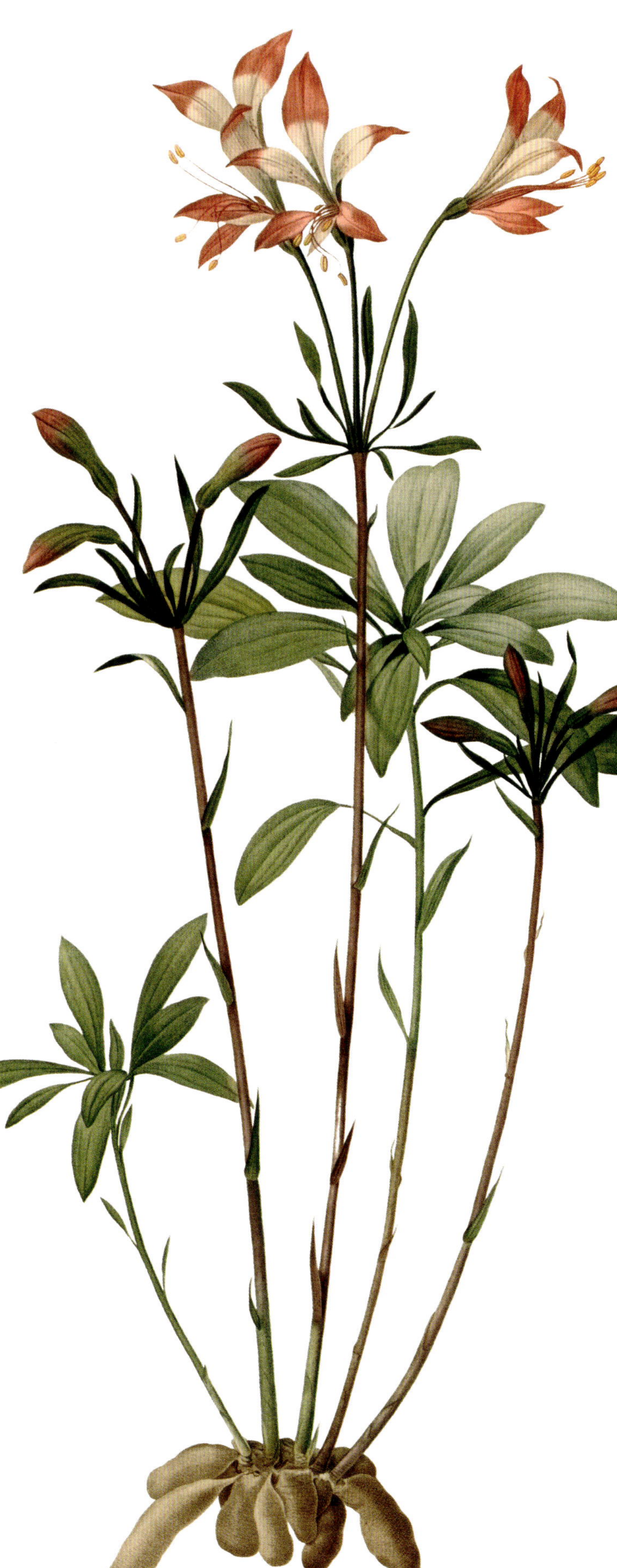

loss felt keenly by Redouté and others who had benefited from his generosity. Work on his planned book, entitled *Geraniologia*, had been interrupted by the revolution and the text was never finished, although 44 engraved plates (31 of which were based on paintings by Redouté) were published in around 1792.

Amid the turbulence, the two Redouté brothers continued to paint, and were both fortunate in being appointed to work on the *vélins*, by now declared state property. Pierre-Joseph drew the illustrations of Algerian and Tunisian plants for Desfontaines' *Flora Atlantica* (1798–99), and continued to draw succulent plants, a project in which he had been encouraged by L'Héritier, eventually producing, with the botanist Augustin de Candolle, *Plantarum succulentarum Historia, ou Histoire naturelle des plantes grasses, avec leurs figures en couleurs* in 28 parts from 1798 to 1805, and a further series with another botanist, J. B. Antoine Guillemin in 1829–37. This book on succulent plants was the first in which Redouté used coloured stipple engraving, which became his speciality, and he engraved and supervised the printing of the plates himself.

The *Plantarum succulentarum Historia* helped to secure Redouté's reputation, and he embarked upon one of his most famous works, eventually amounting to eight volumes, of paintings of bulbous plants, which he entitled *Les Liliacées*, published between 1802 and 1816. Here, the use of stipple gave a very soft effect, ideal for delicate flowers such as irises.

Another of Redouté's projects had been to paint some of the newly introduced plants grown in the garden of J.-M. Cels, a friend of L'Héritier, and also of the botanist Etienne-Pierre Ventenat; the resulting book was published in ten parts between 1800 and 1802. Ventenat had been employed by the Empress Joséphine to study and describe the plants in her garden at Malmaison, and he now introduced Redouté to his employer. This was to be the start of a happy partnership and the most prosperous period in Redouté's life, and it was during these years of salaried employment that Redouté was able to concentrate on producing his most sumptuous works. The first of these was the two-volume *Les Jardins de la Malmaison* (1803–5), with accompanying text by Ventenat, followed by the *Description des plantes rares cultiviées à Malmaison et à Navarre* (1812–17), with text by Aimé Bonpland and 54 particularly fine plates by Redouté.

Thanks to Joséphine, and the success of his work, Redouté was finally able to enjoy a degree of financial security, buying a large

LEFT: Alstroemeria pelegrina *by Pierre-Joseph Redouté from* Les Liliacées.

OPPOSITE: Agapanthus umbellatus *by Pierre-Joseph Redouté from* Les Liliacées.

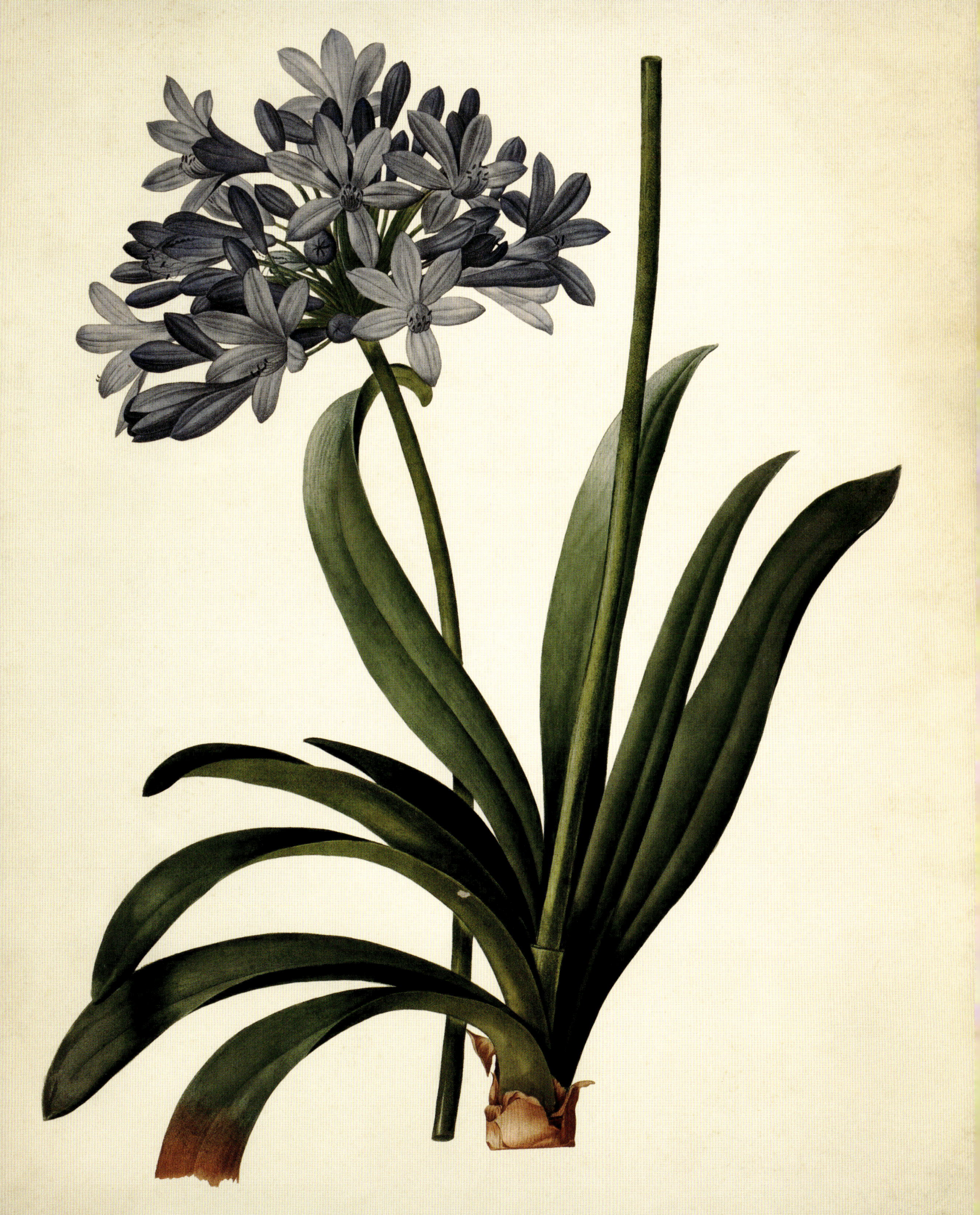

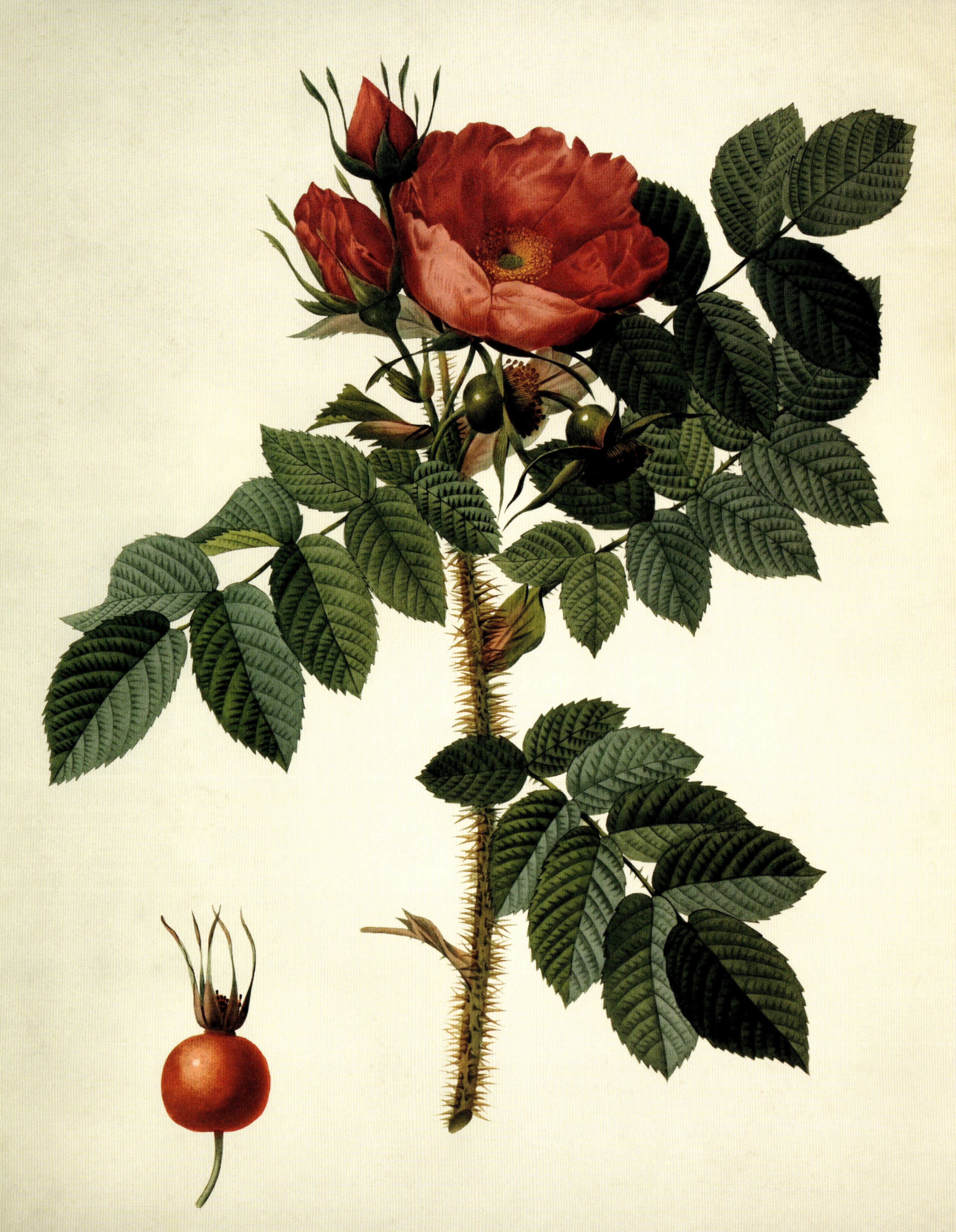

country house and garden for himself, his wife and family, and this appears to have brought him great happiness.

By 1809, Joséphine's marriage to Napoleon had been annulled, and although she retained the estate of Malmaison, she found it politic to move away from Paris, to another chateau, that of Navarre, in Normandy; she continued to employ Redouté until her death in 1814. Earlier in that same year, after a long series of military disasters, including the retreat from Moscow, Napoleon had abdicated, allowing a cautious return to France of exiled French aristocrats, led by Louis XVIII, who ruled as a constitutional monarch for most of the next decade (although interrupted by Napoleon's return from exile in Elba for 100 days in 1815). Redouté, meanwhile, decided to attempt to illustrate as many of the roses grown in France as he possibly could, helped by the fact that Joséphine had amassed an enormous collection in her garden. Even after her death, when much of the estate had been sold by her children to pay her debts, Redouté continued to visit Malmaison, as well as many other gardens in France, in search of roses to paint, a massive project culminating in the publication entitled simply *Les Roses*. This, probably his most famous work of all, contained 169 plates, accompanied by a text written by a friend, the botanist Claude-Antoine Thory and was published in parts between 1817 and 1824. Ironically, given its subsequent success, the first volume was undersubscribed, partly due to lack of the patronage to which he had become accustomed, and throughout the production of the work Redouté struggled to find financial backing. While working on

OPPOSITE: Rosa kamtschatica *by Pierre-Joseph Redouté from* Les Roses.

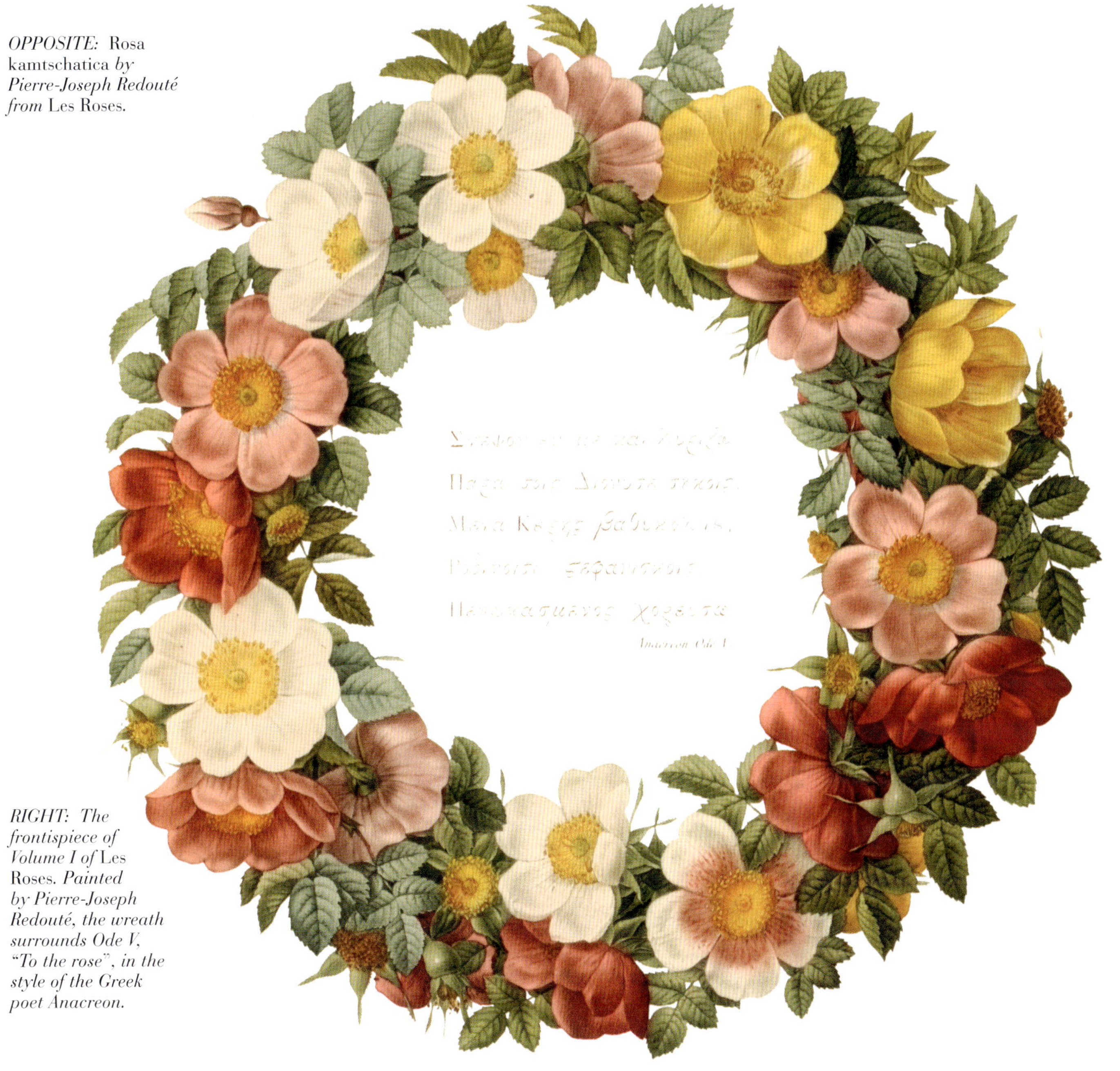

RIGHT: The frontispiece of Volume I of Les Roses. *Painted by Pierre-Joseph Redouté, the wreath surrounds Ode V, "To the rose", in the style of the Greek poet Anacreon.*

RIGHT: "Dahlia double" *by Pierre-Joseph Redouté from* Choix des plus belles fleurs.

OPPOSITE: Various auriculas by Pierre-Joseph Redouté from Choix des plus belles fleurs.

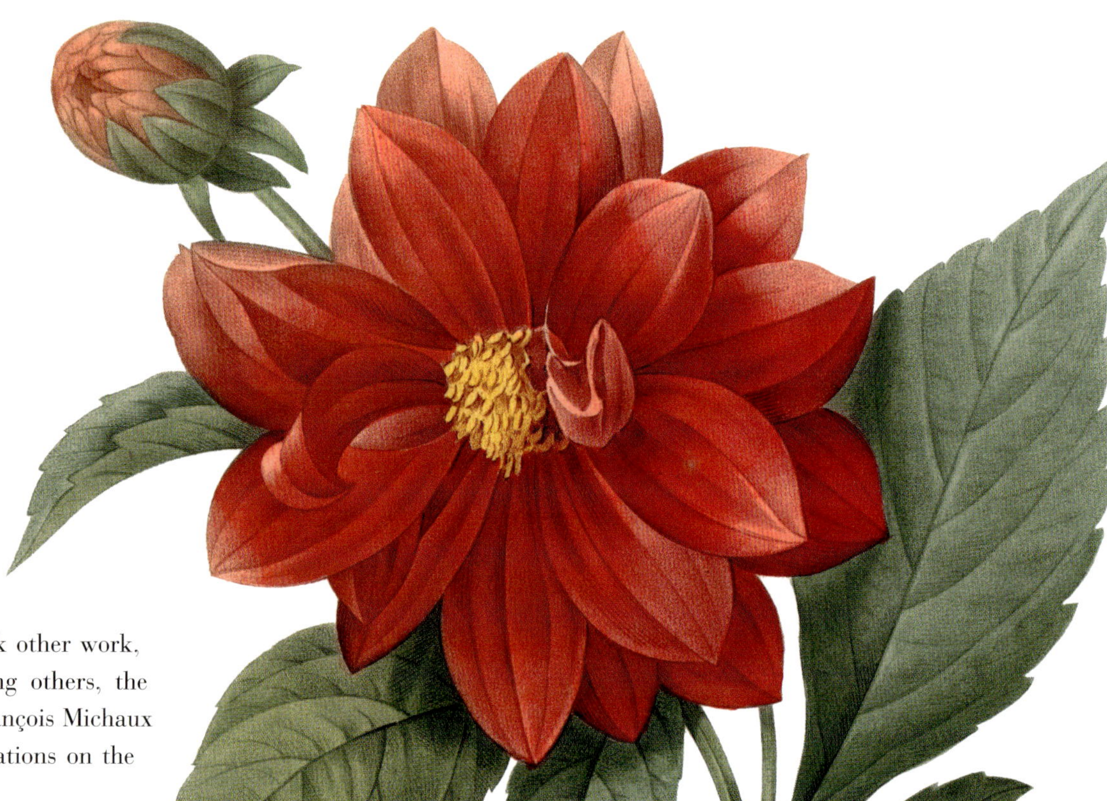

Les Roses, Redouté undertook other work, and collaborated with, among others, the father and son André and François Michaux (*see* page 55) in their publications on the trees of North America.

Life now became more difficult for Redouté. In 1822, his younger daughter Adelaide, a promising flower painter, died, as did his friend and teacher, Gerard van Spaëndonck, who had been probably the greatest artistic influence in his career. Van Spaëndonck's death, however, conveniently resulted in a vacancy at the Jardin des Plantes, and Redouté was appointed as one of two Maîtres de dessin au Muséum d'Histoire naturelle. His work included teaching flower painting to amateurs, as well as the art of botanical illustration to more serious students, and he was also responsible for maintaining, and adding to, the collection of *vélins*.

When Louis XVIII died in 1824, his brother, Charles X, ascended the throne, becoming the last Bourbon King of France and ruling until his abdication in 1830. Charles recognized Redouté's exceptional talent, investing him as a Chevalier of the Légion d'Honneur. Redouté had become friendly with the King's daughter-in-law, the Duchesse de Berry, who was a pupil and supporter of Pancrace Bessa (*see* page 138). In 1824, Redouté dedicated his *Album de Redouté*, a selection of paintings from *Les Roses* and *Les Liliacées*, to her, and this led indirectly to the purchase for her, by her father-in-law, in 1828, of the original watercolours for *Les Roses*.

Redouté, by now financially straitened, also attempted to generate some income by producing a smaller and more popular work between 1827 and 1833, under the title of *Choix des plus belles fleurs*. This contains pictures of popular garden flowers, and is interesting as a record of what was cultivated at the time, and the state of development of some flowers such as pansies and sweet peas.

On the resignation of Charles X in 1830, Louis Philippe, Duc d'Orleans, succeeded to the throne, and by a twist of fate, Redouté's career came full circle with his appointment to the Queen, Marie-Amelie, as Peintre de Fleurs du Cabinet de la Reine, a position he had previously held under Marie-Antoinette. Without this support, Redouté, who seems to have had little business acumen, would almost certainly have become bankrupt. As it was, he continued to paint until the penultimate day of his life, which ended suddenly on 19 June 1840, leaving a truly great legacy of flower paintings.

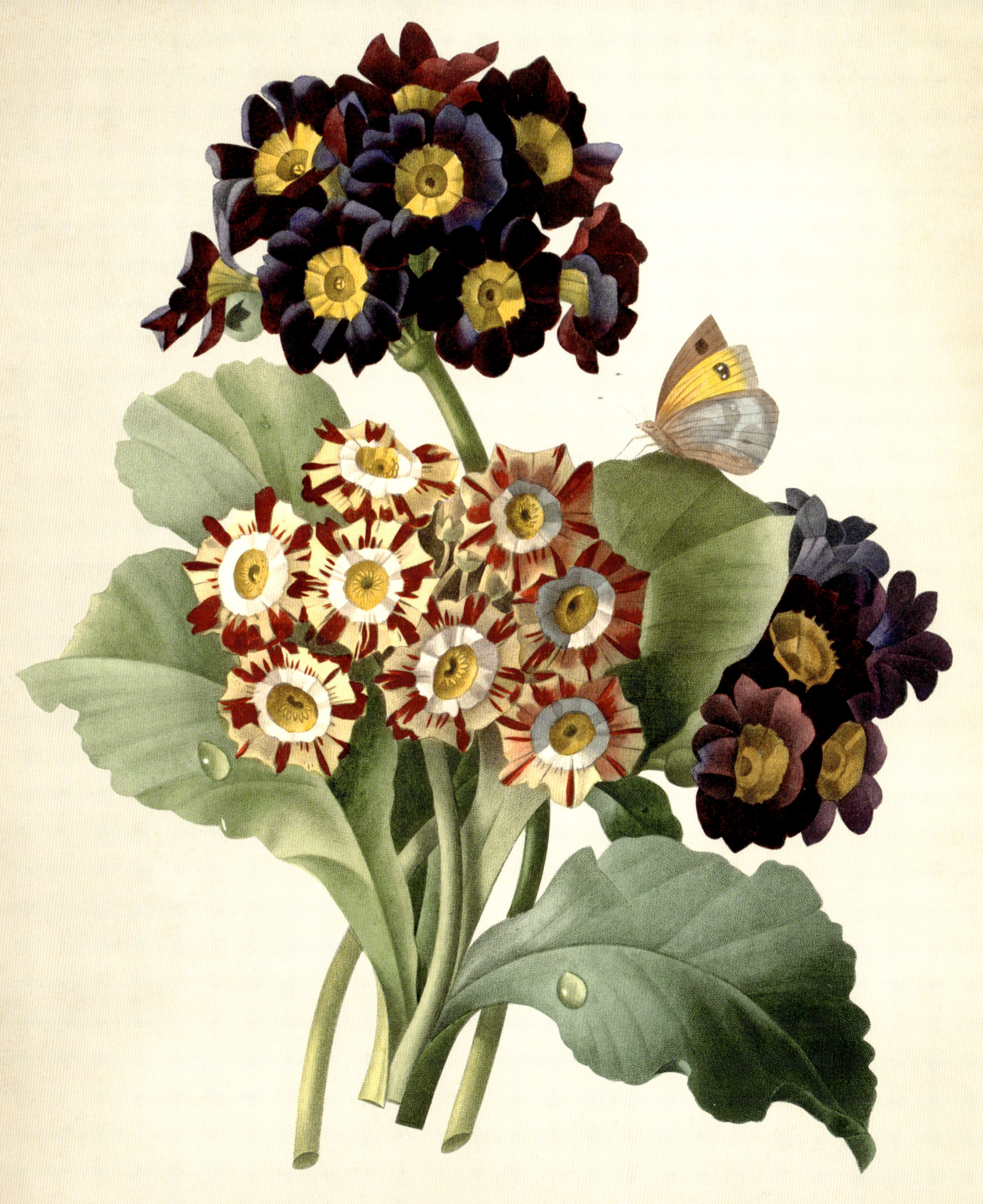

Oreilles d'Ours. *Primula Auricula* Var.

P. J. Redouté Victor

In a moment of serendipity, six great botanical illustrators found themselves working together towards the end of the eighteenth century – van Spaëndonck, the two Redouté brothers, Pancrace Bessa, Pierre Jean François Turpin and Pierre Antoine Poiteau. At the same time, natural historians and explorers such as Alexander von Humboldt and Aimé Bonpland were undertaking exciting expeditions, and the specimens of plants, rocks and animals they collected came flooding into the Paris Museum. In addition to their own work, the artists frequently collaborated in various combinations to illustrate a number of publications documenting the finds – for example Poiteau and Turpin worked together on the drawings for Bonpland's plant volumes of *Le Voyage aux régions équinocxiales du Nouveau Continent* (1805–18), particularly *Monographie de Melastomacées* (1806–23) and, inspired by Linnaeus, *Flora Parisiensis secundum systema sexuale disposita...* (1808).

Pancrace Bessa was another pupil of Gerard van Spaëndonck, working with him at the Jardin du Roi at around the same time as Pierre-Joseph Redouté. Bessa's output was immense, his style influenced by that of both van Spaëndock and Redouté, and he collaborated on several works with the latter. Books illustrated by Bessa and Redouté included *Choix de Plantes: dont la plupart sont cultivées dans le jardin de Cels* by Etienne Pierre Ventenat (published in Paris between 1803 and 1808), the second edition of Henri-Louis Duhamel du Monceau's *Traité des Arbres et Arbustes que l'on cultive en France en plain terre* (Paris, 1800–19) and François and André Michaux's *Histoire des arbres Forestiers de l'Amérique septrentionale* (Paris 1810–13), which described the majority of the trees native of eastern North America.

Some of the plants cultivated in the gardens of Redouté's patron, the Empress Joséphine, were described in Aimé Bonpland's *Description des plantes rares cultivées à Malmaison et à Navarre* (Paris, 1812–17), with a later publishing spin-off of 12 of Bessa's plates under the title *Almanach de flore...* being produced in 1817. The first published work solely by Bessa was *Fleurs et fruits...* (1808) with 24 fine folio stipple engravings, a technique involving etching by dots rather than lines, which had been developed in France during the eighteenth century, and enabled finer gradations of tone to be produced. He also illustrated other books on fruit, a subject he appears to have liked, including *Le Jardin fruitier* (1813, with later editions to 1839) by Louis Claude Noisette (and Etienne Michel's *Traité du Citronier* (1816).

While Redouté enjoyed the patronage of a powerful woman in Joséphine, Bessa was fortunate in having the encouragement of another – Marie Caroline Ferdinande Louise, Duchesse de Berry, and daughter-in-law of Charles X. The Duchesse was one of Bessa's pupils, and the 572 original watercolours on vellum for *Herbier général de l'amateur* (1810–27) were presented to his daughter-in-law by Charles in 1826. These paintings had been engraved and hand coloured for the *Herbier*, which was produced in eight volumes.

Bessa had been appointed teacher of flower painting at the Jardin des Plantes in 1823, and in addition to more serious botanical work, he illustrated several charming small flower books written for amateurs; he also exhibited at the Paris salons from 1806 onwards. Other work included the preparation of drawings for the Commission of Science that accompanied Napoléon on his Egyptian campaign, as well as illustrations of grasses from specimens collected in the Antipodes for Louis-Isidore Duperrey's *Voyage autour du monde* (Paris 1826–29).

In 1832, Bessa's patroness the Duchesse de Berry went into exile for a second time (the first had been during the July Revolution in 1830), and at this point Bessa left Paris and retired to nearby Écouen, where he died in 1846.

LEFT: Chinese aster, Callistephus chinensis, *by Pancrace Bessa.*

OPPOSITE: Swainsona galegifolia, *an Australian pea, whose flowers may be pink, white, red or yellow, by Pancrace Bessa.*

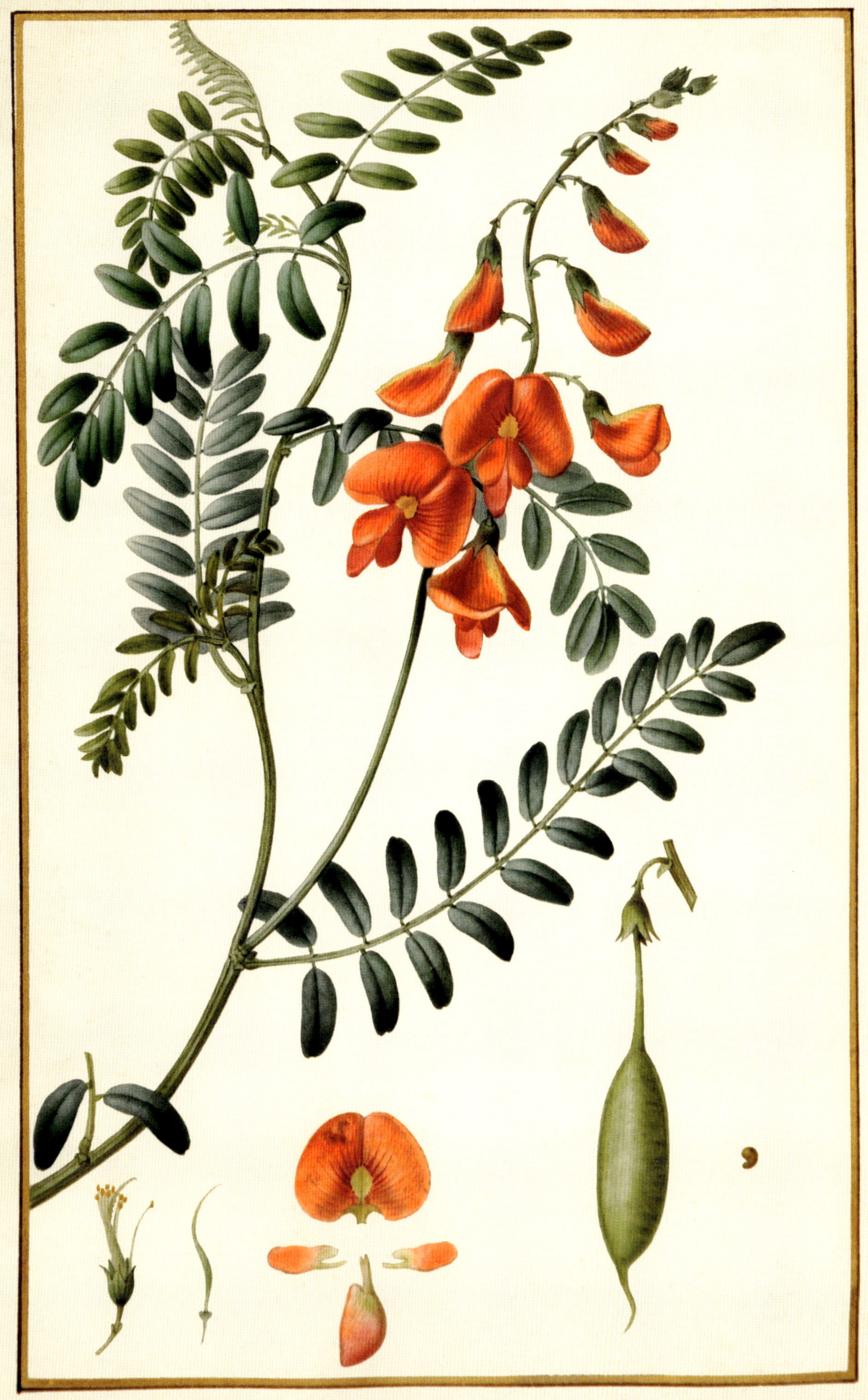

P. Bessa.

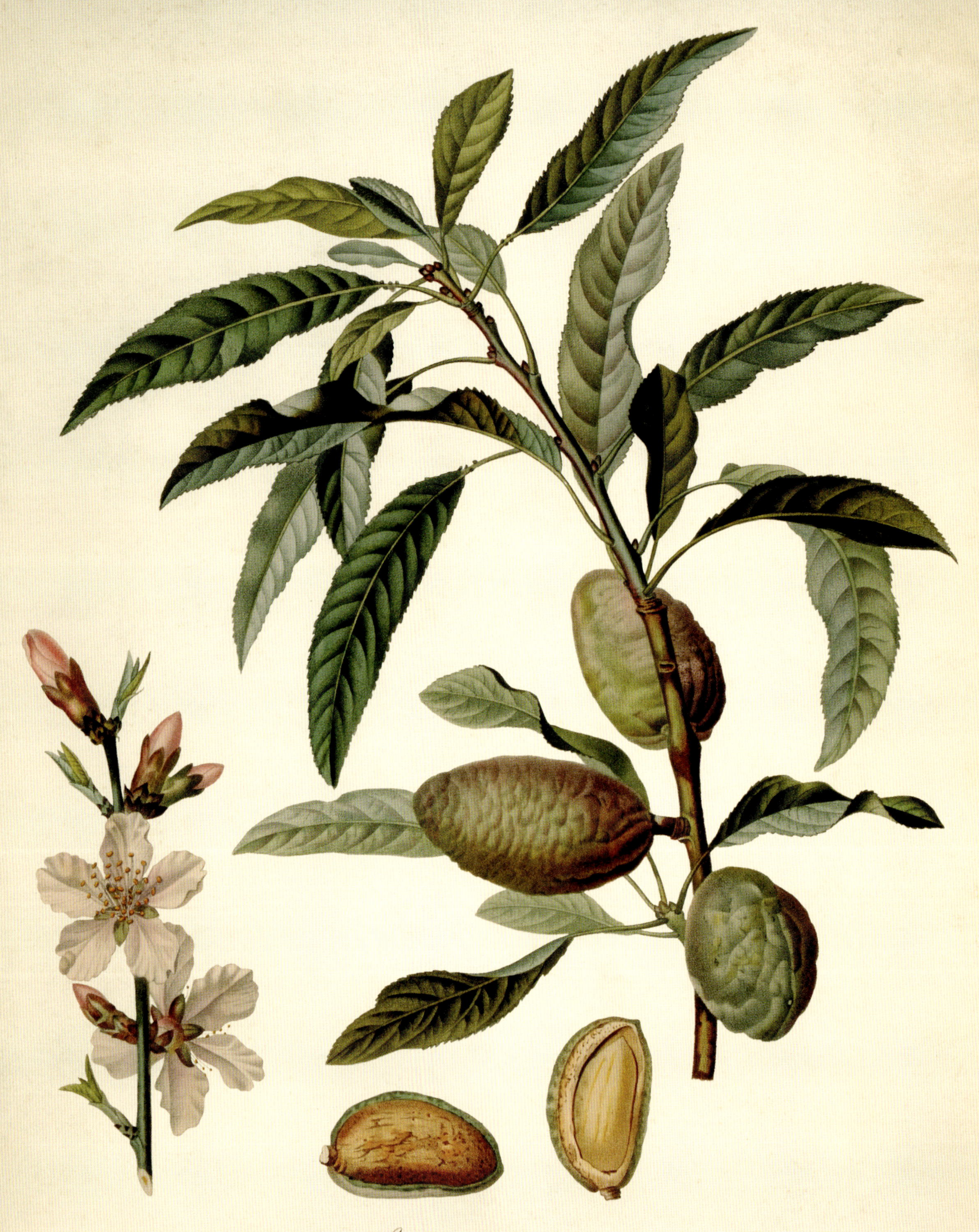

Amande sultane. 325.

De l'Imprimerie de Langlois *Bocourt sculp[t]*

Pierre Antoine Poiteau was a botanist and gardener, who, from 1790 onwards worked on and off at the Muséum d'Histoire naturelle in Paris. Here he came under the influence of both Gerard van Spaëndonck and Pierre-Joseph Redouté. In 1796, he travelled to the West Indies, where he met Pierre Jean François Turpin, returning to France in 1802 with an enormous collection of seeds, which he subsequently described and named.

Pierre Jean François Turpin (1775–1840) was born in Normandy, into a poor family, but rose to become one of the foremost botanical painters of his day. At the age of 14 he joined the army and was sent to San Domingo, where he met Poiteau, and later von Humboldt. These chance meetings were to prove the fortunate catalyst for an outpouring of work by all three men.

In 1808, Poiteau and Turpin worked on drawings for Augustin de Candolle's *Icones plantarum Galliae rariorum* and *Flore*

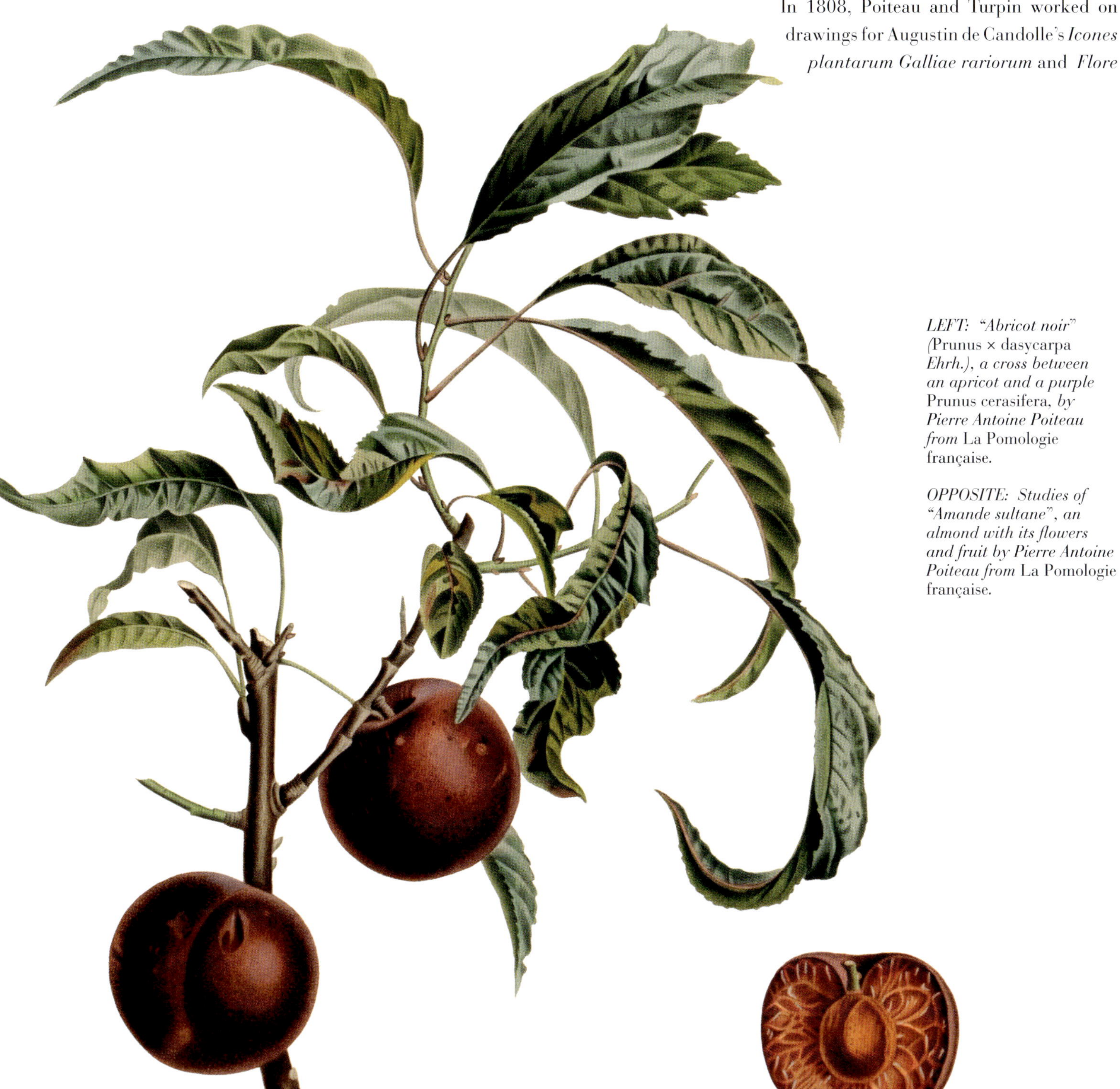

LEFT: "Abricot noir" (Prunus × dasycarpa *Ehrh.), a cross between an apricot and a purple* Prunus cerasifera, *by Pierre Antoine Poiteau from* La Pomologie française.

OPPOSITE: Studies of "Amande sultane", an almond with its flowers and fruit by Pierre Antoine Poiteau from La Pomologie française.

parisienne, contenant la description des plantes qui croissent naturellement aux environs de Paris, which was published in eight parts, between 1808–1813, but never completed. In 1815, Poiteau worked at Versailles, with its famous collection of citrus trees, and in 1818 he illustrated the *Histoire naturelle des orangers*, dedicated to the ubiquitous Duchesse du Berry, with text by the naturalist Antoine Risso. Later that year, he travelled to French Guiana, and on his return in 1822 he moved to work at the chateau of Fontainebleau.

Meanwhile, between 1808 and 1835 Turpin and Poiteau had collaborated on another project, this time in the form of a fine, new, and greatly enlarged, edition of Duhamel du Monceau's *Traité des arbres fruitiers*, Poiteau following this with another book on fruit, *La Pomologie française*, in 1846, and in 1848 and 1853 two volumes of a practical work on horticulture entitled *Cours d'horticulture*. The collaboration between these two fine artists ended with Turpin's death in 1840, the same year in which Redouté died.

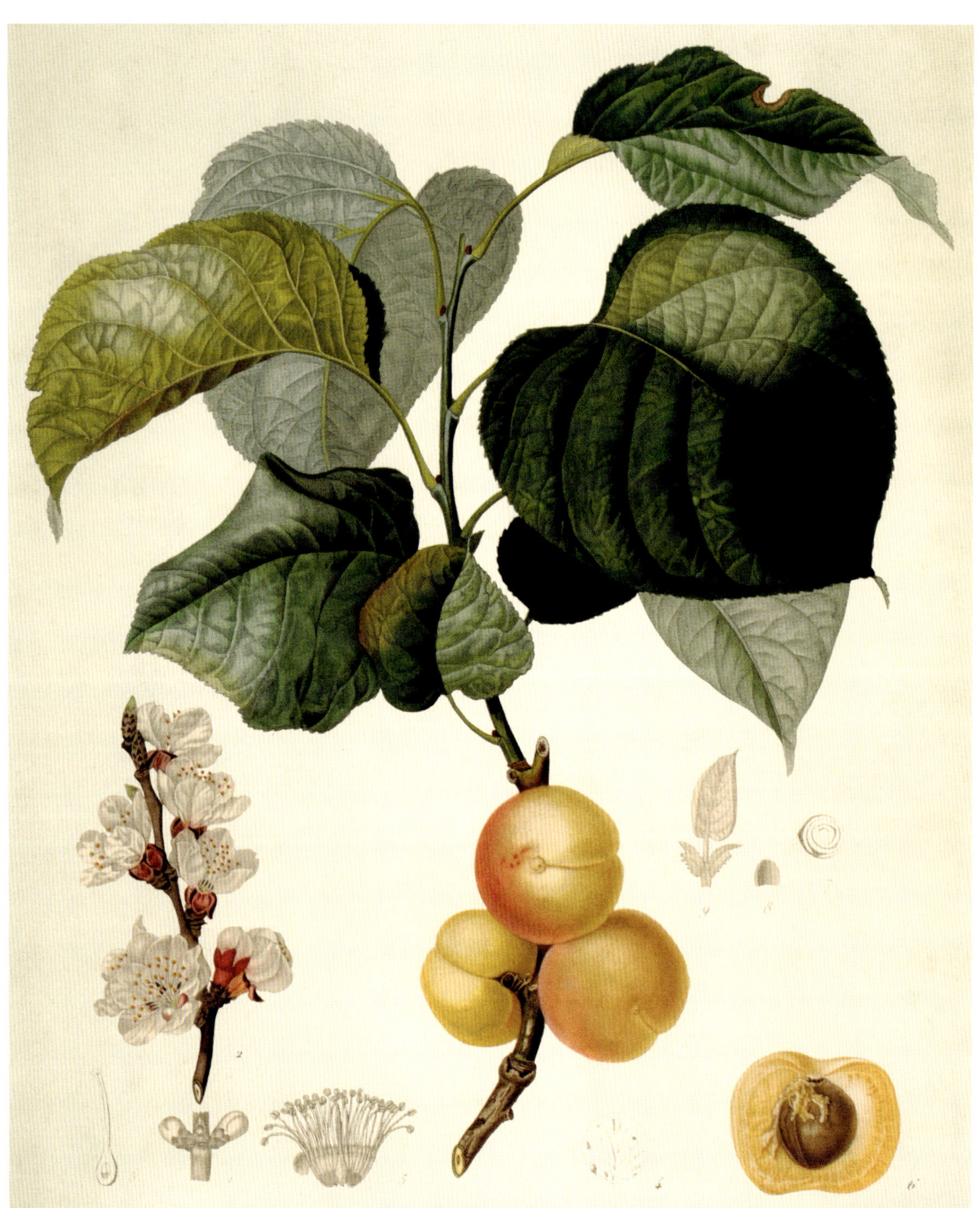

LEFT: Studies of an apricot, its flowers and fruit by Pierre Antoine Poiteau from La Pomologie française.

OPPOSITE: A study of tulips made in 1839 by Félice Cronier who was heavily influenced by Redouté's style.

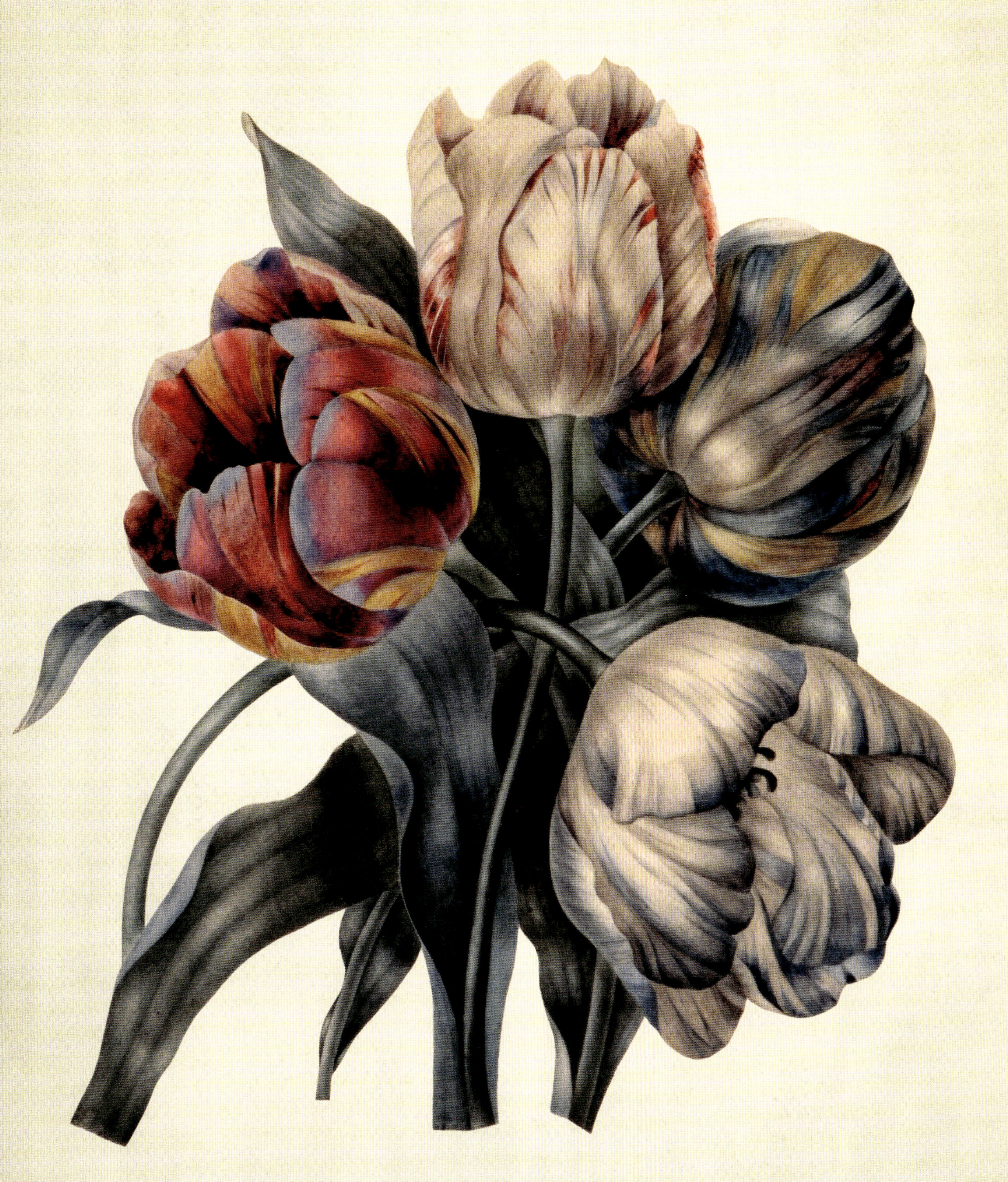

Empress Joséphine

The Empress Joséphine (1763–1814), born Marie Rose Joséphine Tascher de la Pagerie, was the daughter of a sugar-planter on the island of Martinique in the West Indies. At the age of 16, she travelled to France, where she married the Vicomte Alexandre François Marie de Beauharnais, with whom she had two children, Eugène and Hortense. The couple parted in 1785, but Joséphine, because of her association with the nobility, was imprisoned alongside her ex-husband for several months during the Terror in 1793.

BELOW: Ixora chinensis *by Pierre-Joseph Redouté from* Description des plantes rares cultivées à Malmaison et à Navarre.

Despite his republican sympathies, Beauharnais was guillotined, and Joséphine narrowly escaped the same fate thanks only to the fall of Robespierre. The following year, Joséphine, by then the mistress of a leading republican, Paul Barras, met the young Napoleon, who immediately fell in love with her, and they married in 1796, on the eve of Napoleon's Italian campaign.

By 1799, Napoleon had overrun Italy, invaded Egypt, and become the powerful First Consul of France, and Joséphine had acquired the run-down estate of Malmaison just to the west of Paris. While Napoleon plotted, Joséphine gardened extravagantly, turning the garden at Malmaison into a fashionable *"jardin paysager"* and building a hothouse for the rare and exotic plants which reminded her of her youth. In 1804, Napoleon became Emperor of the French, and Joséphine his Empress; money was lavished on the garden and estate at Malmaison. Joséphine genuinely loved flowers, and corresponded with nurserymen both in France and England, as well as with plant collectors such as Sir Joseph Banks.

In order to document the plants grown at Malmaison, Joséphine employed the botanist Etienne-Pierre Ventenat (1757–1808), and he in turn introduced Pierre-Joseph Redouté to her. Between them, the two produced a folio work entitled *Les Jardins de la Malmaison*, published in 20 parts between 1803 and 1805, with descriptions by Ventenat and 120 plates painted by Redouté. The book included an illustration of a rather obscure plant, *Josephinia imperatricis*, named after his patroness by Ventenat in 1804; another plant, *Lapageria*, after Joséphine's maiden name, had been published by the botanists Ruiz and Pavon two years previously.

Joséphine was a generous patron, allowing Redouté the chance to develop his considerable talent; in addition to another illustrated book on the garden, *Description des plantes rares cultivées à Malmaison et à Navarre* (1813) this time in partnership with the botanist Aimé Bonpland,

ABOVE: Liparia sphaerica *(now* Liparia splendens*) from South Arica, by Pierre-Joseph Redouté from* Description des plantes rares cultivées à Malmaison et à Navarre.

OPPOSITE: Rosa redutea glauca, *which was named after Redouté, and painted by the artist for* Les Roses.

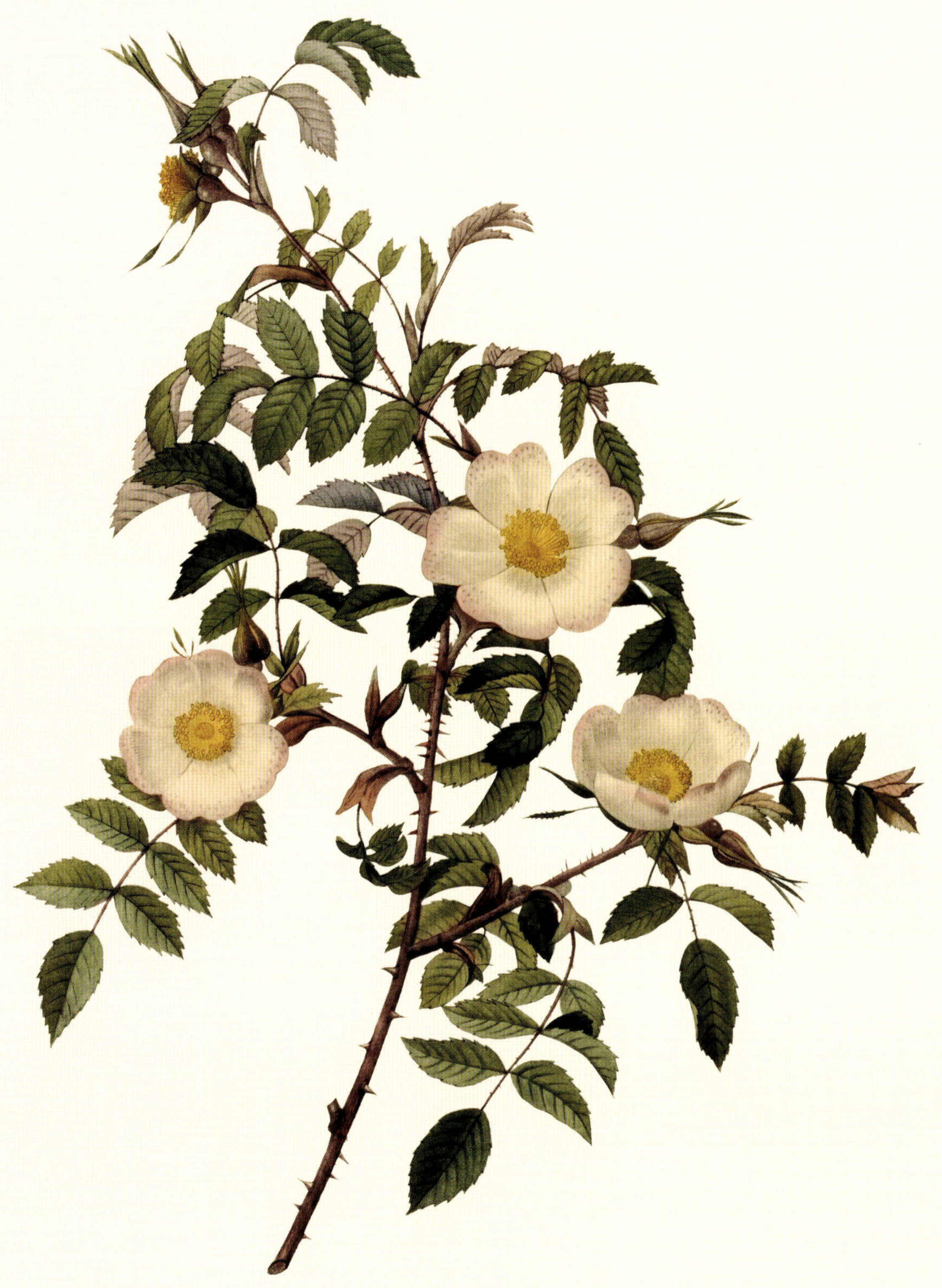

Redouté produced the plates for the eight volumes of *Les Liliacées* (1802–-16) and *Les Roses* (1817–24).

Meanwhile, due to her inability to produce a child, Joséphine's marriage to Napoléon had been annulled in 1809, and she moved further away from Paris, her ex-husband and his new wife (who had borne him a son) to the chateau and estate of Navarre. Here, in receipt of a generous allowance, she continued to garden, and to employ Redouté, but she was not to enjoy her independence for long, dying on 29 May 1814, aged only 51. After her death, the house and gardens were sold by her children to pay off her debts.

11 Botanical and Horticultural Illustrated Journals

CURTIS'S BOTANICAL MAGAZINE

From the end of the eighteenth century onwards, partly in response to the advances being made in scientific knowledge, and partly due to the flood of new plants coming in from abroad, there was an increase in the number of specialized societies in Britain; the Linnean Society was founded in 1788 and the Horticultural Society (which was to become the Royal Horticultural Society) was founded in 1804.

These societies published books, journals and occasional papers on a variety of subjects, and were chiefly aimed at botanists, gardeners and nurserymen. There was also, however, a rise in the number of private individuals who enjoyed growing the "new" plants, which often came into the established Botanic Gardens at Oxford, Cambridge, Chelsea and Kew, but were then distributed either privately or commercially. These keen, intelligent amateurs craved knowledge of the origins (and therefore cultivation requirements) of their new plants, and this desire was satisfied by a wealth of illustrated books, often issued in instalments, and periodicals such as *Curtis's Botanical Magazine*. There are many instances of failure to complete a proposed series of books published in parts (for example *Flora Londinensis*, *see* below) and also a significant number of short-lived journals, but *Curtis's Botanical Magazine* is the exception and still flourishes today, 225 years after its inception.

William Curtis (1746–99), founder of the eponymous magazine, was born on 11 January 1746 in the market town of Alton, Hampshire, where his father was a tanner. Curtis showed a precocious interest in natural history, and spent much time reading herbals and books on the subject. Thanks to his grandfather, a surgeon-apothecary, Curtis served apprenticeships with two apothecaries in London, before studying anatomy at St Thomas's Hospital and assisting in practical botanical demonstrations there.

After qualifying as an apothecary, Curtis practised medicine in London for some years, but was always more interested in natural history, and through this became friendly with the brothers of the naturalist the Reverend Gilbert White, of Selborne. In 1771, the White brothers helped Curtis realize his scheme of establishing a garden of native British plants on a plot of land in Lambeth, and in the same year he published a pamphlet containing instructions for collecting insects, bravely sending a copy to Joseph Banks (*see* pages 98–99) who replied with an encouraging letter. The following year, Curtis produced another, larger, work on the same subject, and when the post of Demonstrator of Botany at the Chelsea Physic Garden became vacant he was recommended to it by the preceding Demonstrator and appointed in 1773.

Curtis stayed at the Physic Garden for four years, but appears to have neglected his duties somewhat, being preoccupied with the production of his *Flora Londinensis*, and he left the garden in 1777, in order to pursue his writing and publishing interests.

The *Flora* was initially conceived as part of a far larger and more comprehensive work, covering the whole country, but, largely due to a lack of funds, only the first section was published, covering all the plants grown within 16 kilometres (ten miles) of London. Curtis was determined to produce a work of the highest standard, and commissioned the best artists and engravers

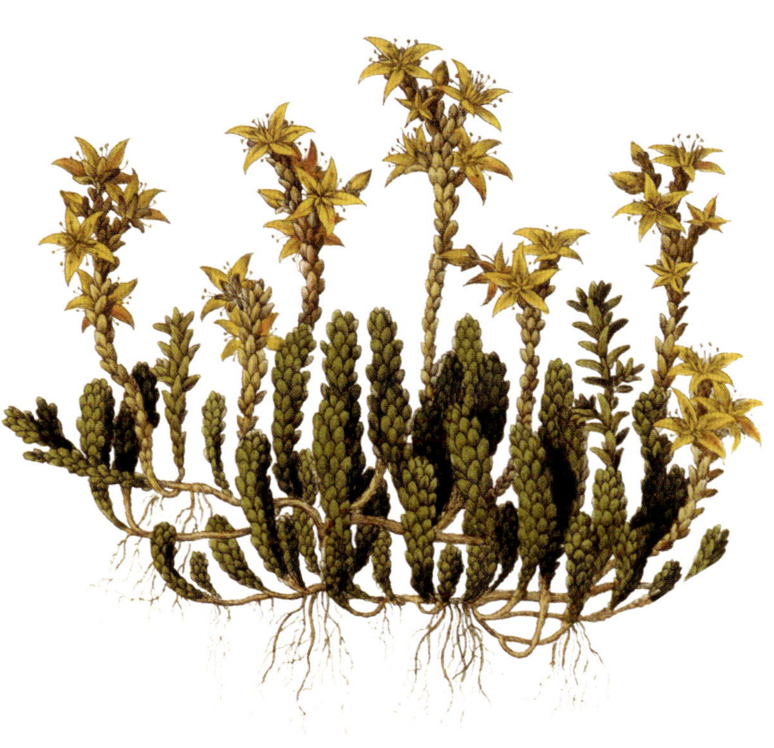

that he could afford to draw life-size illustrations of plants. A total of 72 parts, each containing six plates by artists such as William Kilburn, James Sowerby and Sydenham Teast Edwards, were published between 1775 and 1798 and were sold at two prices, uncoloured or hand coloured. By 1781, Curtis, who had given up his work as an apothecary, was running short of funds, and was fortunate in securing the generous patronage of Lord Bute, former adviser to Princess Augusta at Kew. Curtis gratefully dedicated the first volume to Bute, and the second to a friend, Dr John Lettsom, who likewise gave him a non-repayable loan. For various reasons, including the rather unreliable dates of publication of the various parts which irritated the subscribers, Curtis had to abandon his *Flora* in 1798; ironically, thanks to its high standards of production, it is still one of the finest English illustrated floras, and stands as a magnificent memorial to Curtis and his artists.

Despite the difficulties encountered with *Flora Londinensis*, Curtis kept up work on his private botanic garden at Lambeth, and issued a catalogue of the many plants grown there. He managed to attract a small group of loyal subscribers who, in addition to entry to the garden, were given the opportunity to acquire seeds, to have the run of a small library and to hear Curtis's course of lectures on botany; the text of these was later published in *A Companion to the Botanical Magazine* (1788–89). In 1789, Curtis moved his garden to a more favourable site covering 10 acres in West London, and the Brompton Botanic Garden was opened to subscribers and the public.

To try to recoup his losses on *Flora Londinensis*, Curtis launched the journal which was to preserve his name to the present day, *Curtis's Botanical Magazine*. He had been told by friends and

FAR LEFT: Sedum acre *from Volume I of* Flora Londiniensis.

LEFT: Orchus fusca, *now* Orchus purpurea, *the lady orchis, from Volume II of* Flora Londiniensis.

OPPOSITE: Cardamine pratensis *from Volume I of* Flora Londiniensis.

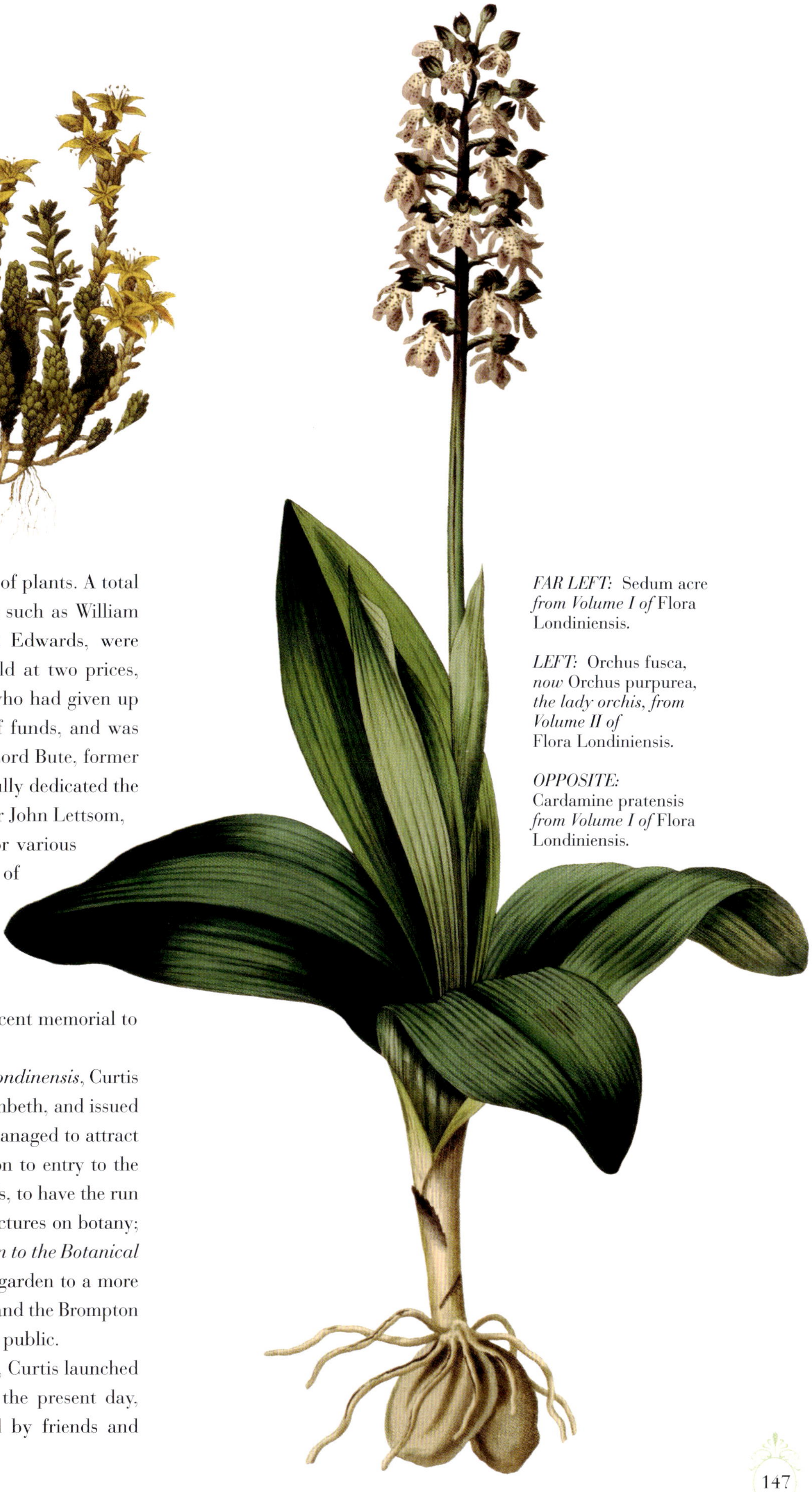

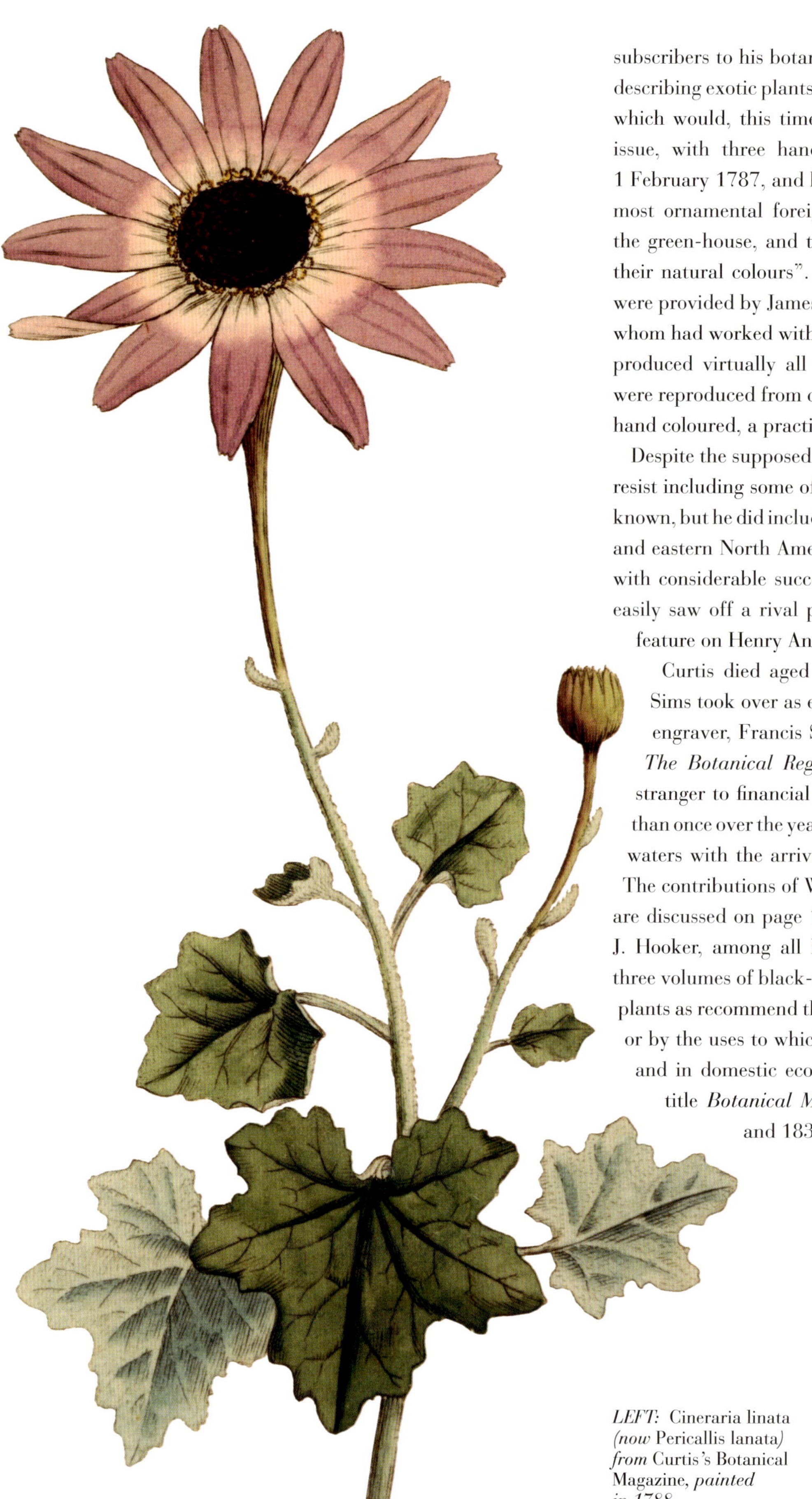

subscribers to his botanic garden that there was a need for a work describing exotic plants, and he decided to risk another publication, which would, this time, appear promptly every month. The first issue, with three hand-coloured engravings, was published on 1 February 1787, and boasted of contents that would include "the most ornamental foreign plants, cultivated in the open ground, the green-house, and the stove, will be accurately represented in their natural colours". The illustrations for the first five volumes were provided by James Sowerby and Sydenham Edwards, both of whom had worked with Curtis on *Flora Londinensis*, and Edwards produced virtually all other illustrations until 1815. The plates were reproduced from copper engravings of the paintings and were hand coloured, a practice that continued until 1948.

Despite the supposed emphasis on exotic plants, Curtis could not resist including some of his favourite garden flowers, already well-known, but he did include many new plants from the Cape, Australia and eastern North America, too. Fortunately, this publication met with considerable success, both critical and financial, and Curtis easily saw off a rival publication, *The Botanist's Repository* (*see* feature on Henry Andrews, pages 156–57).

Curtis died aged 53 in July 1799, and his friend Dr John Sims took over as editor in 1801. Sydenham Edwards and the engraver, Francis Sansom, started a rival monthly magazine, *The Botanical Register* in 1815. Curtis's magazine was no stranger to financial crises, being threatened with closure more than once over the years, but eventually it sailed into rather calmer waters with the arrival of William J. Hooker as editor in 1826. The contributions of William and his son Joseph, to the magazine are discussed on page 150, but it is worth recording here that W. J. Hooker, among all his other projects, found time to produce three volumes of black-and-white "figures and descriptions of such plants as recommend themselves by their novelty, rarity, or history, or by the uses to which they are applied in the arts, in medicine, and in domestic economy…"; these were published under the title *Botanical Miscellany* by John Murray, between 1830 and 1833.

LEFT: Cineraria linata *(now* Pericallis lanata*) from* Curtis's Botanical Magazine, *painted in 1788.*

OPPOSITE: Passiflora laurifolia *drawn and painted by Sydenham Edwards for* The Botanical Register.

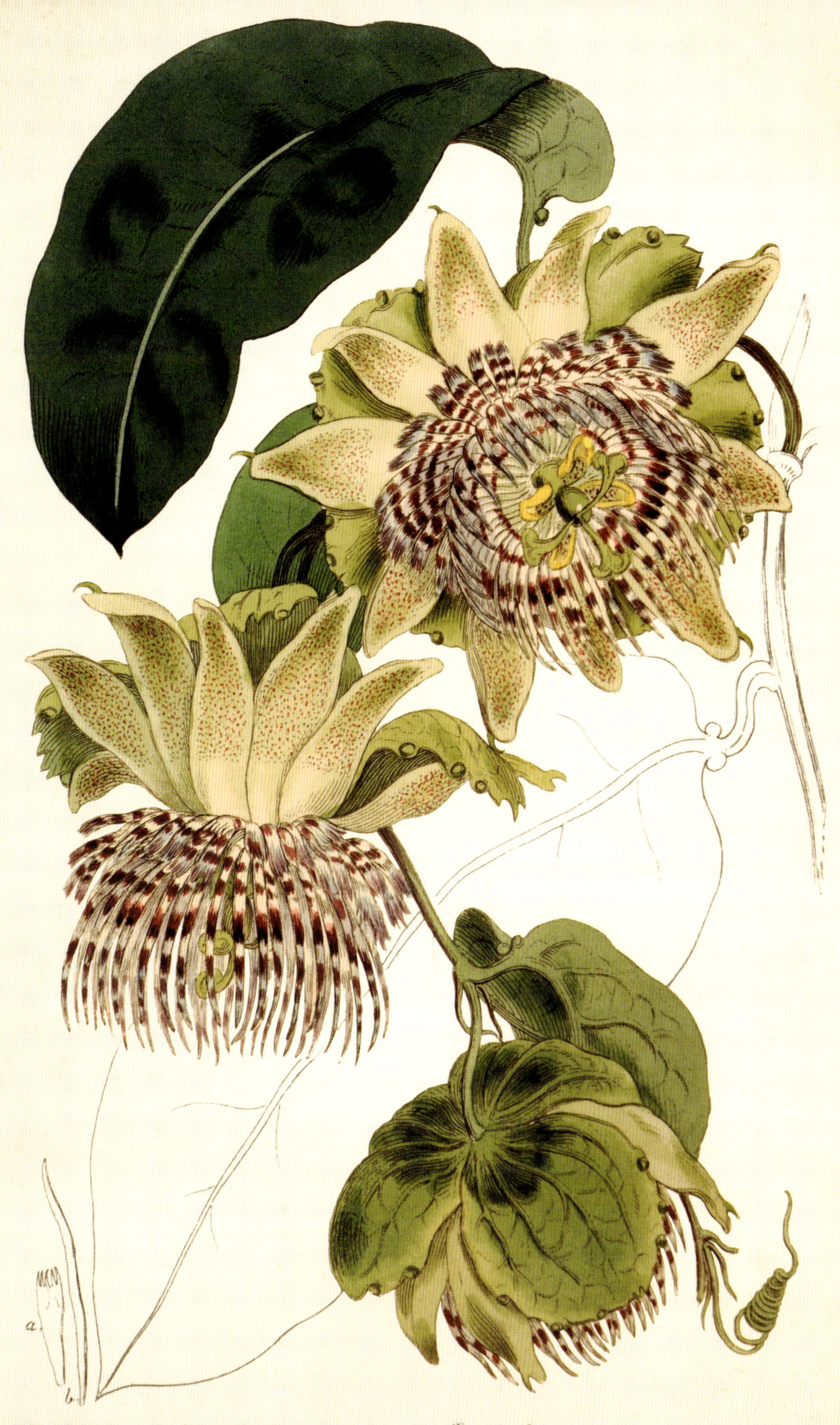

Sydn. Edwards Del. *Pub. by J Ridgway. 170 Piccadilly. April 1 1815.*

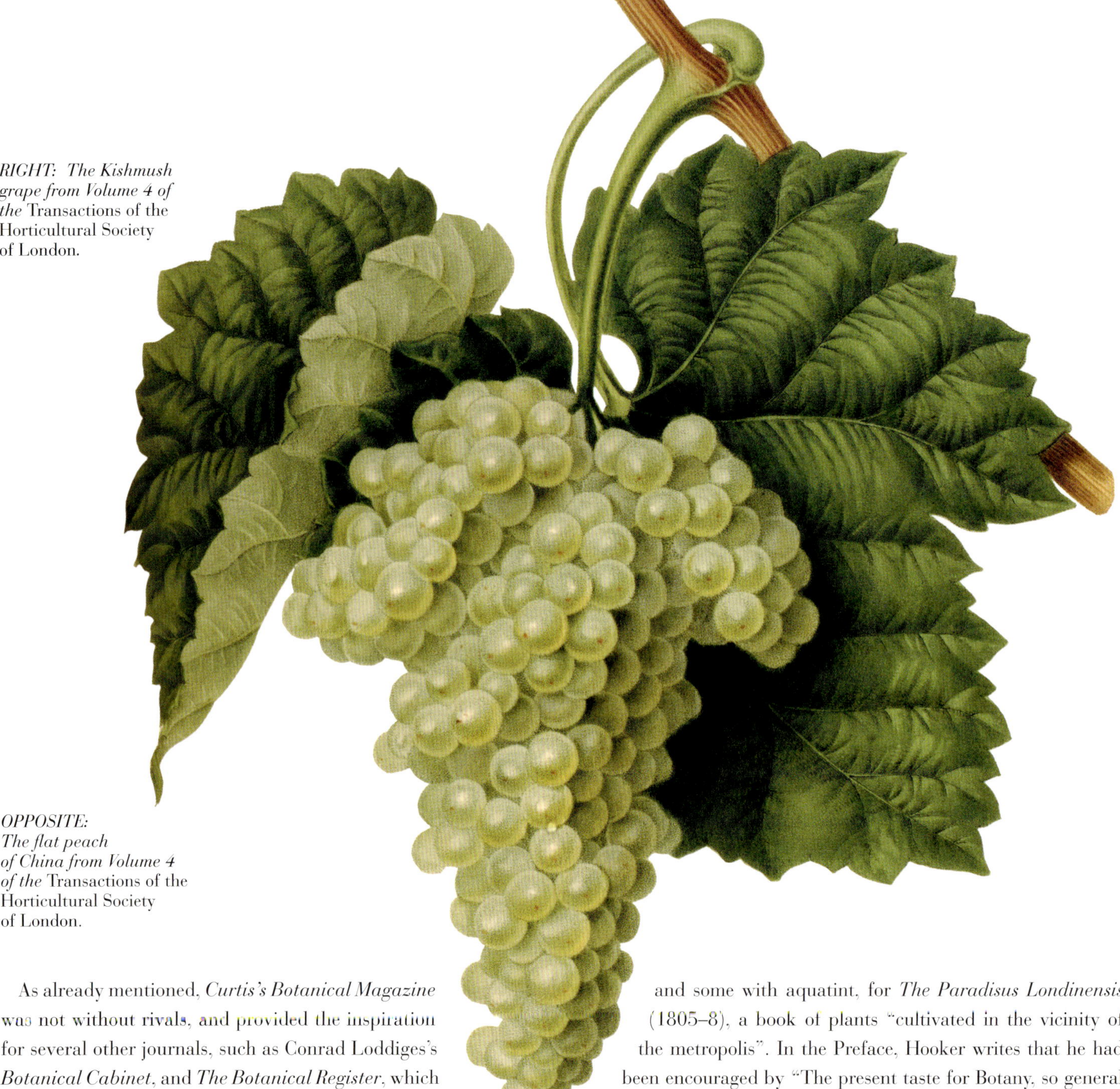

RIGHT: The Kishmush grape from Volume 4 of the Transactions of the Horticultural Society of London.

OPPOSITE: The flat peach of China from Volume 4 of the Transactions of the Horticultural Society of London.

As already mentioned, *Curtis's Botanical Magazine* was not without rivals, and provided the inspiration for several other journals, such as Conrad Loddiges's *Botanical Cabinet*, and *The Botanical Register*, which Edwards and Sansom had founded in 1815. John Lindley (*see* page 107) took over as editor of the latter in 1829 and employed Sarah Drake as his chief illustrator until it ceased publication in 1847. By this time Lindley had become involved in founding and editing *The Gardeners' Chronicle* and also edited the *Transactions of the Horticultural Society of London*, which was published in ten volumes, in two series, between 1805 and 1848. A feature of the engravings illustrating this work was that, after about 1820 they were done on steel, rather than copper. Artists involved with this project included Sarah Drake and Mrs Augusta Withers, and the little known Barbara Cotton; many of the plates, particularly those featuring fruits, in the earlier volumes were drawn by William Hooker (1779–1832), not to be confused with his contemporary, William Jackson Hooker.

William Hooker had been a pupil of Francis Bauer (*see* page 104) and drew 119 plates, some of which were coloured by hand, and some with aquatint, for *The Paradisus Londinensis* (1805–8), a book of plants "cultivated in the vicinity of the metropolis". In the Preface, Hooker writes that he had been encouraged by "The present taste for Botany, so general among all ranks..." to produce a book illustrating plants which are "... new, uncommonly beautiful, or incompletely figured by others..."; the accompanying text was written by botanist R. A. Salisbury, who was one of the founders of the Horticultural Society. In addition to his paintings of fruit for the Horticultural Society, Hooker also illustrated T. A. Knight's *Pomona Herefordiensis* (1811) and *Pomona Londinensis* (c.1816–18).

Another journal seeking to gain a share of the market was Benjamin Maund's *Botanic Garden*, which set out to depict "hardy ornamental flowering plants, cultivated in Great Britain...". This was published in monthly parts, each part consisting of a coloured plate illustrating four flowers, with four pages of accompanying descriptive text. Benjamin Maund (1790–1863) was a pharmacist, botanist and bookseller who had his own press at Bromsgrove in Worcestershire and printed the journal himself. The *Botanic Garden*, which was published from 1825 to 1850, was illustrated

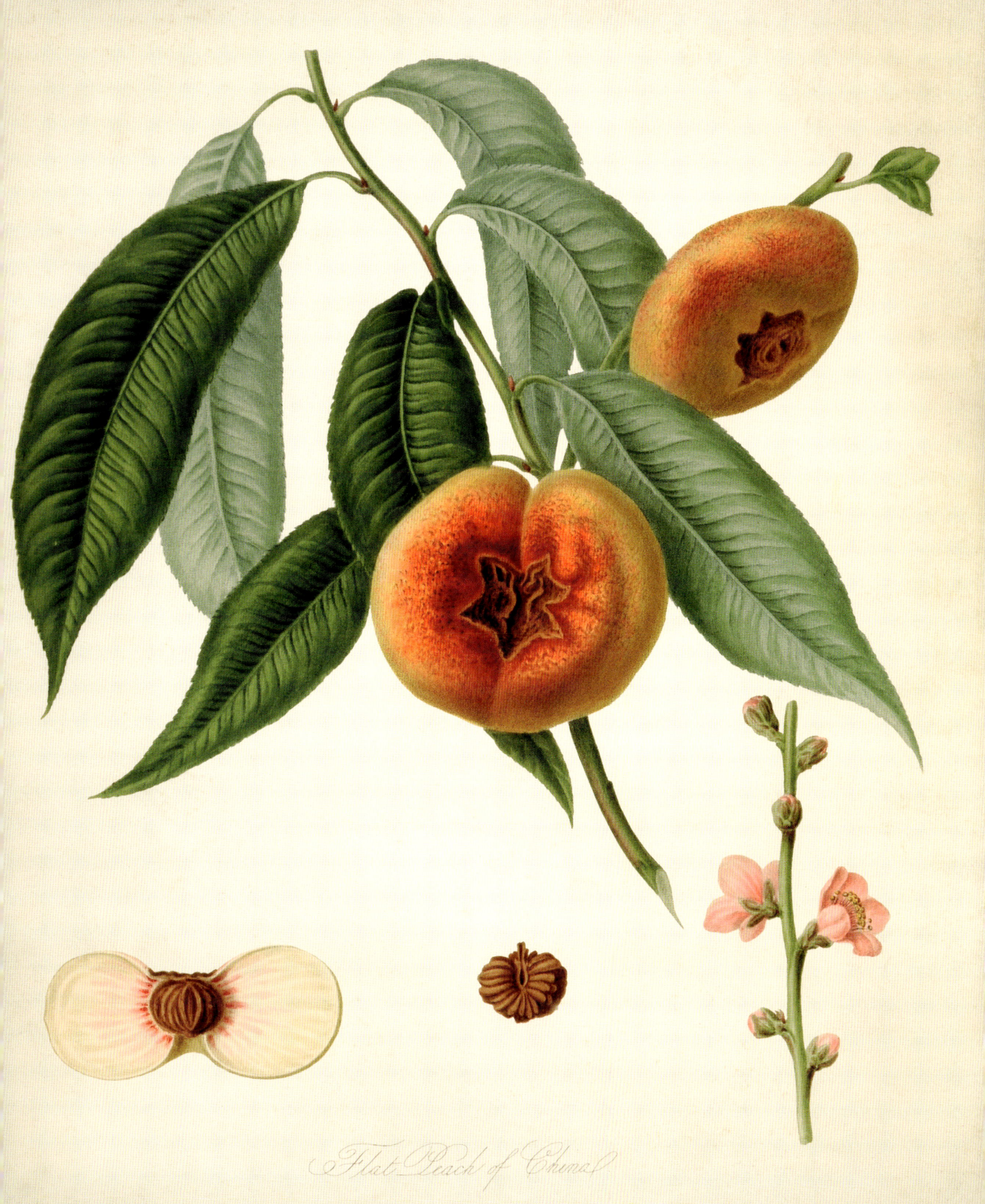
Flat Peach of China

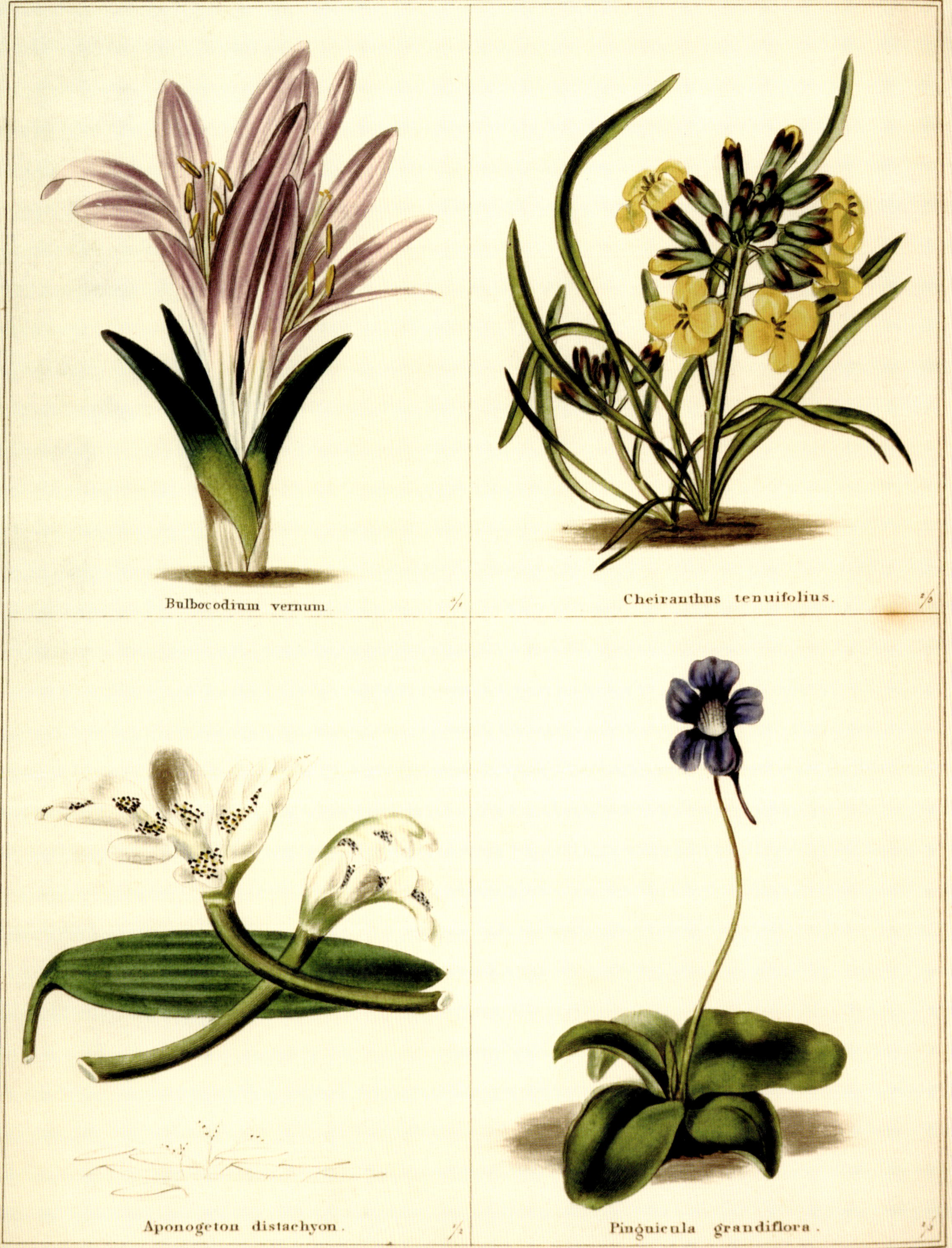

E. D. Smith delt. S. Watts sculp.

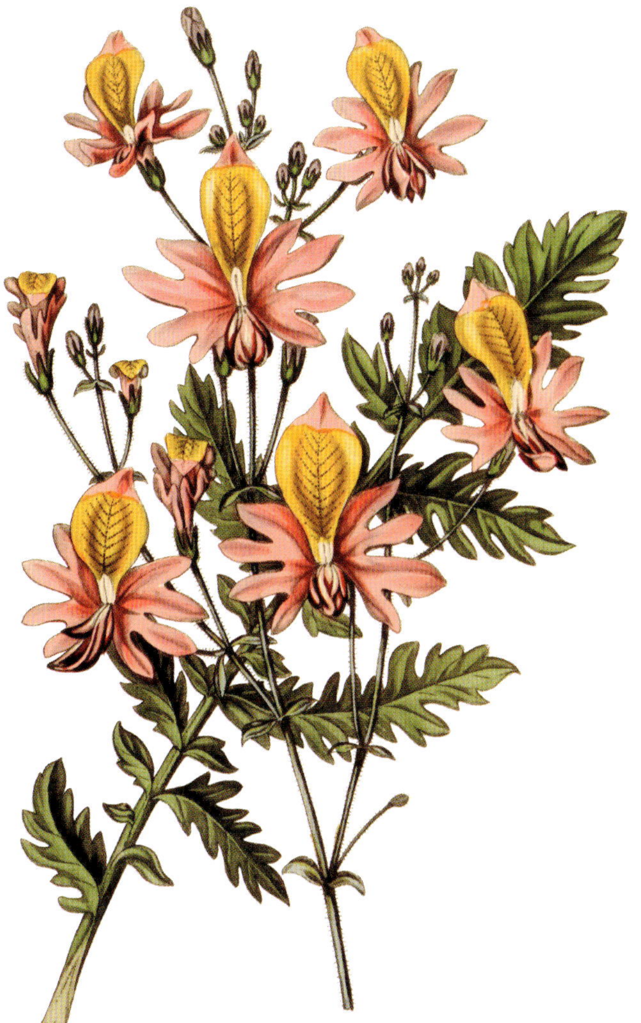

LEFT: Schizanthus retusus *from the* Magazine of Botany.

RIGHT: Neomarica caerulea *from* Paxton's Magazine of Botany.

OPPOSITE: A page of illustrations from Maund's Botanic Garden, *which include (clockwise from top left):* Bulbocodium vernum, Cheiranthus tenuifolius *(now* Erysimum tenuifolium*),* Pinguicula grandiflora *and* Aponogeton distachyon.

by a variety of botanical artists including Mrs Augusta Withers, Mrs Priscilla Bury, Edwin Smith and Maund's own daughters, Eliza and Sarah. For the last few years of its life, the *Botanic Garden* was accompanied by a supplement, issued in parts, consisting of series of 70 hand-coloured engravings entitled *The Fruitist: a treatise on orchard and garden fruits, their description, history and management*.

Maund was obviously not short of ideas and energy, and, in addition to the works described above, also contrived to collaborate with the Reverend John Stevens Henslow, clergyman, geologist and Professor of Botany at Cambridge, to produce another journal entitled *The Botanist*, which was published in five volumes between 1837 and 1846. The artists once again included the hard-working Mrs Withers and Mrs Bury, as well as Sarah Maund and Miss R. Mills, the latter becoming the chief illustrator for G. B. Knowles and F. Westcott's *The Floral Cabinet and Magazine of Exotic Botany*, published in three volumes from 1837 to 1840.

Joseph Paxton's greatest claim to fame is probably the construction of the Crystal Palace for the Great Exhibition held in London in 1850; it was for this that he was knighted, but his extraordinary achievements also included a substantial body of published work. Born in 1803, Paxton began his career as a gardener for the Horticultural Society at Chiswick, leaving after three years to become head gardener to the Duke of Devonshire at Chatsworth, where he remained until 1858. While there, Paxton worked with boundless energy, building several glasshouses including a water-lily house, in which he successfully flowered the giant *Victoria amazonica*, and another one measuring 83 metres (272¼ feet) in length, known as the Great Stove; he also installed a spectacular fountain and established an arboretum.

In between times, he worked on the design of several public parks, became a MP and wrote and edited books and journals. From 1834 to 1849, he oversaw the publication of 16 volumes of *Paxton's Magazine of Botany, and Register of Flowering Plants*, with very attractive, hand-coloured engravings from drawings by F. W. Smith and S. Holden, and containing practical planting advice, and he was one of the founders of the *Gardener's Chronicle* in 1841. Paxton also collaborated with John Lindley in the production of *Paxton's Flower Garden* (1850–53) in three volumes with over a hundred hand-coloured plates. He died a wealthy man in Sydenham, where his Crystal Palace had been re-erected after the end of the Great Exhibition, in 1865.

Illustrated horticultural and botanical journals and magazines were not, of course, confined to Britain, and during the nineteenth century a whole array of such periodicals was published in Europe and America. Some of these were short-lived, while others are still published today. Some of the better known of these included *Flore des Serres et des Jardins de l'Europe*, a monthly journal published by the nurseryman Louis van Houtte, and edited by him with assistance from the botanists Charles Lemaire and M. Scheidweiler, which was published in Ghent from 1845 to 1883, with text in French, German and English. This fine work, which acted also as van Houtte's nursery catalogue, contained descriptions and over 2500 illustrations (mostly coloured) of rare and good plants "newly introduced to the continent or England". The high-quality lithographed plates were mainly based on paintings by the artist P. de Pannemaeker, who was a flower painter and landscape artist of some standing.

Charles Lemaire had previously edited *Jardin Fleuriste*, another gardening journal published in Paris from 1851 to 1854, and worked in Ghent until 1870 on *Illustration Horticole*, in which interesting and ornamental plants were described and illustrated with engravings. Ghent was then a thriving centre for the horticultural trade, with many nurseries selling exotic plants, and ambitious breeding of new varieties of azaleas, begonias and other ornamentals.

The *Revue Horticole*, a journal of practical horticulture, was founded in 1829 and continued until 1920. It had a wide remit, covering gardening principles and practice, new tools and other garden equipment, as well as ornamental plants and fruit, and was illustrated with engravings until 1851, when these were superseded by chromolithographs, an early form of colour printing. The illustrations of fruit by Alfred Riocreux (1820–1912), who also drew the plates for the botanist Joseph Decaisne's *Le Jardin fruitier du Muséum* (the museum in question being the Muséum d'Histoire Naturelle in Paris) are particularly good, and were lithographed with great expertise by the Belgian G. Severeyns.

Gartenflora, the foremost German illustrated garden magazine, ran from 1852 to 1922. It was started by Eduard August von Regel, while he was head of the Botanischer Garten in Zurich and Professor of Botany at the University. In 1854, he was appointed head of the Imperial Botanic Garden in St Petersburg, but continued to publish the journal in Germany, describing and illustrating over 2000 new species from the Russian Empire, which was then expanding across central Asia. Regel remained in St Petersburg until his death in 1892.

RIGHT: Rubus crataegifolius *from* Gartenflora.

OPPOSITE: Iris kolpakowskiana Rgl *from* Gartenflora.

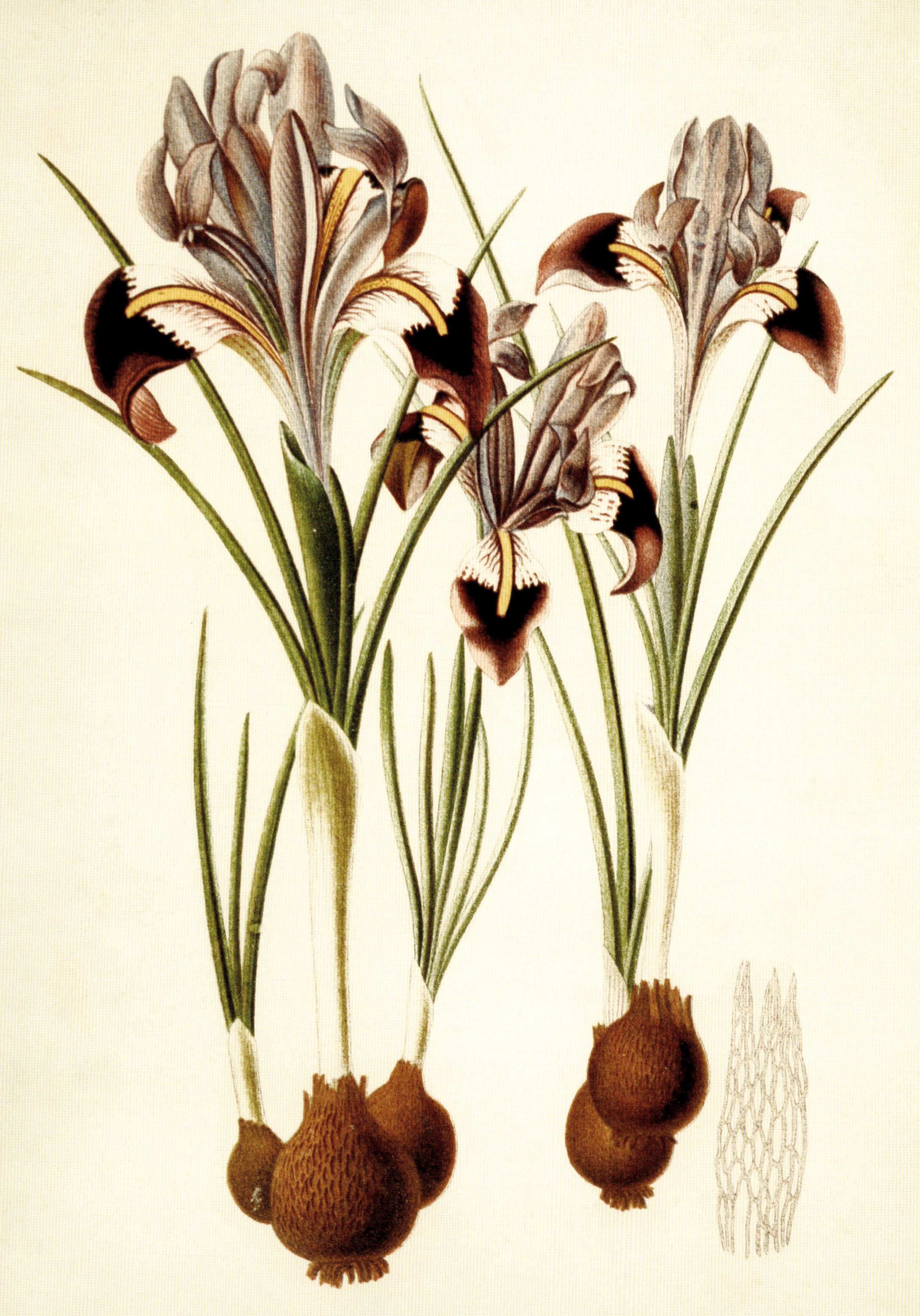

Henry C. Andrews

Henry C. Andrews (fl.1794 –1830) was a botanical artist, engraver and horticultural publisher, whose fame rests chiefly on his well-illustrated volumes on roses, heathers and geraniums. The nineteenth century saw the rise in popularity of books illustrating a single group of plants, known as monographs, and Andrews, able to both draw and engrave his own plates, was well placed to produce these.

His books were generally informative, with good clear illustrations, and were reasonably priced, putting them in a different league from the great, illustrated florilegia of the previous century.

Little is known of Andrews, except that he lived in London, and was the son-in-law of John Kennedy, a partner in the famous Vineyard nursery of Lee & Kennedy in Hammersmith. The nursery, which had been established in the mid eighteenth century, specialized in supplying "exotic" plants from countries such as South Africa, and these often tender novelties were particularly popular among the aristocracy.

Andrews's first venture was to establish and edit a monthly journal, *The Botanist's Repository*, as a rival to the older *Curtis's Botanical Magazine*. This seems to have been a fairly successful venture, with Andrews producing ten volumes between 1797 and 1814, and it was followed by a modest work, entitled *Engravings of Heaths*, with 30 plates drawn and engraved by the author, and published in 1800. A greatly expanded version was the next to be published, entitled *Coloured Engravings of Heaths* (1802–09), with 288 hand-coloured engravings, followed in 1804–06 by the five-volume *The heathery; or, A monograph of the genus Erica*, with 300 plates.

From this can be seen the amazing popularity of the Cape heaths, or Ericas, new species of which were pouring into the European nurseries (including Lee & Kennedy) from plant collectors and agents in the Cape. The love of these plants became a craze, with the great garden owners vying with each other for the best collection. Many of Andrews' specimens came from the Marquis of Blandford at Whiteknights, Reading, and another great collection belonged to the 6th Duke of Bedford at Woburn, who had a special glasshouse constructed to provide the correct growing conditions for his heaths. This latter collection was described by the Duke's gardener, George Sinclair, in *Hortus Woburnensis* (1833), who also noted that a window in the heath house was decorated with "about 50 of the most beautiful flowering species… executed by Mr. Andrews, and so accurately done, that they can

LEFT: Erica conspicua *from Andrews'* The heathery; or, a monograph of the genus Erica

OPPOSITE: Protea canaliculata *from the same work.*

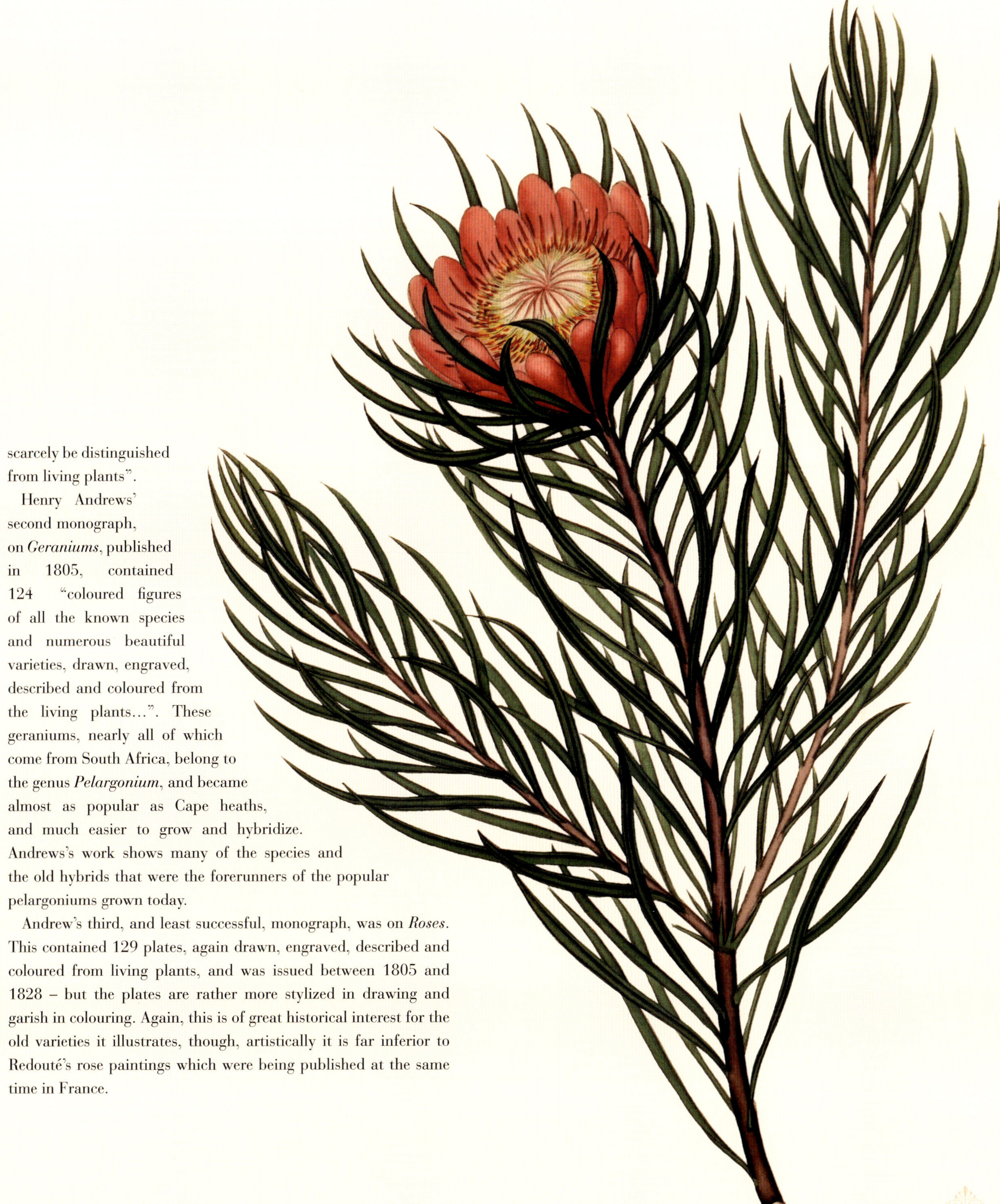

scarcely be distinguished from living plants".

Henry Andrews' second monograph, on *Geraniums*, published in 1805, contained 124 "coloured figures of all the known species and numerous beautiful varieties, drawn, engraved, described and coloured from the living plants…". These geraniums, nearly all of which come from South Africa, belong to the genus *Pelargonium*, and became almost as popular as Cape heaths, and much easier to grow and hybridize. Andrews's work shows many of the species and the old hybrids that were the forerunners of the popular pelargoniums grown today.

Andrew's third, and least successful, monograph, was on *Roses*. This contained 129 plates, again drawn, engraved, described and coloured from living plants, and was issued between 1805 and 1828 – but the plates are rather more stylized in drawing and garish in colouring. Again, this is of great historical interest for the old varieties it illustrates, though, artistically it is far inferior to Redouté's rose paintings which were being published at the same time in France.

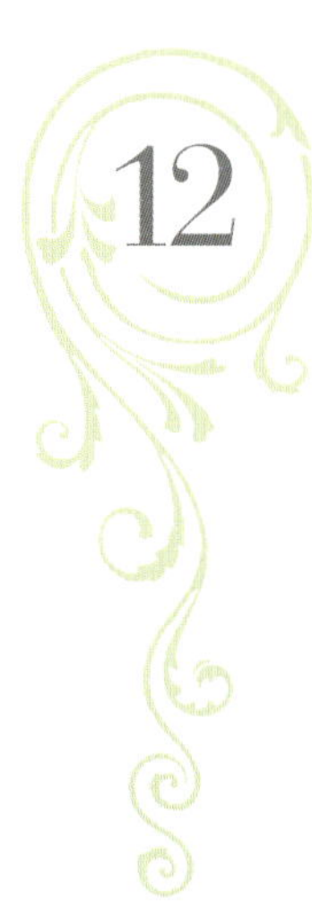

12

EARLY CHINESE PLANT DRAWINGS

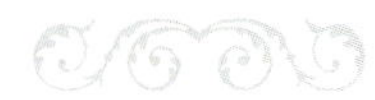

CHINESE PLANT PAINTING BEFORE EUROPEAN INFLUENCE

During the Song Dynasty (960–1279) Chinese art reached a high state of naturalism in plant and animal portraiture. The Emperor Huizong, who reigned during 1101–26, is said to have been a great aesthete, and a skilful painter. He taught the court painters personally, and awarded prizes for those who achieved the greatest realism.

His favourite style was bird-and-flower, and usually consisted of a branch with leaves, flowers or fruit, and attendant birds and insects. Some examples of painted hand scrolls or silk hanging scrolls survive from the thirteenth century. One, which is nearly life size, shows a group of lotus leaves and flowers with a pair of ducks swimming beneath them, and another shows flowering gardenias with a pair of warblers. The leaves in both these are painted with great skill in the handling of perspective, particularly of the flat lotus leaves viewed from the side at eye level. The lotus, being an important symbol of Buddhism, was often painted, and several ancient examples were preserved in temples until the nineteenth century when they were acquired by collectors or museums.

The winter-flowering *Narcissus tazetta* was also a popular subject in China, associated with the rebirth of nature after winter. Zhao Mengjiang (1199–before 1267), a member of the Song imperial family who was active around 1250, made a speciality of drawing it: a long scroll, painted in ink, shows a field of elegant narcissus with their waving leaves and bowing flower stems, marching across the page. Half of the original, but still measuring over 3.5 metres (11 feet) long, is now in the Metropolitan Museum of Art in New York. In later centuries, some of this early naturalism was lost, and the skill of the brushwork and economy of line became more valued than the careful study of nature. Beautiful calligraphy as well as skill in painting with a brush became important accomplishments for a scholar, and flowering plum branches, chrysanthemums, pine needles and particularly bamboo leaves became common subjects for black-ink painted scrolls.

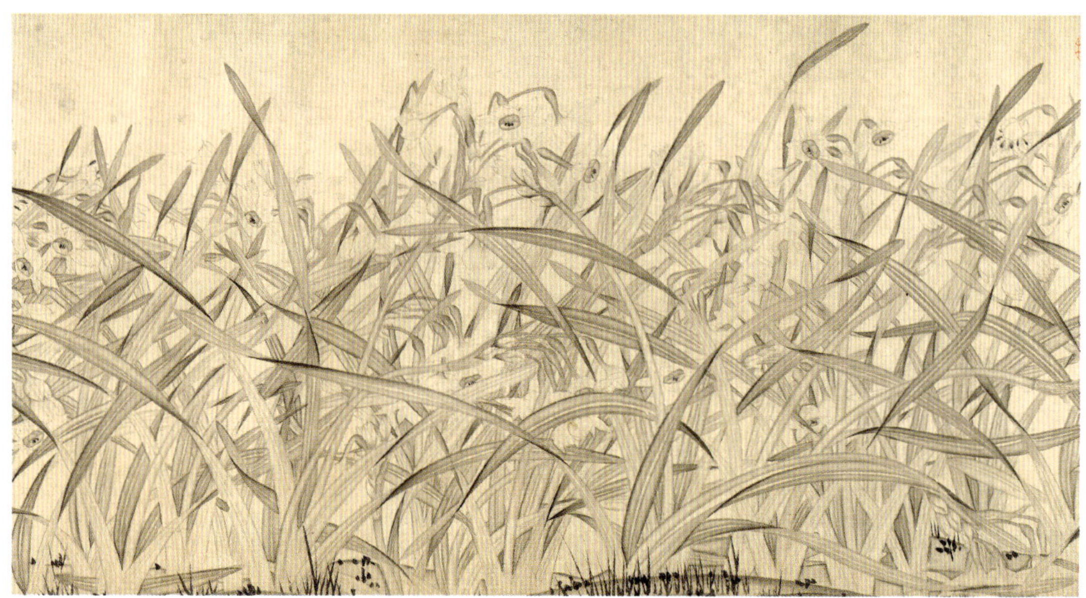

LEFT: Made in ink on paper sometime during the fist half of the thirtenth century, this scroll by the Chinese artist Zhao Menjian depicts Narcissus tazetta.

OPPOSITE: Thespia tiliacea *(now* Hisbiscus tiliaceous*) from John Reeves's Botanical collection from Canton, China.*

PAINTING FOR EUROPEANS

From the beginning of the sixteenth century onwards the Chinese were as unwelcoming to foreigners as were the Japanese, arresting those who tried to trade with them, and forbidding foreigners to leave their designated offices in the trading ports. After several unsuccessful attempts, the Portuguese managed to establish their factory at Macao in 1557, but were only allowed to trade on sufferance in the port of Canton and a short distance up the Pearl River. Only the Jesuits, led by Matteo Ricci, who brought western science and Christianity to Peking in 1601, were allowed to remain in the country.

The first English traders began to deal with China under the auspices of the East India Company in the early eighteenth century and were allowed to set up their first factory in Canton in 1711, trading tea for silver. The East India Company's warehouses in Canton became the centre of the export of tea to England, as well as silk, porcelain and other Chinese products. Opium, which was grown in India, became the means of paying for these goods, taking the place of silver. In England, there was a craze for anything Chinese, from wallpaper, dinner services and furniture, such as that designed by Thomas Chippendale, to garden design and garden buildings; as

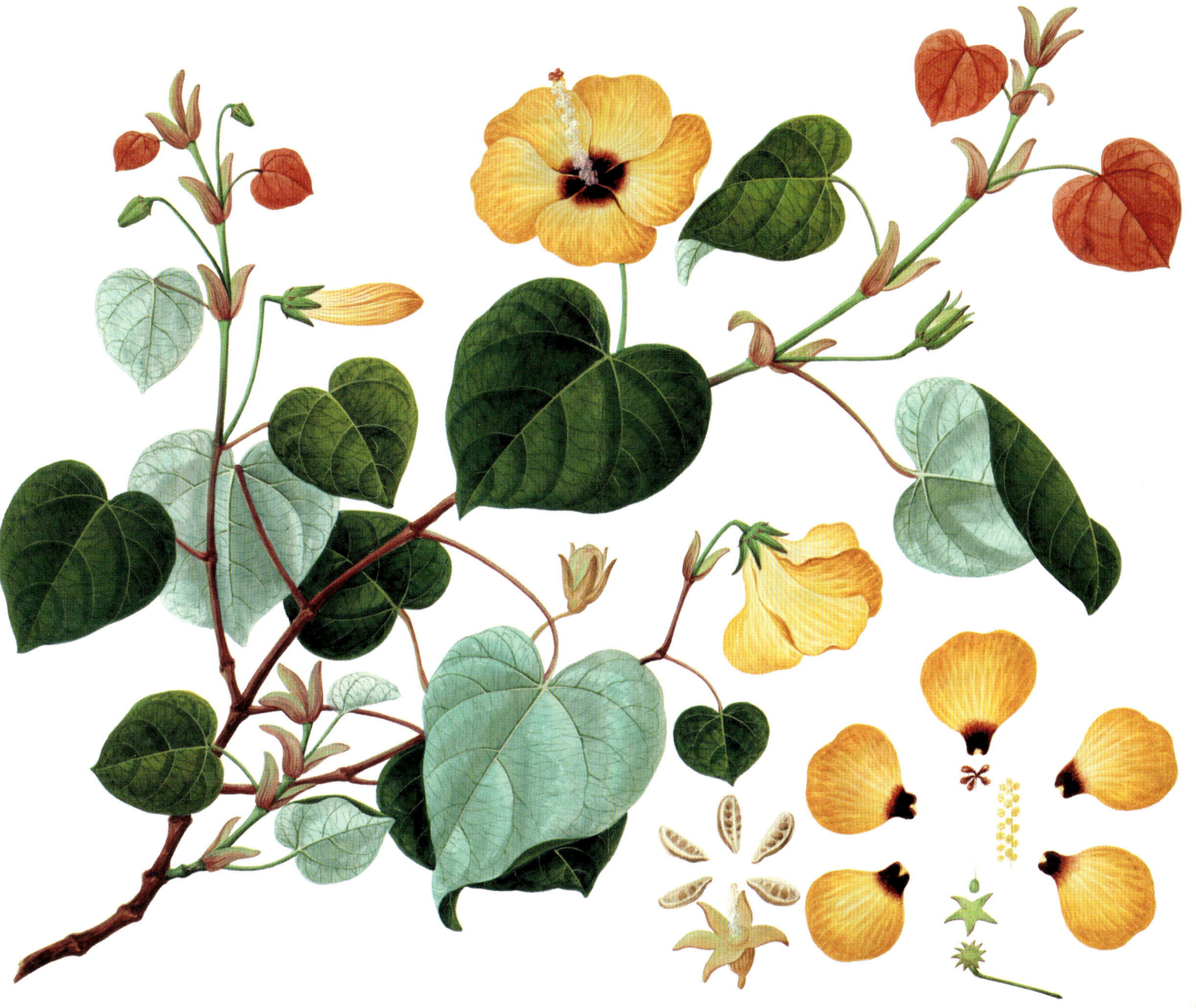

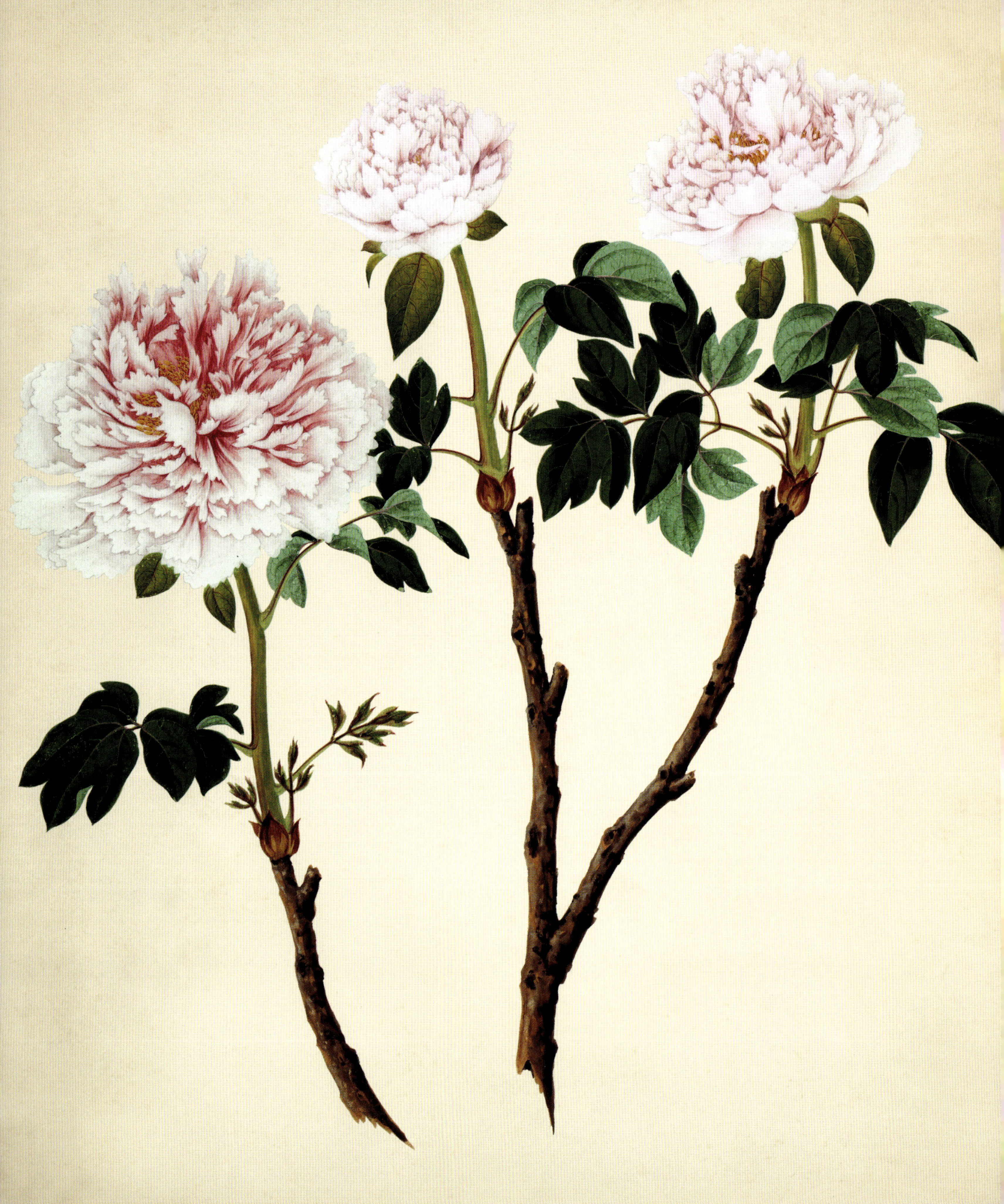

early as 1685, the diplomat and keen gardener, Sir William Temple, whose knowledge of China was distinctly hazy, had coined the term "Sharawadgi" to describe the informal or irregular planting thought to be favoured by the Chinese. In contrast, the English architect Sir William Chambers (1723–96), who had actually visited China twice, published his *Designs of Chinese buildings* in 1757, and designed a Chinese temple and ten-storey pagoda (completed in 1762) for Princess Augusta's garden at Kew.

Lord Macartney's embassy from George III to the Manchu Emperor Chien Lung in 1792, to try to gain further concessions and expand trade, seemed to have been a complete failure, but it intensified the fashion in England for everything Chinese and so fuelled the demand for more trade with China. A few Chinese garden plants came back on the tea ships, but they were obviously only a fraction of those grown by the Chinese; for example, perpetual-flowering roses arrived in around 1750, *Magnolia denudata*, one of the parents of the common *M.* × *soulangiana*, and the first moutan or tree peony were planted at Kew by Sir Joseph Banks in the 1780s, and chrysanthemums reached France in 1789.

In 1803, Sir Joseph sent William Kerr, a gardener in the Royal gardens at Kew to collect plants in China and send them back to England. He lived in Macao and Canton for about nine years, but is said to have fallen into bad company: he did, however, send back to Kew the tiger lily, *Lilium tigrinum*, white double *Rosa banksiae*, named after Lady Banks, as well as the egg-yolk-flowered shrub named after him, *Kerria japonica*. He visited Java and the Philippines and ended up as a superintendent of the Slave Island and King's House gardens in Colombo, dying in 1814.

The greatest contribution to the knowledge of Chinese plants and animals in the early nineteenth century was made by John Reeves, a tea inspector for the East India Company. Reeves was born in Essex in 1774, joined the company in 1808 and worked in China from 1812 to 1831. He lived for much of the time in the foreign colony in Macao, only visiting the factory in Canton during the winter tea-shipping season, where he used his contacts to acquire plants and animals from all over eastern China, and arranged to have them painted by Chinese artists. The Horticultural Society commissioned Reeves to send back plants and paintings of flowers to London. These eventually filled five volumes and, although sold when the Society was facing bankruptcy in 1859, have mostly since been restored to the Lindley Library. They show mainly Chinese garden flowers, azaleas, peonies, camellias, chrysanthemums and wisteria. The style of the paintings is an interesting combination of Chinese and European styles. The paintings are scientifically more accurate than the usual Chinese work of the period, but the choice of specimen and love of any twist or frill in the petals show distinct Chinese influence.

Reeves appears to have kept many paintings for his own collection and these were given to the Natural History Museum by his family. They include over 2000 paintings of animals, birds and fish, as well as flowers and fruit. Many were named after Reeves, notably the beautiful golden, black-and-white Reeves's Pheasant, whose tail feathers can reach 1.6 m (63 inches) long, and the flowering tree, *Reevesia thyrsoidea*, named by John Lindley in 1827.

There were others, like Reeves, who commissioned Chinese artists to draw and paint plants; one of these was Sir Thomas Stamford Raffles (1781–1826) who was employed by the East India Company, and is best known as the founder of present-day Singapore. Raffles was fascinated by natural history of all types, and made great collections of specimens; sadly, on the way back to England in 1824, the ship on which he was travelling caught fire, and many of his specimens, papers and drawings were lost. Even so, a Chinese and a French artist worked night and day during the two months in which he waited for the next ship, and managed to replace a number of the drawings, which are now in the collection of the British Library.

Raffles's name is commemorated by the parasitic plant *Rafflesia*, the flowers of which look and smell like rotting flesh; one species, *R. arnoldii*, has the largest known flower in the world. The eponymous Arnold, was a British botanist who, with a Malay servant, collected a specimen of the plant in Sumatra but who died of a fever soon afterwards. Raffles's wife, who, with her husband, had been present at the discovery, apparently finished Arnold's incomplete drawing of the plant, and it was sent to Sir Joseph Banks, along with the dried specimens. Banks passed the picture on to Francis Bauer (*see* page 104), at that time working as botanical artist at Kew, and he produced an amazingly detailed painting of the plant, which was published in the *Transactions of the Linnean Society* in 1822.

OPPOSITE: Paeonia, *an old moutan cultivar from John Reeves*'s *Botanical collection from Canton, China.*

EUROPEAN PAINTINGS OF CHINESE PLANTS

The Chinese tea trade continued to provide a trickle of plants for English gardens. The difficulties of getting live plants back was increased by the slowness of the ships, the two crossings of the tropics that the ships had to make, as well as the journey around the Cape. John Lindley was the energetic secretary of the Horticultural Society of London (as it then was), and apart from being the leading authority on orchids, was becoming increasingly interested in the possibly of acquiring new plants from China. In 1827, he published the new camellia, *C. reticulata*, in his periodical, *The Botanical Register*. This huge-flowered camellia had been introduced by Captain Rawes in 1820, and still survives in some old gardens. A specimen at Chatsworth was planted on the glassed-in wall in 1850, and is still thriving.

After the discoveries of John Reeves through the merchants in Canton, Lindley encouraged the council of the Horticultural Society to send their own collector to China. Robert Fortune was born near Edrom, Berwickshire in 1812, and trained at the Royal Botanic Garden in Edinburgh in 1839, before going to the society's garden

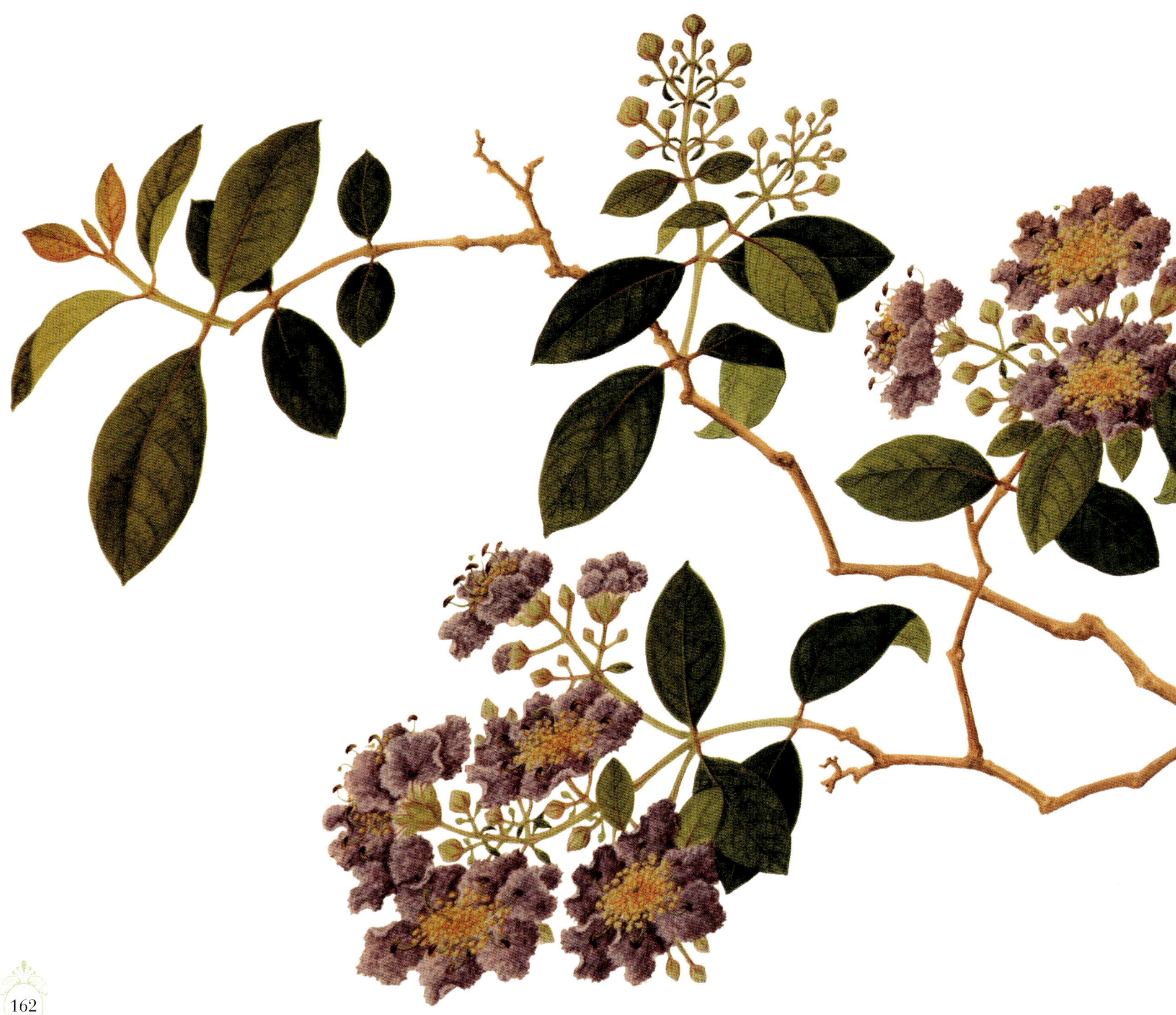

RIGHT: This painting of Enkianthus quinqueflorus *is part of a collection of works by Chinese artists held in the collection at the Royal Botanic Gardens at Kew.*

BELOW: Lagerstroemeria speciosa *is part of the same collection. The artist is unknown.*

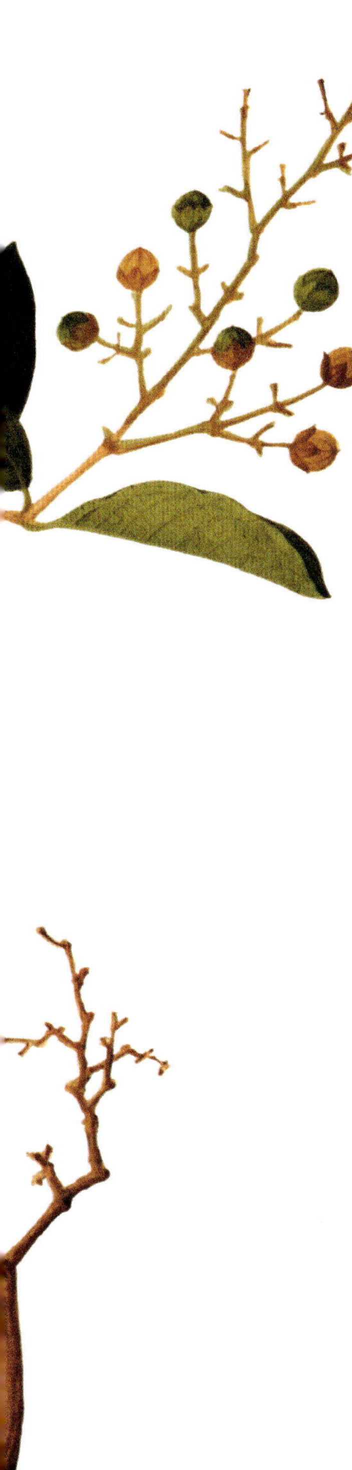

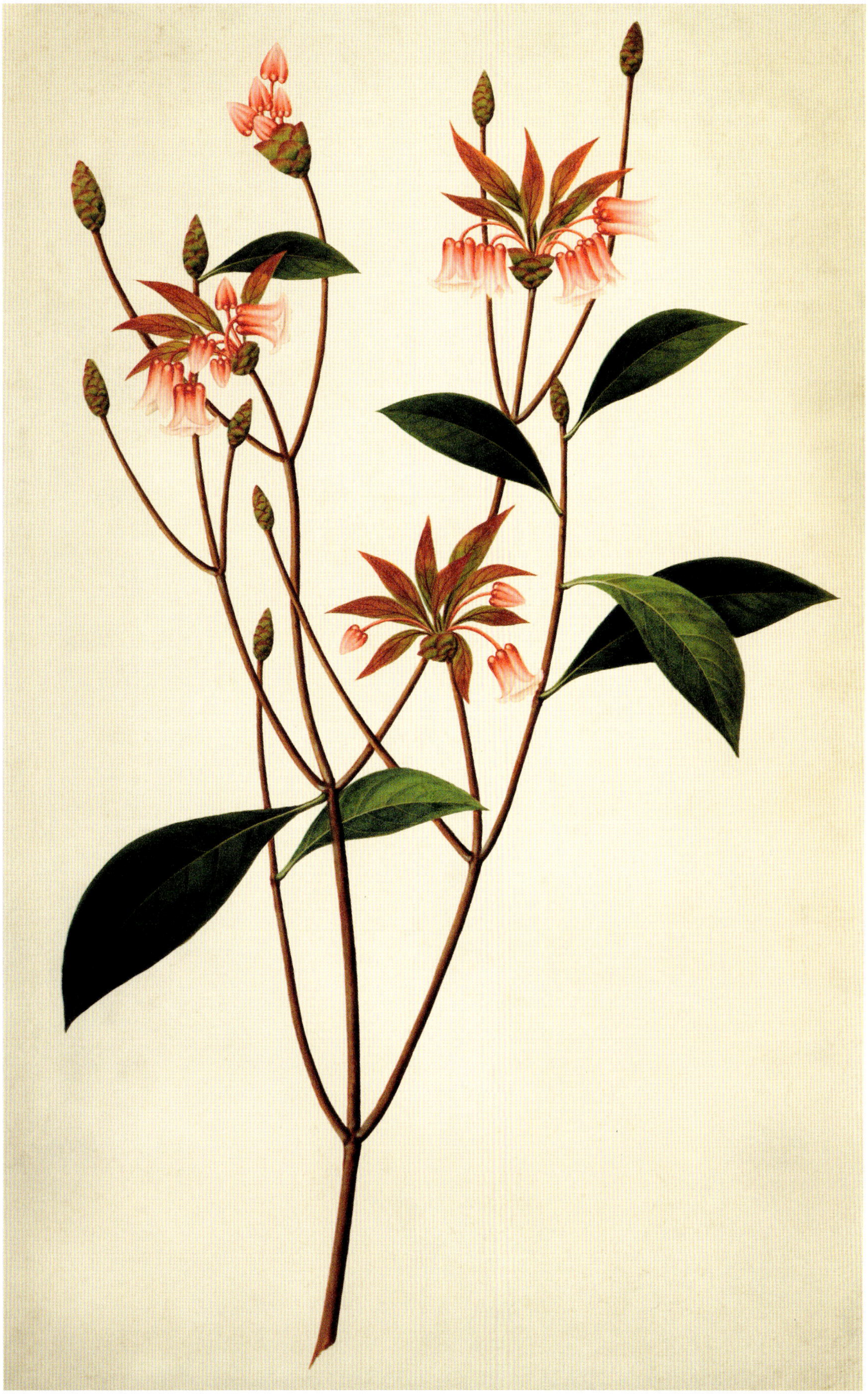

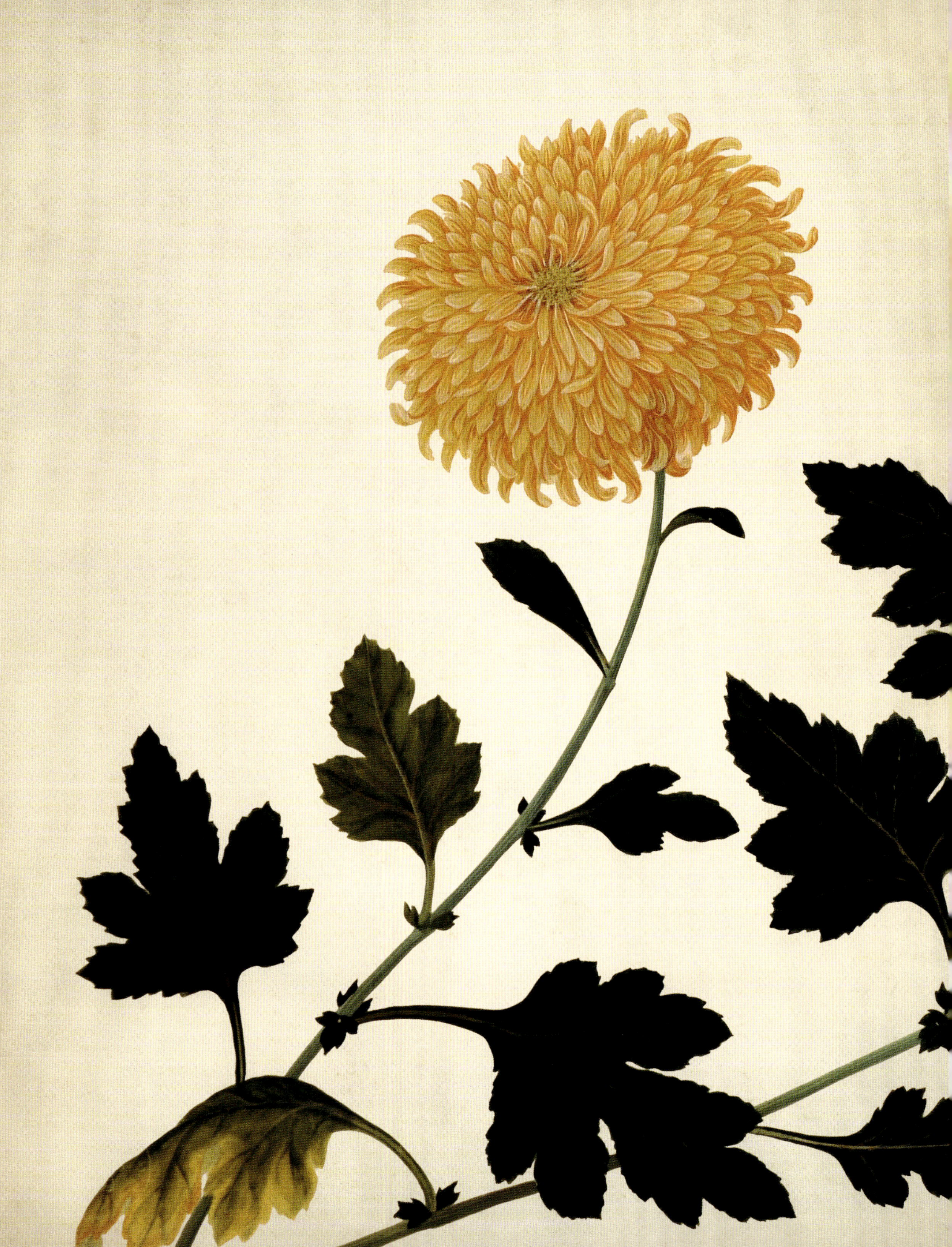

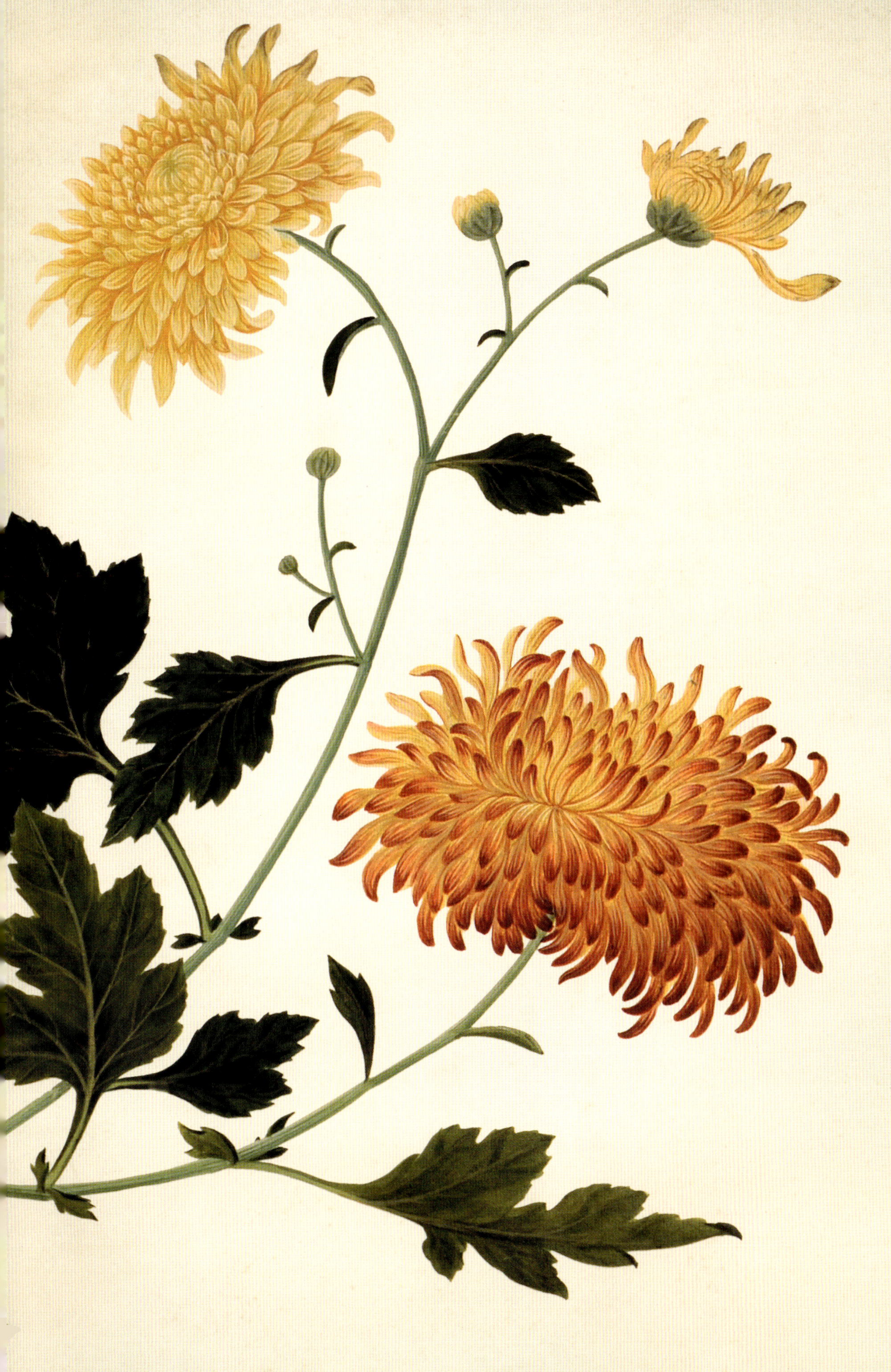

A study of Chinese Chrysanthemum *cultivar by an unknown Chinese artist from the collection at the Royal Botanic Gardens at Kew.*

at Chiswick as superintendent of the hothouses. Between 1843 and 1861, he made at least five journeys to China, some for the society and some for the East India Company. In 1842, the end of the first Opium War eased restrictions on travel, but most Chinese were still deeply antagonistic to foreigners. In spite of this, Fortune, having learnt Chinese, managed to penetrate further afield than Kerr and Reeves, particularly in the area around Ningbo, where he collected many new plants to send back to Europe, most of which were hardier than the ones growing around Canton. Later, he also visited Japan. Many of his plants are still grown today: "Fortune's Double Yellow" is a beautiful climbing rose "found in the garden of a mandarin in Ningpo [sic]" and from the mountains of Hong Kong, he introduced *Ixora chinensis*, whose rounded heads of intense scarlet flowers are now universal in subtropical gardens. Winter-flowering jasmine, *Jasminum nudiflorum*, and the scented honeysuckle *Lonicera fragrantissima* and *Anemone japonica* are among the many now common garden plants that Fortune introduced. A number of Fortune's plants, such as *Leycesteria formosa* (1839), *Rehmannia chinensis* (1839) and *Anemone japonica* (1857) were illustrated in *Curtis's Botanical Magazine*.

It was not only ornamental plants that came to Europe – Chinese fruits were also brought back; peaches were often imported, and a painting by the French artist Jean Gabriel Prêtre (fl.1800–40) in the art collection at the Royal Botanic Gardens, Kew, shows details of the loquat *Eriobotrya japonica*, which was introduced in 1787. Prêtre, who had been a member of Napoleon's Egyptian Commission, painted not only plants, but also birds and insects, and was employed by the Muséum national d'Histoire naturelle in Paris; he also worked for Napoleon's wife Empress Joséphine, painting animals and birds in her menagerie. He was fortunate in being at the Paris museum at a time of great artistic activity, and collaborated with other artists such as Bessa, Poiteau and Turpin and the Redouté brothers (*see* pages 130–45).

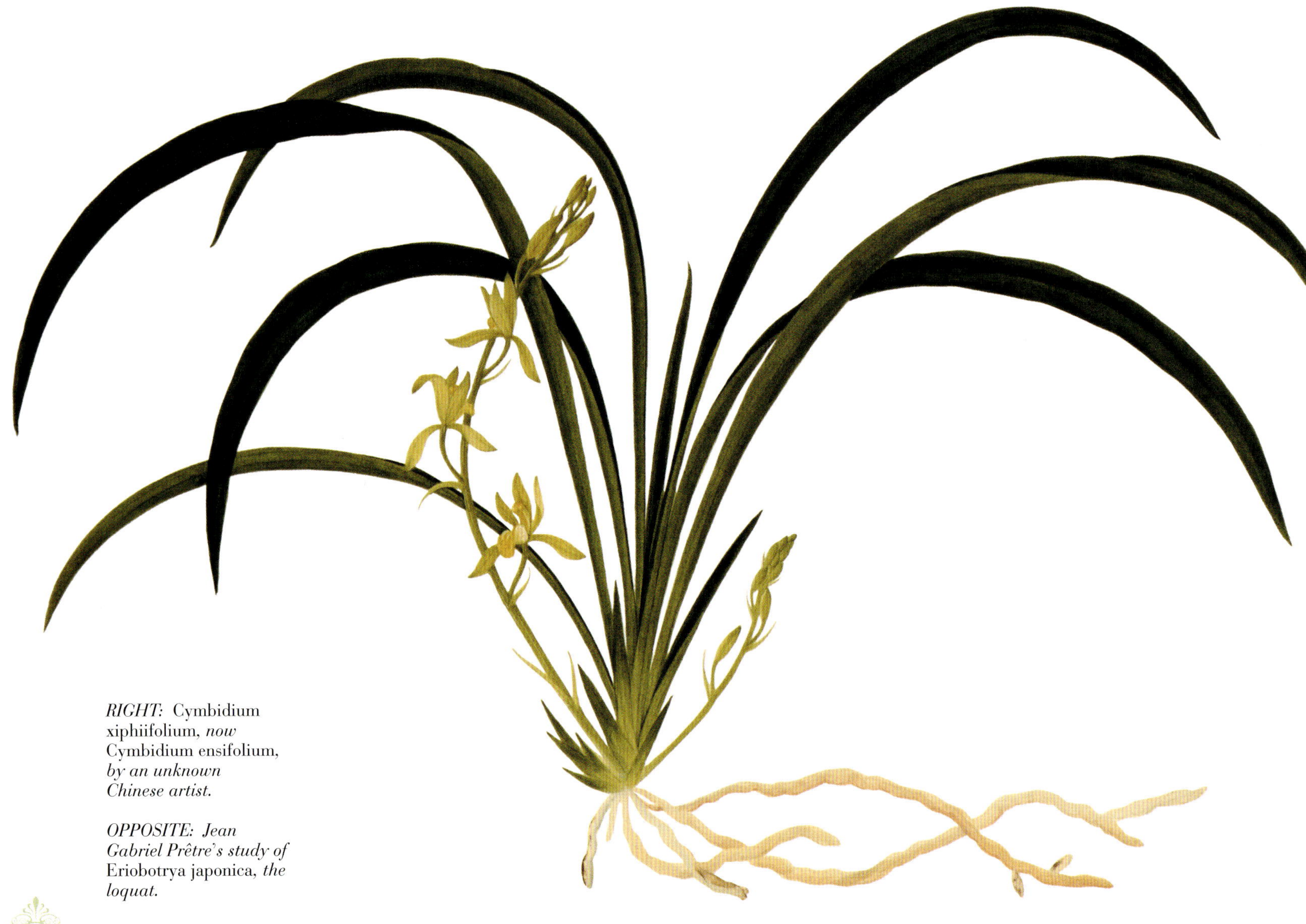

RIGHT: Cymbidium xiphiifolium, *now* Cymbidium ensifolium, *by an unknown Chinese artist.*

OPPOSITE: Jean Gabriel Prêtre's study of Eriobotrya japonica, *the loquat.*

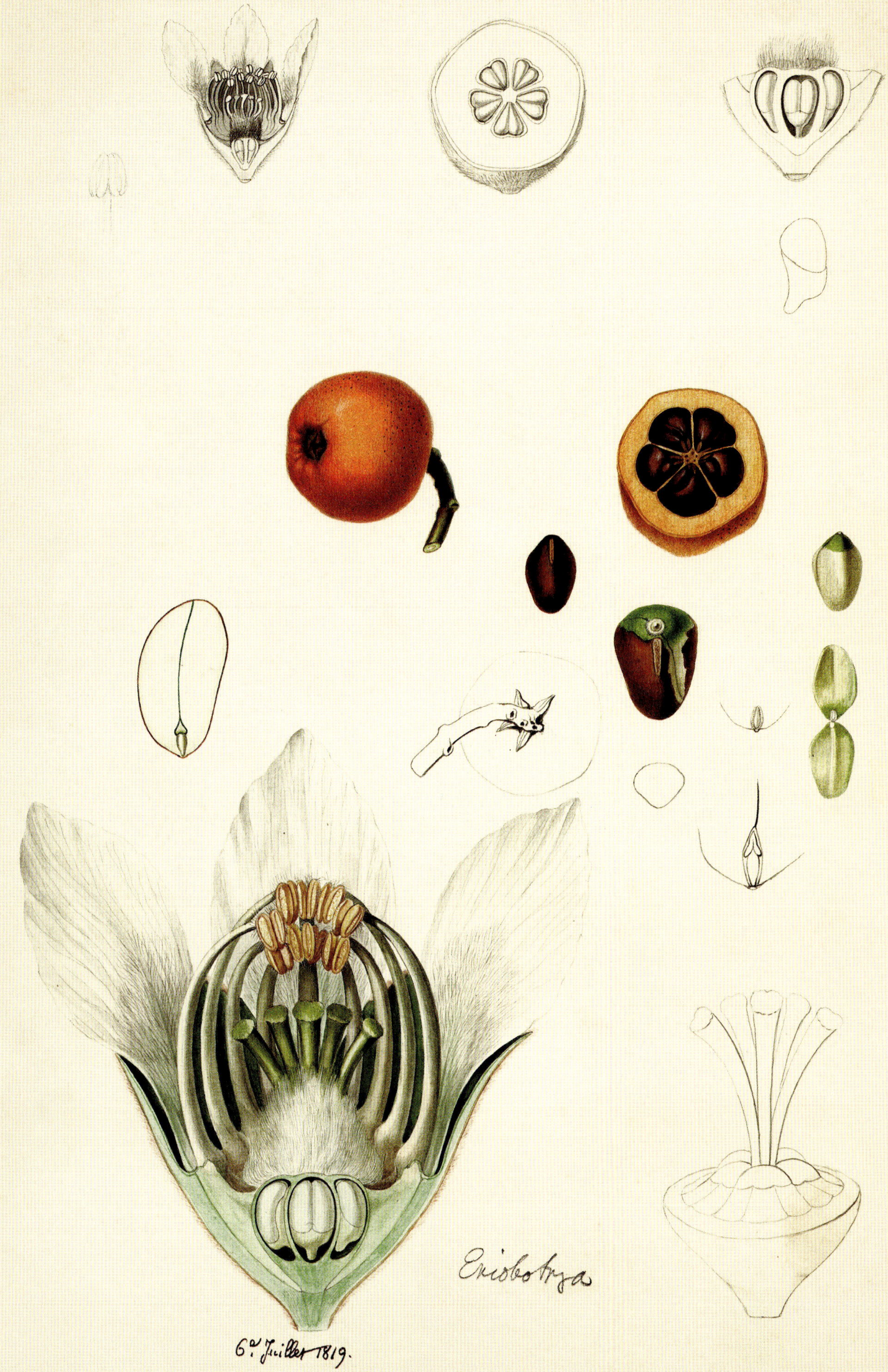

Eriobotrya
6e Juillet 1819.

Père David and the French Missionaries

The opening up of China to the outside world in the mid nineteenth century, after the end of the second Opium War in 1860, allowed subjects of the Treaty powers (Britain, France, Russia and the USA) to travel more freely within the country. Christian missionaries were also allowed to work in inland China, and this, combined with the fact that many of them, as well as the expatriate consular and customs staff, were interested in natural history, enabled large areas of the country to be explored scientifically for the first time.

There had been earlier attempts at trade and exploration: the Portuguese had monopolized China's trade with Europe during the sixteenth century, a few Jesuit missionaries managed to travel in the country during the seventeenth century, and the English finally managed to gain a foothold through the East India Company.

The work of the great British plant collector Robert Fortune, who worked for the East India Company, is mentioned elsewhere (*see* pages 162–66) but for the most exciting discoveries of the nineteenth century we must look to the French, and in particular to the missionaries, some of whom lived, worked and died in the Chinese hinterland.

Père Jean Pierre Armand David (1826–1900) is probably the best known of these, and was a prolific collector of animals, birds and rocks, as well as plants. He arrived in China in 1862 and made three great expeditions, which he described in detail in his diaries. His first trip, to Mongolia, yielded little of botanical importance, but the second, which lasted from 1868 until 1870 and took him into the border area of Tibet, as well as to Chengdu and Moupine (now Baoxing), resulted in an amazing collection of plants and other natural-history specimens, including the first skins of the giant panda.

In Moupine, David based himself for several months in a small convent run by members of the Missions Etrangères, and made forays into the mountains from there. He saw many plants in flower that were later brought into cultivation in Europe, and among these the handkerchief tree, *Davidia involucrata*, is one of the most unusual. Other plants named after him, and now well-known to gardeners, include *Lilium davidii*, *Rhododendron davidii*, and *Viburnum davidii*. David sent back dried specimens of over 2000 species of plants to the Muséum nationale d'Histoire naturelle in Paris, and the botanist Adrien-René Franchet published descriptions of these in a book entitled *Plantae Davidianae*. Incidentally, this book provides a striking example of a problem encountered by botanical artists when drawing a plant only from a dried specimen: the "handkerchiefs" of the Davidia are shown pointing upwards instead of hanging down!

Père Jean Marie Delavay (1834–95) was another member of the Missions Etrangères who lived and worked in China, for many years between Dali and Lijiang in Yunnan, and collected plants even more assiduously than David; it is estimated that he sent 200,000 dried specimens back to the museum in Paris. Among the plants named after him are *Clethra delavayi*, *Magnolia delavayi*, *Paeonia delavayi* and *Osmanthus delavayi*. Many are illustrated in Franchet's *Plantae Delavayanae* (1889–90).

Paul Guillaume Farges (1844–1912) overlapped with Delavay, but collected plants in eastern Sichuan; he appears to have devoted much time to improving the lot of the miserably poor people among whom he lived, but still found time to collect many good plants, including *Rhododendron fargesii*.

About 20 years later the last of the great missionary collectors, Jean-André Soulié (1858–1905), arrived; like many missionaries he was a doctor, and also a keen botanist, who sent thousands of dried plants back to Paris. His name is remembered in *Rhododendron souliei* and *Primula souliei*, among others. He worked in the China–Tibet borderlands in Yunnan and Sichuan, lastly in Xinlong, and was one of several missionaries who were tortured and killed by Tibetan lamas in the unrest which followed Francis Younghusband's invasion of Lhasa in 1903 and 1904.

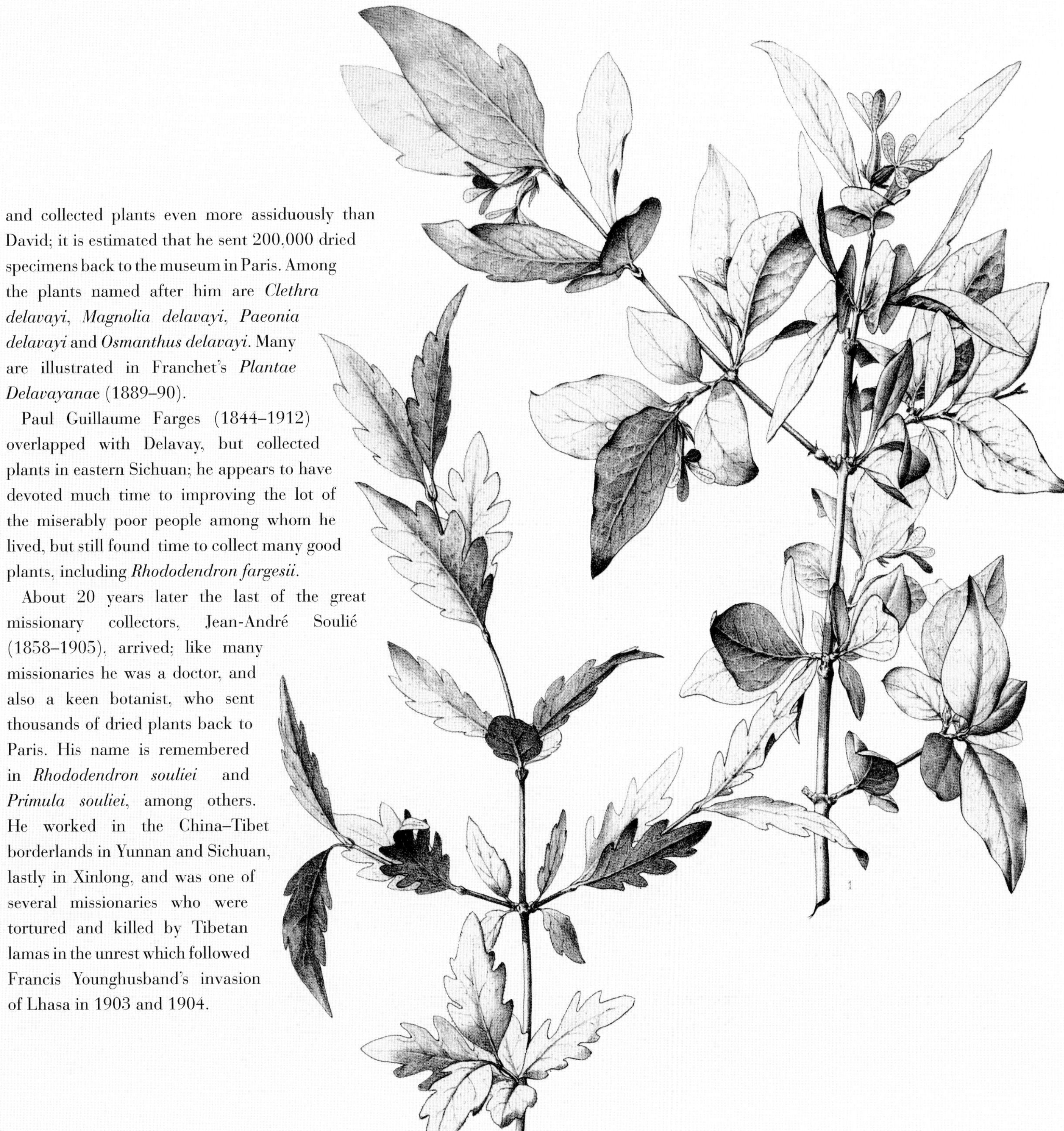

RIGHT: Abelia biflora *from* Plantae Davidianae.

13

THE COMPANY SCHOOL IN INDIA

The East India Company was founded in 1600 during the reign of Queen Elizabeth I, and was granted a monopoly of trade with the East Indies and anywhere east of the Cape of Good Hope and west of the Straits of Magellan.

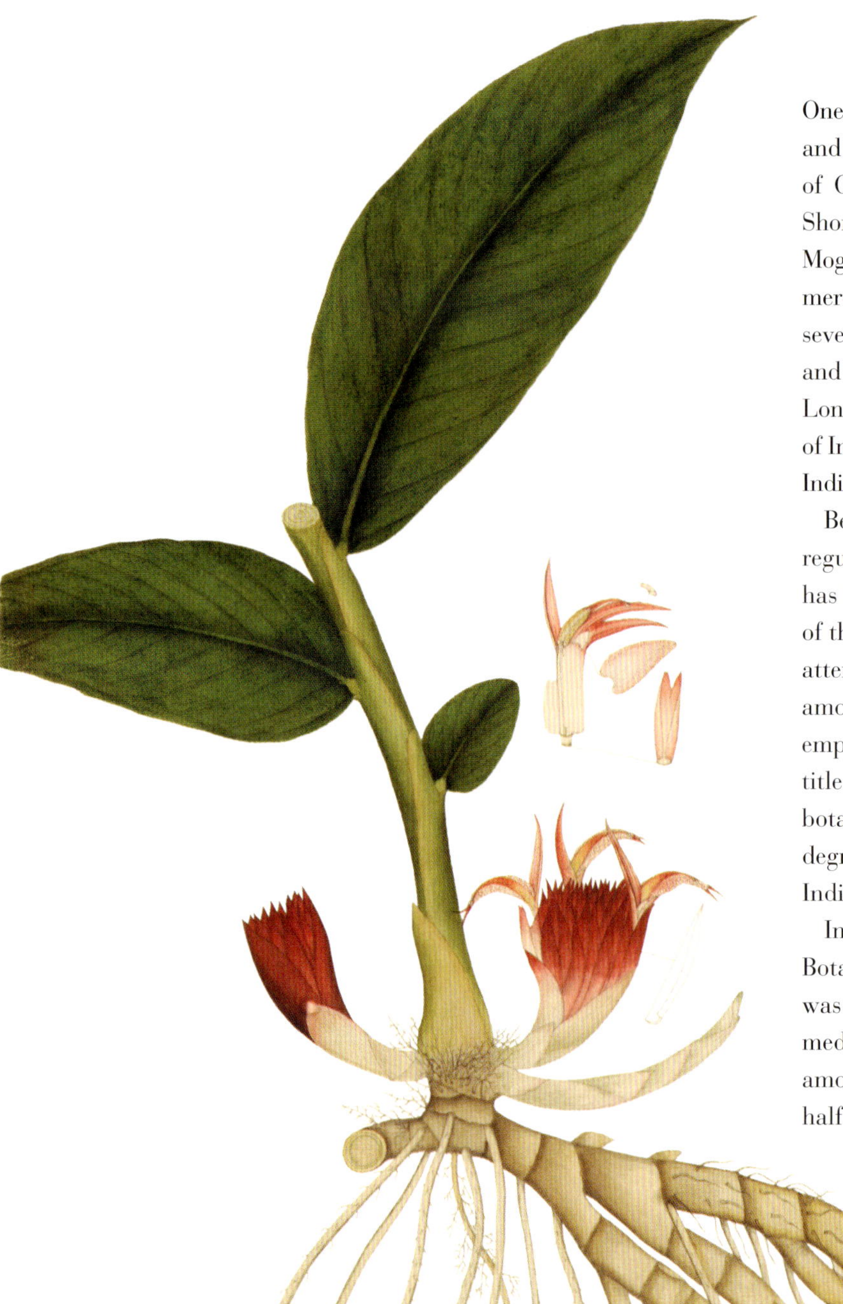

One of its earliest bases was at Surat near Bombay (now Mumbai), and circa 1610 it set up a new base at Machilipatnam on the coast of Coromandel, as the southeast coast of India was then called. Shortly afterwards, in 1612, Sir Thomas Roe took an embassy to the Mogul Emperor, Jahangir and obtained permission for the company merchants to trade freely around the coast of India. Through the seventeenth and eighteenth centuries, the company became more and more powerful, with the encouragement of the government in London, and eventually raised a large army, finally annexing much of India. It behaved as a ruling power until 1858, when following the Indian Mutiny, the colonial Government of India took control.

Before the introduction of quinine to control malaria, and with regular outbreaks of cholera, India was an unhealthy posting, and it has been calculated that between 1736 and 1834, only 10 per cent of the officers of the company survived to return to England. Great attempts were made to reduce the incidence of disease and death among Europeans, and while it was in its heyday, the company employed medical graduates, usually from Edinburgh, with the title of Assistant Surgeon; many of them were also interested in botany, which was, at that time, an integral part of any medical degree course, and it was these who began the systematic study of Indian plants.

In 1778, the company set up a post called "Naturalist and Botanist" or "Naturalist to the Madras Government", whose job it was to investigate and test any plants that might be exploited for medicine or trade. Apart from food crops, indigo and spices were among the chief exports from India to England. Patrick Russell, half-brother of Alexander Russell who wrote the *Natural History*

LEFT: Zingiber rubens, *Bengal ginger, commissioned by William Roxburgh.*

OPPOSITE: Kaempferia rotunda *and a dissection of the flower commissioned by William Roxburgh.*

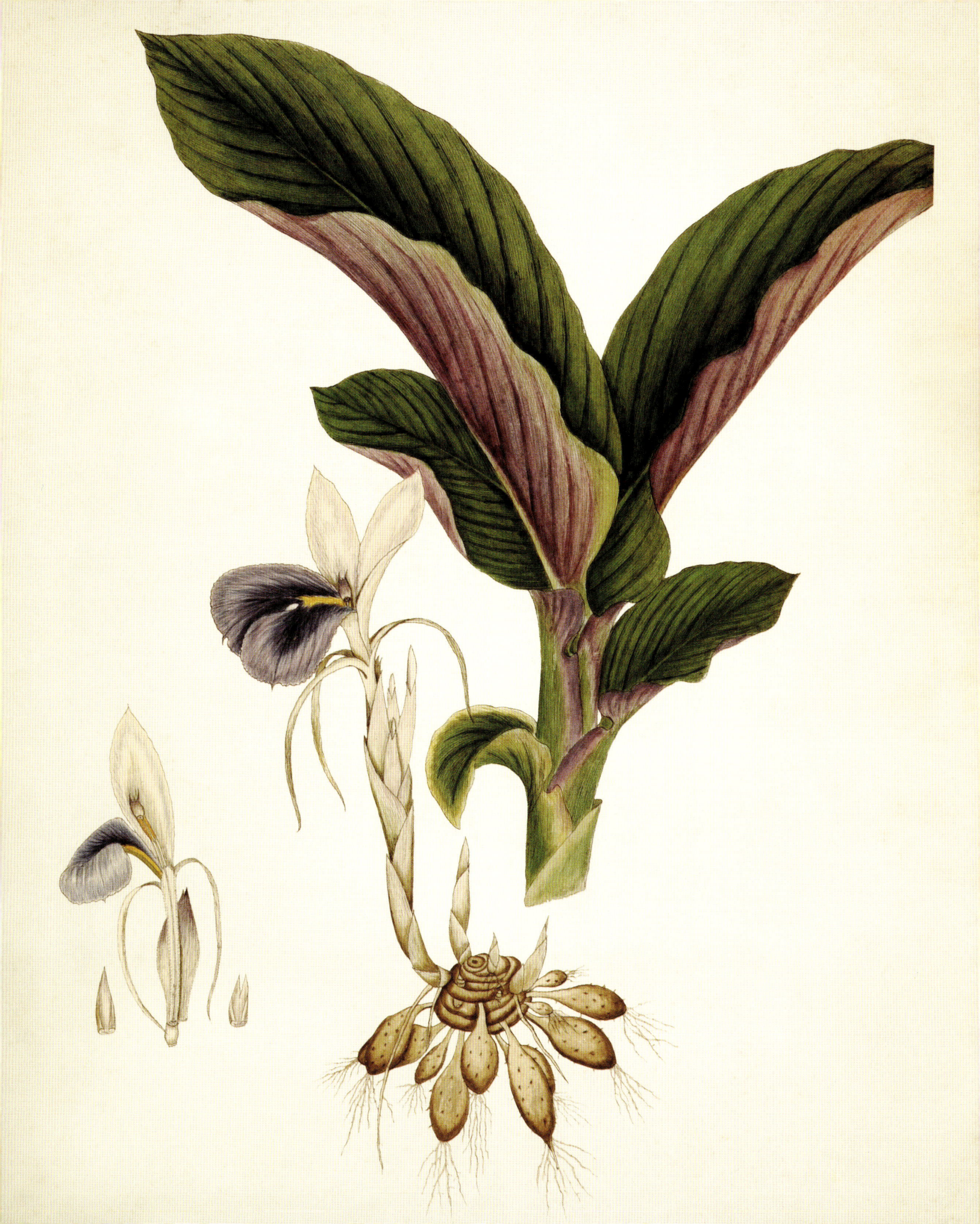

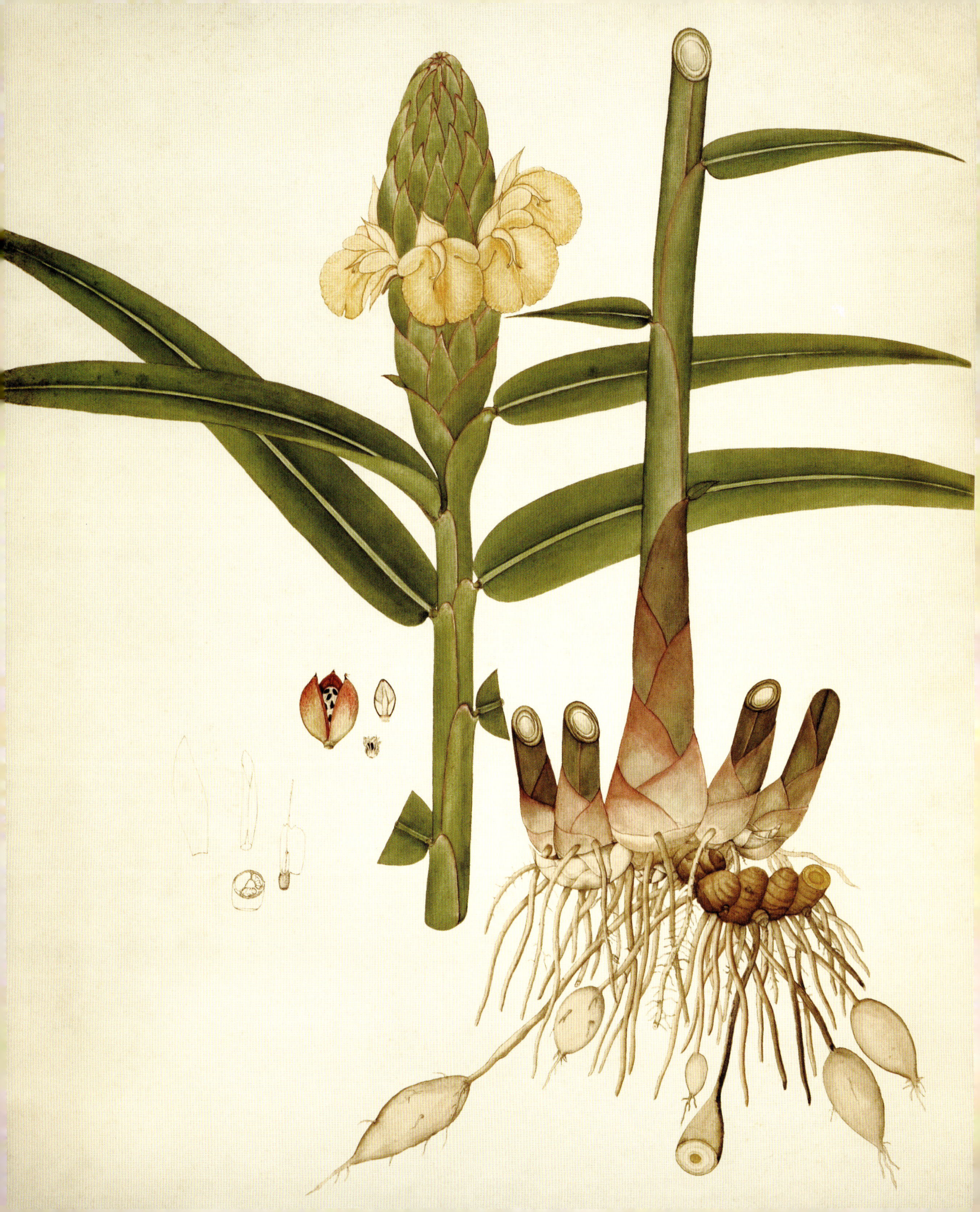

of Aleppo (*see* page 66), was Naturalist from 1785, and while he was on leave he persuaded Sir Joseph Banks and the company to publish William Roxburgh's *Plants from the Coast of Coromandel* (1795–1818), a handsome folio with hand-coloured engravings copied from illustrations by Indian artists, some of which had been made under Roxburgh's supervision. Roxburgh had arrived in Madras in 1776, after studying medicine in Edinburgh, and succeeded Russell in the post until 1793 when he went to Calcutta, to be the first Superintendent of the Honourable Company's Botanic Garden, which had been founded by the Company in 1786. While at Calcutta, Roxburgh continued the custom that he had started in Madras of training and employing Indian artists to record plants grown in the garden and other items of interest. Under Roxburgh and his successors, the Calcutta Botanic Garden became very influential, both for the study of Indian plants and food crops, and as a staging post for plants between China and Europe.

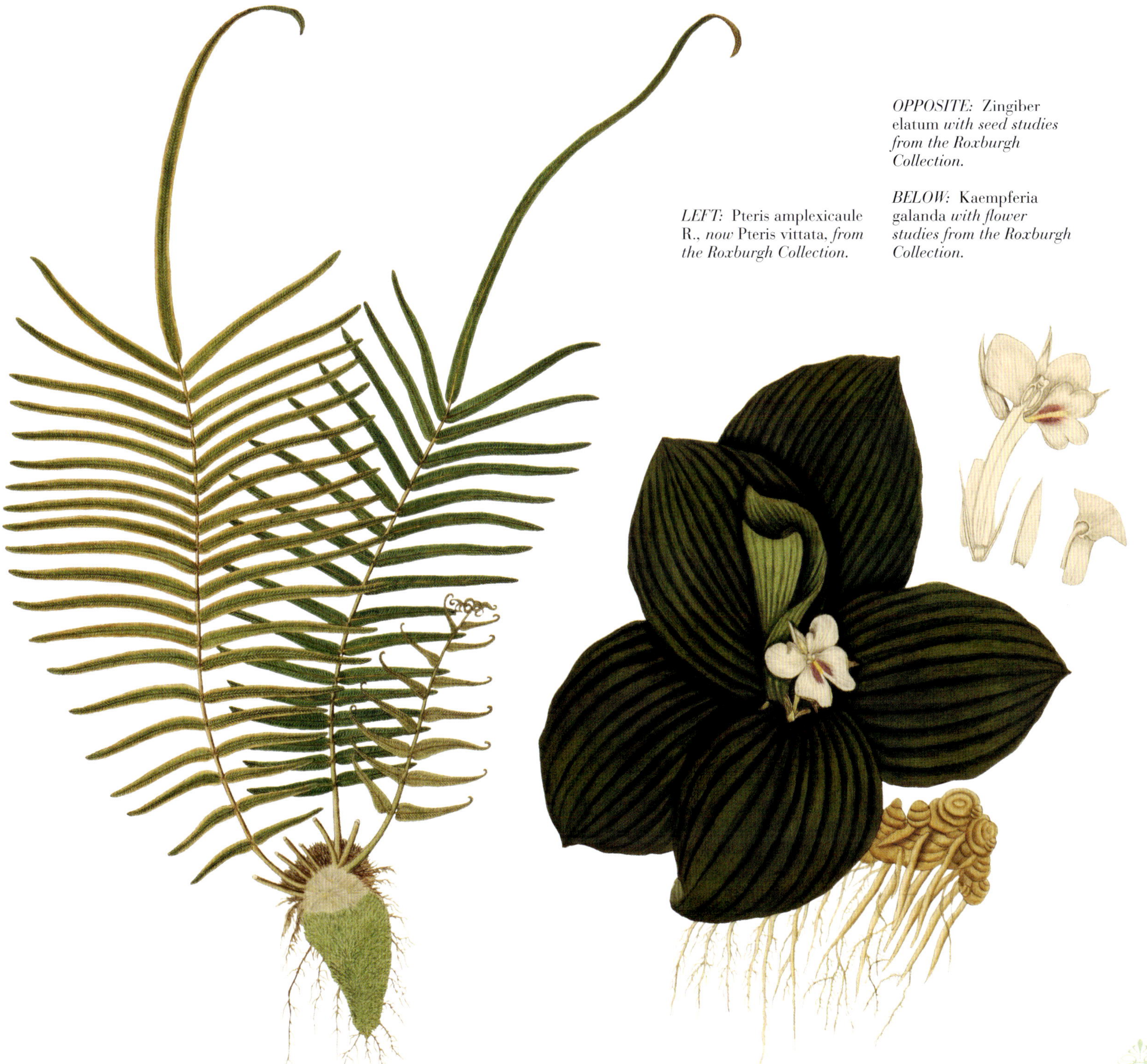

OPPOSITE: Zingiber elatum *with seed studies from the Roxburgh Collection.*

LEFT: Pteris amplexicaule R., *now* Pteris vittata, *from the Roxburgh Collection.*

BELOW: Kaempferia galanda *with flower studies from the Roxburgh Collection.*

While most of the early publications on Indian botany were supported by the East India Company, some were produced by rich individuals in England. One of these was William Roscoe, whose *Monandrian Plants of the order Scimataceae* was published in 1828. Roscoe was an unusual man of humble beginnings (his father ran a tavern and then a market garden), but as a boy he developed a restless and enquiring mind, trying his hand at several trades before becoming articled to an attorney and reading avidly to educate himself, first learning Italian and later classical languages with a group of friends. After becoming a successful attorney, and marrying, he moved with his family to a country retreat in Toxteth, and in 1796, published a study of the Medici. Later, he became a partner in a firm of bankers and developed his interest in botany, becoming one of the founders and first president of the Liverpool Botanic Garden, then at Mount Pleasant. His book covers gingers, cardamoms and related plants and the Himalayan genus *Roscoea* is named after him. His book was restricted to 150 copies, and contains 112 hand-coloured lithographs, which have a simplicity reminiscent of the artists of the Company School.

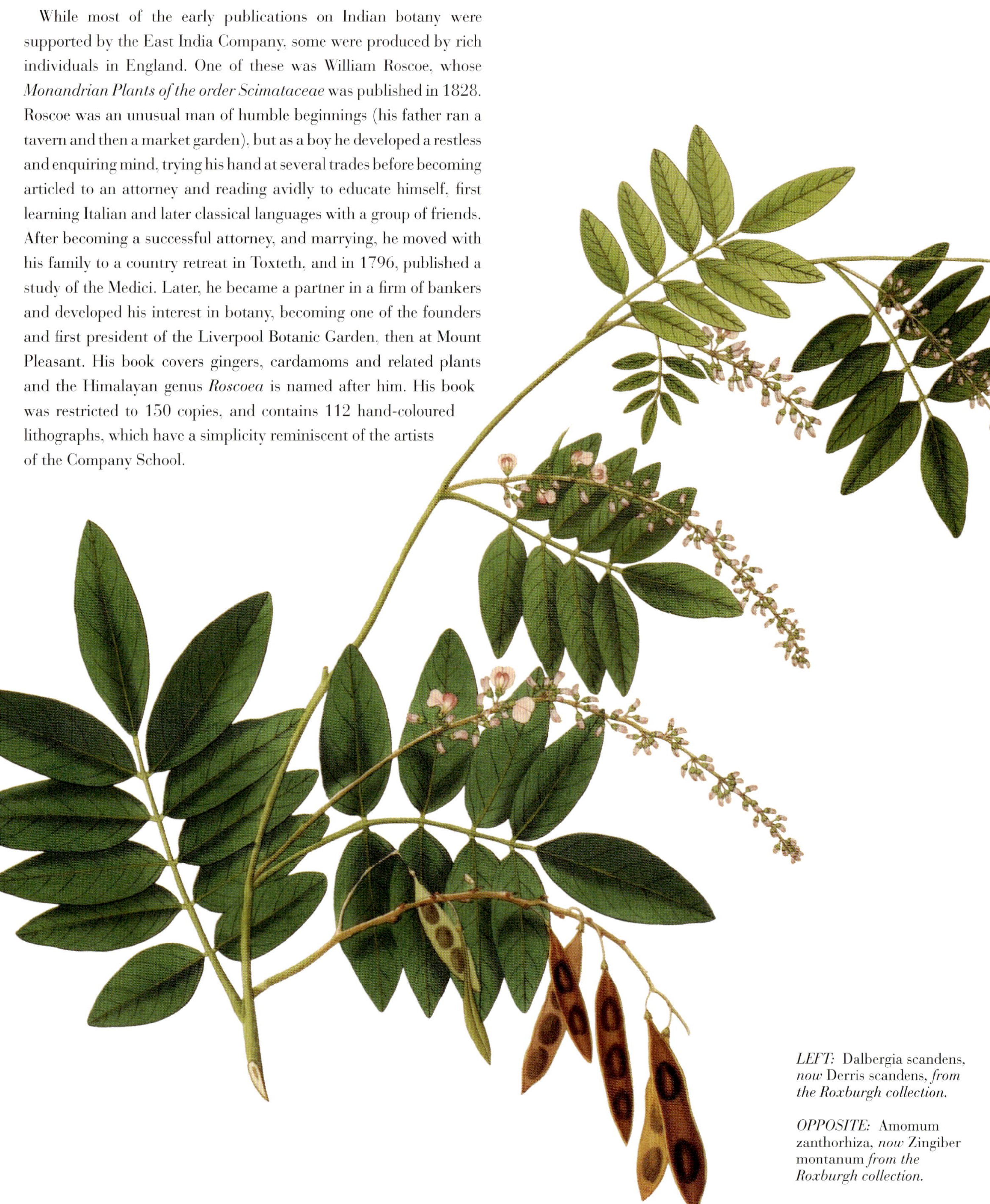

LEFT: Dalbergia scandens, *now* Derris scandens, *from the Roxburgh collection.*

OPPOSITE: Amomum zanthorhiza, *now* Zingiber montanum *from the Roxburgh collection.*

N. 1507 Amomum zanthorhiza. Roxb.

Roxburgh's successor as superintendent at the Calcutta Botanic Garden was Nathaniel Wallich (1786–1854). Wallich was born in Copenhagen, and became surgeon to the Danish settlement in India, and then became Roxburgh's assistant in 1809, before joining the East India Company and being confirmed as Superintendent in 1817, a post he held until 1856. The botanic garden in Calcutta was famous for its Indian plants but also as a staging post for plants from China, which could be settled and propagated there before being sent on to Kew. Wallich also employed Indian artists, and their work was published in the sumptuous *Plantae Asiaticae rariores*, a folio of three volumes, published between 1829 and 1831 while he was living in London.

Many of the plates in this work were by the Indian artist Vishnupersaud, who was perhaps the most skilful of the company botanical artists; while many of them concentrated on the detail and arranging the specimen carefully in two dimensions, filling most of the page, Vishnupersaud's work has real depth and perspective, as well as elegance. Other paintings in *Plantae Asiaticae rariores* were by Gorichaud, another Indian artist, and by the well-known English artists Charles M. Curtis and Miss Drake, who also worked for John Lindley. Wallich's choice of plants includes many from the Himalayas, which he visited in 1820 and 1821, about 20 years before Hooker's exploration of Nepal and Sikkim.

The Himalayas were the main subject of *Illustrations of the botany and other branches of the natural history of Himalayan mountains and the flora of Cashmere* by John Forbes Royle, published between 1833 and 1840. Royle was born in India, studied medicine in Edinburgh, and then returned to India in 1820, soon becoming Curator of the Botanic Garden at Saharanpur, north of Delhi. This gave him the opportunity to visit the Himalayas, and include ecological information in his book. Most of the plant illustrations are lithographs after drawings by Vishnupersaud,

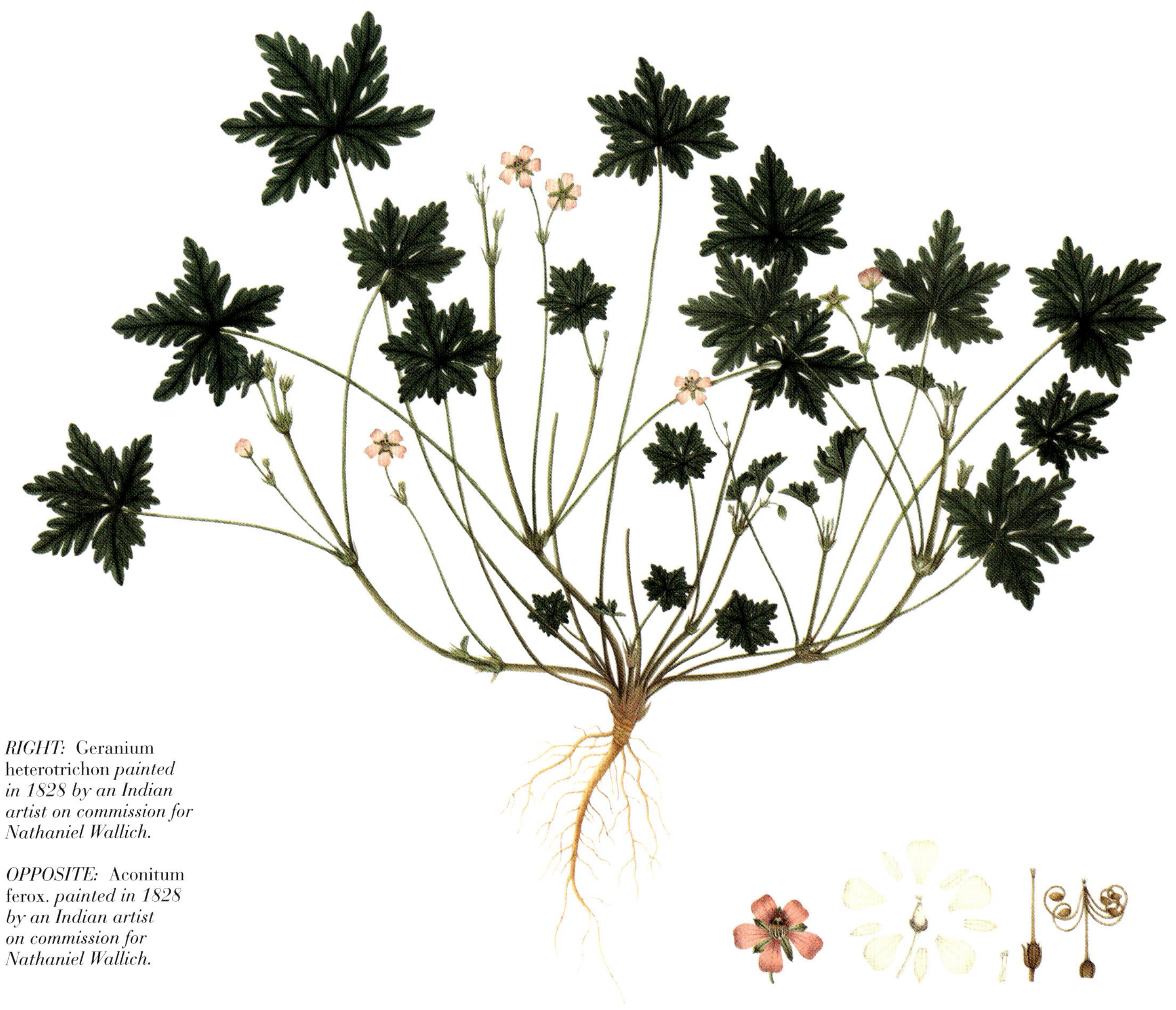

RIGHT: Geranium heterotrichon *painted in 1828 by an Indian artist on commission for Nathaniel Wallich.*

OPPOSITE: Aconitum ferox. *painted in 1828 by an Indian artist on commission for Nathaniel Wallich.*

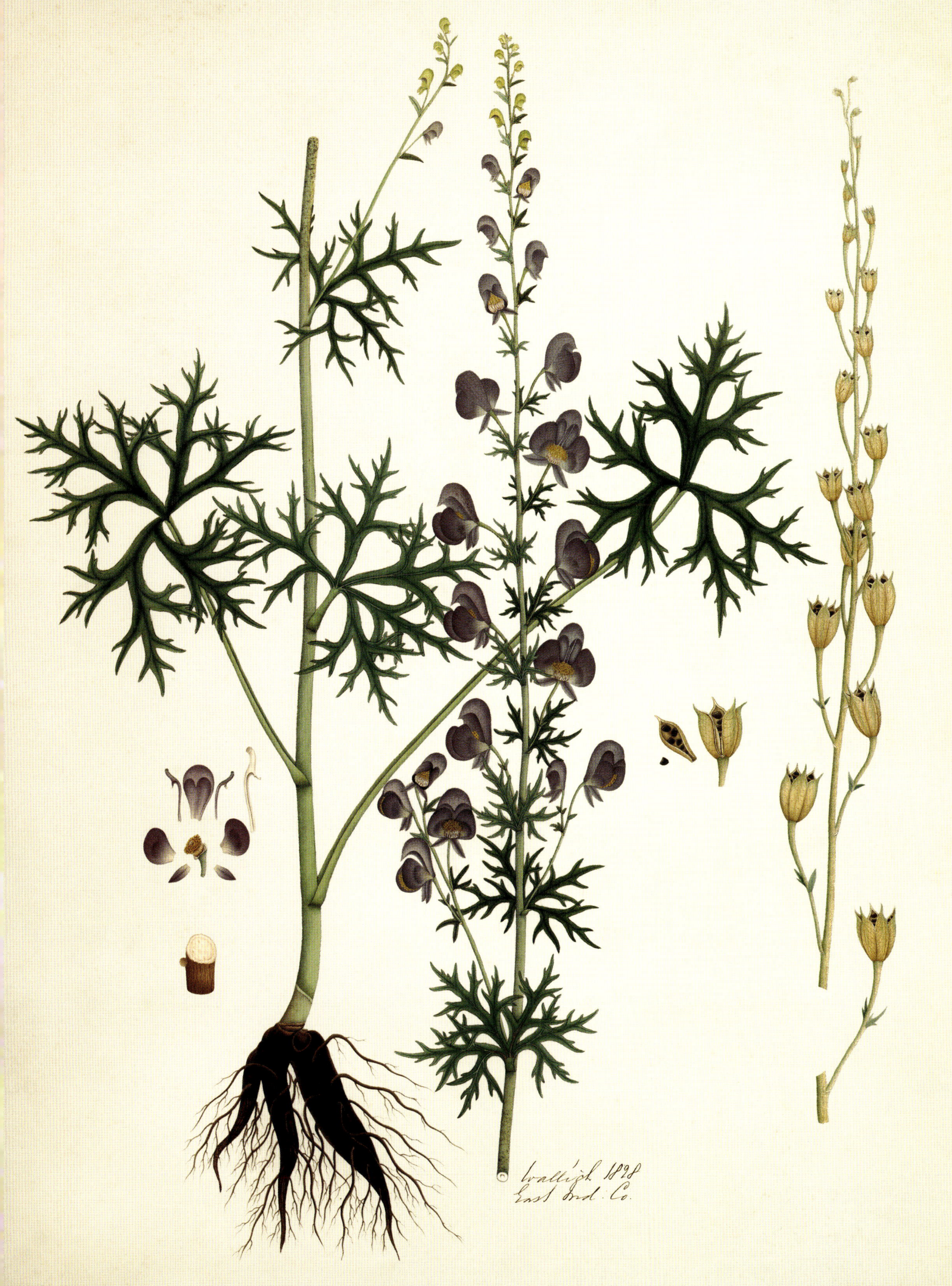
Wallich 1828
East Ind. Co.

but a few are by the same other artists who worked for Wallich.

The publications of Roxburgh and Wallich used only a very small part of the hundreds of plant illustrations drawn for the East India Company. Two copies of each painting were made in Calcutta; one was to be kept there, the other sent to London to the company's library, so that they could be studied by botanists in Europe. When William Hooker was planning to visit Ceylon, he visited London (from Glasgow), and made reduced copies of over 2000 of Roxburgh's drawings, so that he could learn the plants of southern India, and perhaps take his copies to use in place of a textbook. Both Hooker's copies and the originals, passed on from the India Office library, are now at Kew; only a few of them have ever been published. Roxburgh's drawings were also copied for Robert Wight (*see* page 181), so the same image can often be found in several libraries or publications. Copying earlier paintings was and still is a tradition in Indian art, and the copies are not considered inferior to the original. The current writings of Henry J. Noltie have revealed many fascinating aspects of Company School botanical painting.

Though most of the botanical collectors in India were British, or worked for the East India Company, a few botanists from other European countries came to India. One of the most attractive characters was Victor Jacquemont who decided to leave France to escape from an unhappy love affair, and was commissioned by the Jardin des Plantes in Paris to collect in the Himalayas. He first visited Royle in Saharanpur in 1828 and then travelled via Simla up the Sutlej river towards Tibet. He returned to Simla for the winter, where he was invited by the Maharajah of Kashmir to visit his territory. In 1831, Jacquemont explored much of Kashmir, and then returned to Bombay with his collections, but died there of cholera, in 1832, aged only 31. He had collected many species new to science, and some, notably the white-barked *Betula jacquemontii*, were named after him. His collections were returned to France and published with drawings by Alfred Riocreux and Eulalia Delile, who were working at the Jardin des Plantes, as *Voyage dans l'Inde* between 1835 and 1844.

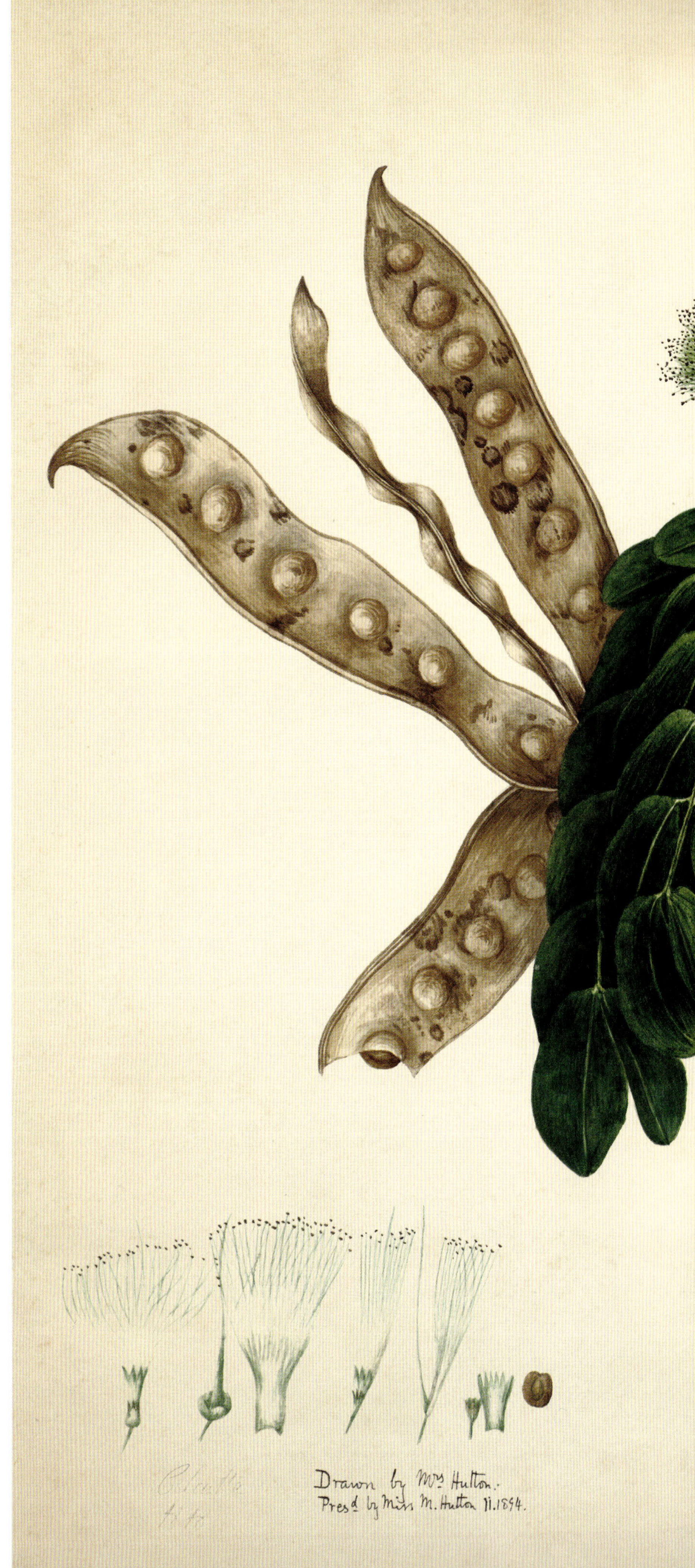

RIGHT: Mimosa sirissa, *now* Albizia lebbeck, *the Siris tree, painted by Mrs Hutton in Calcutta in 1818.*

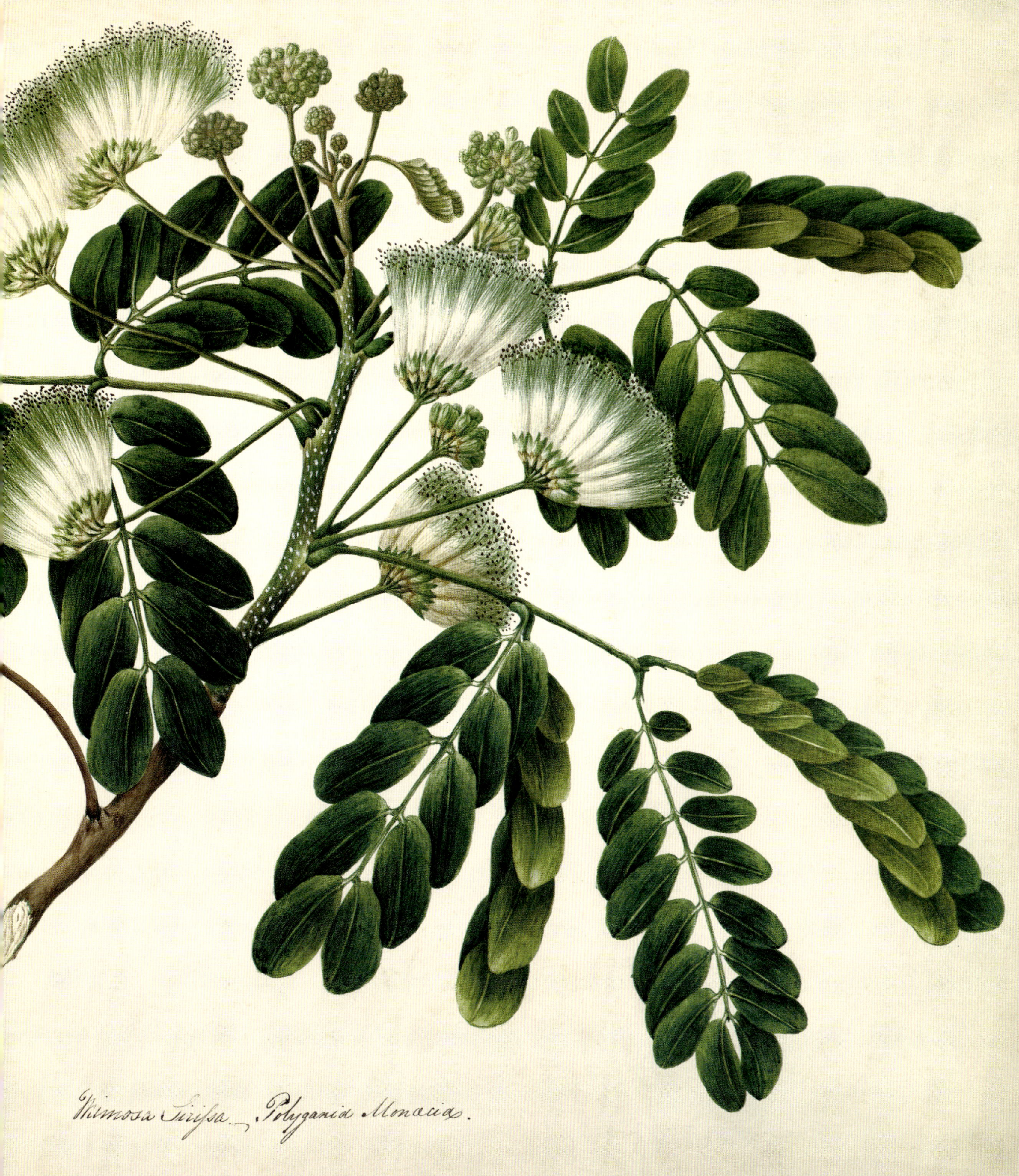
Mimosa Sirissa_ Polygamia Monoecia.

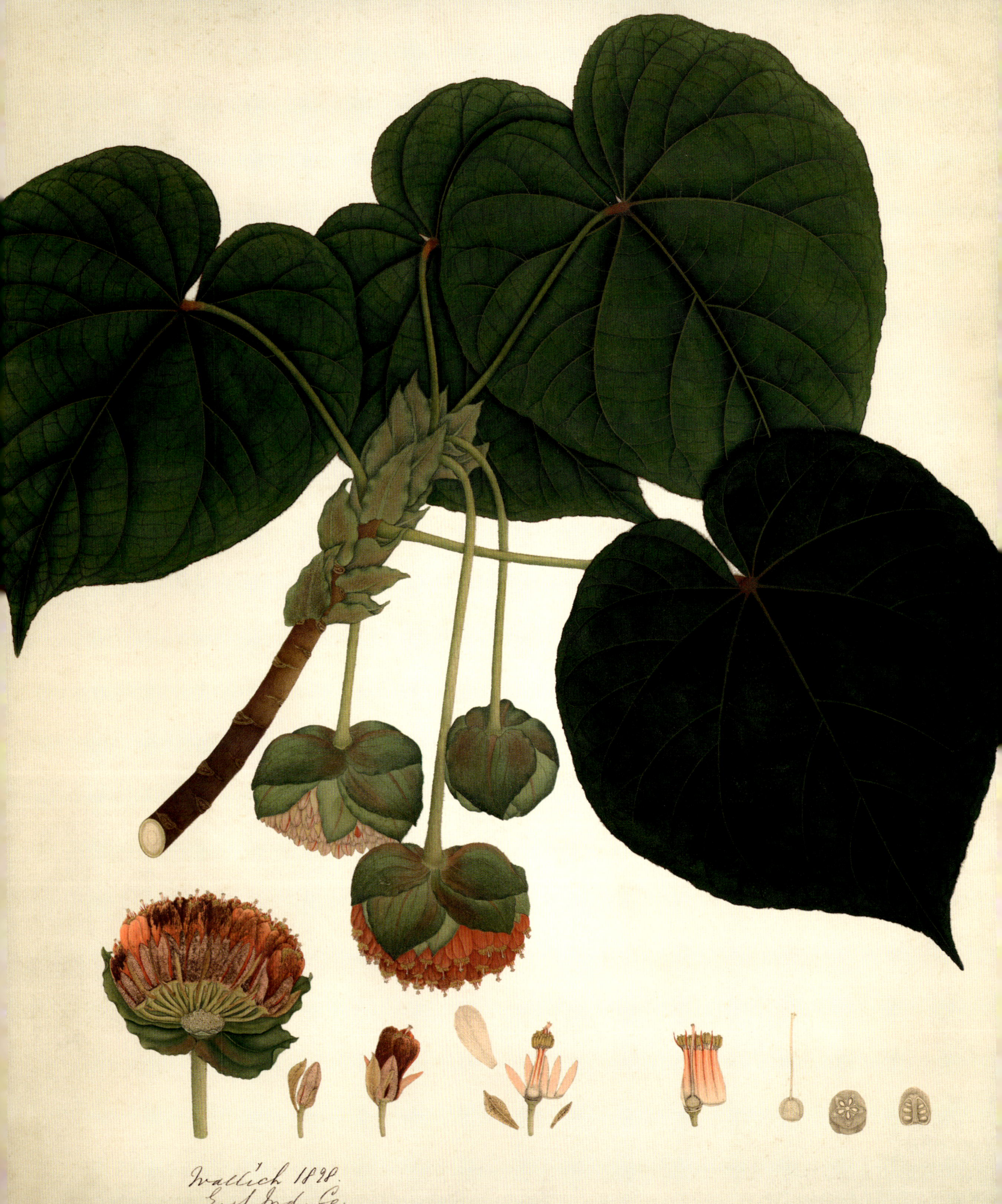
Wallich 1828.
East Ind. Co.

The most prolific publisher of illustrations of Indian plants was Robert Wight (1796–1872). Wight was born in Scotland, and trained in medicine in Edinburgh. He went to India as Assistant Surgeon with the Madras Medical Service in 1819, and started to collect plants in his spare time, employing Indian artists to make paintings of the specimens, as well as drying them for his herbarium. Soon, Wight was sending specimens of new species to William J. Hooker, then Professor of Botany at Glasgow University, and when Wight returned to Scotland for sick leave in 1831, he brought with him over 100,000 specimens of around 4000 species, as well as his collection of paintings.

Of all the botanists working at this time, Wight was one of the most convinced of the value of illustrations. In India these had the great advantage over herbarium specimens that they were easier to transport and also relatively easy to copy. Wight had copies made of Roxburgh's published and unpublished plant illustrations, which he could use for reference in the absence of a written flora. He also saw the need in India, for the production of cheap books for everyday use by botanists, not the hugely costly ones being produced in London for rich men's libraries. While home in Scotland, Wight learnt lithography, and when he returned to India, he began to publish illustrated books, using line drawings after the paintings by his two Indian artists, Rungia and Govindoo. His *Illustrations of Indian botany*, with lithographs, was published in 1831, and covered mainly plants from Peninsula India. In 1834, he published a handbook, with short descriptions, the *Prodromus Florae Peninsulae Indiae Orientalis*, and followed this, in 1840, *Icones plantarum Indiae Orientalis: or figures of Indian plants* with 2115 lithographed line drawings, some by Wight himself. This is still one of the most useful guides to the flora of southern India. Wight also published a larger, more impressive book on plants from the Neilgherry Hills, *Spicilegium Neilgherrense, or a selection of Neilgherry plants, drawn and coloured from nature, with brief descriptions of each*, which has 204 hand-coloured lithographs after paintings by Rungia and Govindoo in a large quarto size, and was printed in Calcutta. Wight retired from India in 1853, and farmed near Reading until his death in 1872, aged 76.

Joseph Hooker's own paintings and those of his father, William Jackson Hooker, are described in the next chapter, but Joseph was instrumental in publishing a beautiful folio, with paintings by Indian artists. His first expedition into the Himalaya was to Darjeeling, a hill station at 2134 m (7000 ft), used by Europeans to recuperate from the unhealthy heat

OPPOSITE: Dombeya wallichii *from the Wallich collection.*

RIGHT: Medinilla rubicunda *from the Wallich collection.*

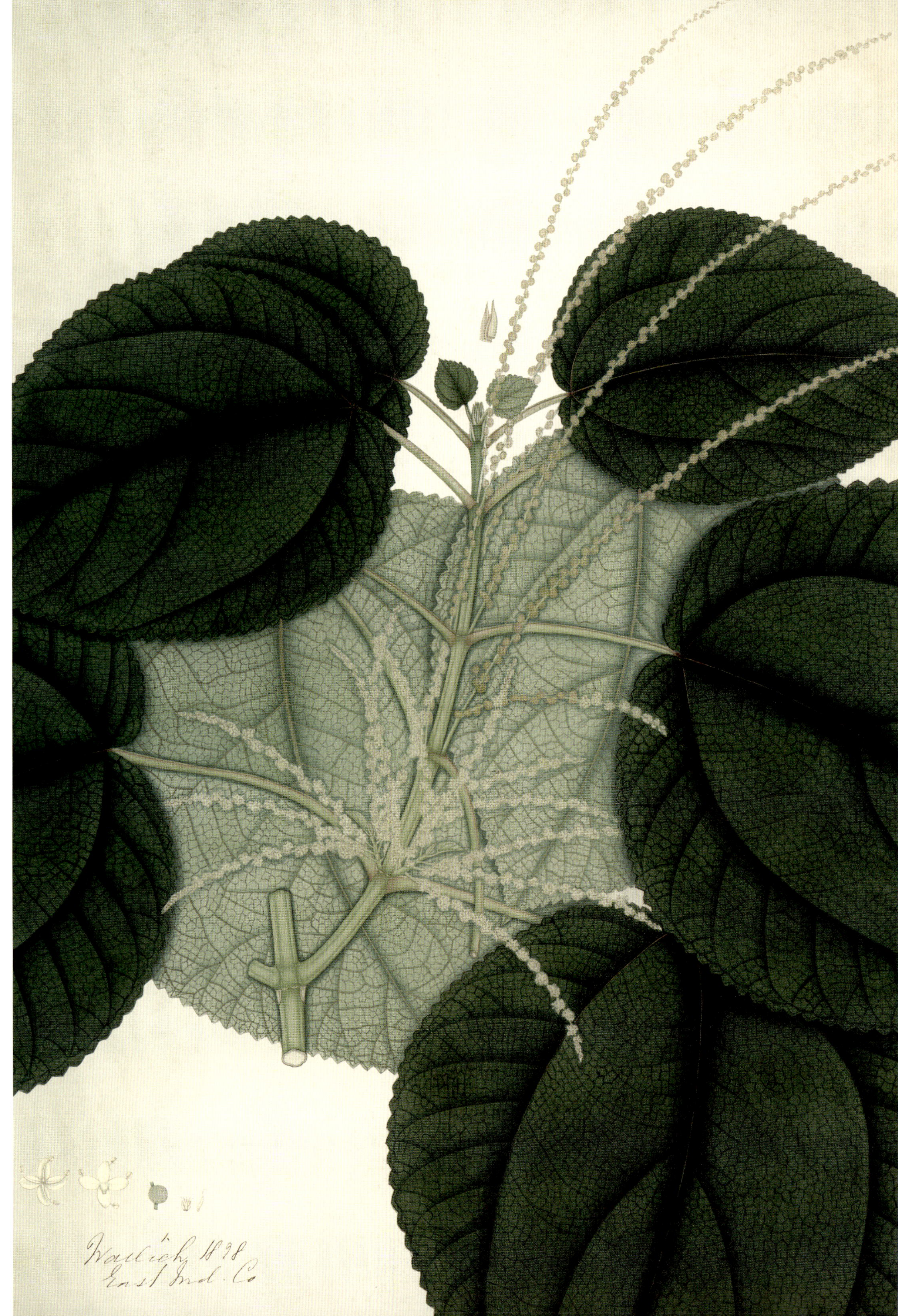
Wallich 1828
East Ind. Co

of Bengal. Here, he met Dr Archibald Campbell, superintendent of the local sanatorium and J. F. Cathcart, a semi-invalid, but keen amateur botanist who employed local artists to paint the flowers brought to him by collectors from the hill villages.

Hooker wrote to his father at Kew in 1850, "Cathcart's drawings… are all now going on with being made at or near Darjeeling, of Darjeeling plants. He has five artists at work, who turn out about three plants a week – it costs him more than all my pay together." There were eventually more than a thousand drawings, and Hooker promised to have them published. Cathcart died in Switzerland on his way home from India, but Hooker was true to his word, and the paintings became the basis for one of the most interesting and beautiful of all botanical books: *Illustrations of Himalayan Plants, selected from drawings made for the late J. F. Cathcart Esq. of the Bengal Civil Service*. Published in London in 1855, it contains large hand-coloured lithographs, redrawn from the originals by W. H. Fitch at Kew. The title page is particularly charming, with a garland of Himalayan flowers, with bamboo, palms, rhododendrons, *Agapetes*, *Dactylicapnos ventii* and *Pleione hookeriana*. The centrepiece is the yellow poppy, *Cathcartia* (or *Meconopsis*) *villosa*. The highlights of the book were a huge plate of a pink *Magnolia campbellii*, drawn by Hooker himself and one of the first illustrations of a blue poppy, *Meconopsis simplicifolia*. There were 176 subscribers to the book, including Charles Darwin and George Canning, Governor-General of India from 1856 to 1862. No finer books on plants than this one have ever been printed, and later attempts were more botanical and less ostentatious.

The Royal Botanic Garden of Calcutta, as it had become, produced a series of monographs in the late nineteenth century, as part of their annals, culminating in *The Orchids of the Sikkim-Himalaya*, with 484 hand-coloured lithographs by Robert Pantling, then superintendent of the government's quinine plantation. Other titles included *The Magnoliaceae of British India*, with text by Sir George King, a native of Peterhead in Aberdeenshire, and illustrations by Indian artists.

OPPOSITE: Boehmeria macrophyllum *from the Wallich Collection, painted in 1828.*

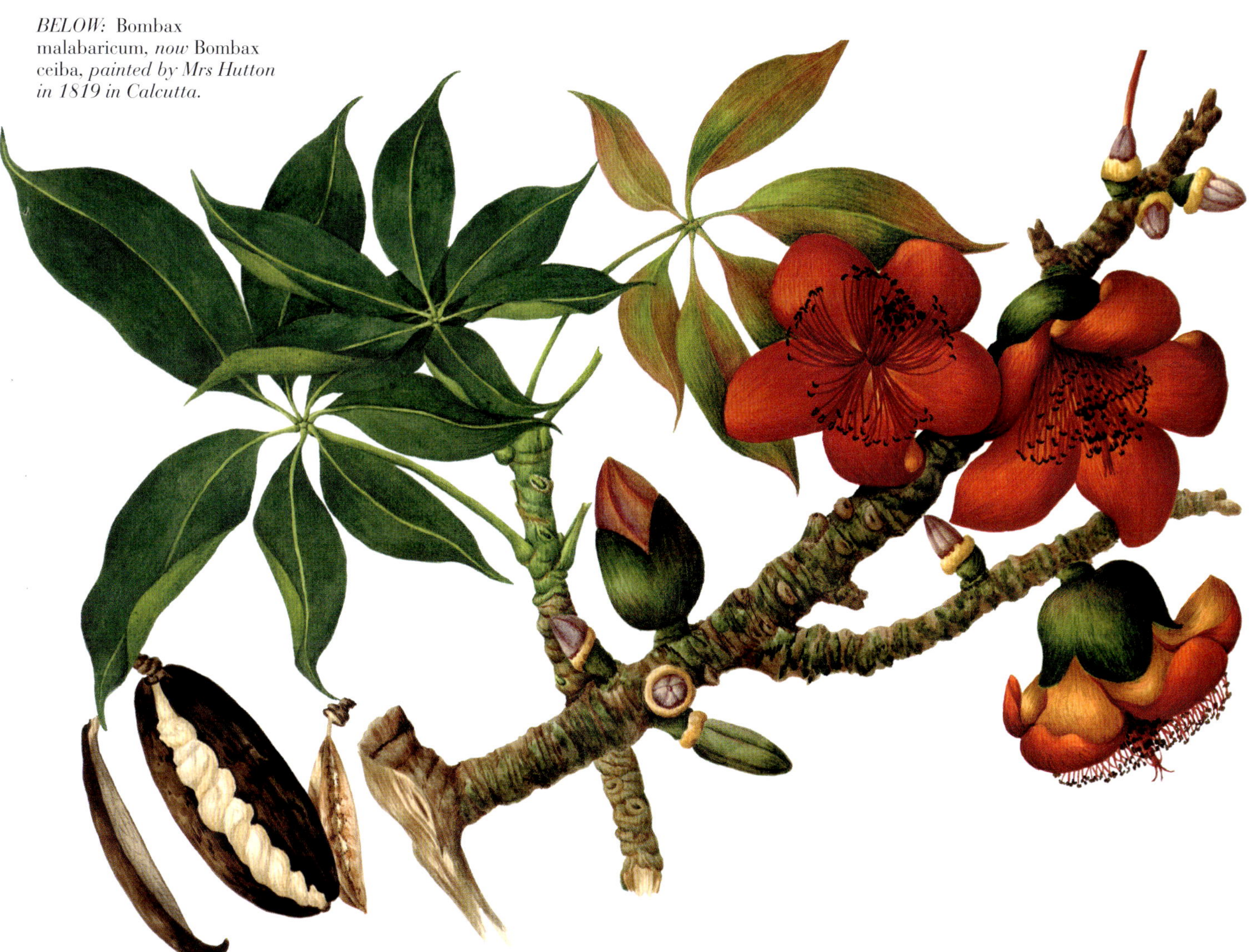

BELOW: Bombax malabaricum, *now* Bombax ceiba, *painted by Mrs Hutton in 1819 in Calcutta.*

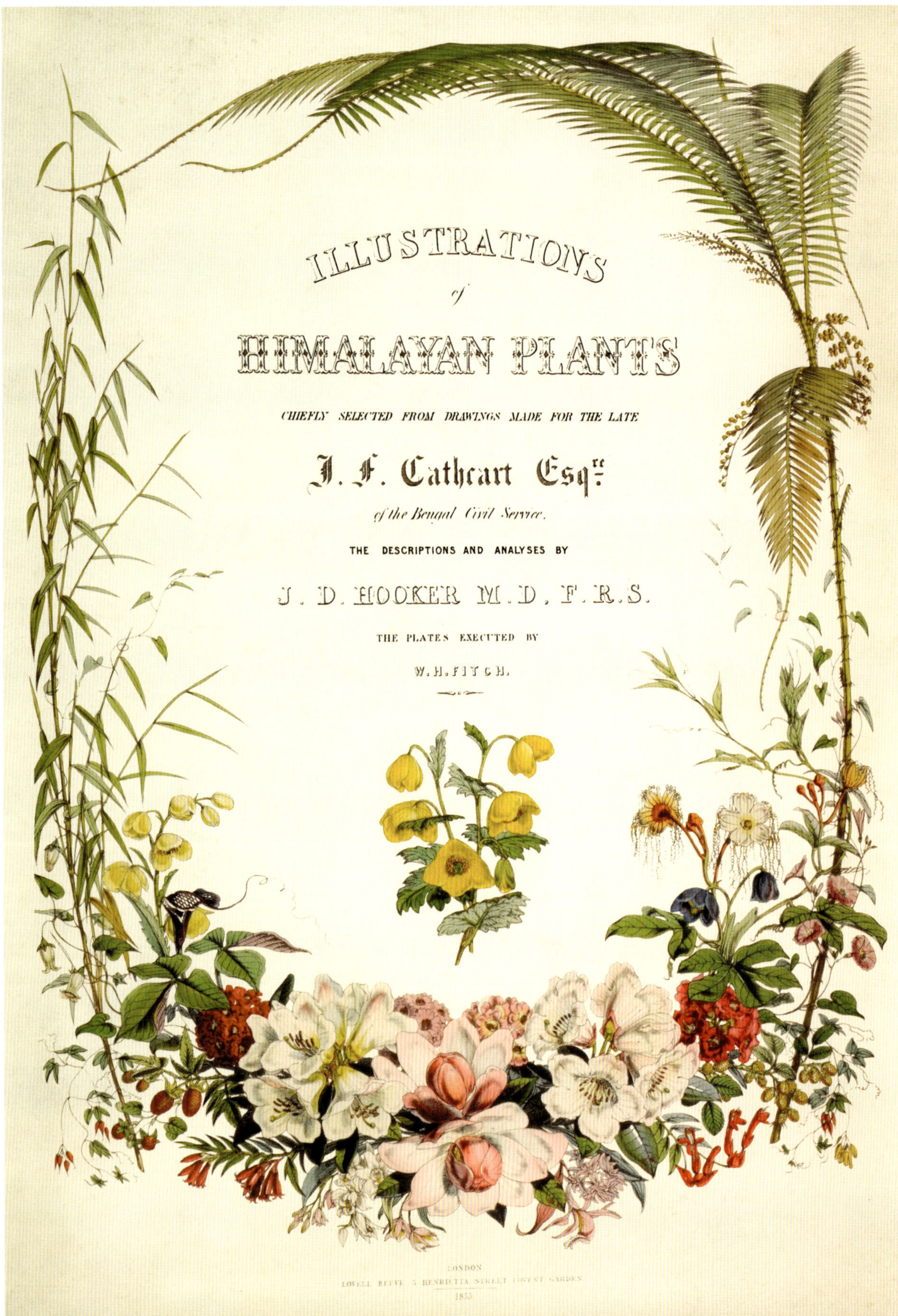

ILLUSTRATIONS
of
HIMALAYAN PLANTS
CHIEFLY SELECTED FROM DRAWINGS MADE FOR THE LATE
J. F. Cathcart Esqr.
of the Bengal Civil Service.
THE DESCRIPTIONS AND ANALYSES BY
J. D. HOOKER M. D. F. R. S.
THE PLATES EXECUTED BY
W. H. FITCH.

LONDON
LOVELL REEVE 5 HENRIETTA STREET COVENT GARDEN.
1855

LEFT: The title page of Illustrations of Himalayan Plants *by Joseph Hooker, with a garland of Himalayan plants and the yellow* Cathcartia villosa *in the centre*

OPPOSITE: Meconopsis simplicifolia *from* Illustrations of Himalayan Plants.

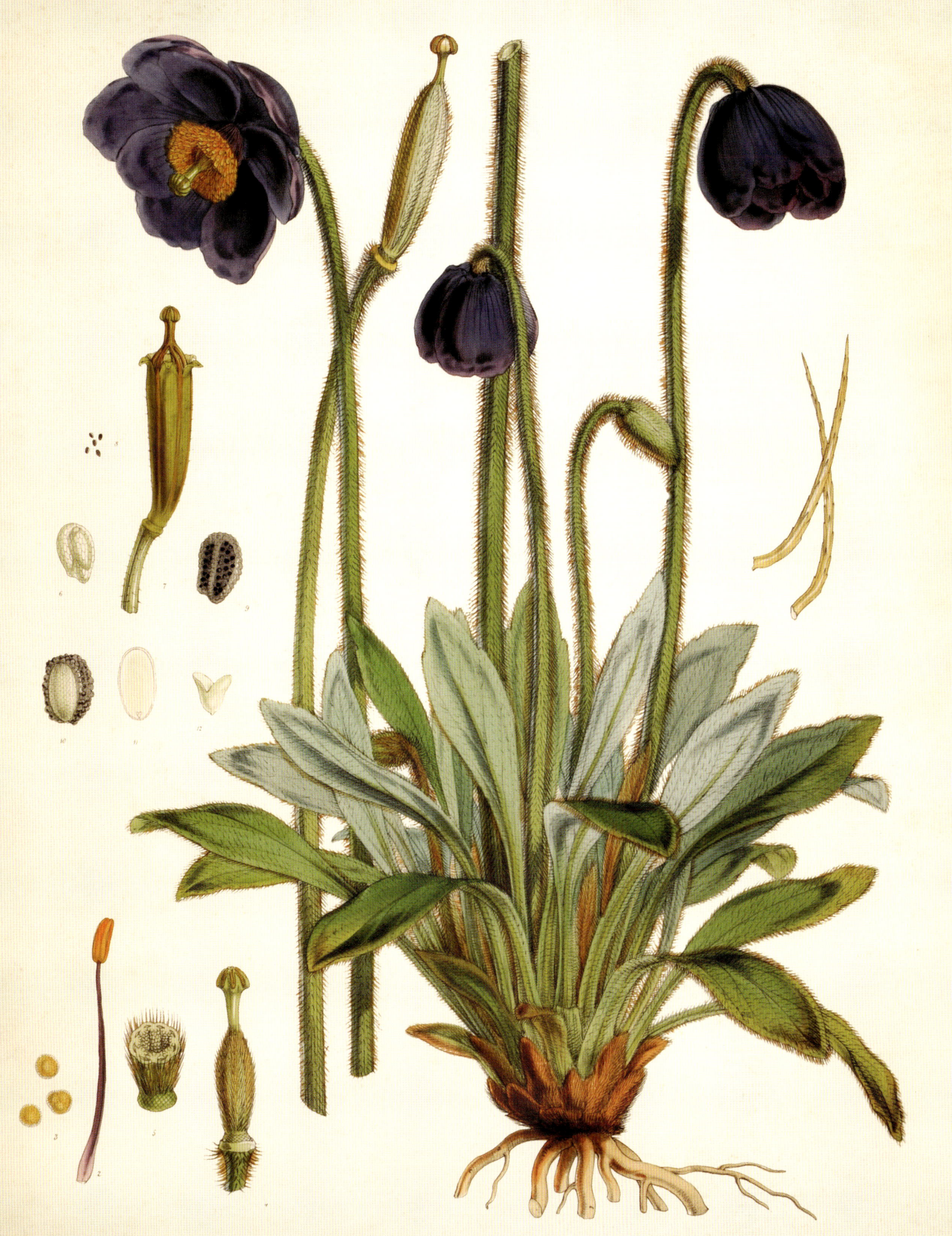

The Story of Flora Danica 1761–1883

The name *Flora Danica*, has been used for two publications, the first of which was a herbal published in 1648 by the doctor and botanist Simon Paulli. The *Flora Danica* to which we refer here, however, was the brainchild of Georg Christian Oeder, who in 1753 proposed a popular publication in which all the native plants of Denmark would be illustrated and described.

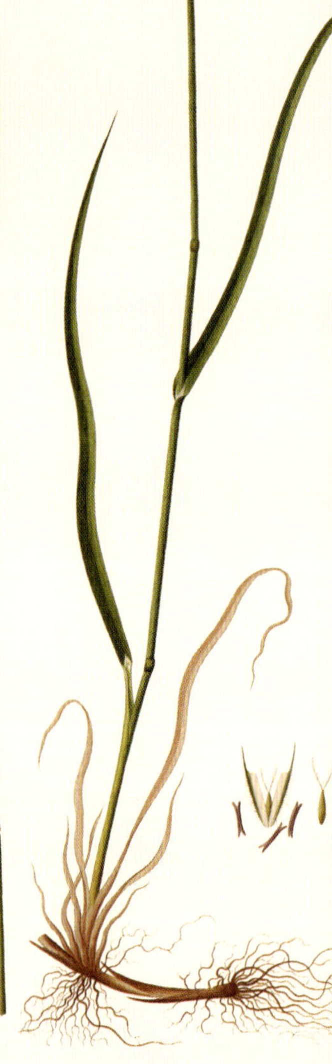

The project was subsidized by the royal family, and the first part published in 1761; Oeder (after whom *Pedicularis oederi* was named by a later editor, Martin Vahl) was editor until 1772, and was succeeded by a series of others. The *Flora* appeared in 51 parts (the last appearing in 1883) plus three supplements, and in two versions: one hand-painted and one uncoloured. Copies of the latter were distributed free, via the clergy, to schools and professional people, who were expected, in return, to provide relevant local information to the Royal Botanical Institution.

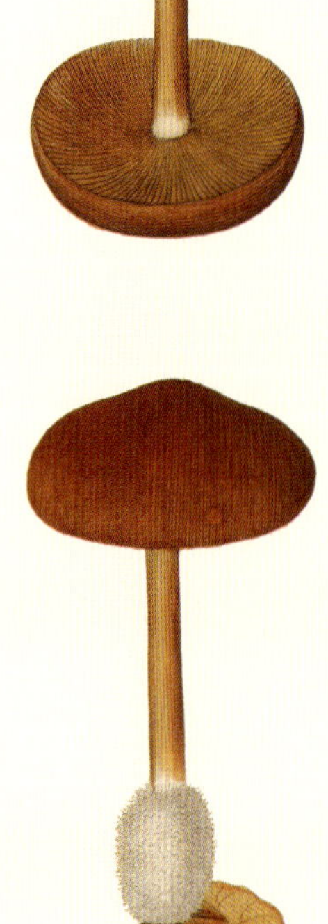

This ambitious project required dedicated botanical artists and engravers, and Oeder found the father and son combination of Michael and Martin Rössler of Nuremberg, who combined these roles. Michael (1705–77) was the engraver, Martin (1727–82) the artist, and the majority of the copper plates produced by them for *Flora Danica* are now in the collection of the Botanical Museum in Copenhagen. The text, in Latin, Danish and German, was never completed, although the first section, an introduction to botany, was published.

For political reasons, the plants figured in the *Flora* before 1814 originally included specimens not only from Denmark, but also from Norway, and until 1864, when the territories of Schleswig-Holstein were lost, a number of German plants. Later still, it was proposed that the work should include plants from all areas of Scandinavia, and Swedish plants were added.

A porcelain dinner service, decorated with almost exact copies of paintings from the *Flora*, was ordered by the King of Denmark in 1790; it is thought to have been intended as a gift to Catherine the Great, but the order was cancelled when Catherine died in 1796. Subsequently, the service was made and delivered to the King in 1802, and its first recorded use was at a banquet celebrating his birthday in January 1803; this set, comprising around 1500 pieces, is still in the hands of the Danish royal family. The pieces were hand made, with leaves and flowers painted by Johann Christoph Bayer (1738–1812) of Nuremberg, who had studied flower painting with Johann Christoph Dietsch (1710–68), a landscape artist and flower painter. Bayer worked on the original *Flora* from 1769 onwards, and was employed from 1776 to 1802 by the Royal Copenhagen Porcelain Manufactory.

In 1862, on the occasion of the wedding of Princess Alexandra, daughter of the Danish King Christian IX, to the English Prince of Wales, later King Edward VII, a further dinner service was produced from the same factory; this set is kept at Windsor Castle. The Royal Copenhagen factory still produces pieces hand painted with plates from *Flora Danica*.

ABOVE: Phleum pratense *from* Flora Danica.

LEFT: Agaricus peronatus, *now* Collybia peronata, *from* Flora Danica.

OPPOSITE: A dandelion, Taraxicum officinale, *from* Flora Danica.

Flora Danica Tab DLXXIV.

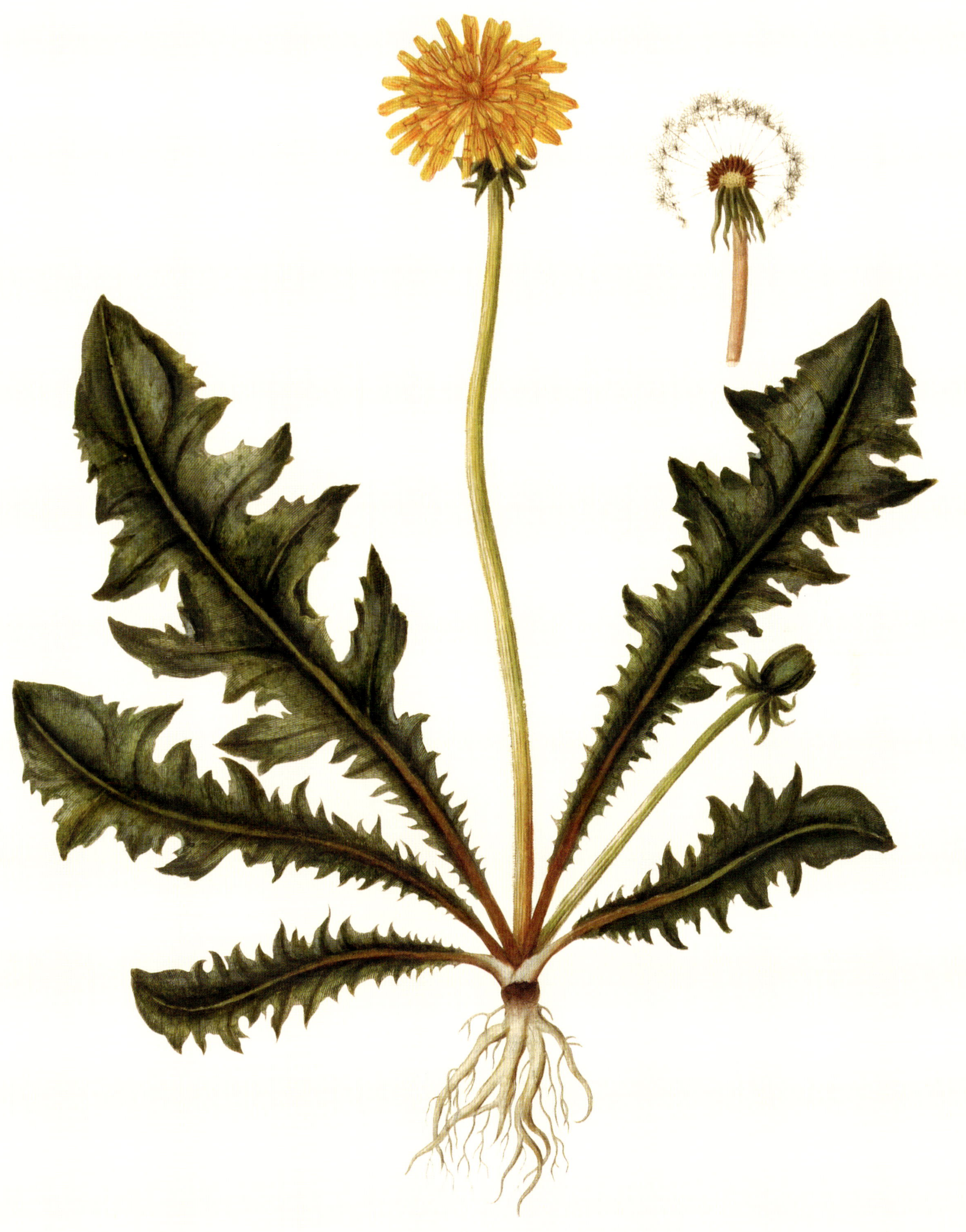

14

A NEW ERA AT KEW

William Jackson Hooker and his son Joseph Dalton were the pre-eminent botanists of the nineteenth century, not only through their running and expansion of the Royal Botanic Gardens, Kew, but also through their friendship and correspondence with other scientists, such as Charles Darwin, T. H. Huxley and the American botanist Asa Gray, and their numerous publications.

William Jackson Hooker was born in Norwich on 6 July 1785, the son of Joseph, a cloth merchant's clerk, and his wife Lydia. William's love of nature was encouraged by his parents, and he was surrounded by gardens. Norwich was a city with a long tradition of horticulture, thanks at least in part to the Dutch and Flemish weavers who had settled there in the sixteenth century, and the Huguenots who had sought refuge in the seventeenth century. Joseph was a grower and collector of succulent plants, while William soon developed a keen interest in mosses and lichens.

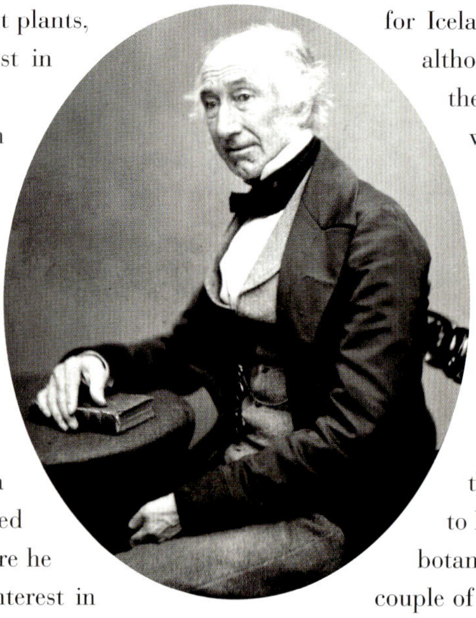

William was fortunate in being encouraged in his botanical interests by a local botanist and friend of Sir Joseph Banks (*see* pages 98–99), Dr James Edward Smith, who had written the descriptions for James Sowerby's *English Botany*, and for Sibthorp's *Flora Graeca* as well as becoming the first President of the Linnean Society. William attended Norwich Grammar School, and, on inheriting some money and property from his godfather, a farmer and brewer in Kent, was later educated privately at Starston Hall, near Norwich, where he was taught about estate management. His interest in natural history – especially botany and entomology – increased, and when he sent a specimen of a rare moss to Dawson Turner, a banker and keen amateur botanist living at Great Yarmouth, he received by return a copy of Turner's *Botanists' Guide*.

Hooker soon became friends with Turner and his family, and was encouraged to join in the family obsession with drawing; Dawson's daughters were fortunate in being given lessons by the artist John Sell Cotman. William learnt to draw seaweeds well enough to illustrate a book, *Fuci*, on which Turner was working. It was eventually published, with 258 fine colour plates, between 1808 and 1819.

Thanks to Turner and Dr Smith, Hooker was given an introduction to Sir Joseph Banks, who gave him the run of his library and herbarium in London. Banks encouraged Hooker to travel, and offered him a place as naturalist on a ship bound for Iceland, which Hooker accepted enthusiastically, although this was nearly the end of his travels as the boat caught fire on the return voyage and was completely destroyed. Fortunately, all the crew were rescued, but Hooker's specimens and most of his notes were lost.

On his return to England, Hooker was advised by Dawson Turner to put the money he had inherited into buying a part share of a brewery in Suffolk, the rest of which was owned by Turner and two business partners. Hooker accepted, and moved into the house that came as part of the deal, and attempted to learn about the business, while keeping up his botanical work as a sideline. In 1814, after a tedious couple of years in which he tried to take an interest in brewing and also to sell the farms he had inherited in Kent, Hooker embarked on an extended tour around the south of France and Switzerland, where he met many of the important botanists of his generation, with whom he was to correspond in later years.

Hooker's association with the Turner family soon became permanent, with his marriage to Maria, the eldest daughter, in 1815, and the next year their eldest son, William Dawson, was born, followed by Joseph Dalton (*see* page 192) the following

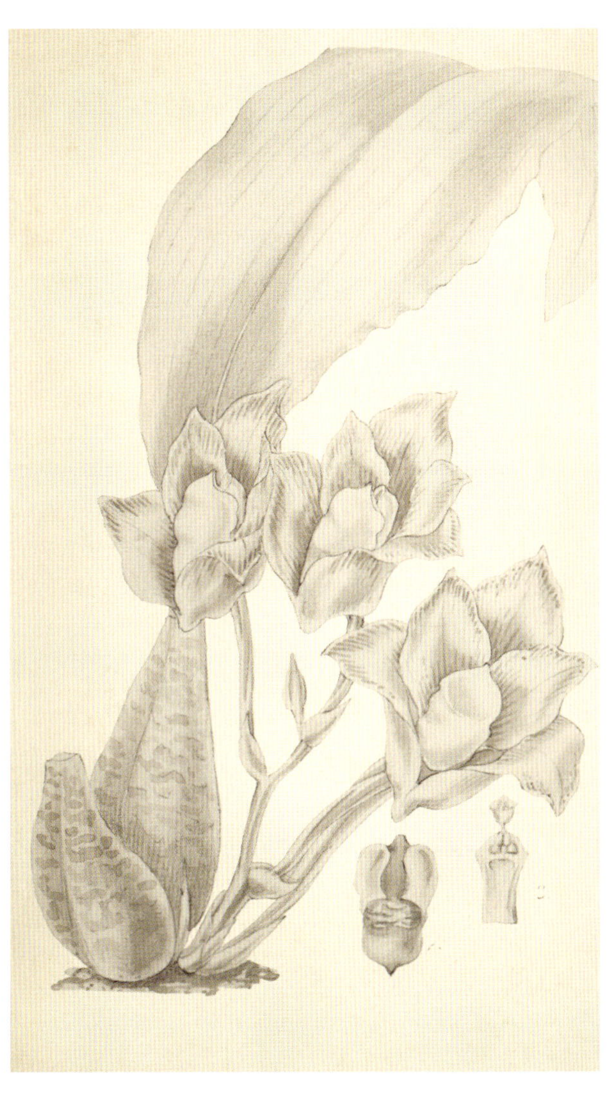

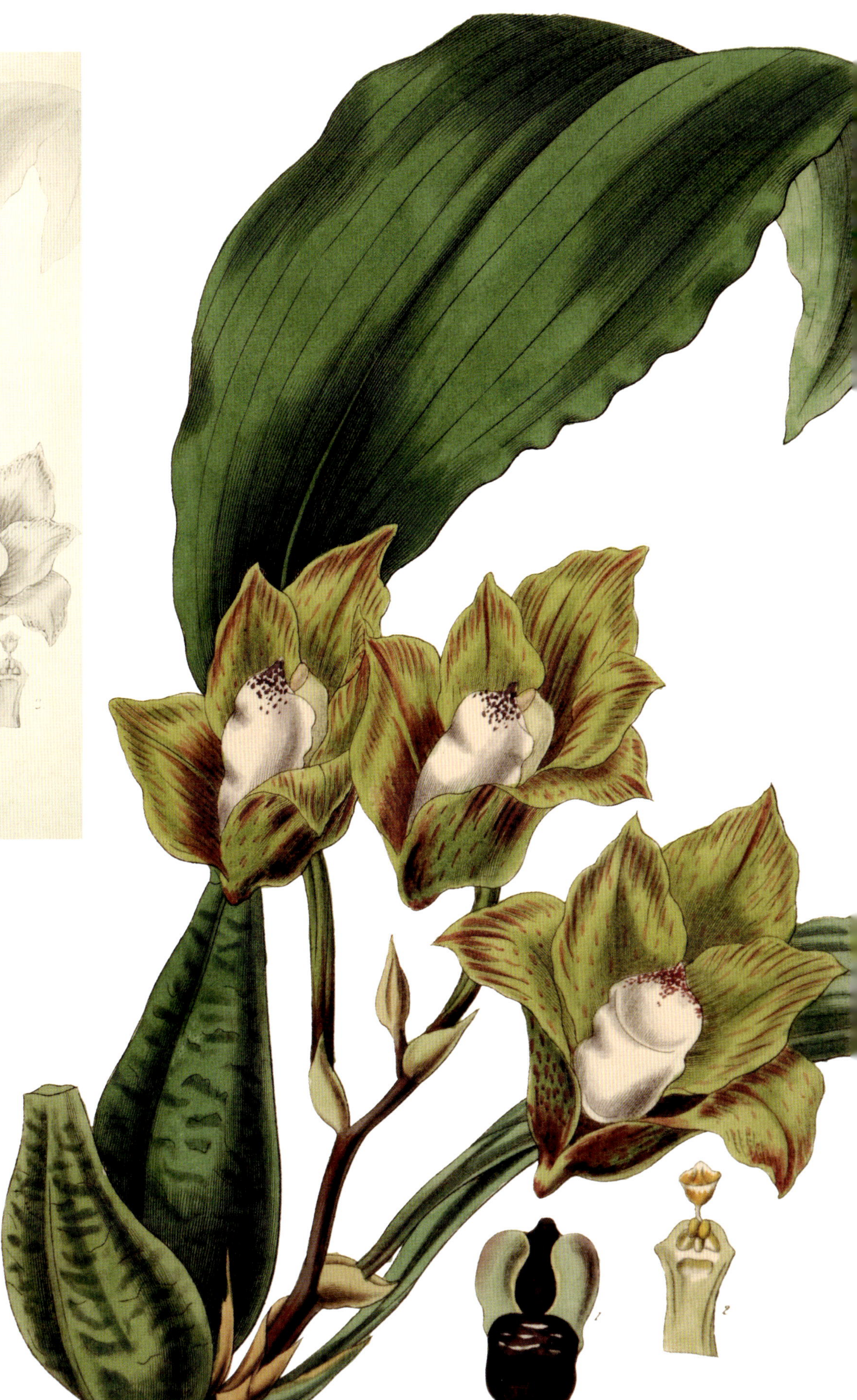

ABOVE: William J. Hooker's original pencil study of Maxillaria tetragona.

RIGHT: The final painted version of Maxillaria tetragona, *now* Bifrenaria tetragona, *as it appeared in* Curtis's Botanical Magazine.

OPPOSITE: William Hooker, who succeeded William Townsend Aiton as Director ot the Royal Botanic Gardens, Kew.

summer. Hooker worked hard at both brewing and botany, but there was a disastrous financial slump at this time, which affected all businesses, the brewery being no exception. Hooker was more or less tied to his failing business and family responsibilities, having by now a third child, but a number of young botanists, amongst them John Lindley (*see* page 107) came to visit him in East Anglia, and William encouraged them to travel, and gave them introductions to Sir Joseph Banks. Meanwhile, Banks was able to help William himself one final time (he died in 1820), by securing for him, with the help of the botanist Robert Brown, the Professorship of Botany at the University of Glasgow.

The Hooker family duly moved to Glasgow, and Hooker arranged for plants from Kew and elsewhere to be sent up to increase the collection at the Botanic Gardens. He also prepared sets of large coloured drawings showing the anatomy of plants for his students, and became a popular lecturer, often taking his students on field trips. He wrote and published books and papers, including a serial publication entitled *Exotic Flora*, and from 1827 onwards, edited and drew the plates for *Curtis's Botanical Magazine*. In addition to his university lectures for medical students, he gave popular lectures on botany to members of the public, and continued to accumulate a large personal herbarium; various students came, in their spare time, to help Hooker organize the specimens, and in due course Hooker took on a young man, Walter Fitch (*see* page 200) to help him.

William J. Hooker was knighted for his services to botany in 1836, and in the summer of 1841 was appointed the first Director of the Royal Garden at Kew, which was at that time quite small, extending to only about 11 acres. Hooker set about running the garden with huge energy, improving it, enlarging it, and building new greenhouses, including the famous Palm House. He continued in this vein for the next 20 years, but gradually gave over much of the responsibility to his son, Joseph.

Sir William died on 12 August 1865. Fitch's last work for Sir William was to design a garland of wheat and grasses for his Wedgwood memorial plaque, which still sits on the wall of St Anne's Church, Kew Green.

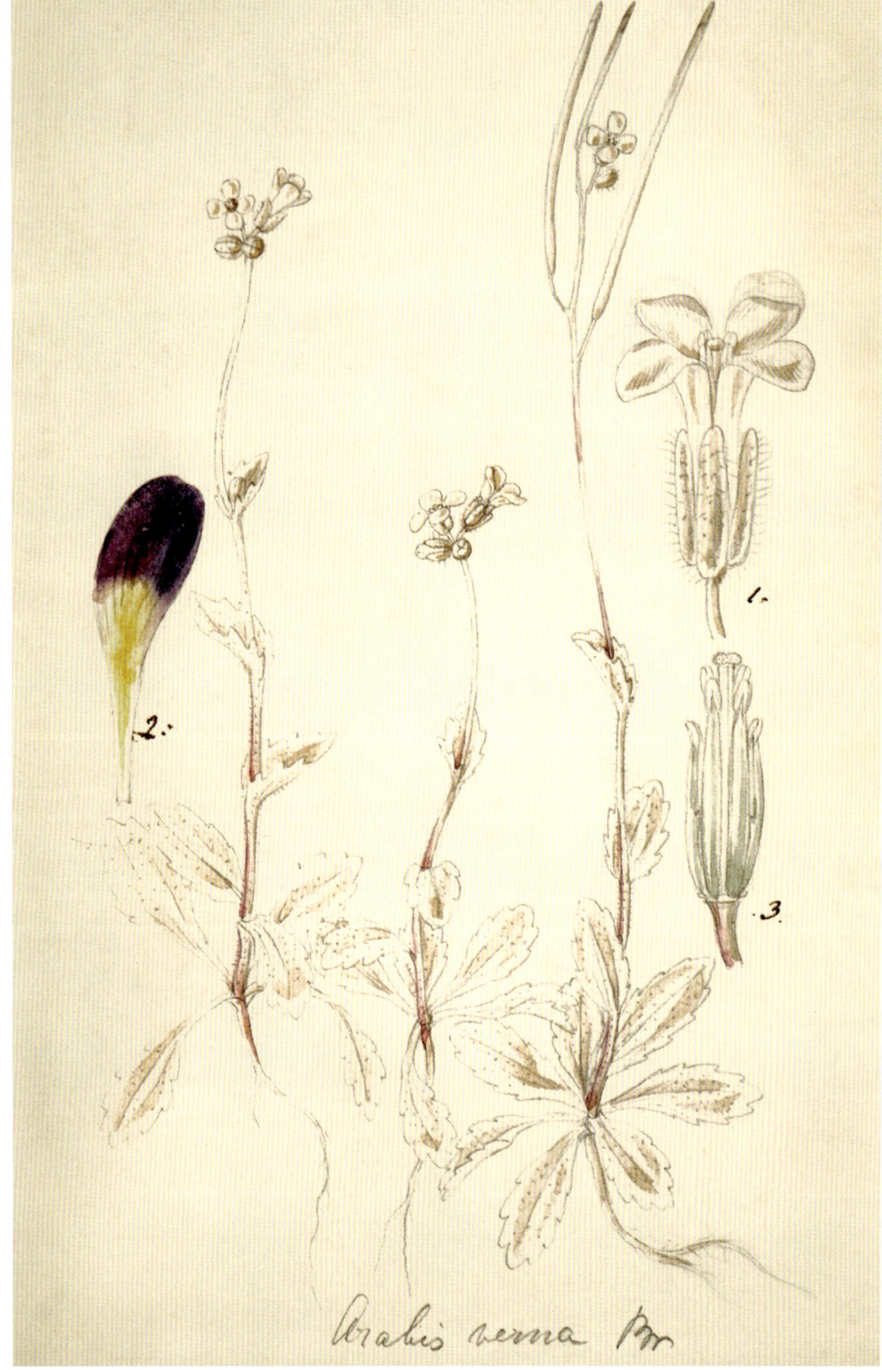

RIGHT: William Hooker's original study of Arabis verna.

RIGHT: The final version of Arabis verna, *which appeared in* Curtis's Botanical Magazine.

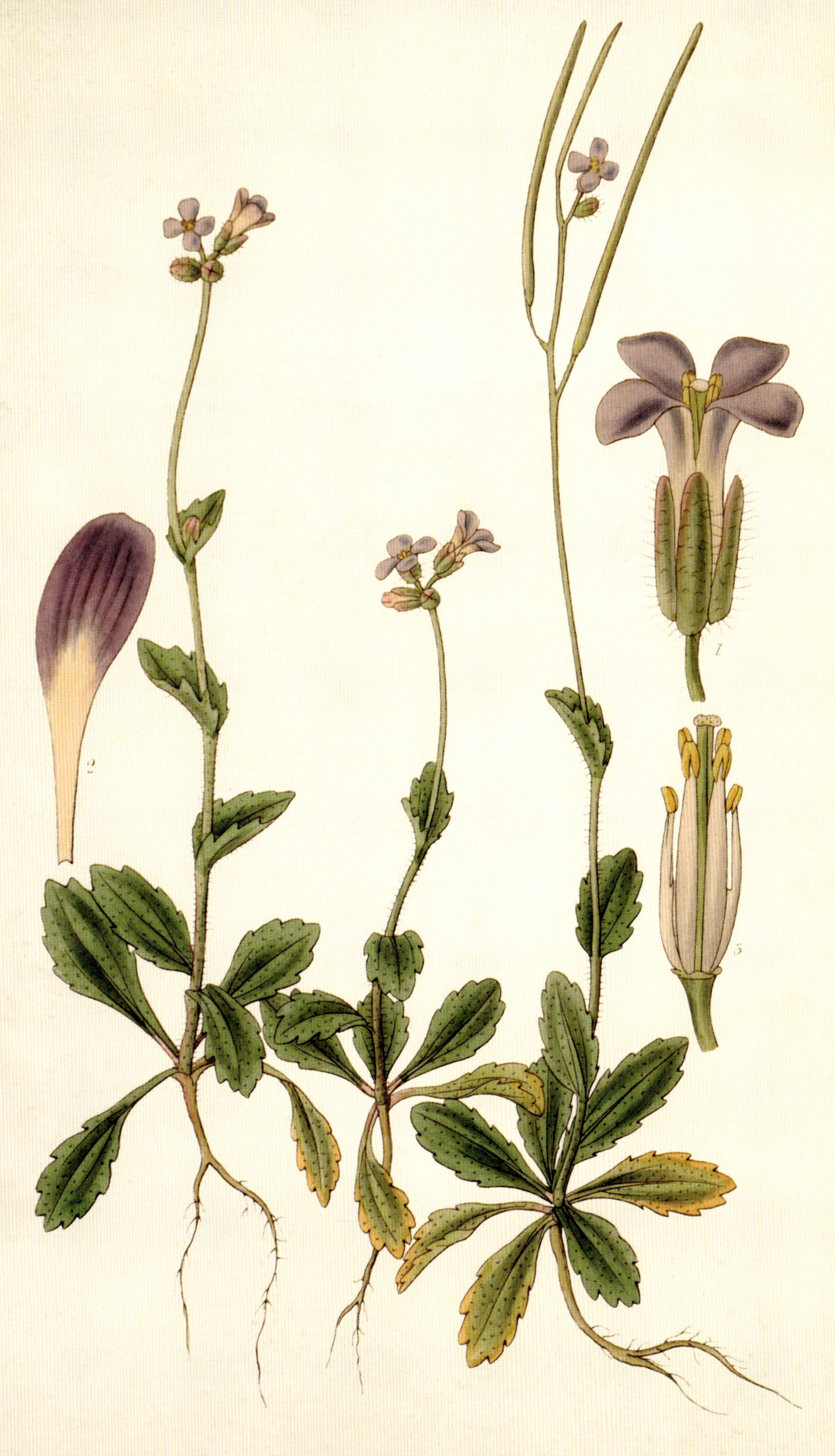

Arabis verna

JOSEPH HOOKER

Sir William's son, Joseph Dalton, was born in Suffolk on 30 June 1817, but spent most of his childhood in Glasgow, where he trained in medicine. He spent four years, from 1839 to 1843, as assistant ship's surgeon and botanist on board HMS *Erebus*, which was sent to explore the Antarctic. Although officially a doctor, Joseph's greatest interest was in plants, and he persuaded the commander of the ship, Sir James Clark Ross, to appoint him as botanist to the expedition. He was thus able to collect flowering plants in Tierra del Fuego and the Antarctic islands, and lichens on the Antarctic continent itself. In southern Argentina he found several species very similar to the arctic alpines with which he was familiar in the Scottish mountains, and this sparked his interest in the geography of plants, encouraging him to study the similarities and differences between arctic flowers and those which grow on high mountains in the tropics, and taking a particular interest in island floras. Hooker arrived back in Britain with herbarium specimens representing over 1500 species of plants.

Thanks partly to his father, and partly to his own interests, he had good connections in the scientific world, and showed an intelligent curiosity in all matters concerning natural history, botany in particular. He was also a good artist, making numerous pencil and watercolour sketches on his travels, both of the landscape and of plants, and many of these are now stored in the archives at Kew.

Joseph had many friends and correspondents, particularly Charles Darwin, to whom he was first introduced by his father, and they became close friends. Darwin was working on his theory of evolution, and the two men had a shared interest in the geography of plants. Hooker looked up to the older man (though there was a difference of only eight years between them), although he was not at first convinced by Darwin's theory, and Darwin valued Hooker's opinions and expertise in botany, humbly describing himself in a letter written in 1843, as a "Botanical Ignoramus".

Despite having qualified as a doctor, which at that time was the only scientific training available for a botanist, in 1845 Joseph applied for the Chair of Botany at Edinburgh University which would have allowed him to study botany full time, but although the university had received a glowing reference from Darwin, by then a well-known personality, he was unsuccessful.

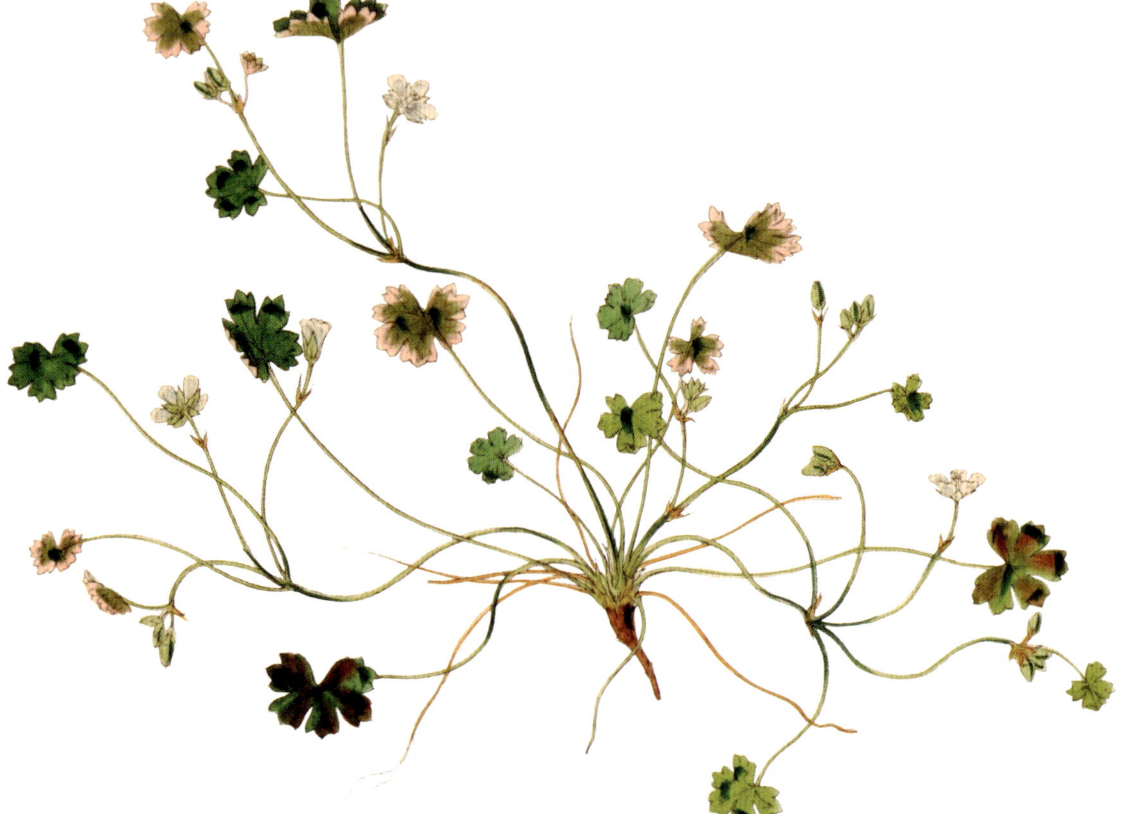

ABOVE: Joseph Hooker, who travelled all over the world to discover and record new plant species.

LEFT: Geranium microphyllum, *drawn by W. H. Fitch, which appeared in Hooker's* Flora Antarctica.

OPPOSITE: Anisotome latifolia, *from Auckland and Campbell Islands, also from* Flora Antarctica.

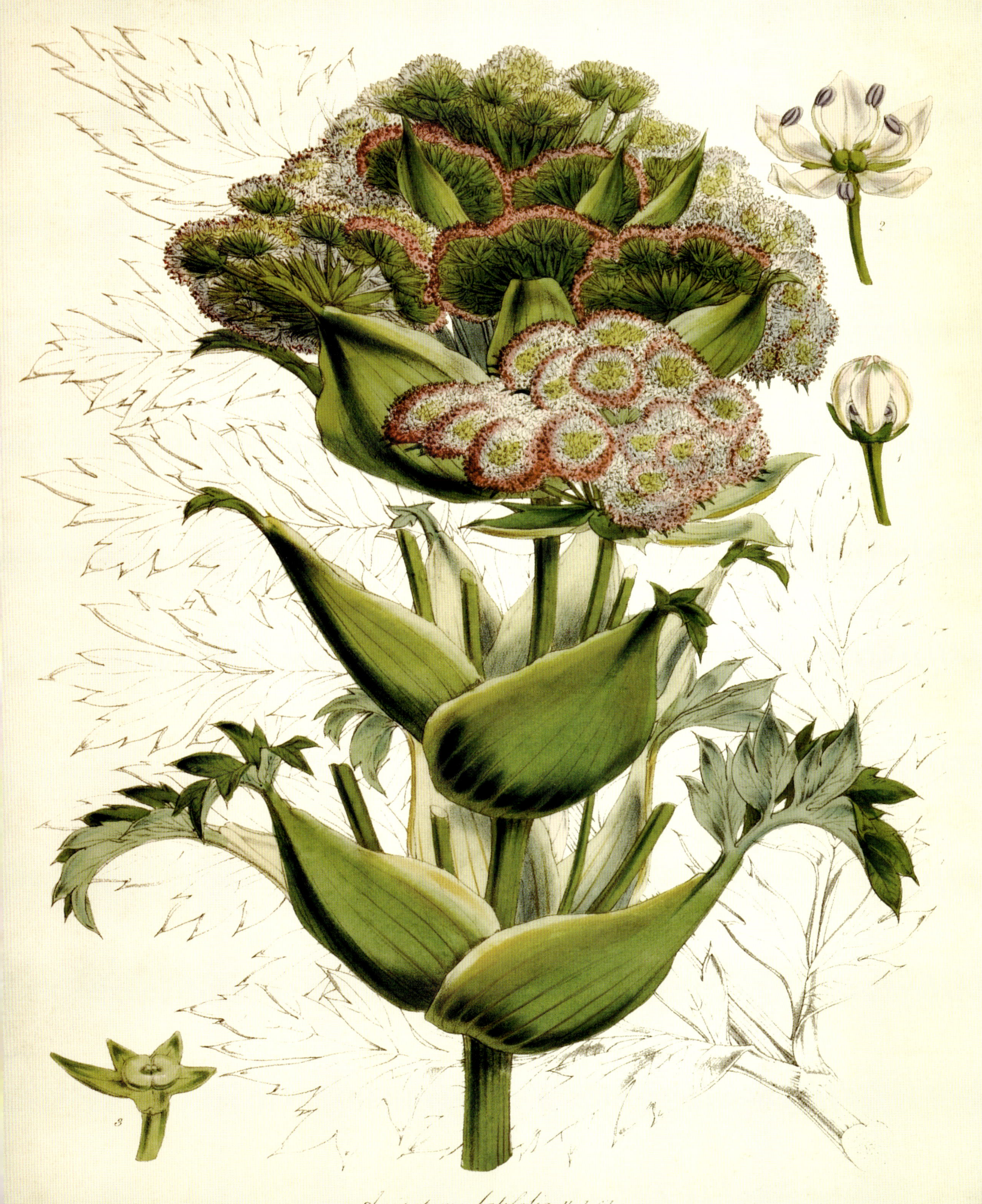

Anisotome latifolia Hook. fil.

Meanwhile, Joseph continued to write up his Antarctic journey – the first volume of *Flora Antarctica* had been published in 1844, and the second in 1847 – and he spent the year of 1846 studying fossil botany at the Geological Survey. In 1847, he travelled to India and the Himalayas (where he was briefly imprisoned by the Rajah of Sikkim) and managed to amass thousands of plant specimens, including 25 new *Rhododendron* species. The introduction of these rhododendrons was to make a profound impression on British gardeners, leading to a mania for collecting and hybridization, and the planting of large gardens on acid soils in the Weald of Sussex and Surrey, on the Berkshire sands, in Cornwall and in Scotland. Hooker's Indian expedition gave rise to several publications, notably *The Rhododendrons of Sikkim-Himalaya* (1849–51) and *Illustrations of Himalayan plants* (1855, *see* page 183), and also produced plants such as *Rhododendron hookeri* and *Impatiens hookeriana* which were later illustrated in *Curtis's Botanical Magazine*.

In 1851, Joseph married Frances Harriet Henslow, daughter of the clergyman, geologist and Professor of Botany at Cambridge, John Stevens Henslow. The couple seem to have enjoyed a happy family life, having four sons and three daughters, to whom they were devoted; sadly, though not unusually for the time, one of their daughters died at the age of six.

In 1854, Joseph was appointed by the government to be Assistant Director of the gardens at Kew, which now covered an area in excess of 300 acres. Working with his father, who was still Director, Joseph became increasingly busy in the day-to-day running of the garden, and he also took on much of the organization and editorial work of *Curtis's Botanical Magazine*. As with subsequent editors, the nature of the plants portrayed changed according to both personal taste and gardening fashions, and, while Joseph continued to describe and illustrate plants of economic interest, he also added new species, often recently collected by the various plant hunters sent abroad by the big nurseries. During his time as editor, he included many rhododendrons from the Himalayas, as well as bulbs new to western gardeners, such as the striking *Lilium auratum* which had been collected by the nurseryman John Gould Veitch in Japan. The year 1860 was a typically busy one for Joseph – he went on an expedition to Syria and the Lebanon, and also, in collaboration with his fellow botanist George Bentham, published the first part of *Genera Plantarum*, an attempt at a system of classification for

BELOW: Joseph Hooker's preliminary field sketches and studies of Rhododendron dalhousiae.

OPPOSITE: W. H. Fitch's finished painting of Rhododendron Dalhousiae *from Hooker's* Rhododendrons of the Sikkim-Himalaya.

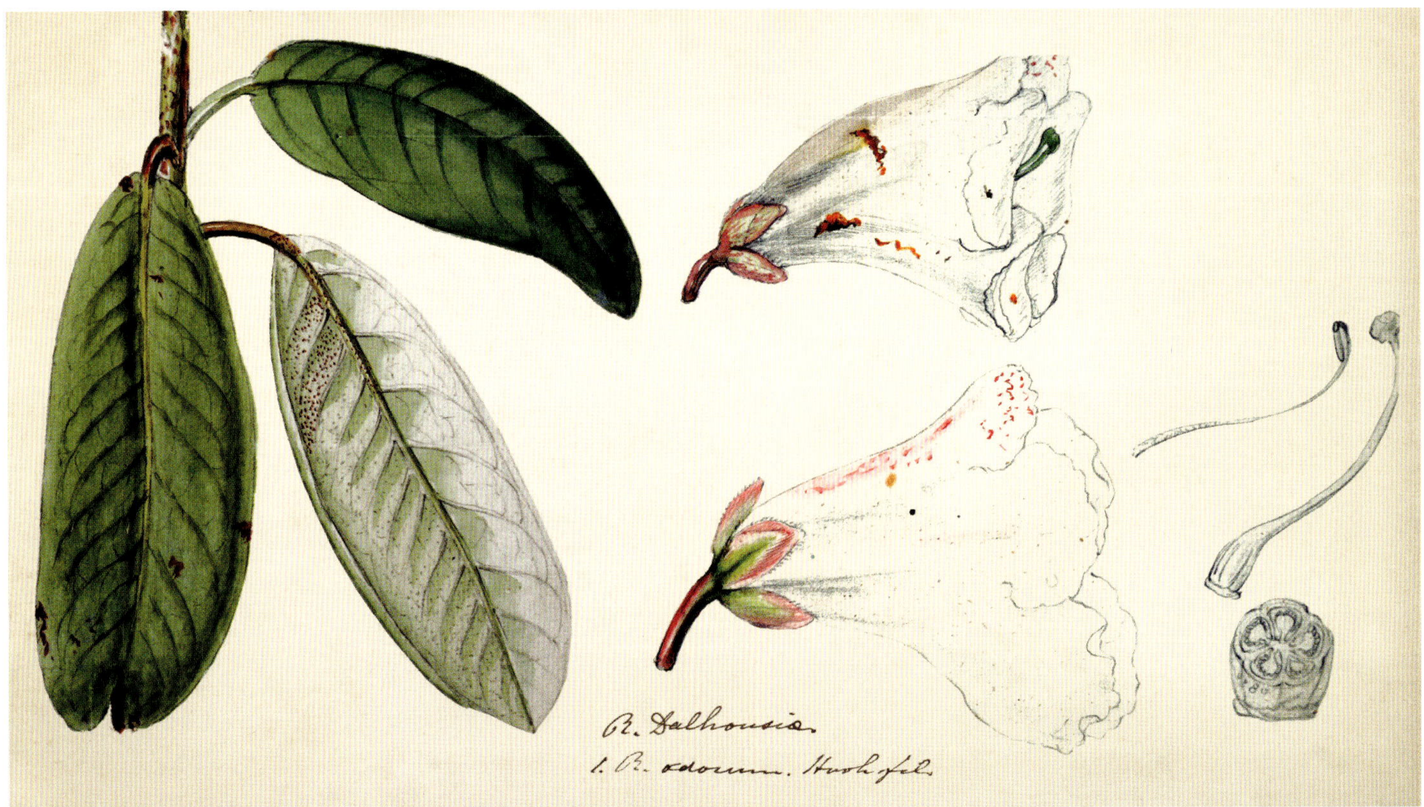

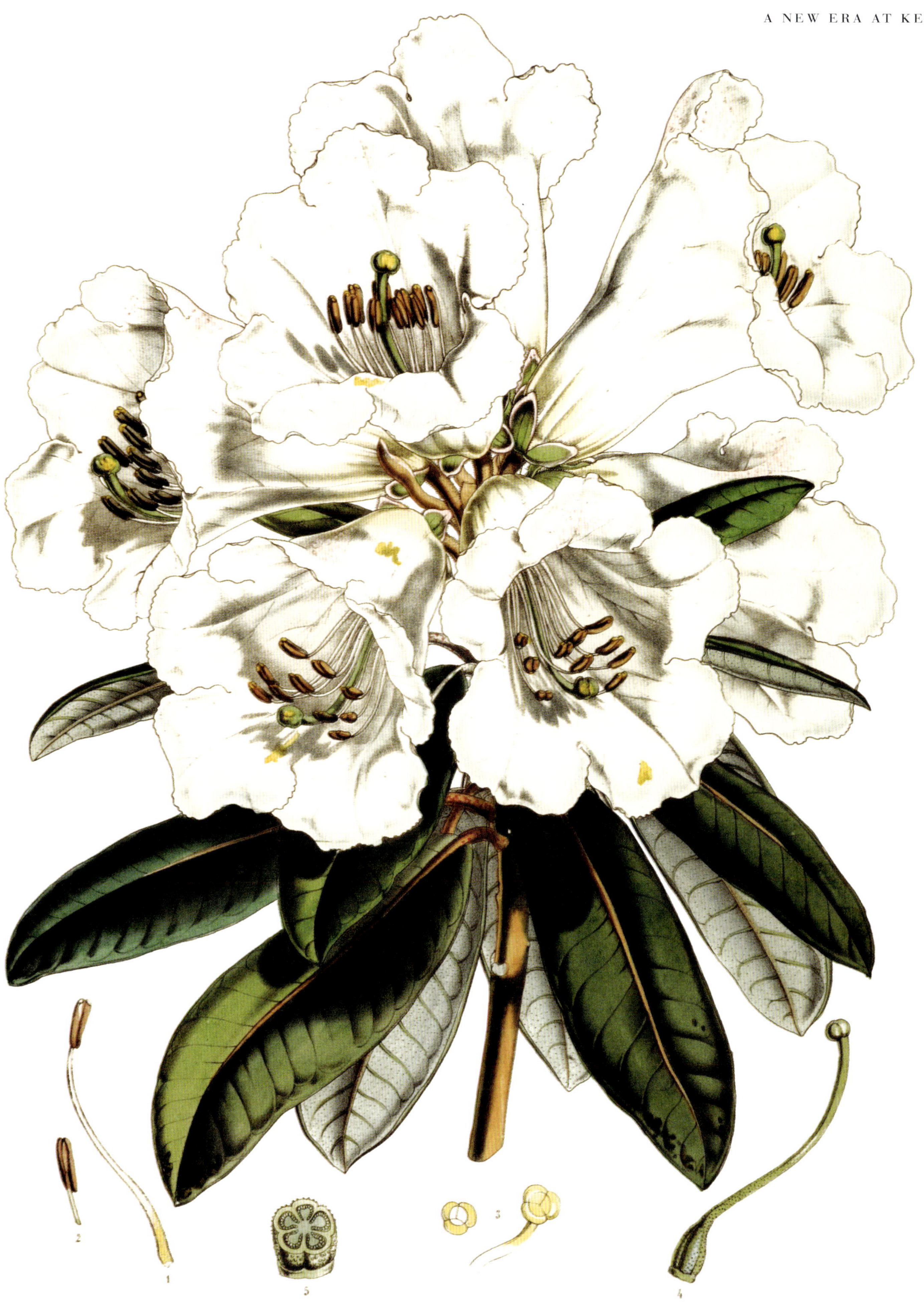
2
1
5
3
4

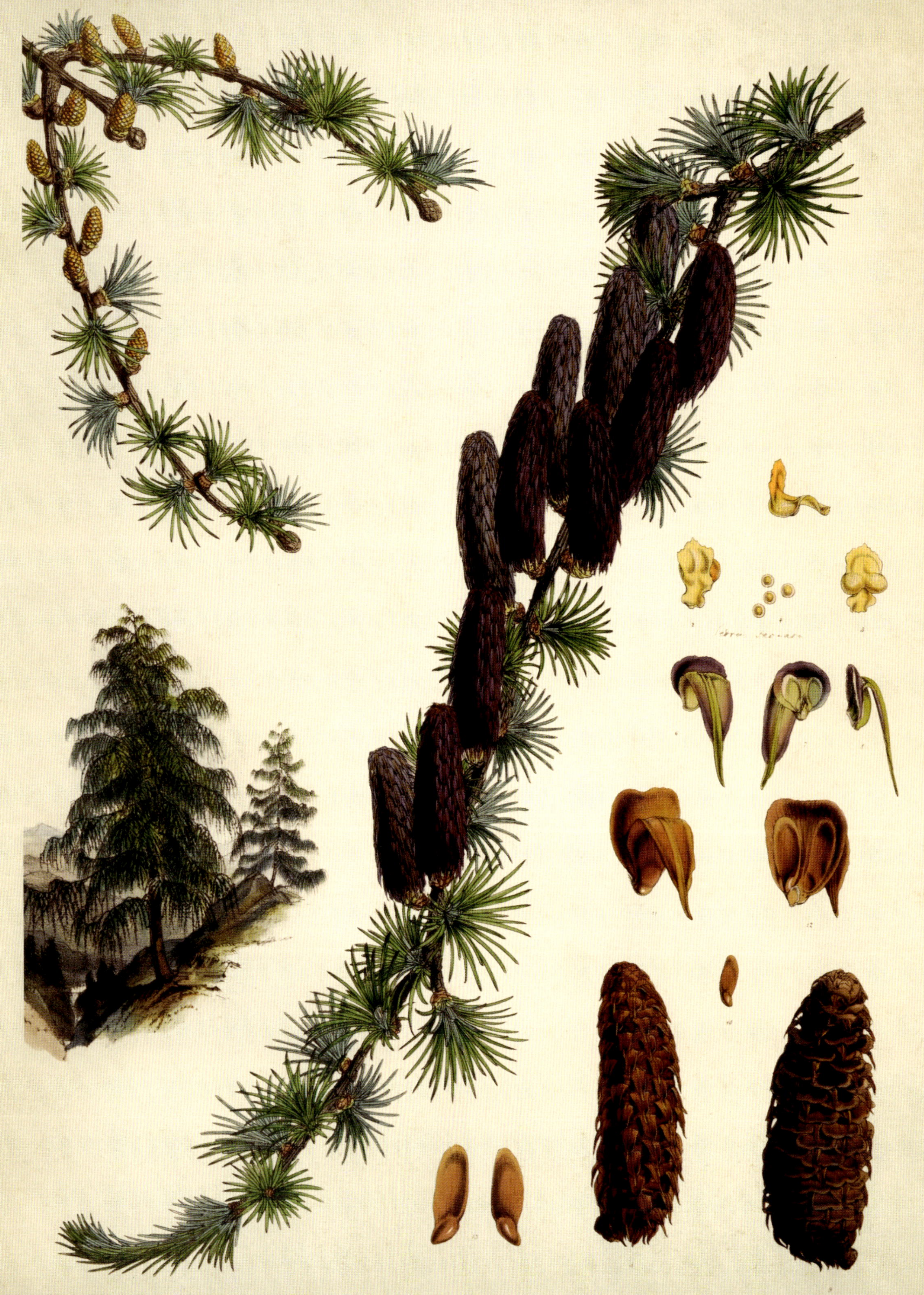

all known plant genera, a massive project that was not completed until 1883.

On the death of his father in 1865, Joseph was offered the post of Director, although he had already been more or less running the garden at Kew for some time. He was now answerable to the government for the management of the gardens, the herbarium, and research, and he felt, in contrast to those who viewed it as a pleasure garden, that Kew existed primarily as a place of scientific study. This difference in ideas led to some disagreements, but in general Hooker seems to have been happy with his lot, writing to Darwin in 1866 that he was "... very busy, out of doors 6-hours a day, & delighting in my occupation. I can make even Kew 50 percent better than it is".

Joseph, like his father, lived a busy life, working hard, and continuing to travel when possible, and he was sufficiently respected by his peers to be knighted and to serve as President of The Royal Society. Sadly, in 1874, his devoted wife Frances died unexpectedly, but Joseph recovered from the shock sufficiently to remarry two years later; his new wife was a widow, Hyacinth Jardine, with whom he had two sons.

The following year, Joseph was knighted, and from July to September he and Hyacinth undertook an 8000 mile-long expedition across the USA with his friend, the botanist Asa Gray. On their return, the Hookers built a house, The Camp, and made a garden at Sunningdale, where the acid, sandy soil was ideal for rhododendrons, and in the autumn of 1885, having been Director for 20 years, Joseph retired from Kew, although he did not give up work. He continued to edit, with assistance from others, *Index Kewensis*, *Flora of British India*, a new edition of George Bentham's *Handbook of the British Flora* and *Curtis's Botanical Magazine*, the latter until 1904.

Sir Joseph died on 10 December 1911 at the age of 94. His memorial plaque, decorated with plants modelled by the botanical artist Matilda Smith, sits next to that of his father at Kew.

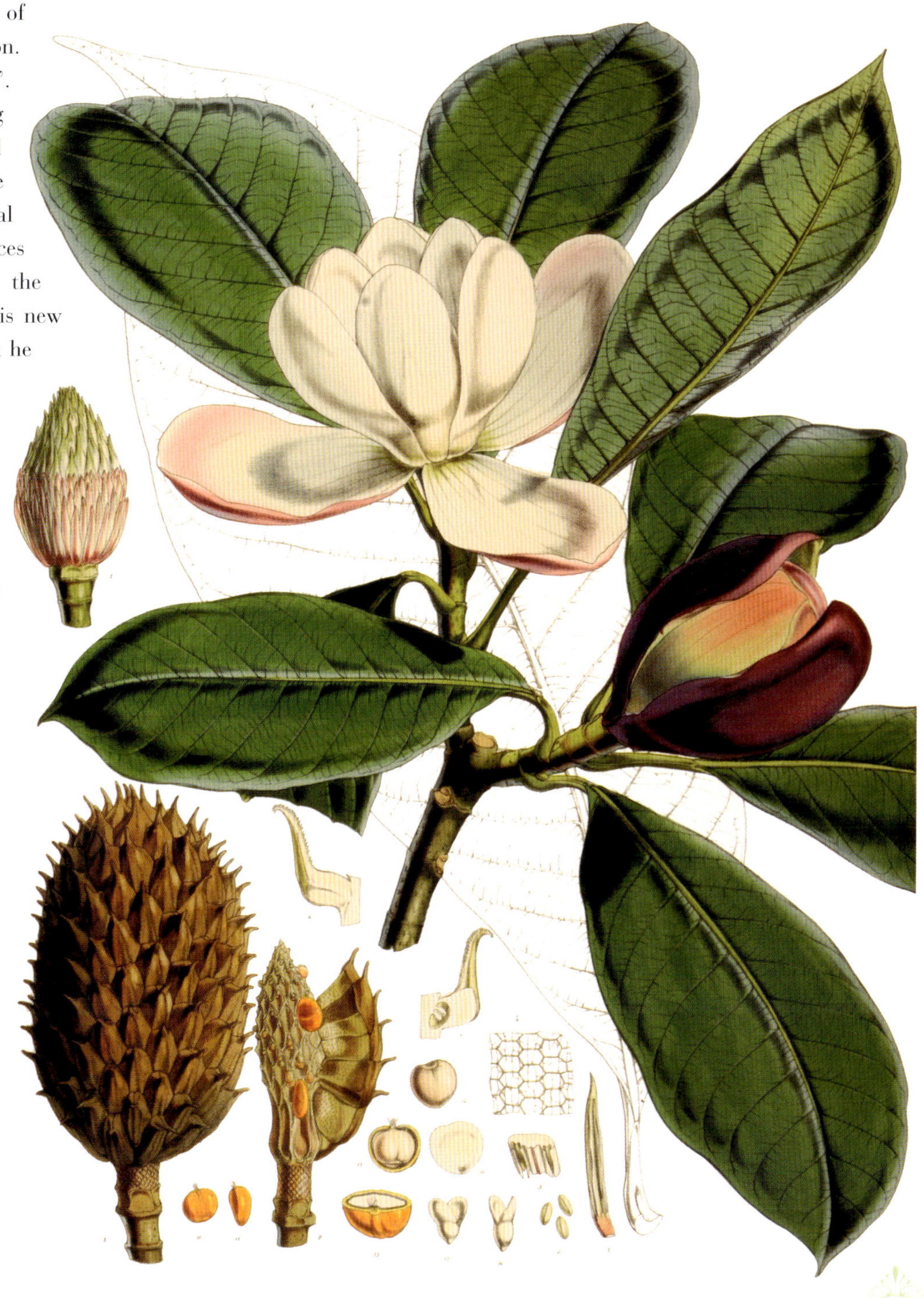

RIGHT: Talauma hodgsoni, *now* Magnolia hodgsonii, *from Joseph Hooker's* Illustrations of Himalayan Plants.

OPPOSITE: Studies of Laryx griffithii *from Joseph Hooker's* Illustrations of Himalayan Plants.

FOLLOWING PAGES (LEFT): W. H. Fitch's painting of Lilium auratum *for Elwes'* The genus Lilium.

FOLLOWING PAGES (RIGHT): W. H. Fitch's painting of A: Codonopsis gracilis*; B:* C. javanica*; C:* C. inflata.

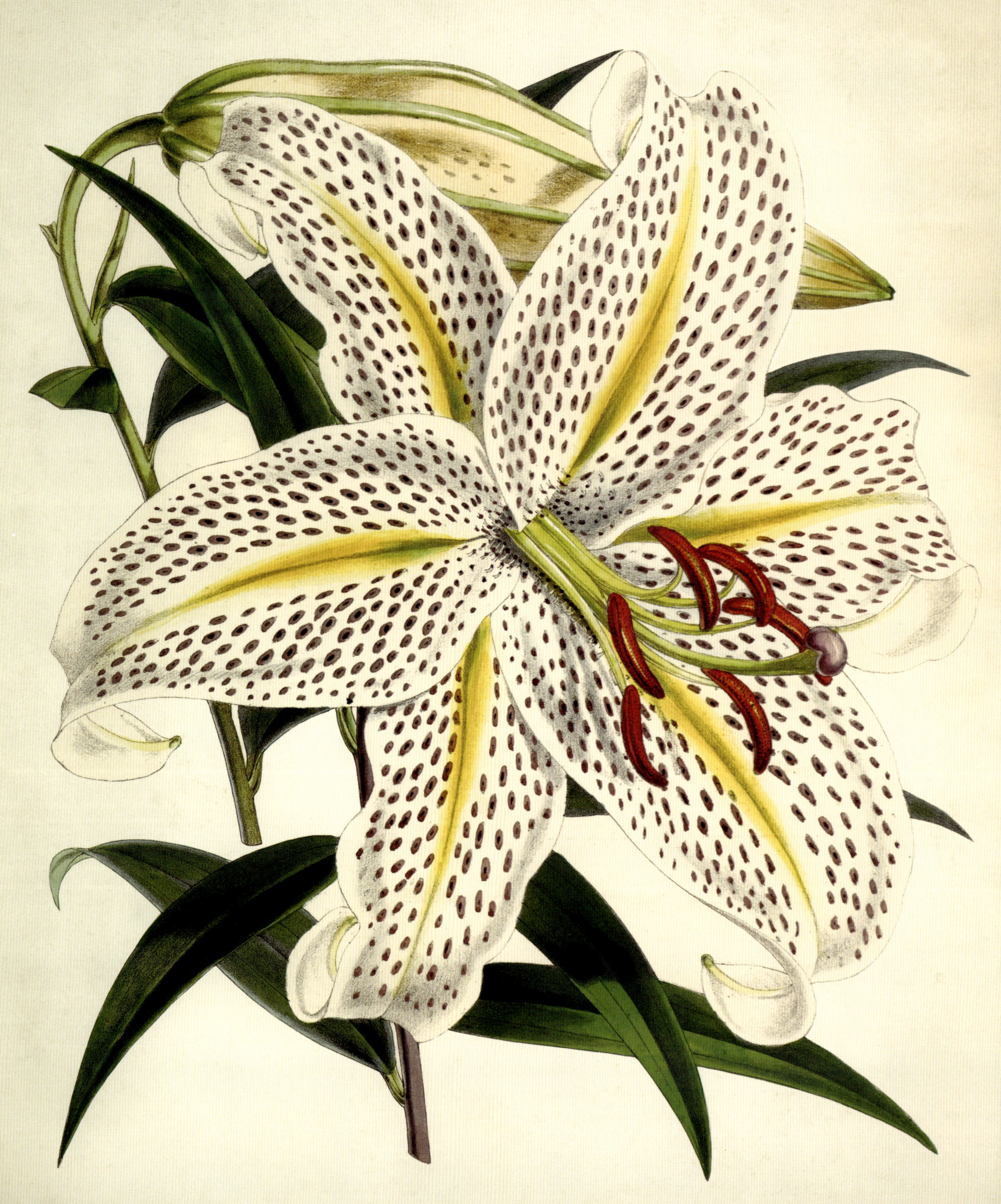

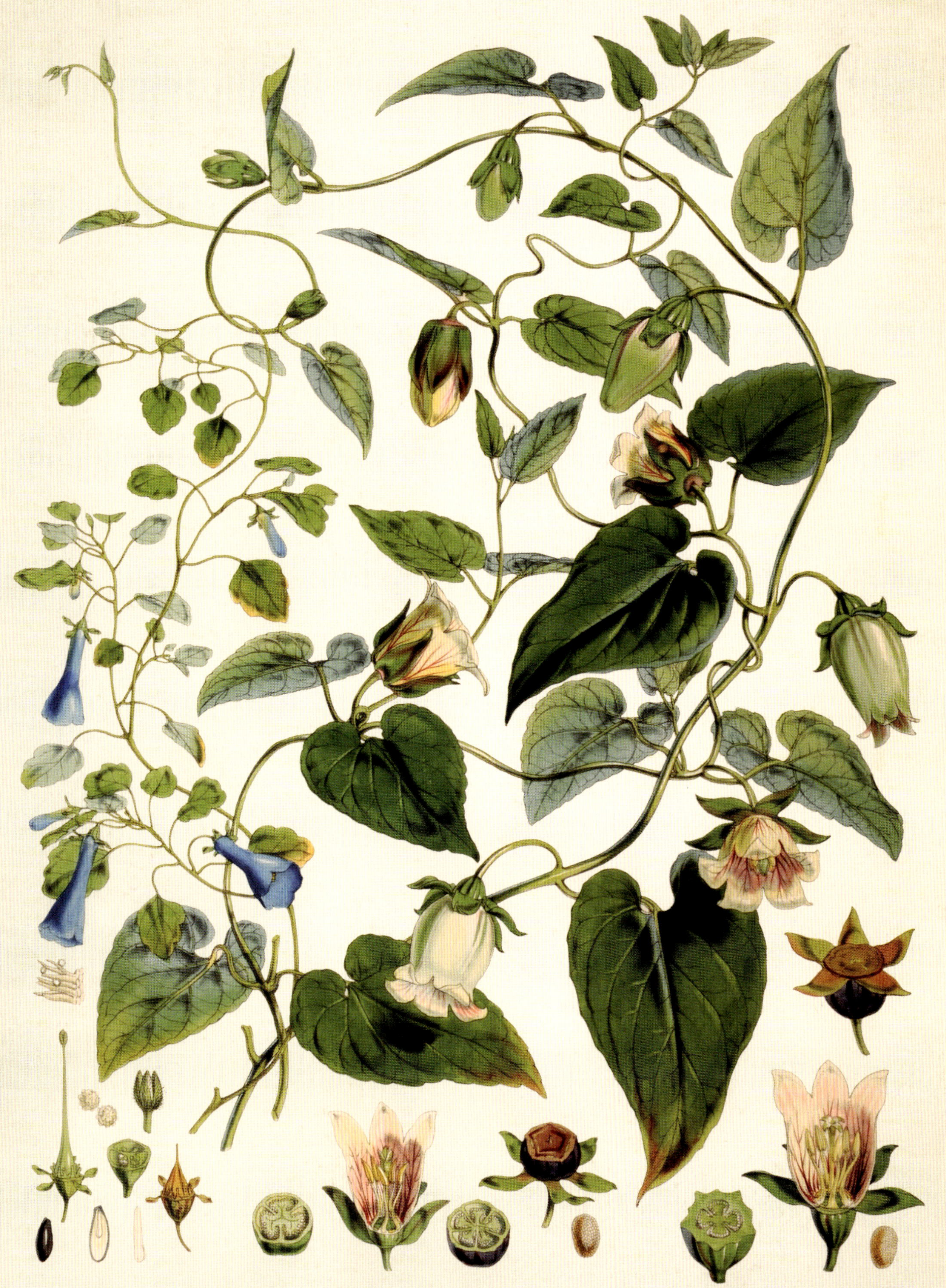

The Hookers could never have achieved their pre-eminence in publishing botanical illustrations without a first-class artist, W. H. Fitch. Walter Hood Fitch was born in Glasgow on 29 January 1817. His father worked in the cloth trade, and the family moved to Leeds, at that time a centre of the industry. Walter spent two years at the Grammar School, where he received a good classical education, but by the time he was 11, the family had returned to Glasgow, and his education ended. He showed promise as a draughtsman, and became assistant to a topographical artist, Andrew Donaldson, before moving on to learn the art of lithography.

In 1830, he was apprenticed to a pattern drawer in a calico mill, owned by a friend of William J. Hooker, and when Hooker came to look for an assistant, Fitch was recommended to him. At first, Fitch spent his spare time taking lessons from Hooker in how to mount dried plant specimens and how to make accurate botanical drawings, skills he soon mastered, and Hooker bought him out of his apprenticeship and took him on full time. One of Fitch's first jobs was to assist Hooker in preparing large-scale drawings for use in the lectures that Hooker gave at Glasgow University, and Fitch also attended these lectures alongside Hooker's two sons, William and Joseph.

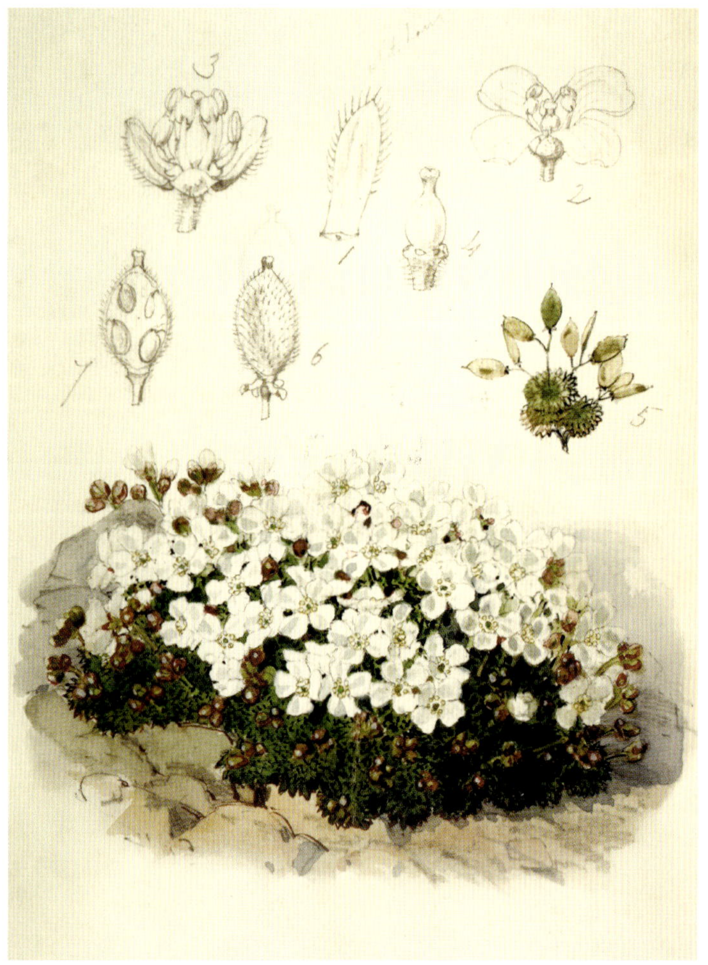

In addition to his academic post, Hooker was the editor, and sole artist, of *Curtis's Botanical Magazine*, and Fitch's work was soon of a sufficiently high standard to produce drawings for the magazine, which were then reproduced by means of engraving on copper. From the autumn of 1834 onwards, Fitch became indispensable to William Hooker, and in addition to the drawings for lectures and almost all the plates for the magazine, he also produced 392 figures for Hooker's book *Botanical Illustrations...* (1837), and numerous plates for a new, uncoloured periodical *Icones Plantarum*.

In addition to being a skilled artist, who worked at a prodigious rate, Fitch also became an excellent lithographer, and was able to draw straight onto the stone, and he was responsible for the excellent reproduction of Francis Bauer's plates of ferns for Hooker's *Genera Filicum* (1838–42). In 1841, when William became Director of Kew, Fitch moved with the Hooker family from Glasgow to London. At first, Fitch lived in the Hooker household, and seems to have worked almost continuously on *Curtis's Botanical Magazine* and other books for both William and his son Joseph. Despite his prolific output, he was never officially on the staff at Kew, but was employed directly on a freelance basis by William Hooker; father and son did their best to accord Fitch the recognition due to him, and Joseph arranged for his salary to be increased. Joseph also named a Polynesian plant, *Fitchia nutans*, for him, describing it in the *London Journal of Botany* (1845) as "A very noble plant... I have named it in honour of one who is well known as a most accurate and elegant Botanical artist, Mr Walter Fitch, to whose pencil are due the plates of this work, of the *Icones Plantarum*, of the last twelve volumes of the Botanical magazine, and of the greater part of *Flora Antarctica*".

Both William and Joseph relied heavily on Fitch, who by now had a large family to support. Despite Sir William's efforts, the government refused to employ him as the official artist at Kew, but he was paid directly for his work on *Curtis's Botanical Magazine* by the publisher. When Joseph took over as Director of Kew, Fitch remained sole artist for the *Curtis's Botanical Magazine*, and in December Hooker dedicated Volume 95 to him, with the words: "To Walter Fitch Esq, FLS, The Accomplished Artist and Lithographer of upwards of 2500 plates already published of the Botanical Magazine this volume is dedicated by his faithful and sincere friend Jos.D Hooker".

Although Fitch and Joseph Hooker had been friends for many years, and worked together on numerous botanical projects, and

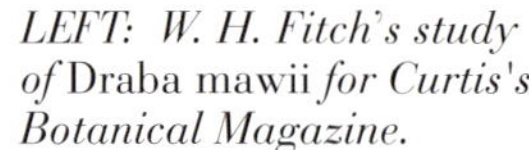

LEFT: W. H. Fitch's study of Draba mawii *for Curtis's Botanical Magazine.*

OPPOSITE: W. H. Fitch's study of Adianthus reniforme *Wild. for Joseph Hooker's* Filices Exoticae.

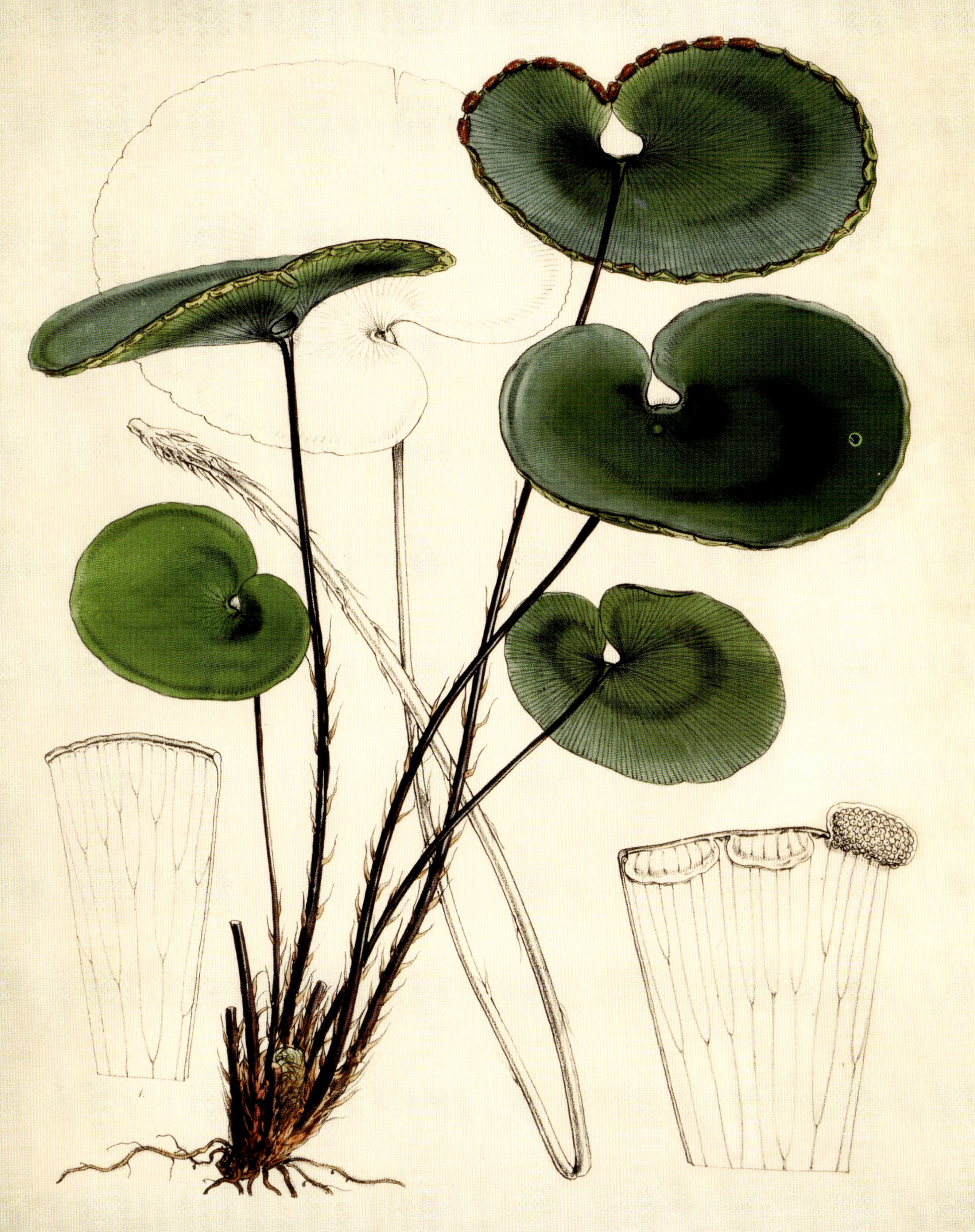

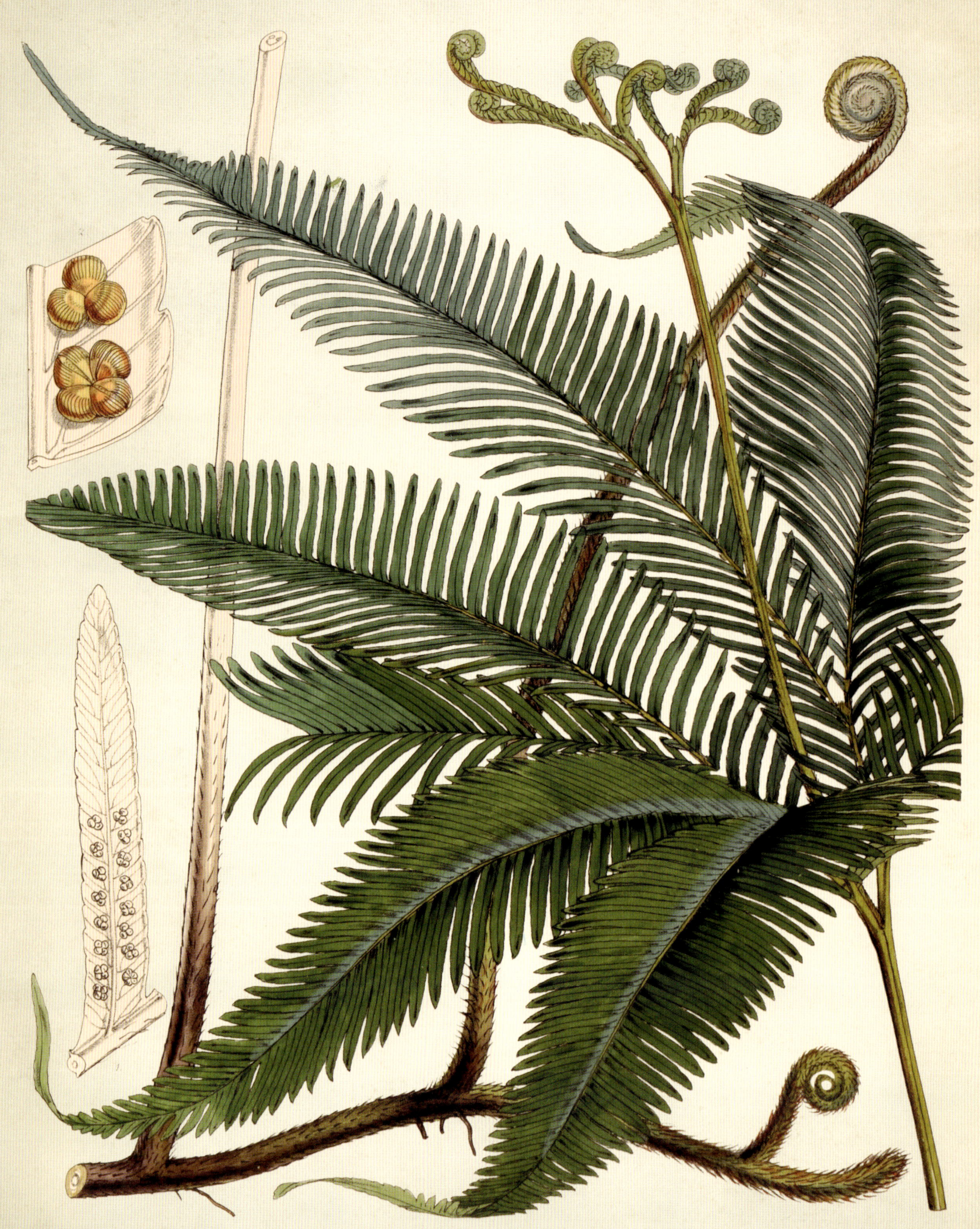

Fitch had produced beautiful landscape paintings from Joseph's travel sketches, the two men eventually fell out over money, which was not entirely Joseph's fault, but rather that of the authorities. In 1877, Fitch wrote:

> Dear Sir Joseph,
> I forward you the three sketches of the 3 Botanical Magazine plates which finish this year, and it is probable they will be the last.
> Yours respectfully, Walter Hood Fitch.

For some reason, the two were unable to find a solution to the problem, and Joseph was forced to find other artists at short notice. He was fortunate in being able to employ Fitch's nephew, John Nugent Fitch, as lithographer. As for artists, his eldest daughter, Harriet, who, ironically, been taught to draw by W. H. Fitch, and Joseph's sister-in-law Anne Barnard, were able to contribute drawings, and he also trained his second cousin, Matilda Smith, who, many years later, became the first official botanical artist employed by Kew.

Sir Joseph subsequently worked hard to obtain a government pension for Fitch, successfully procuring him a Civil List pension in consideration of his services to Botanical Science, and he and Fitch made up their differences. Fitch continued to paint, but did no more work for *Curtis's Botanical Magazine* and he died at Kew in 1892, aged 74.

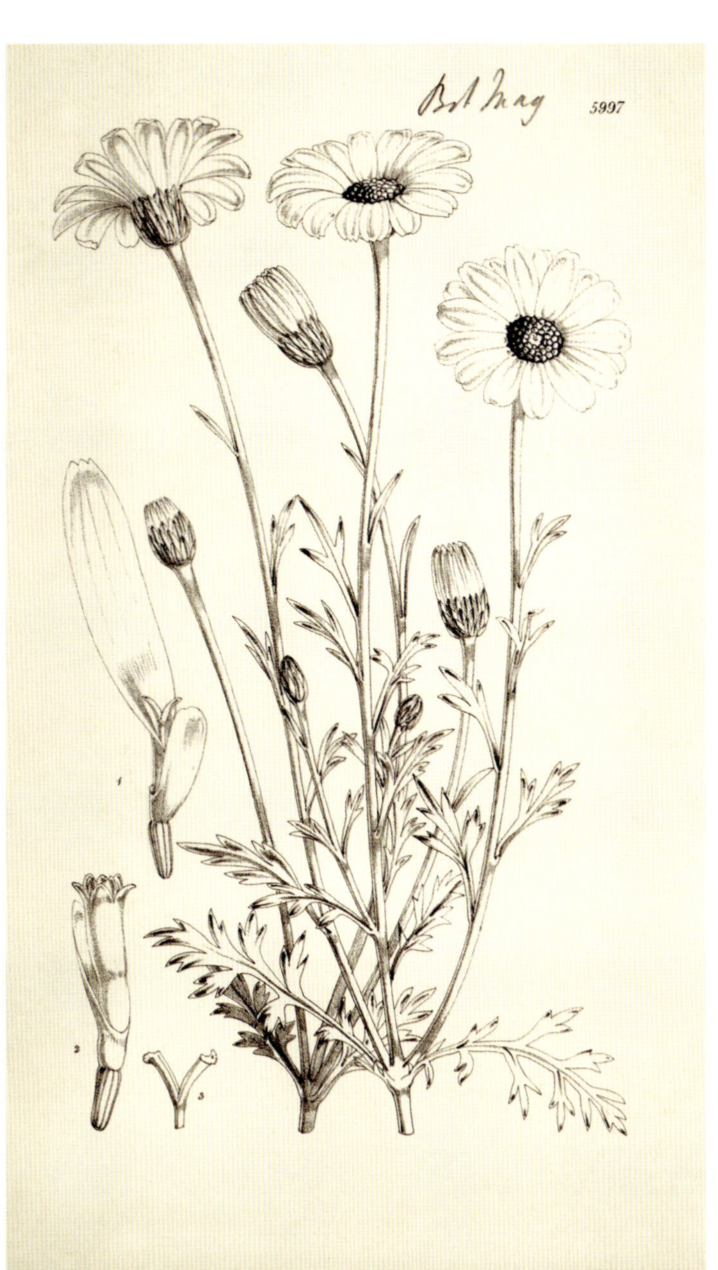

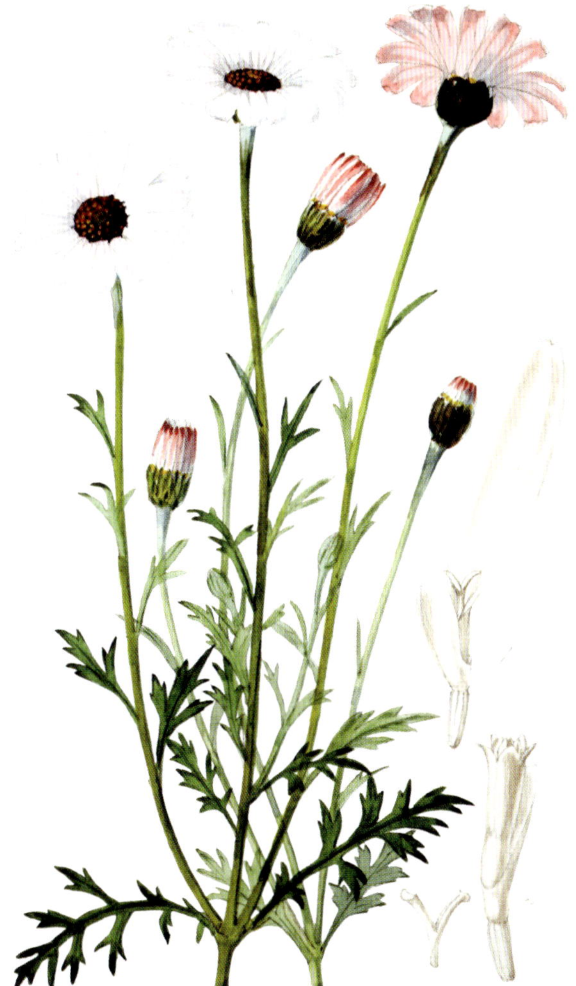

LEFT: W. H. Fitch's preliminary study for Chrysanthemum mawii.

ABOVE: The final artwork of W. H. Fitch's painting of Chrysanthemum mawii, *which was published in* Curtis's Botanical Magazine.

OPPOSITE: W. H. Fitch's study of Gleichenia flabellata *(now* Sticherus flabellatus*) for Joseph Hooker's* Filices Exoticae.

George Maw

George Maw (1832–1912) was a successful businessman, a traveller, keen amateur archaeologist, geologist and botanist. His greatest botanical achievement was the magnificent book which he wrote, illustrated and published, *The Genus Crocus*, the original paintings for which are today in the library at the Royal Botanic Gardens, Kew.

George was born in London, the son of John Hornby Maw, who was the owner of a family pharmaceutical business, and a keen artist, who moved his family first to Hastings, and later to Bideford in Devon. George, who was a good draughtsman and designer, assisted his father in the designing of decorative tiles, which were fired by a local potter.

In 1850, the Maw family moved to the Midlands, where they took over an old tile factory in the heart of the Potteries, and produced floor and wall tiles in the mock-mediaeval "Hispano-Moorish" and Persian (or Turkish) styles. The business prospered, and a new pottery works was constructed. Meanwhile, Maw took a long lease on Benthall Hall at Broseley, a substantial sixteenth-century house, above the Severn River, in the Ironbridge Gorge, Shropshire, and set about making a garden.

By 1862, Maw & Co had developed the process of producing majolica ware and exhibited some of their tiles at the International Exhibition in the Crystal Palace. The resulting publicity enabled the company to flourish, and George was able to travel abroad more, combining his professional interest in collecting examples of ceramics with his amateur interest in searching for plants when visiting southern Europe and Morocco.

While abroad, he kept detailed notes and sketches of anything and everything of interest to him, whether ceramics, rocks or plants, and passed information and plants on to his friends Joseph Hooker and Charles Darwin.

By now Maw had conceived a grand plan to write and illustrate a monograph on *Crocus*, and asked Hooker for his opinion of the project. Hooker, who had experience of publishing, replied cautiously, "The coloured plates are most exquisite, but I fear that the process is far too expensive for Botanical works of which so very few copies are required that Hand colouring is the cheapest plan by far."

In the spring of 1871, Maw and Hooker set off with John Ball, another amateur botanist and friend of Hooker, on a trip to study the flora of the High Atlas Mountains of Morocco, and the results of this expedition were published in 1878 under the title *Journal of a tour in Marocco and the Great Atlas*, with geological notes by Maw. Hooker and Maw were in constant communication, and Maw frequently sent plants to Kew to be illustrated for *Curtis's Botanical Magazine* by the long-suffering W. H. Fitch. Among the plants illustrated are *Salvia dichroa*, from the Atlas, *Sedum dasyphyllum* var. *glanduliferum*, *Bellis rotundifolia* var. *caerulensis*, *Andryala mogadonensis*, *Sempervivum tectorum*

ABOVE: An engraved portrait of George Maw.

RIGHT: Crocus imperati *from* The Genus Crocus.

OPPOSITE: W. H. Fitch's painting of Sempervivum tectorum *var.* atlanticum *from a plant which was sent back from Maw and Ball's expedition to the High Atlas Mountains of Morocco.*

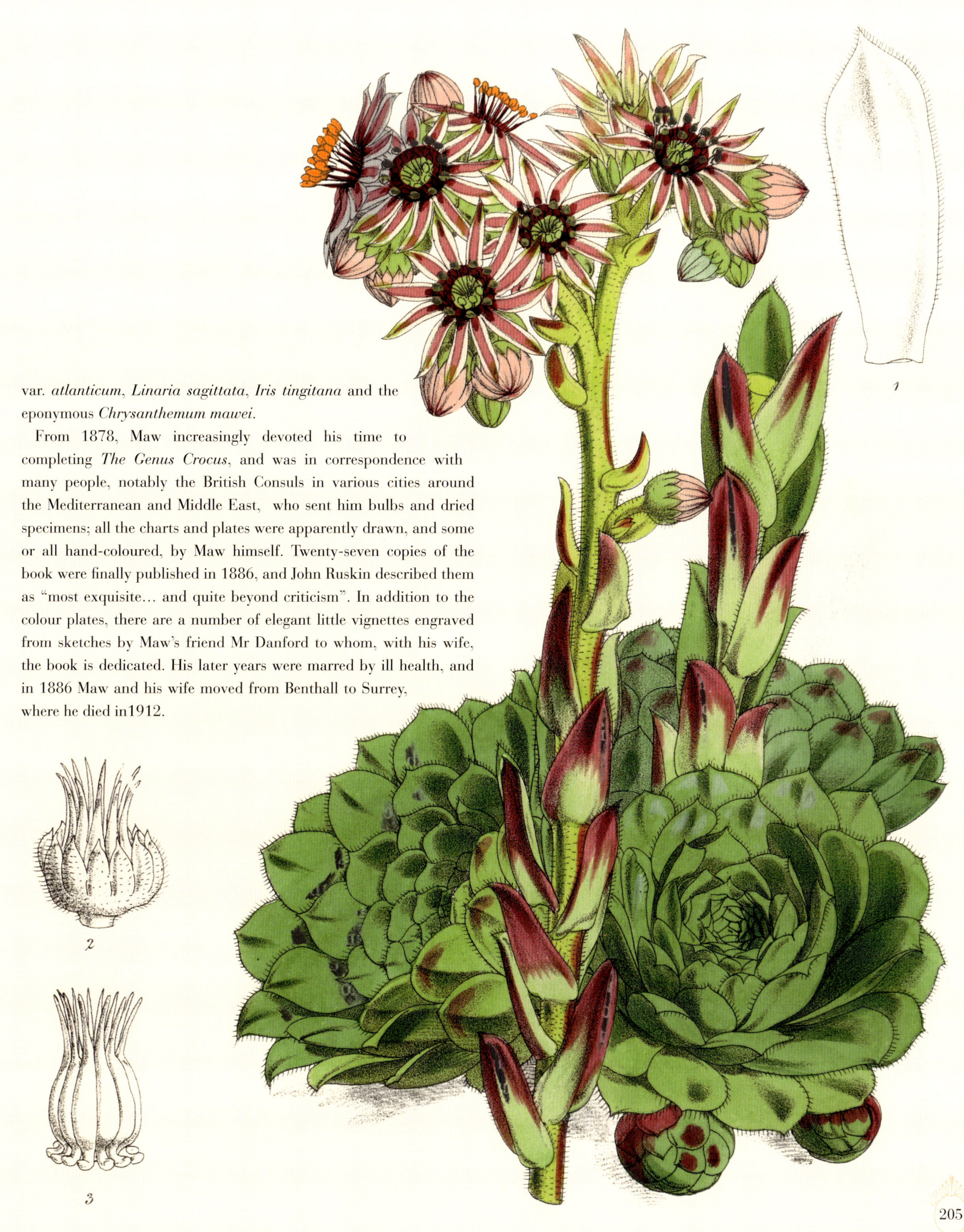

var. *atlanticum*, *Linaria sagittata*, *Iris tingitana* and the eponymous *Chrysanthemum mawei*.

From 1878, Maw increasingly devoted his time to completing *The Genus Crocus*, and was in correspondence with many people, notably the British Consuls in various cities around the Mediterranean and Middle East, who sent him bulbs and dried specimens; all the charts and plates were apparently drawn, and some or all hand-coloured, by Maw himself. Twenty-seven copies of the book were finally published in 1886, and John Ruskin described them as "most exquisite… and quite beyond criticism". In addition to the colour plates, there are a number of elegant little vignettes engraved from sketches by Maw's friend Mr Danford to whom, with his wife, the book is dedicated. His later years were marred by ill health, and in 1886 Maw and his wife moved from Benthall to Surrey, where he died in1912.

15

VICTORIAN TRAVELLERS

With the expansion of the British Empire in the mid nineteenth century, both in the Far East and in the new colonies, adventurous travellers visited exotic places, or were posted abroad, and became enthusiastic amateur botanists, discovering new species, sending home specimens and in a few cases making detailed paintings of the local wild flowers.

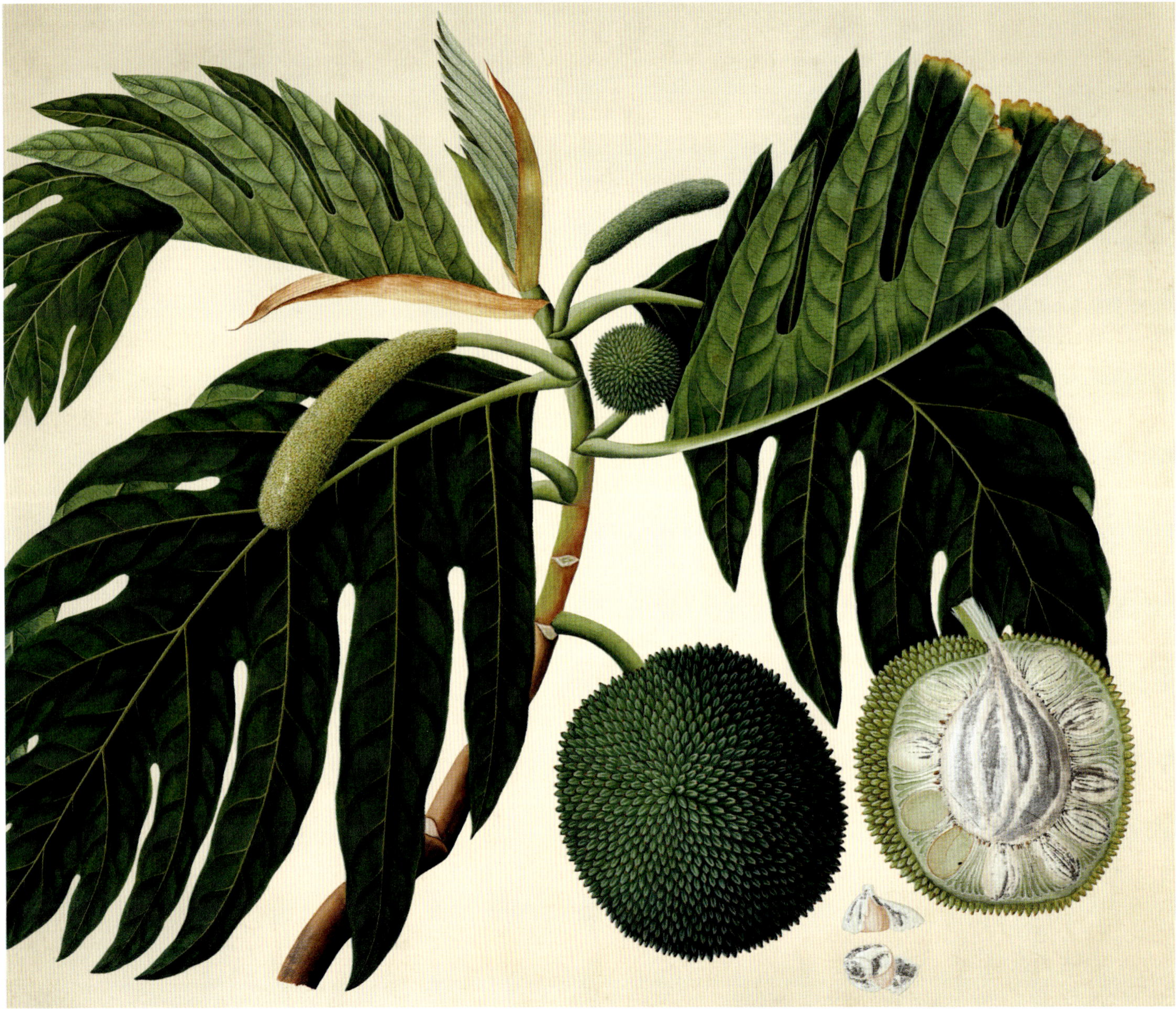

Many wives kept their husbands company and ran their households, having ample spare time to indulge other interests, such as painting flowers, which had long been regarded as a suitable accomplishment for a lady. Some of these collections went to the library at the Royal Botanic Gardens at Kew, where they have been studied alongside the herbarium specimens. Some of them are well known, but others have never been published.

Janet Hutton (fl. 1800–20s), née Robertson, was one of the earliest of these. Her husband was an East India merchant, and they lived first in Penang, from 1802 until 1808, and after that in Calcutta from 1817 until 1823. Mrs Hutton drew many large watercolours of plants, particularly edible fruits and useful plants, very much in the style of the Company School, and there has been some discussion about whether she painted the plants herself, or collected paintings by local artists. Alternatively, she may have made copies of paintings by local artists, especially while she was in Calcutta, where there was a thriving school of Indian flower painters already working for the East India Company botanists. In 1894, around 200 of her paintings were presented to Kew by her daughter.

A very different collection of paintings, also now at Kew, were made in Hong Kong by a Royal Artillery officer, Lt. Colonel, later General, John Eyre, who died in Chichester in 1865. It was only in 1842 after the first opium war, that Hong Kong was established as an English base, so the area was still very wild and botanically unexplored. General Eyre was stationed in Hong Kong from 1849 to 1851, and while he was there discovered and painted many new species. He explored with his fellow officer, Major John George Champion (1815–54), who fell wounded at the head of his men at the Battle of Inkerman in the Crimean War and died of wounds in the hospital in Scutari. They were there with the botanist and government official, H. F. Hance (1827–86), who suffered from malaria, and was therefore unable to make so many strenuous excursions. Champion named the genus *Eyrea* (Sapindaceae), after his friend, and a German botanist, Berthold Seemann, named the winter-flowering *Camellia hongkongensis*, both discovered by Eyre. The beautiful *Rhododendron championae* was named by William Hooker after Mrs Champion, who usually accompanied her husband on his botanising. It is now very rare, but its beautiful scented, white flowers can still be found hanging over the jogging track on Victoria Peak. Champion's herbarium and his drawings are now at Kew, and were used by George Bentham for his *Flora Hongkongensis*, published in 1861; Eyre's paintings are now at Kew, and include may cultivated plants as well as the new, wild species he collected.

The Victorian craze for botany and gardening inspired not only amateurs but a very pioneering and successful nursery business, that of James Veitch and Sons, which started in Killerton near Exeter, before moving to the City in London in 1832 and then to the King's Road, Chelsea in 1853. Their first collectors were the brothers Thomas and William Lobb, who visited South America and Java, collecting many new plants, which were successfully cultivated in the nursery. By

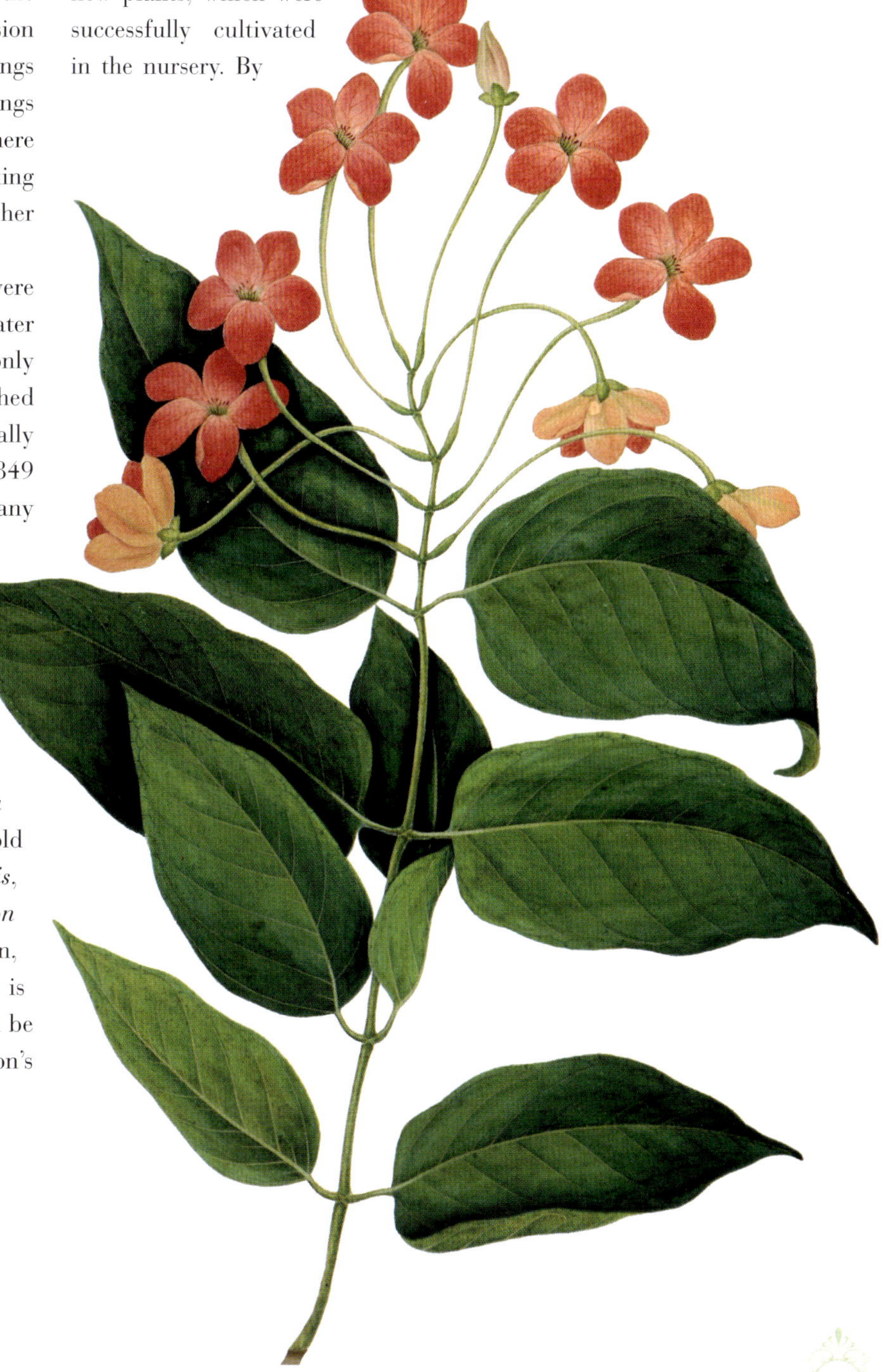

OPPOSITE: Janet Hutton's painting of Artocarpus incisa, *the breadfruit tree.*

RIGHT: General John Eyre's painting of Quisqualis indica, *made in Hong Kong in 1850.*

the late nineteenth century, following gardening fashions, they were concentrating more on hardy plants, and Charles Maries and E. H. Wilson visited Japan and China under their auspices. In all, over 400 of their new introductions were illustrated in *Curtis's Botanical Magazine*. One of Veitch's collectors was Frederick William Burbidge (1847–1905), who started his career as a gardener at the Horticultural Society's Garden at Chiswick and then at the Royal Gardens at Kew, where he started painting flowers. After two years, he joined the *Garden* magazine, and then in 1877 went to Borneo as a collector for Messrs Veitch. Burbidge's collections there are mostly remembered for the successful introduction of the spectacular pitcher plant, *Nepenthes rajah*, from Mount Kinabalu, but he also discovered and introduced approximately 20 new ferns. After returning from Borneo, he became curator of the Trinity College, Dublin Botanic Garden in Ballsbridge. Burbidge not only painted flowers, but also wrote about painting, notably the *Art of Botanical Drawing* (1873). Some of his drawings are at Kew, others in Trinity College, Dublin.

Charlotte Lugard (1859–1939) painted flowers collected by her husband, Edward James Lugard, an army officer. She accompanied him on an expedition to Bechuanaland, Basutoland (now Lesotho) and Ngamiland (in Botswana) in 1897 and 1898. Her drawings are now at Kew, and the succulent *Monadenium lugurdae* N.E. Br. is named after her.

Charlotte Williams was born in 1867 in Wimbledon, of Irish antecedents, and at the age of 30 she married Sir Otway Wheeler-Cuffe, a civil engineer from County Kilkenny. Immediately after their marriage in 1897, they went to live in India, moving to Burma in 1913, where Lady Wheeler-Cuffe studied the flora and painted hundreds of flowers, principally orchids. They were based for some time in Maymyo, a hill station named after its founder Colonel May, and from there she travelled round the country with her husband. The botanical garden which she set up on 340 acres round a lake still exists.

Lady Wheeler-Cuffe sent seeds and specimens to Sir Frederick Moore in the Botanic Garden at Glasnevin in Ireland, and discovered, on Mount Victoria, the large-flowered, scented epiphyte *Rhododendron cuffeanum*, and she also painted the climbing rose, *Rosa longicuspis*, and the blue orchid, *Vanda caerulea*. Several hundred of her delicate watercolour paintings were presented to Glasnevin in 1926, when she and her husband retired to Ireland. She died in County Kilkenny in 1967.

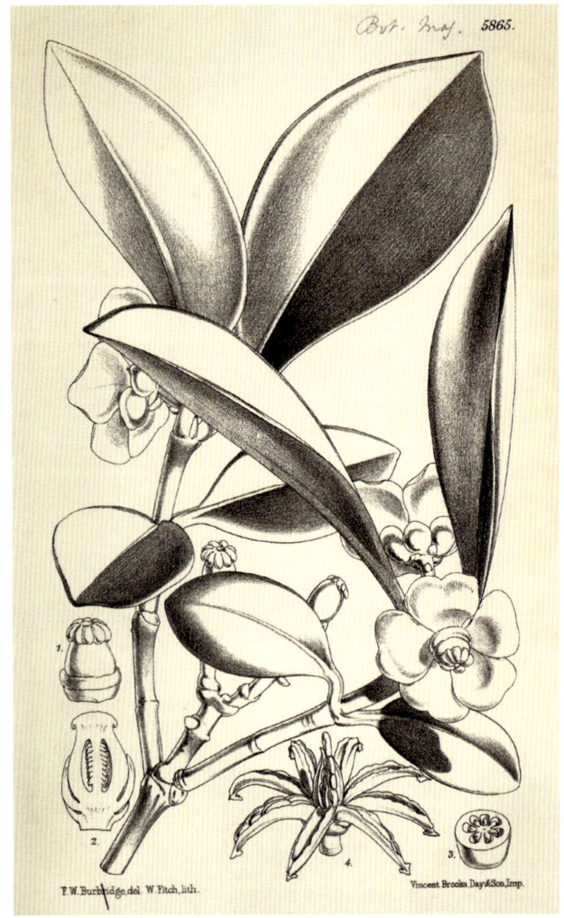

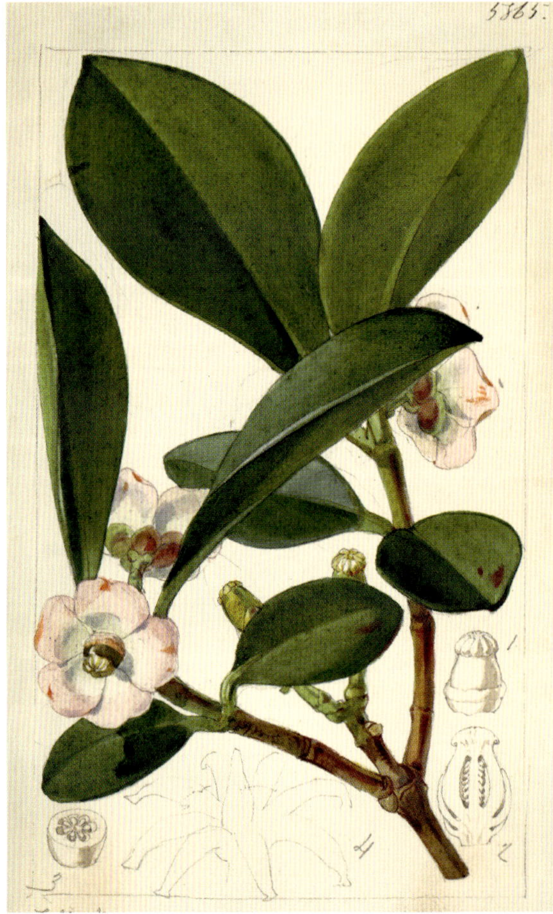

LEFT: Frederick William Burbidge's preliminary and final artwork of Clusia odorata.

OPPOSITE: Charlotte Lugard's study of Fugosia digitata, *now* Cienfuegosia digitata, *which she made in the Kwebe Hills in 1898.*

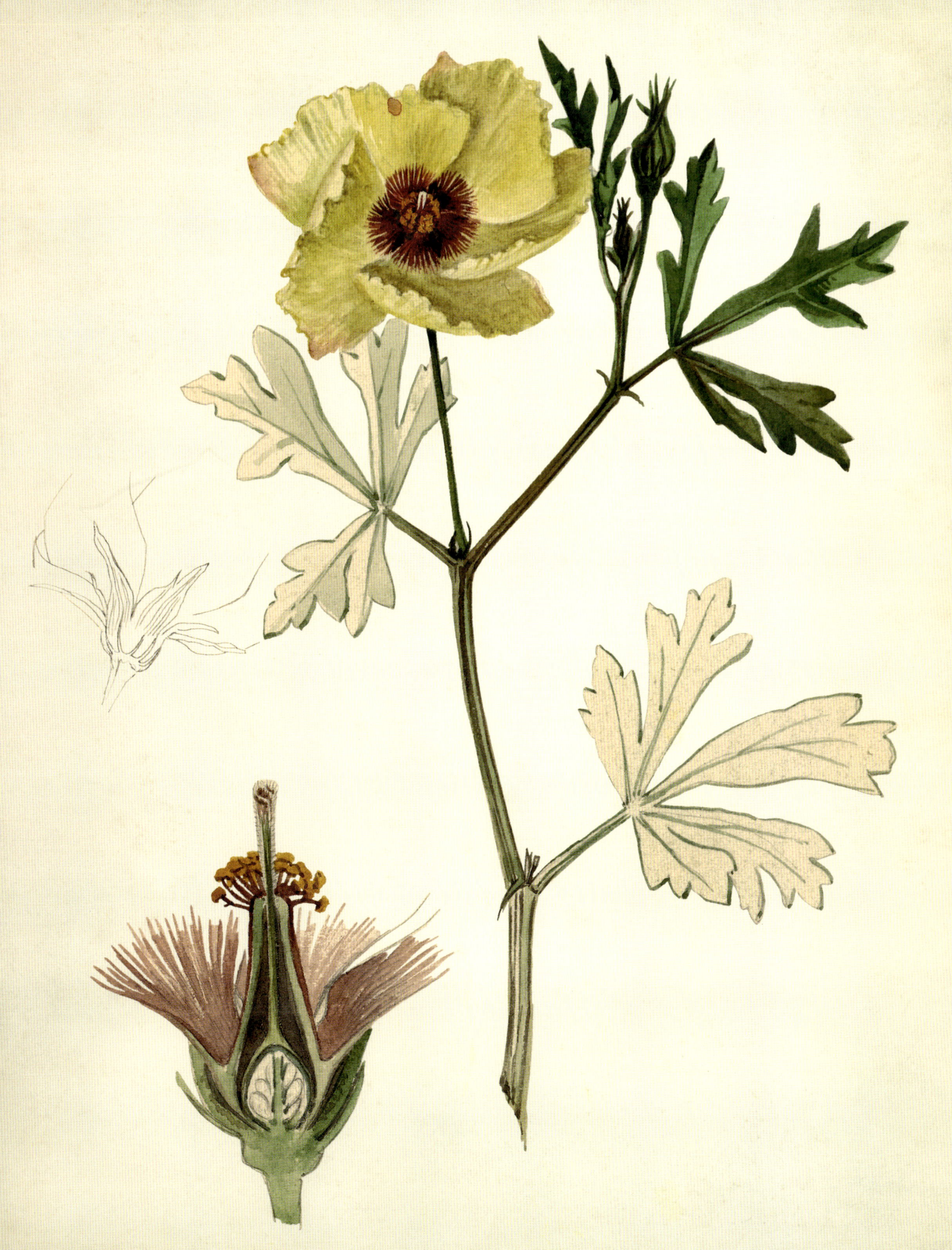

MARIANNE NORTH

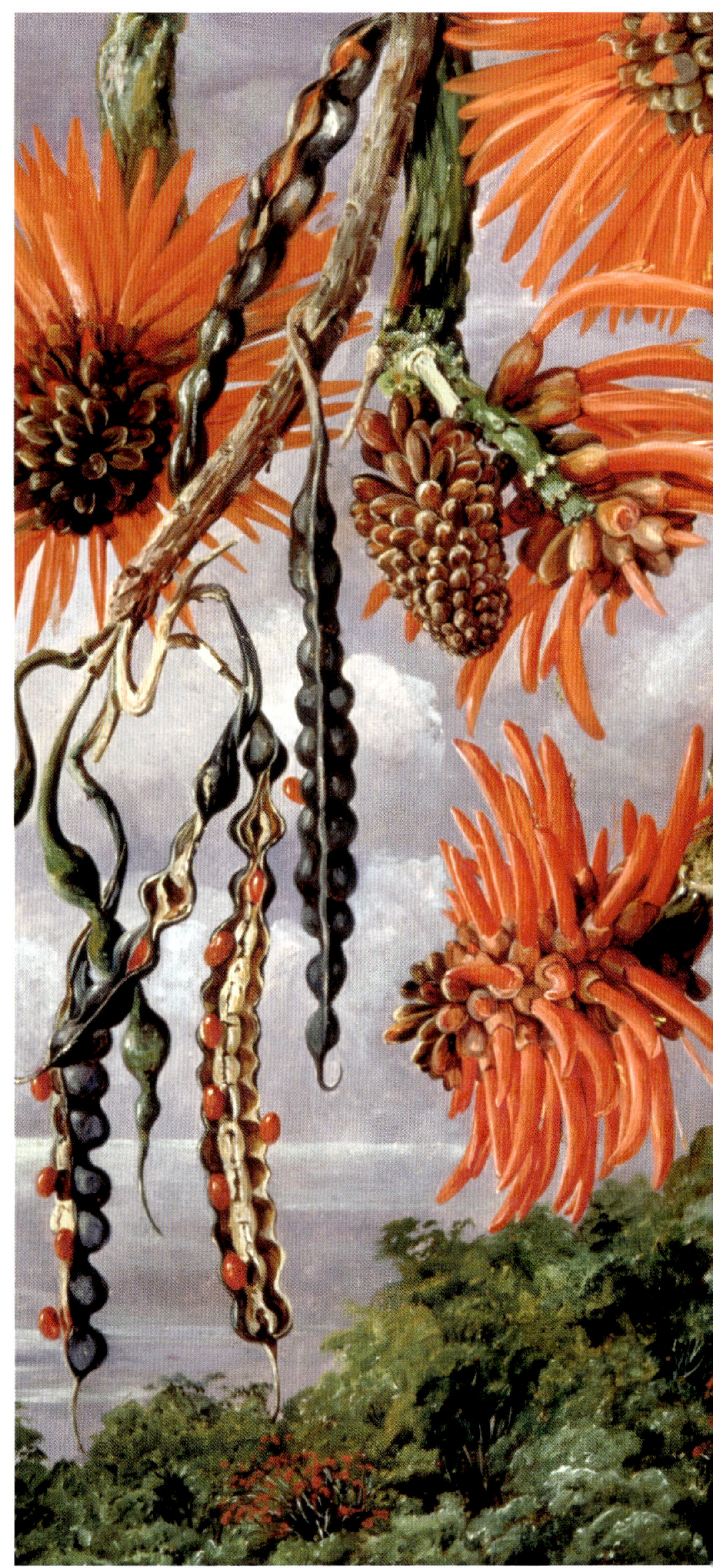

Of all the Victorian lady travellers and flower painters, Marianne North (1830–90) is the most famous, and her paintings have become familiar through the gallery she presented to Kew. She was born on 24 October 1830 – to Frederick North, the Member of Parliament (MP) for Hastings, whose family originally came from Norfolk, and his wife Janet, the daughter of another MP, Sir John Marjoribanks of Berwickshire, who had been left a widow with a young daughter. The Norths were well connected and comfortably off, dividing their time between Hastings, London, Norfolk and Lancashire, as well as travelling round Europe. They had a large circle of friends and acquaintances, among them the poet Edward Lear, who was later to encourage Marianne North in her work, and Sir William Hooker (*see* pages 188–90).

Frederick and his wife were devoted to each other, and Marianne seems to have had a thoroughly pleasant life, taking lessons in music and art, and, in 1850, flower painting from a Dutch artist, Miss van Fowinkel, and the following year from Valentine Bartholomew, flower painter to Queen Victoria.

In January 1855, Marianne's mother died, and during her last illness extracted a promise from her daughter that she would never leave her father. Marianne carried out this request faithfully, and accompanied him everywhere, including trips to the Horticultural Society's garden at Chiswick, and Kew, where she met Sir William Hooker who showed her some of the brilliantly coloured tropical plants growing in the new glasshouses. Marianne's sister Catherine married in 1864, and the following year Frederick lost his seat in the general election, leaving more time for their ever-increasing travels. The pair visited Italy, Turkey, Greece, Egypt and Palestine among other destinations.

In 1867, Marianne took lessons in oil painting from Robert Dowling, an Australian artist, and this (unusually) became her preferred medium for flower painting. By 1868 Frederick was back in parliament, but in declining health. The following year, Marianne took him to Austria to recuperate, but he became worse, and died shortly after their return to Hastings. Marianne wrote, "For nearly forty years he had been my one friend and companion, and now I had to learn to live without him, and to fill up my life with other interests as I best might… I went straight to Mentone to devote myself to painting from nature, and try to learn from the lovely world which surrounded me there how to make that work henceforth the master of my life."

In 1871 Marianne received an invitation from a friend to spend the summer in the USA, so she sold her house and set off; she had introductions to other friends, and was able to spend the next few months touring North America before travelling alone to Jamaica, where she arrived on Christmas Eve. She hired a house, near the

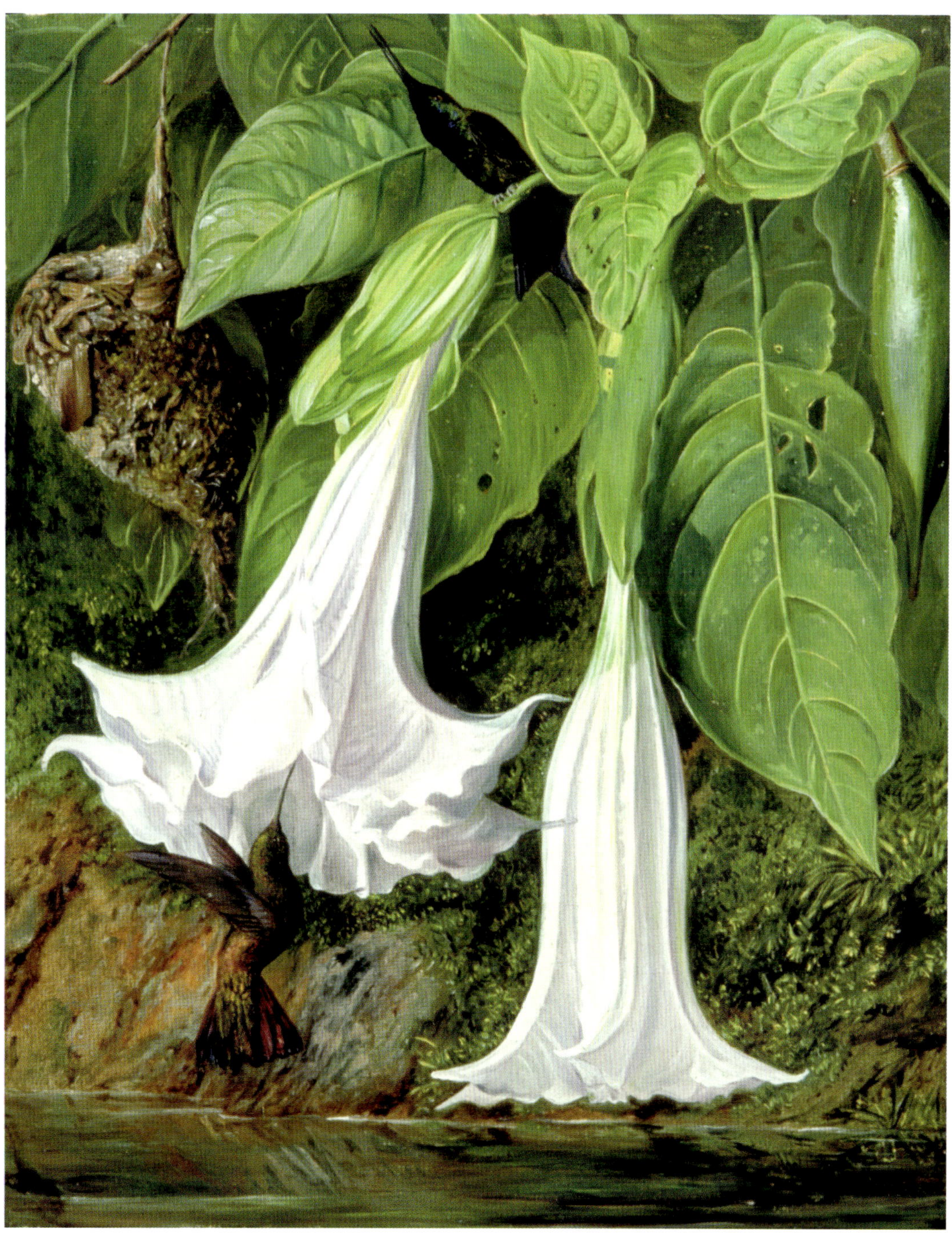

OPPOSITE: Marianne North's painting of an Erythrina *in Brazil.*

RIGHT: Marianne North's painting of Brugmansia suaveolens, *visited by a humming bird.*

deserted botanical gardens of Kingston, and here she had her first taste of the lush tropical vegetation which she had yearned to see since her visits to Kew. She was so excited by the wealth of plants that she declared herself to be "...in a state of ecstasy, and hardly knew what to paint first".

This visit to Jamaica marked the beginning of many years of intrepid travelling and a prodigious output of paintings. She returned home for just long enough to meet up with friends and give them an account of her travels, before embarking for Brazil, where she spent a year. Here she stayed at first near the botanic garden in Rio de Janeiro, painting there every day, and then began to explore the surrounding countryside by railway train, mule train, carriages and on foot. During 1875–77 trips to Tenerife, California, Japan, Singapore and Ceylon (Sri Lanka) followed, and she also visited Java and Sarawak (Borneo), where she was the guest of Charles Brooke, the Rajah of Sarawak, and his wife. Marianne was extremely keen on orchids, and this was a paradise for her – she travelled along the forest rivers by canoe, sometimes with a local guide who would cut the epiphytic orchids from overhanging trees so that they fell into the canoe. She also painted pitcher plants here, and one, a new species, collected from the limestone hills, was named *Nepenthes northiana* after her by Joseph Hooker in 1881.

Even when returning to England after months abroad, Marianne's accounts of her travels are breathless – for example, she wrote, "…to Rome… to see Mr Lear at San Remo with his cosmopolitan gallery of sketches… and my niece at Cannes. I went straight through from Cannes to London in thirty-six hours, arriving at midnight on the 25th February…." Back in London, Marianne was asked to loan some of her paintings to the Kensington Museum, so she "employed the last few weeks of my stay in England in making a catalogue as well as I could of the 500 studies I lent them, putting in as much general information about the plants as I had time to collect, as I found people in general woefully ignorant of natural history, nine out of ten of the people to whom I showed my drawings thinking that cocoa was made from the cocoa-nut".

From September 1877 Marianne spent a year in India, travelling around Simla, Darjeeling, Delhi, Amritsar, Lahore and elsewhere, and here she concentrated on building up a collection of paintings of plants considered sacred by the Indians. She arrived home in the spring of 1879 with about 200 paintings, some of which she later exhibited in a room in Conduit Street, London. A review of this exhibition, published in the *Pall Mall Gazette*, suggested that the paintings would be better exhibited at Kew, so Marianne wrote to the director of the gardens, Sir Joseph Hooker, asking him "…if he would like me to give them to Kew gardens, and to build a gallery to put them in, with a guardian's house. I wished to combine this gallery with a rest-house and a place where refreshments could be had – tea, coffee, etc".

Sir Joseph accepted her offer, though he refused the tea-house on the grounds of practicality, and Marianne immediately selected the site at Kew and commissioned the architect. Meanwhile, she had been summoned to meet Charles Darwin, who suggested a visit to Australia, and Marianne "determined to take it as a royal command and to go at once". She travelled via Borneo once more, and from there to Australia, reaching Brisbane in August 1880. She proceeded to travel around Australia, painting *Eucalyptus* and banksias. Tasmania, and tree ferns, were next, before she movd on to New Zealand, America, and then home.

LEFT: Marianne North's painting of a coconut palm.

BELOW: Marianne North's painting of the Taj Mahal at Agra.

OPPOSITE: Marianne North's painting of various flowers including Clusia rosea *and* Onicidium.

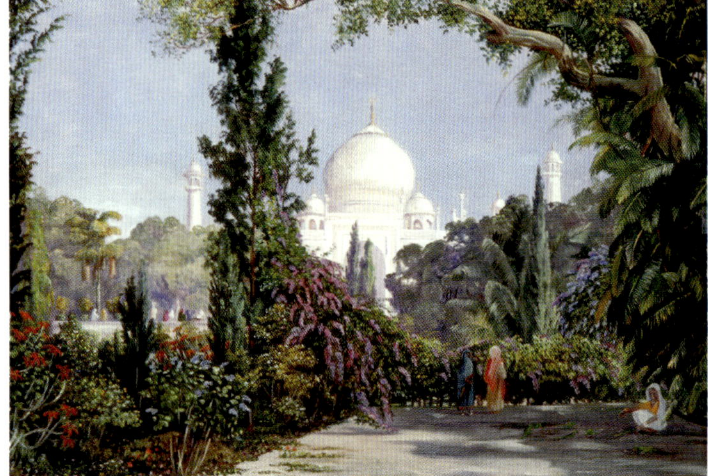

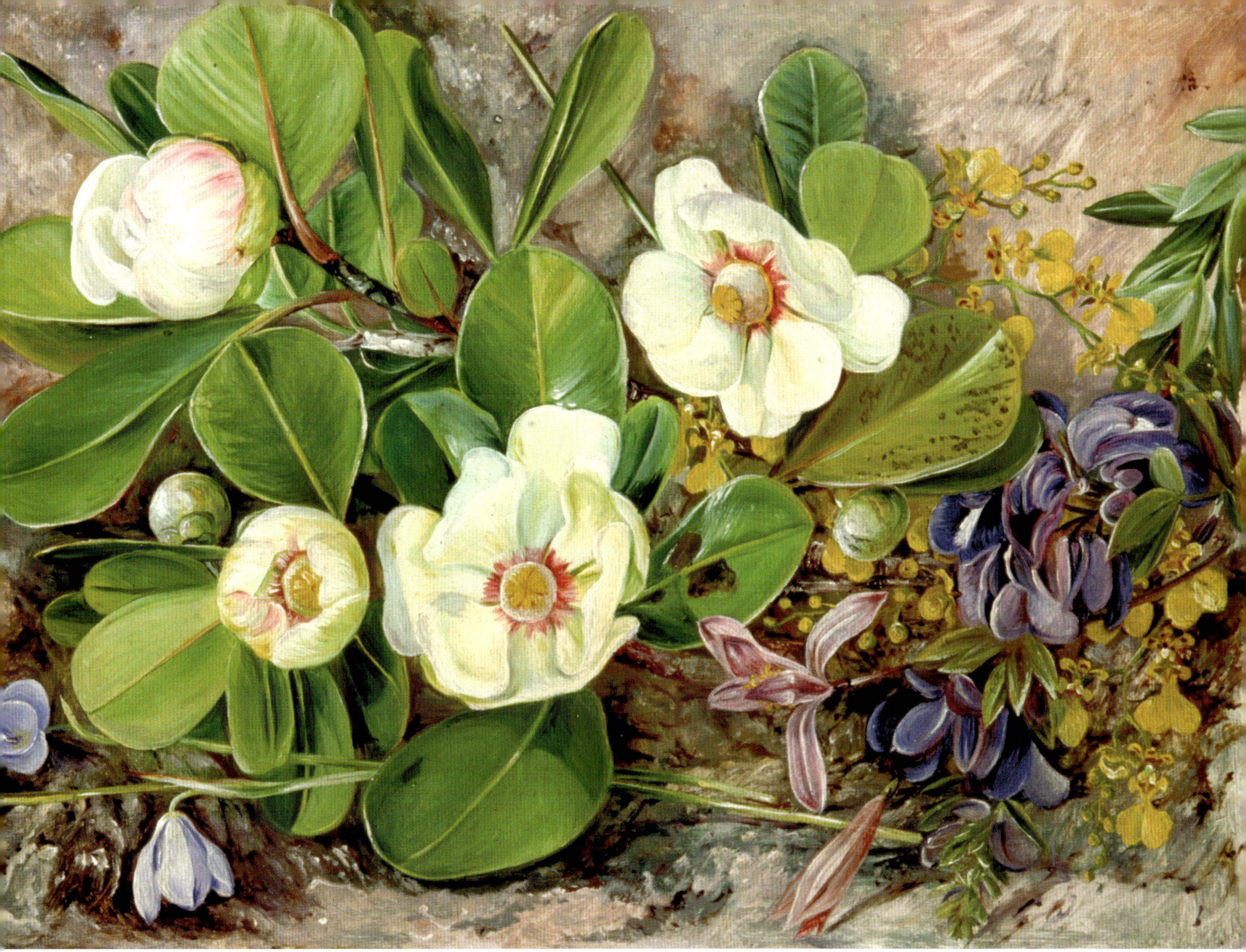

Once back in England again, Marianne went with all speed to see how the new gallery at Kew looked: she was not disappointed: "I found the building finished (as far as bare walls went) most satisfactorily, its lighting perfect. Mr Fergusson kindly arranged about the decorating and painting of the walls. After that I spent a year in fitting and framing, patching and sorting my pictures, and finally got it finished and open to the public on the 7th June 1882."

The gallery had a high room with plenty of wall space for hanging the large number of paintings, with large windows above to provide as much natural light as possible, an artist's studio, a flat for a gardener and, outside, an Indian-style verandah, intended to evoke memories of India. Marianne arranged the hanging of the initial 627 paintings, (a further 205 were added later), and labelled them, with the help of the botanist Mr Hemsley. The general public were amazed by the collection: the bright colours of exotic plants that most could never hope to see in the wild, and the scenes of dramatic landscapes, made a huge impact on people who had never travelled or seen coloured photographs of the tropics. Marianne herself seems to have genuinely delighted in sharing her pictures with other people; she loved displaying the paintings and also enjoyed the whole process of building and decorating the gallery, putting great thought into the siting of the building: "I chose the site … far off from the usual entrance gates, as I thought a resting-place and shelter from rain and sun were more needed there, by those who cared sufficiently for plants to have made their way through all the houses …."

Now that the gallery was open, Marianne decided that, as Africa was the one continent not represented there, she should travel there immediately. She left in August 1882, and after only 18 days sailing arrived at the Cape. As usual, she fell almost immediately among friends, and was soon out painting the proliferation of bulbs, succulents and proteas and a few orchids, including *Disa uniflora* (syn. *D grandiflora*), which she found growing on Table Mountain. (At the time of her visit the plants weren't flowering, so she must have produced her painting from a flower seen at some other time.)

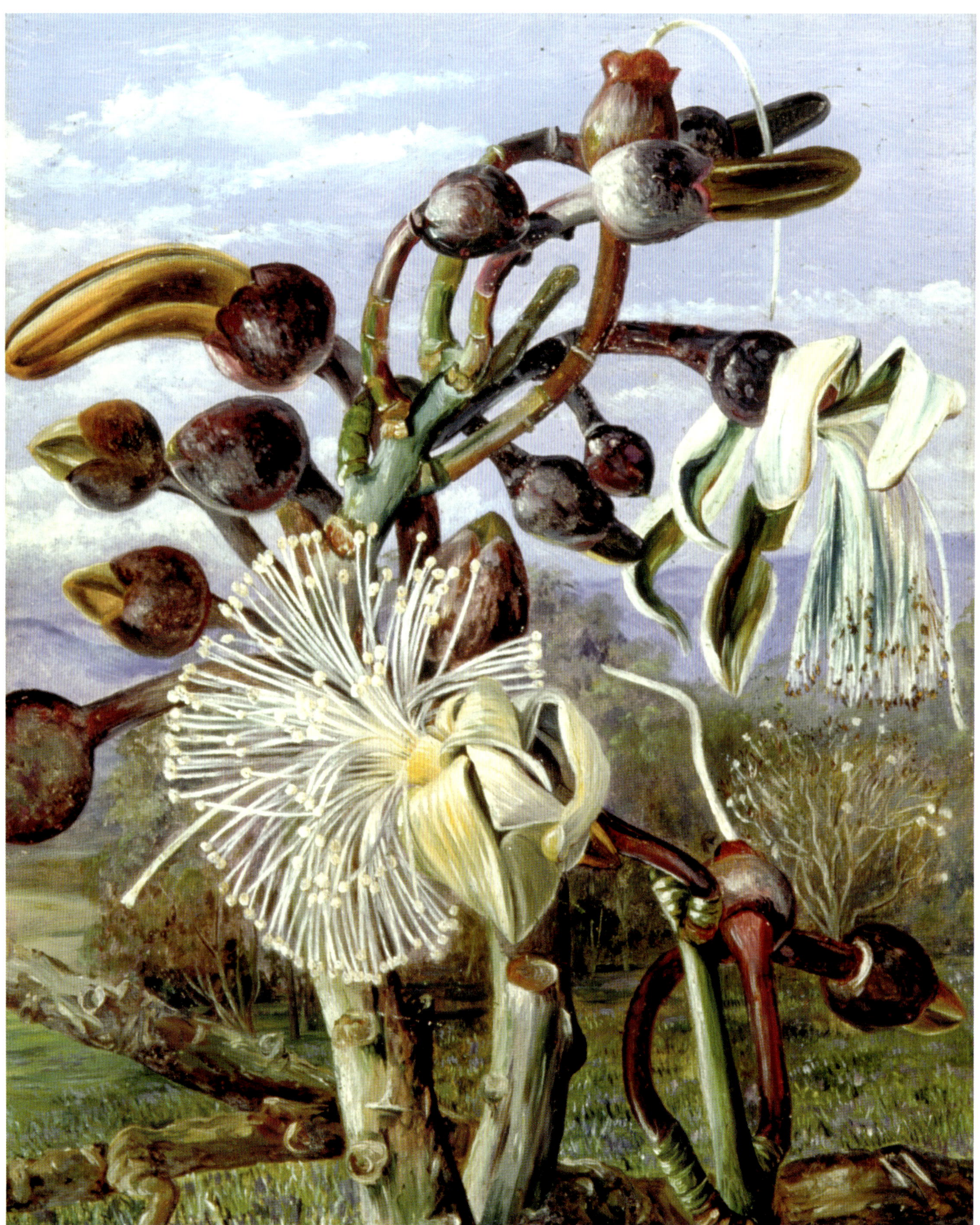

Marianne returned to England once more, and arranged with Mr Fergusson for another room to be added to the gallery, before setting off for the Seychelles in the autumn of 1883. Here, she painted numerous palms, and in particular the peculiar Seychelles coconut *Lodoicea maldivica*, (known as coco de mer, because its seeds were often washed up, and no one knew where they originated), as well as a tree named by Hooker *Northea seychellana*.

By now, Marianne was driving herself to complete her self-appointed project before leaving for England, but when she agreed to go into quarantine for smallpox on a nearby island she suffered such exhaustion that it seems she had a nervous breakdown. She recovered sufficiently to sail back to England, where she recovered, and was able to enjoy finishing and arranging her paintings at Kew.

In November 1884, Marianne was well enough to travel abroad one last time – to Chile, which she wished to visit so that she could paint the blue Puya and the monkey-puzzle tree, *Araucaria araucana*. She rode out into the hills and managed to track down both plants, with the monkey puzzles "covering many miles of hill and valley". After Christmas and New Year at Santiago, she started for home, stopping off for a month with an old friend in Jamaica on the way, and as soon as she got back, returned to Kew to finish and rearrange the pictures in the gallery. She reordered these so as to preserve the geographical groupings of plants, and finally satisfied with the gallery "tried to find a perfect home in the country with a ready-made old house and a garden to make after my own fashion…".

Marianne found a house at Alderley in Gloucestershire, and spent her last years enjoying country life, gardening and writing her memoirs, entitled *Recollections of a Happy Life*, and *Some further Recollections of a Happy Life*, which were published posthumously by her sister Catherine in 1892–93. She died, aged only 59, in 1890.

Marianne North's gallery at Kew, now a Grade II listed building, was refurbished in 2010 and restored to its original state, and the paintings conserved and rebacked onto new boards, by conservators.

OPPOSITE: Marianne North's painting of Pachira marginata *(now* Pseudobombax marginatum*) in Brazil.*

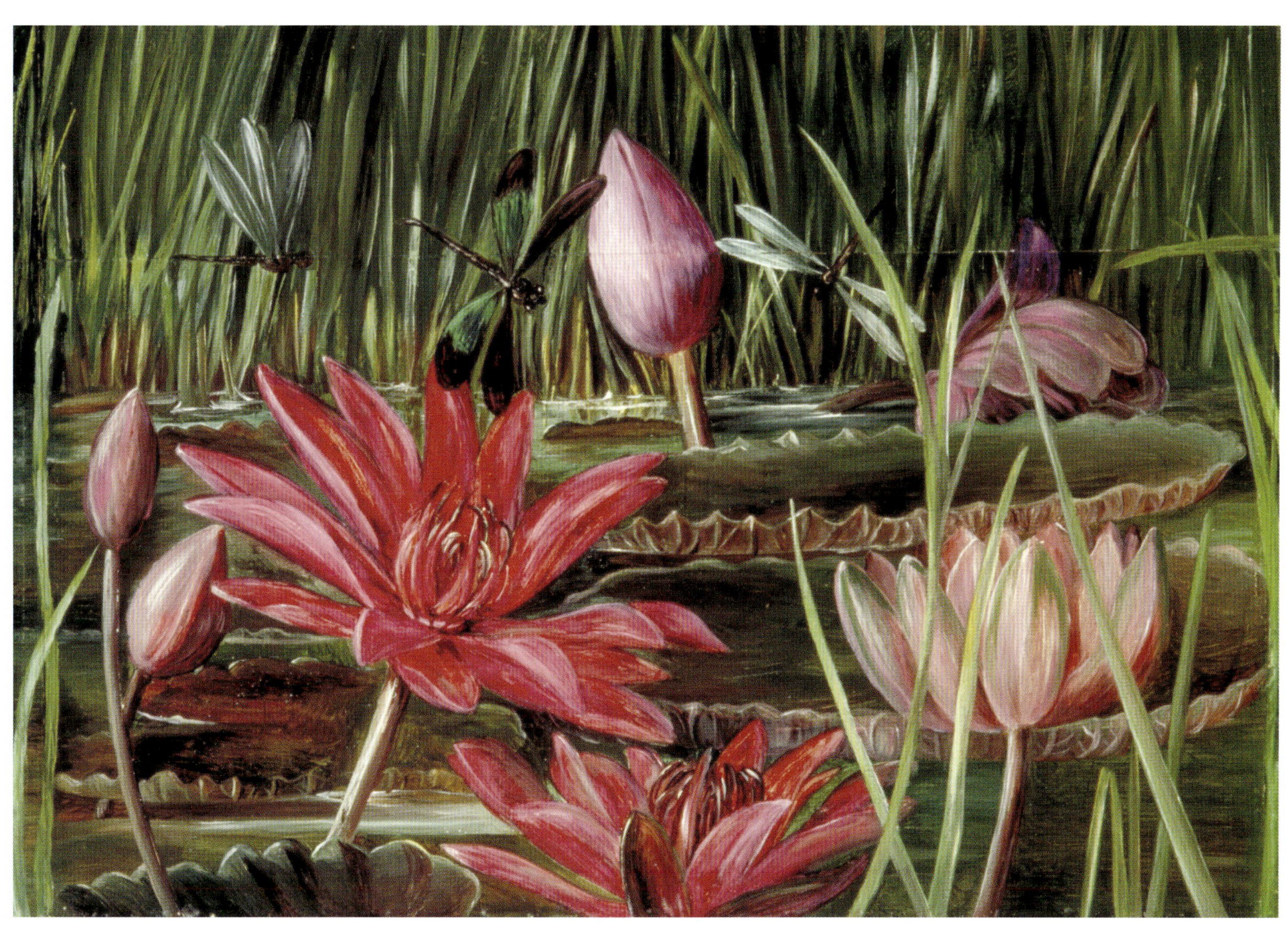

BELOW: Marianne North's painting of Nymphaea rubra.

Elwes and the Genus Lilium

Henry John Elwes was born in 1846 into a wealthy family at Colesbourne Park, Gloucestershire. He was educated at Eton and in Germany, before joining the army in 1865, which gave him the chance to indulge his interests in shooting big game and in natural history, especially ornithology and entomology.

His army career was short lived, and after resigning his commission he was able to travel to Asia, and, inspired by Joseph Hooker's *Himalayan Journals* went to the Sikkim Himalayas in 1870. He was elected a Fellow of the Royal Society for his work on the distribution of Asiatic birds, and also became increasingly interested in entomology and botany. As he remarked to E. A. Bowles in later life, "To think that I spent twenty of the best years of my life catching butterflies."

In 1871, he married Margaret Lowndes, with whom he had two children, and the couple lived and gardened near Cirencester. Elwes continued to travel, and in 1874 visited Turkey, where he collected many bulbs, including a fine snowdrop, subsequently named *Galanthus elwesii* by Joseph Hooker. This was painted by W. H. Fitch (*see* pages 200–03) and published in *Curtis's Botanical Magazine* in 1875.

His interest in bulbs now fired up, Elwes then embarked on his great folio work entitled *The Genus Lilium*, in which he attempted to describe all the species of lilies known at the time. In his self-effacing introduction to the book, which was published in 1880, he writes that the book is "… not the work of a scientific botanist, but is merely the result of a few years horticultural study…". Elwes was helped in his work by J. G. Baker, the botanist at Kew, and numerous correspondents who forwarded information to him. He was also fortunate in securing the services of Fitch, who drew and lithographed the 48 plates, which were then hand coloured. This is a huge folio, with the lilies painted life-size or slightly larger, and the illustrations are as fine as any of Fitch's work, even if they perhaps lack a little subtlety.

By 1933, enough new lilies had been discovered by plant hunters in areas such as western China, only recently opened to travellers, to merit a series of supplemental volumes, published

OPPOSITE: Lilium washingtonianum *from Elwes's* The Genus Lilium.

RIGHT: Lilium superbum *from Elwes's* The Genus Lilium.

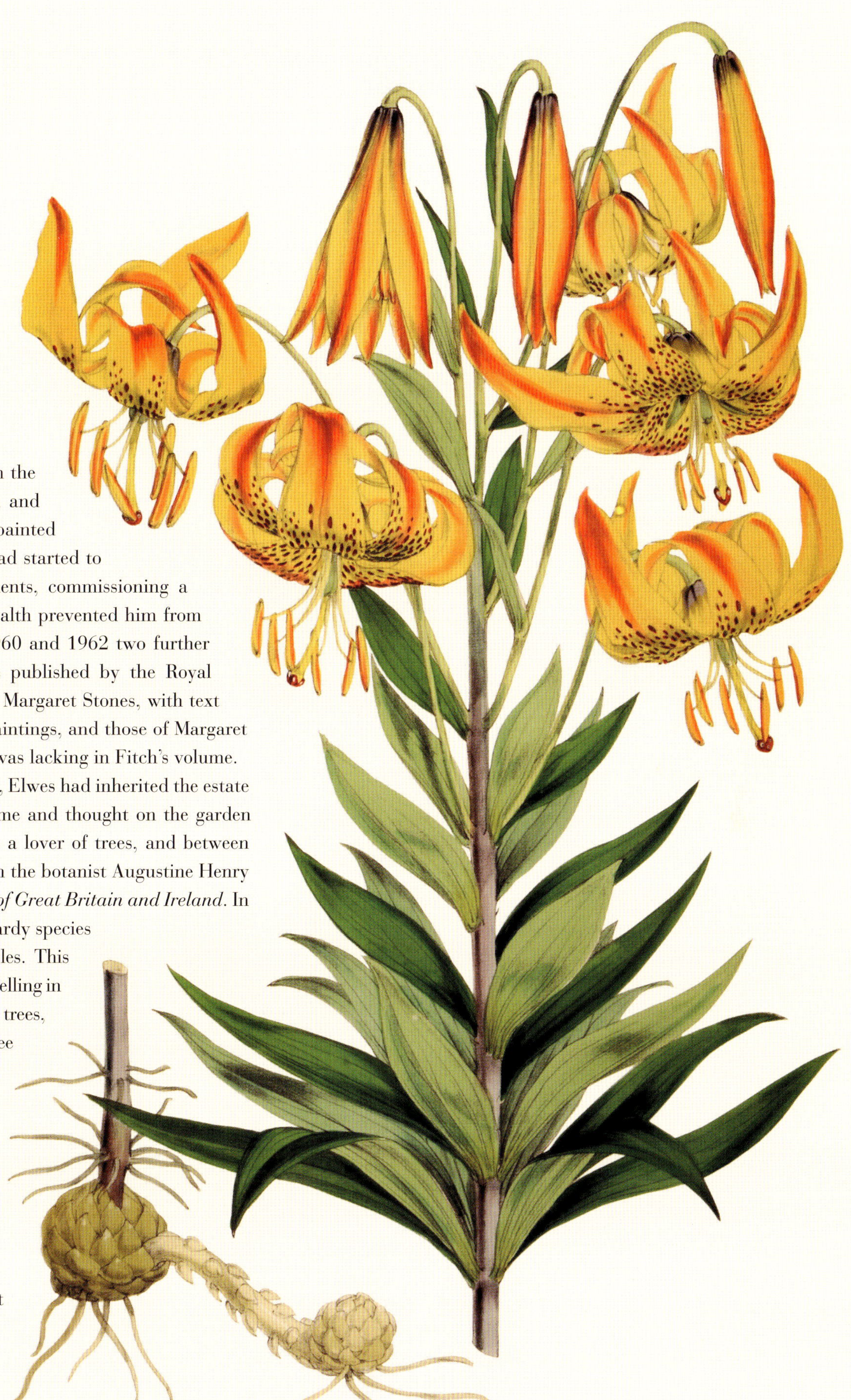

in seven parts until 1940, for which the text was written by Alice Godman and Alfred Grove, and the illustrations painted by Lilian Snelling. Elwes himself had started to prepare the first of these supplements, commissioning a number of plates, but his failing health prevented him from completing the project. Between 1960 and 1962 two further supplements, parts 8 and 9, were published by the Royal Horticultural Society, illustrated by Margaret Stones, with text by W. B. Turrill. Lilian Snelling's paintings, and those of Margaret Stones have some of the grace that was lacking in Fitch's volume.

On the death of his father in 1891, Elwes had inherited the estate at Colesbourne, and spent much time and thought on the garden and wider landscape there. He was a lover of trees, and between 1900 and 1913 he collaborated with the botanist Augustine Henry on a seven-volume work, *The Trees of Great Britain and Ireland*. In this, the two men described every hardy species of tree then grown in the British Isles. This involved an enormous amount of travelling in order to see and measure many of the trees, and Elwes also travelled abroad to see the trees in their native habitats. With the exception of an attractive engraved title page, the book was illustrated with black-and-white photographs of specimens.

Elwes's work was appreciated by his peers; he was elected a Fellow of the Linnean Society in 1874, and was awarded the first Victoria Medal of Honour by the Royal Horticultural Society in 1897. He died at Colesbourne on 26 November 1922.

16

Bringing China to Europe

The influence of plants from western China on European gardens cannot be overestimated, and a series of intrepid collectors sent back seeds and bulbs for subscribers and nurseries in the British Isles. Père David (*see* pages 168–69), gave western botanists a glimpse of the riches of western China in the late 1860s, but the first to introduce large numbers of Chinese wild plants to gardens was Augustine Henry (1857–1930).

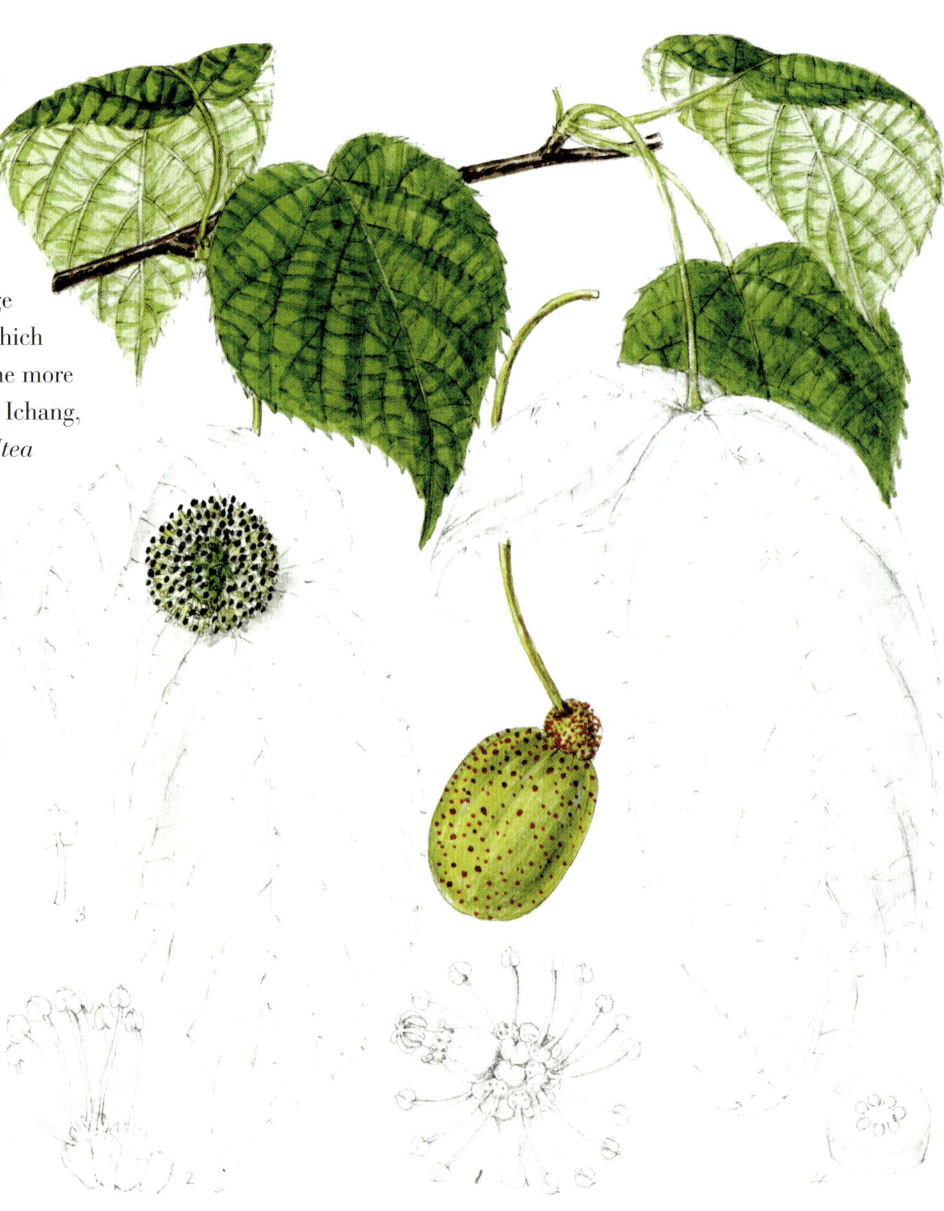

Henry was born in Dundee, of Irish parents, and joined the Chinese Maritime Customs service, then partly staffed by Europeans, in 1880, as a medical officer. Initially, he was stationed in Ichang, where the gorges and rapids (now flooded by a gigantic dam) made a natural border between the Sichuan basin and eastern China. Here, Henry found a large number of ornamental trees, shrubs and lilies which would be hardy in northern Europe, in contrast to the more subtropical plants grown in eastern China. Near Ichang, Henry found *Lilium henryi*, *Hamamelis mollis* and *Itea ilicifolia*, now common garden plants. During his periods of leave, Henry travelled in other parts of China, notably Sichuan and Hubei, and was later stationed in Yunnan, near Mengtze, close to the border with French Indochina, and finally to Szemao further west. Seeds of the dramatic *Rodgersia pinnata* were sent to Kew in 1898, and first flowered there in 1902. Henry collected the seed north of Mengtze in western Yunnan, where plants were found growing on wet cliffs.

On his travels in Hupeh, Henry had reported seeing a handkerchief or dove tree, *Davidia involucrata*, in full flower, and it was to collect seed of this that Messrs Veitch of Chelsea dispatched E. H. Wilson to China.

Ernest Henry Wilson (1876–1930) was born in Gloucestershire, and became a trainee gardener at Birmingham Botanic Gardens, and then at Kew from where he was recommended to James Veitch.

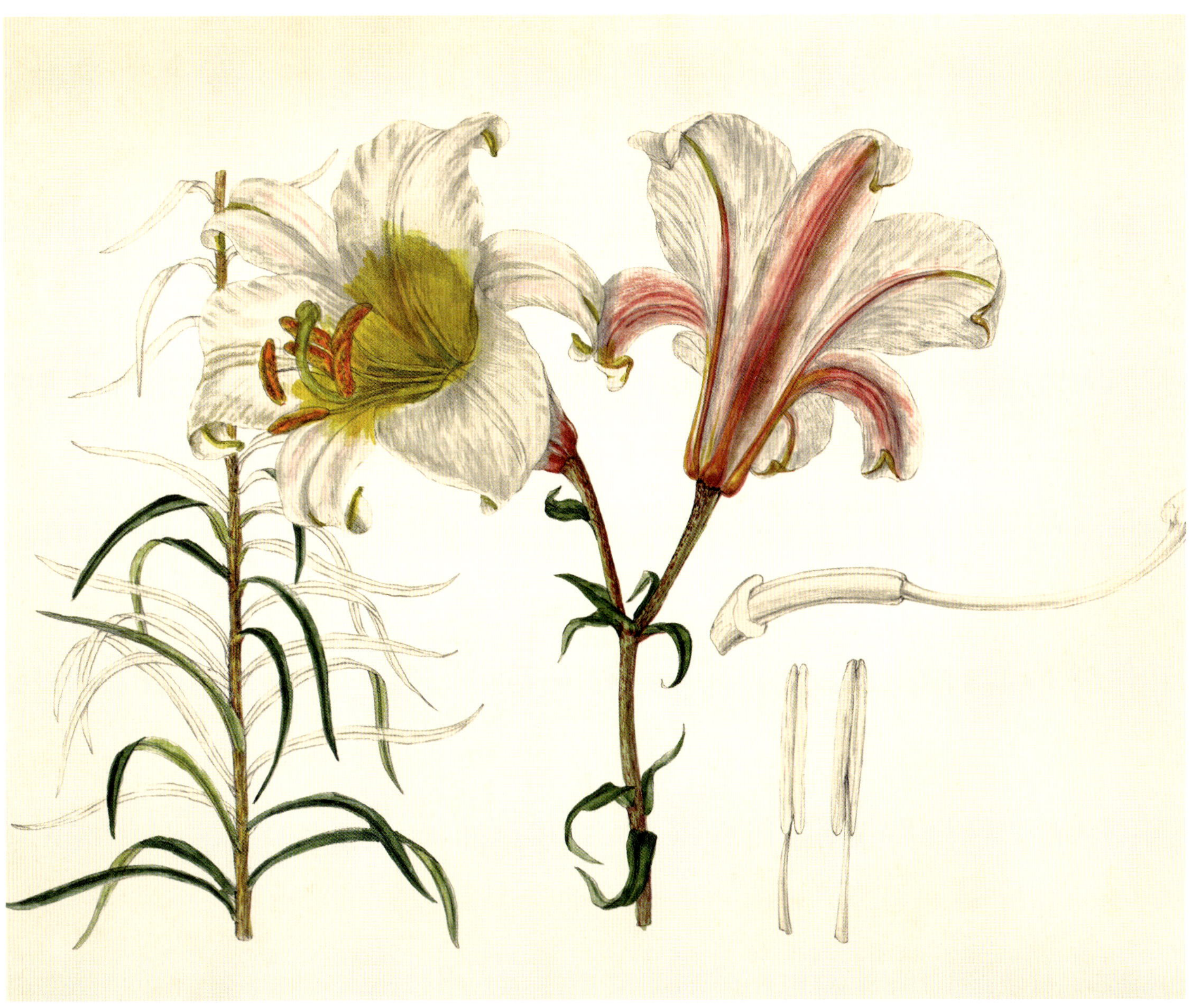

His instructions were not to waste time on other plants until he had met Henry, discovered the whereabouts of the *Davidia* tree and collected the seed. Wilson left England for China in April 1899, crossing America, where he visited the Arnold Arboretum, and the Pacific before landing at Hong Kong. From here, he travelled across China to meet Henry near the Burma border in Szemao. It took him most of the summer to reach Szemao, and he spent the early part of the winter learning about Chinese plants and customs from Henry, who was then about to return to England. Wilson succeeded in introducing the *Davidia*, and numerous excellent shrubs and trees, including *Actinidia chinensis* (the Kiwi fruit), *Clematis armandii* and the evergreen *Magnolia delavayi*. Wilson made a second expedition for Veitch in 1903, and later ones for the Arnold Arboretum, on one of which he was based in a houseboat, nicknamed *The Harvard*, near Lèshān, not far from Mount Omei (Emei). Wilson made his fourth and last expedition to China in 1910, and it was on this trip that he collected *Lilium regale*, the hardy white trumpet lily, which is probably his most popular introduction. Wilson, who was then keeper at the Arnold Arboretum, and his wife were killed in a car crash in Worcester, Massachusetts in the USA in 1930.

OPPOSITE: Matilda Smith's study of Davidia involucrata *var.* Vilmoriniana, *named after the missionary botanist Père Armand David, and the great French nurseryman Maurice de Vilmorin, who first grew the tree in Europe.*

ABOVE: Matilda Smith's study of Lilium regale, *one of E. H. Wilson's most popular introductions from China, painted on 7 and 8 July 1920.*

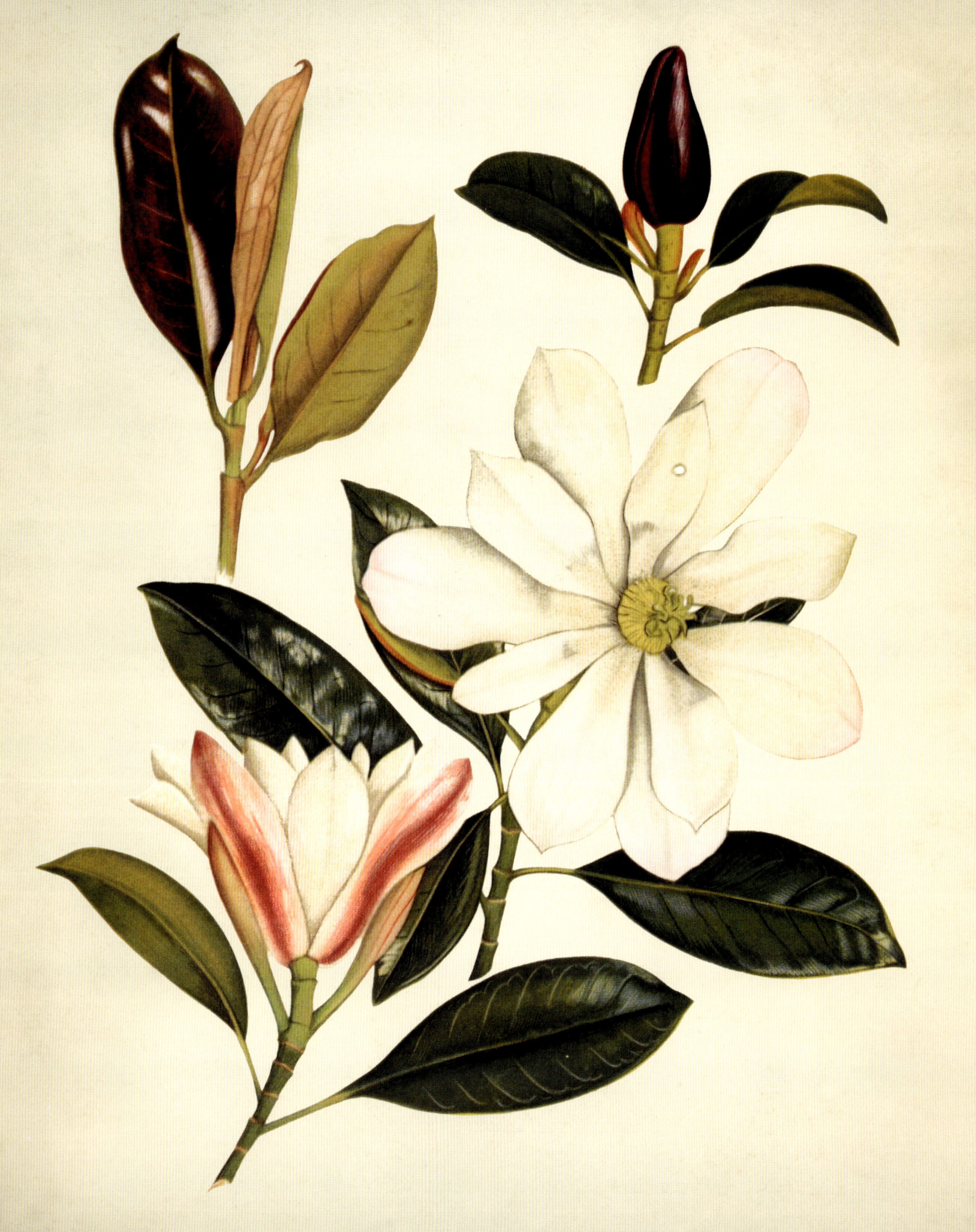

Though Wilson never painted flowers or made sketches of the landscape, he was an accomplished photographer, carrying a full-plate camera with him on the most difficult journeys and producing excellent black-and-white photographs of landscapes, and less often of plants. Most are kept at Harvard University. Numerous coloured paintings of his plants were made once they had flowered in England, and his plants are found in some important books. The lilies, of course, appeared in the first supplement to H. J. Elwes's monograph in 1933 (*see* page 216). Wilson's magnolias, notably *Magnolia wilsonii*, *M. dawsoniana* and *M. sargentiana*, were illustrated in *Asiatic Magnolias in Cultivation* by G. H. Johnstone, when they flowered in gardens in Cornwall; this was not published until 1955, as many of the tree magnolias take 20 or 30 years to flower from seed, and then the Second World War delayed the publication of expensive, illustrated books. In addition to Wilson's introduction, this features the huge, pink-flowered *M. campbellii* subsp. *mollicomata*, and the dramatic purple-flowered "Lanarth", both collected by George Forrest. This book, and the equally beautiful monograph of *Paeonia* by F. C. Stern (1946), were published by the Royal Horticultural Society and illustrated mainly by Lilian Snelling and Stella Ross-Craig. Wilson's collection of *Paeonia obovata* var. *willmottiae* Stapf, was raised by Miss Willmott of Warley Place in Essex, one of the keen gardeners who raised seeds sent from China.

While the collectors were sending back quantities of seed of new plants, gardeners in Britain and Ireland were competing to grow them successfully and be the first to persuade them to flower. The earliest paintings of these new plants were often published in *Curtis's Botanical Magazine*, then published by Lovell Reeve for the Royal Botanic Gardens, Kew. Harriet Thistleton Dyer (1854–1945), the eldest daughter of Sir Joseph Hooker, was one of the chief artists between 1878 and 1899; married to the director of Kew, she was conveniently on hand to paint any new plant that flowered. Thus, in 1899, she illustrated *Rosa ecae*, a small, bright-yellow rose which had been discovered by Surgeon-Major Aitchison while on the Afghan border delimitation commission in the 1880s and named after his wife, E. C. Aitchison.

A second artist, Matilda Smith (1854–1926), was more prolific, and contributed paintings to *Curtis's Botanical Magazine* between 1878 and 1923, including many of the new rhododendrons collected by Augustine Henry and E. H. Wilson. One of these was the bluest of all rhododendrons, *R. augustinii*, which Henry had discovered in Hupeh, and had been introduced by Wilson.

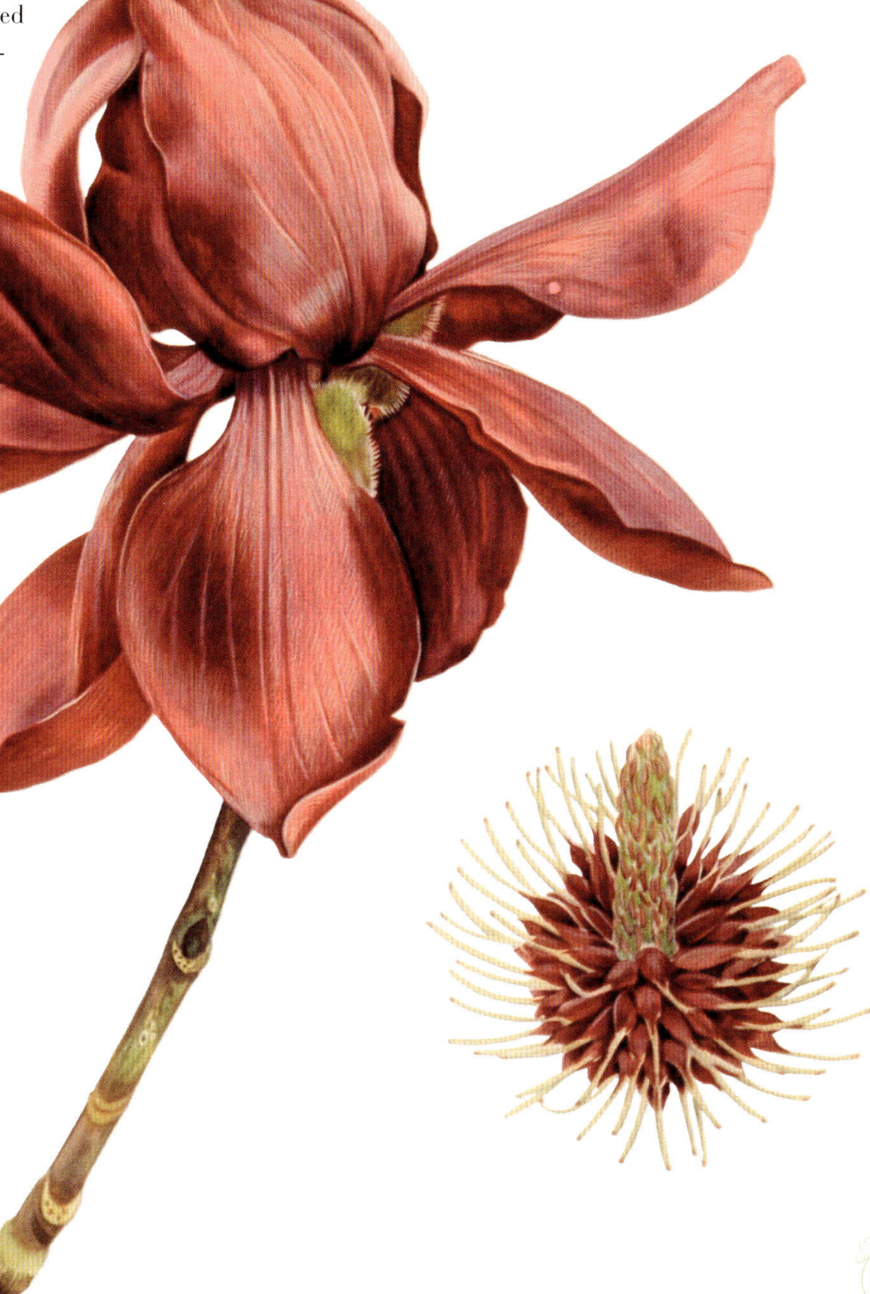

RIGHT: P. Ciccimara's painting of Magnolia Cambellii *subspecies* mollicomata *convar.* Williamsiana *from G. H. Johnstone's* Asiatic Magnolias in Cultivation.

OPPOSITE: Ann Webster's painting of Magnolia nitida *from G. H. Johnstone's* Asiatic Magnolias in Cultivation.

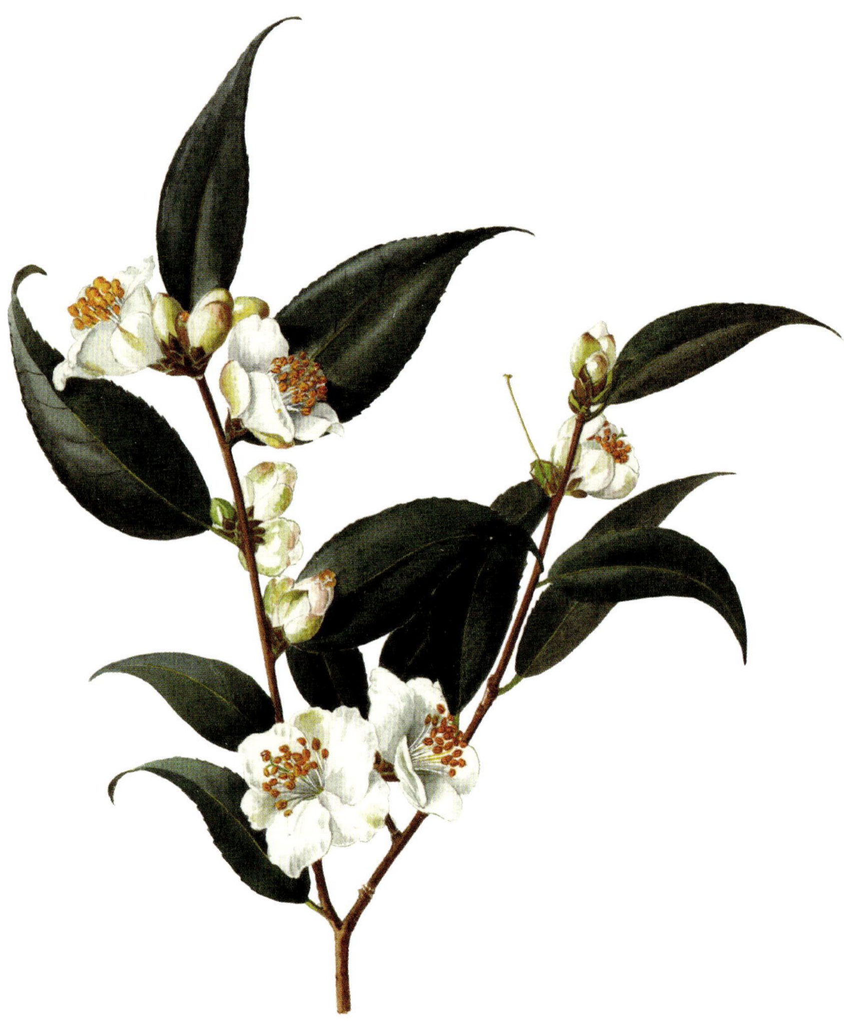

LEFT: A study of Camellia cuspidata *by Lilian Snelling, painted in 1928.*

OPPOSITE: A field sketch of Daphne petraea *by Reginald Farrer.*

Lilian Snelling (1879–1972) was one of the most elegant botanical artists of the twentieth century: her work is accurate, delicate and artistic and, in its use of transparent watercolour, was in line with the fashion of the early part of the century. She was born in Kent, but spent her early career as a botanical artist, painting flowers from the garden at Colesbourne for Henry Elwes, and then as official artist at the Royal Botanic Garden, Edinburgh, where she worked from 1916 until 1921. This was a great era for the garden in Edinburgh, as new rhododendrons and *Primula* were being grown and studied from seed sent back from China and the Himalayas; the primulas especially grow better in the cooler climate of Scotland than they do around London. Snelling proved an expert painter of both these genera, and while working as chief artist for *Curtis's Botanical Magazine* between 1922 and 1952, she painted 41 species of *Primula* and no less than110 species of *Rhododendron*.

Reginald Farrer (1880–1920) was one of the most influential, if eccentric, gardeners of the early twentieth century. He was a keen amateur, with enough private means to indulge his passion for plants, and especially for alpines, which he grew in the family garden near Ingleborough in Yorkshire. Farrer was an enthusiastic and colourful writer, and his books, especially *The English Rock Garden*, published in 1919, were noted for their florid descriptions of plants. Two short quotes give a flavour of the book; here he describes a pink-flowered, dwarf alpine shrub, *Daphne petraea*: "haunting hot and terrible cliff-faces of rose-grey limestone fronting the full radiance of the Italian sun… the sight of those sheer, awful faces blotted with scabs of living pink flat to the cliff… is one that amply repays the distance, difficulties, dangers and despairs that sometimes wait on the worshipper of *D. petraea*." The preface says that "this book was written in 1913, and corrected for press at Lanchou-fu, Kansu, China, during the winter of 1914", but publication was delayed by the war. In 1914, Farrer was engaged on his first expedition to China, with William Purdom, exploring the cold limestone mountains of central Gansu, which, he thought, would produce hardier and more alpine plants than the trees and shrubs other collectors had gathered further south. On this expedition, they found many new plants, often named *farreri* or *purdomi*, including a *Geranium*, a *Buddleia*, a gentian and a rose.

Farrer's second expedition, 1919–20, was to the mountains of northern Burma, and on part of it he was accompanied by Euan Cox of Glendoick. They were based in a bungalow in Maymyo, and from there travelled about a thousand miles into the unexplored mountains along the border with China. After the first season, Cox returned to Scotland, and Farrer continued for the next year. He was a difficult companion and was happier alone; his dramatic pictures, painted in the field, show alpine plants in their wild habitats, the crimson *Notholirion hyacinthinum* with ranges of peaks, or blue *Primula sonchifolia* growing out of the bare ground from which the snow has melted, or the pale bells of *Primula agleniana* against icy peaks. Dwarf rhododendrons hug the tops of rocky ridges and the graceful *Nomocharis pardanthina* var. *farreri* emerges from a grassy slope. These sketches are rather crude, but atmospheric and informative, and were sent back to England after Farrer's death.

In her book *A Rage for Rock Gardening*, Nicola Shulman describes Farrer's tortured life and his death. In October 1920, waiting in a mountain hut for the rain and mist to clear and to begin the season's seed harvest, he became ill with chest pains and a cough; he ceased to eat, but drank whisky with various medicines and died quietly on 17 October 1920. Six workers carried his body from Nyitadi down to the Nung village of Kawnglang Po, and they buried him in a hollow on the hillside above the village: "His seeds and books were abandoned, but they sent his diaries… back to his mother, who cut them up with scissors."

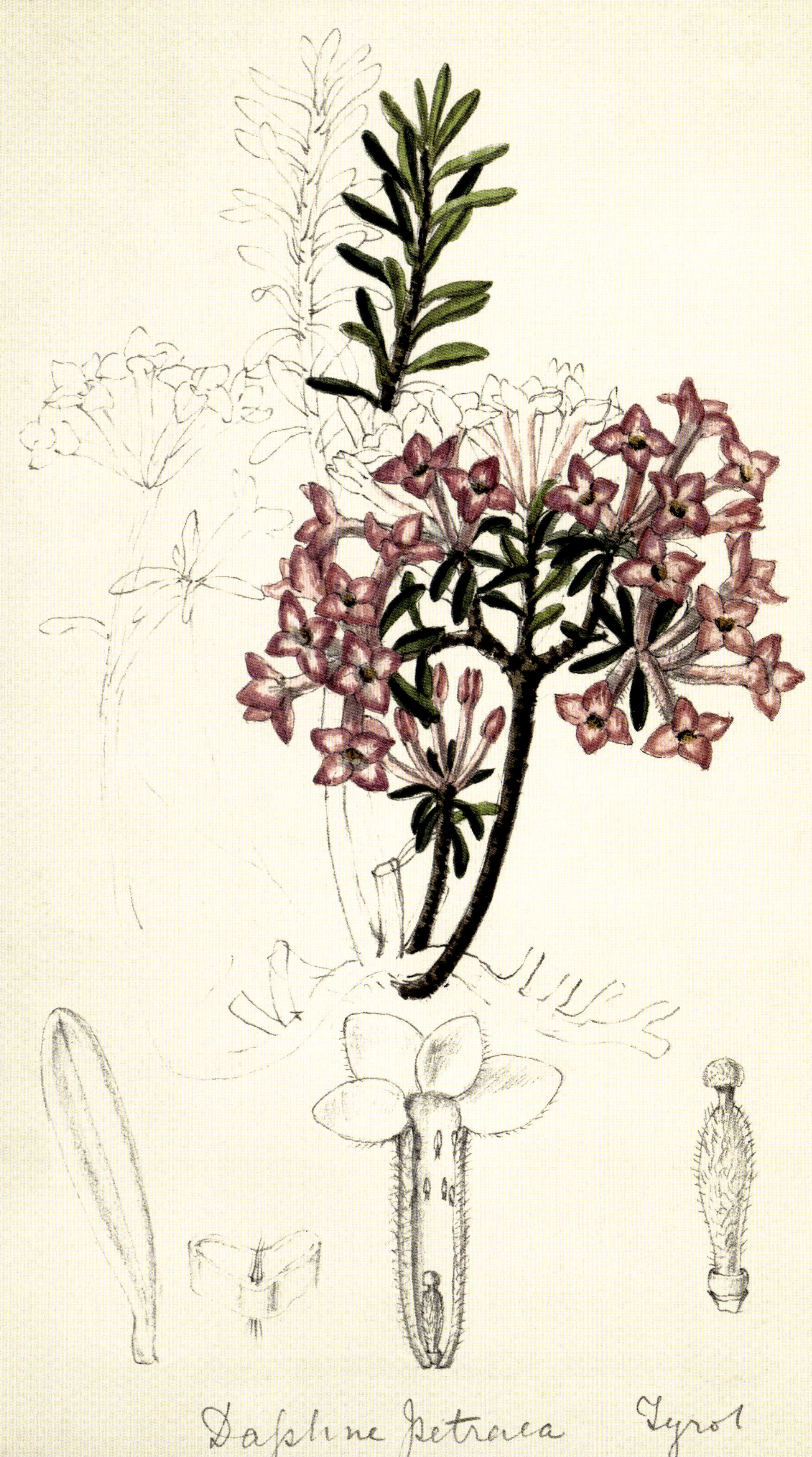
Daphne petraea Tyrol

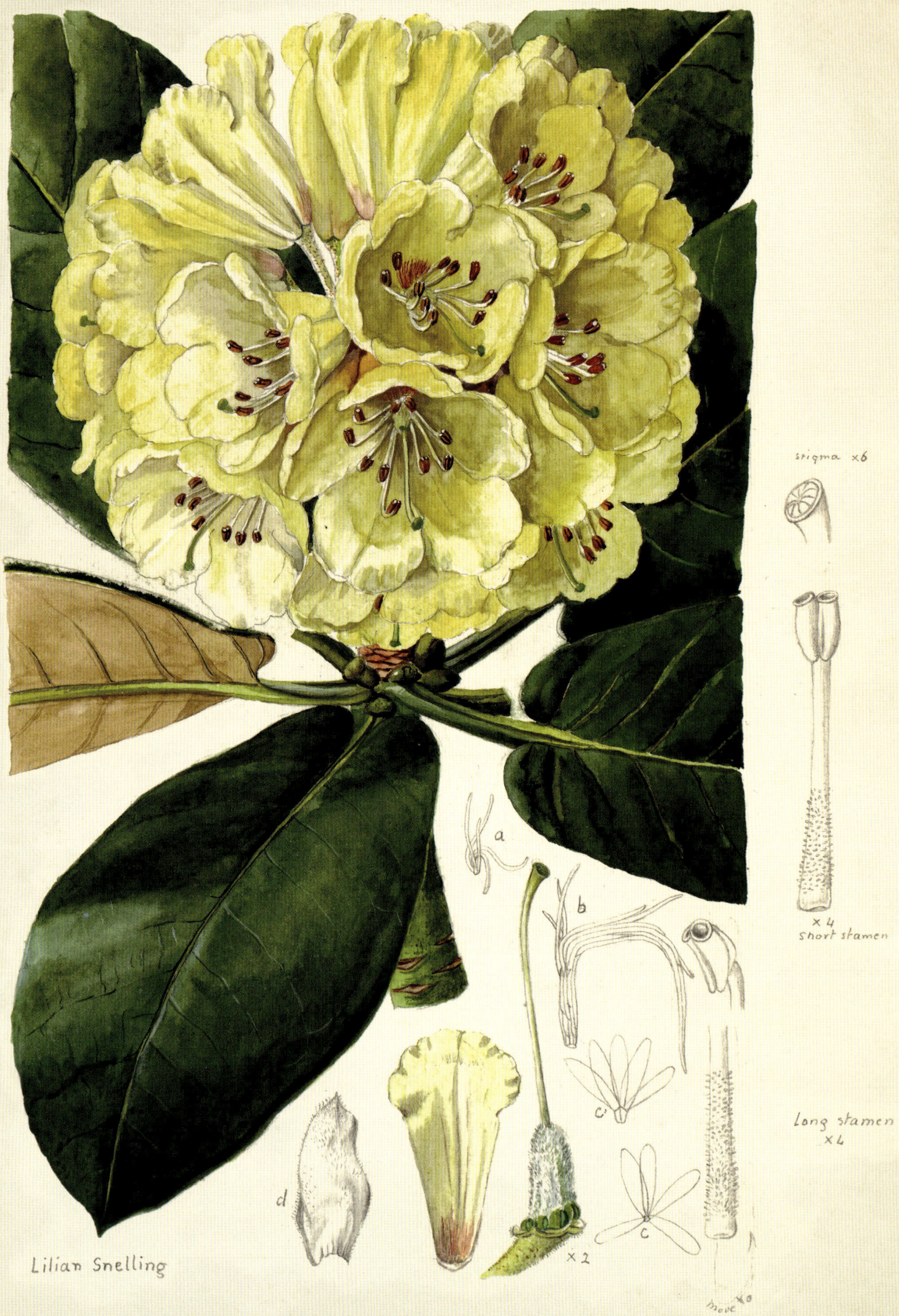

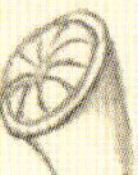
stigma x6

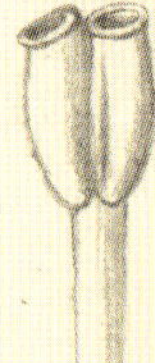
x4
short stamen

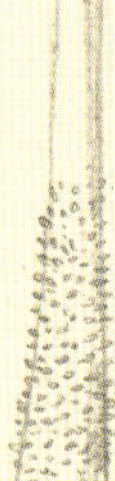
Long stamen
x4

a
b
c
c
d
x2
Lilian Snelling

The Kew style of botanical painting, with its precise drawing and total accuracy was not the only style used for botanical publications in the early twentieth century. Other artists, not primarily botanical illustrators, illustrated important books. Alfred Parsons (1847–1920), who was primarily a landscape painter, painter of gardens and garden designer, also produced delicate watercolours for Ellen Willmott's monograph *The Genus Rosa*, published between 1910 and 1914. Parsons was president of the Society of painters in watercolour, and his paintings are delicate and show well the characteristic habit of the rose species. The reproduction used was not good, but the original paintings, which are in the Lindley Library, are beautiful. Parsons also provided illustrations for William Robinson's *The Wild Garden*, and A. B. Freeman-Mitford's *Bamboo Garden*.

A somewhat similar project to Willmott's Roses was Dykes' *The Genus Iris* (1913). W. R. Dykes (1877–1925) was a master at Charterhouse School near Godalming from 1903 to 1919, and then secretary of the Royal Horticultural Society until 1925. He grew and studied irises, and his monograph was illustrated by the art master at Charterhouse, Frank Round (1878–1958). The paintings are delicate, as befits their subject, and accurate. This seems to be the only botanical work he did, and his original paintings are now in the Natural History Museum and the Lindley library.

After a pause during the First World War, gardening and plant collecting continued as before, with the eastern Himalayas being the focus of collectors such as George Forrest (1873–1932) and Frank Kingdon-Ward (1885–1958). George Forrest was born in Scotland, and when not travelling in China, lived near Edinburgh. He was based in western Yunnan, either near Lijiang when exploring the Yulong Shan, or near Tenchong when collecting in the ranges which run between the headwaters of the Yangtze, Salween and Mekong rivers. Kingdon-Ward, whose father was a mycologist and Professor of Botany at Cambridge, explored the boundary region of India, Burma and Tibet and especially the valley of the Tsangpo, which rises in Tibet and passes through the Himalayas, emerging as the Brahmaputra in Assam, the greatest canyon in the world, nearly 500 km (310¾ miles) long and 6000 m (19,690 ft) deep at its deepest point.

Both these collectors sent back thousands of different seeds to England, to the groups of subscribers who financed the expeditions, and the trees grown from their seeds can still be seen in many large gardens, such as Caerhays Castle and Trewithen Gardens in Cornwall. Both collectors specialized in rhododendrons, magnolias and primulas which were in vogue at the time, but also sent back anything they thought would grow well. Two of Forrest's most beautiful rhododendrons were illustrated by Lilian Snelling: the yellow *Rhododendron lacteum* and the very dwarf, crimson-flowered *R. forrestii*, which was grown in County Meath in Ireland by Lord Headfort, one of Forrest's subscribers. The dwarf white *Rhododendron leucaspis* was one of Kingdon-Ward's collections from the Tsangpo gorge in 1924; it flowered in 1940, and was painted by Stella Ross-Craig.

LEFT: Alfred Parsons' study of the rose "Celestial" from Ellen Willmott's The Genus Rosa.

OPPOSITE: Lilian Snelling's study of Rhododendron lacteum *for* Curtis's Botanical Magazine.

Botanical exploration has sometimes been considered a cover for spying, but Captain F. M. Bailey was first a spy, a master of disguise and only by chance a discoverer of new plants. While travelling near the Tsangpo River in 1913, he found the fabled blue poppy of Tibet and pressed a piece in his pocket book: this turned out to be a new species, named *Meconopsis baileyi* in his honour, and painted by Lilian Snelling from seeds brought back by Kingdon-Ward.

Lilian Snelling continued as the main illustrator for *Curtis's Botanical Magazine* until 1949, but in 1933 she was joined by Stella Ross-Craig (1906–2006), who was based at Kew, which she had joined at the age of 23. Ross-Craig was married to Robert Sealy, a botanist on the staff at Kew, and her botanical line drawing was superb. Her main work was *Drawings of British Plants*, which covered all the wild flower found in the British Isles, and was published in parts between 1948 and 1973. Her clarity of line and consistence of shading is remarkable. Ross-Craig remained as chief botanical artist at Kew until 1958, when her post was filled by Margaret Stones.

Margaret Stones was born in Victoria, New South Wales, in Australia in 1920. She worked as a nurse in the Second World War, and when she contracted tuberculosis, she started painting the wild flowers brought to her bedside. Stones came to England first in 1951, and started work for botanists at Kew and the Natural History Museum. Her first major commission was to paint illustrations of Tasmanian plants for Lord Talbot de Malahide, an English diplomat who had inherited a castle and estate near Dublin and an estate in Tasmania that had been settled by the Talbots in 1824. Milo Talbot had been in the embassy in Ankara from 1943 to 1945, and was later posted to Lebanon and finally, as ambassador, to Laos. He commissioned Margaret Stones to paint the endemic wild flowers growing in Tasmania, and in 1962 she spent some months there, studying and painting the flora. This project was the beginning of *The Endemic Flora of Tasmania*, published in six parts between 1967 and 1978. Many of the plants were sent to Kew from Tasmania by air, others grown by Lord Talbot in his garden at Malahide Castle, or in gardens in England with collections of Tasmanian plants, such as Nymans and Wakehurst Place. Lord Talbot initiated and underwrote the cost of publication, and after his unexpected death in 1973, this was taken over by his sister, the Hon. Rose Talbot. The illustrations are simple and clear, but are sometimes taken from rather small specimens; the later ones benefit by having the enlarged dissections painted on the plate.

From 1976 Margaret Stones produced a set of 244 fine drawings of wild flowers for Louisiana State University at Baton Rouge, as part of the E. A. McIlhenny Natural History Collection. This is a wonderful group of paintings, the equal of anything done in the late eighteenth or early nineteenth century; they are large, life size, very detailed and well embellished with dissections of the important details of flower or leaf, all drawn from wild specimens growing in the state. A book about the project, with numerous illustrations, was published as *Flora of Louisiana* in 1991.

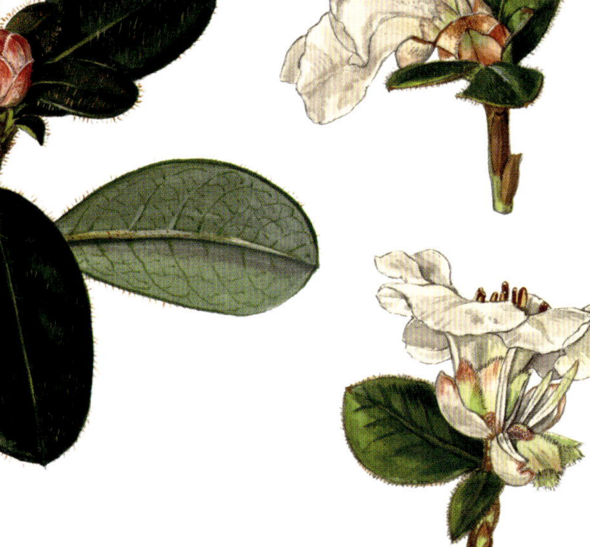

LEFT: Stella Ross-Craig's study of Rhododendron leucaspis for Curtis's Botanical Magazine.

OPPOSITE: Prionotes cerinthoides *by Margaret Stones from Winifred Curtis's* The Endemic Flora of Tasmania.

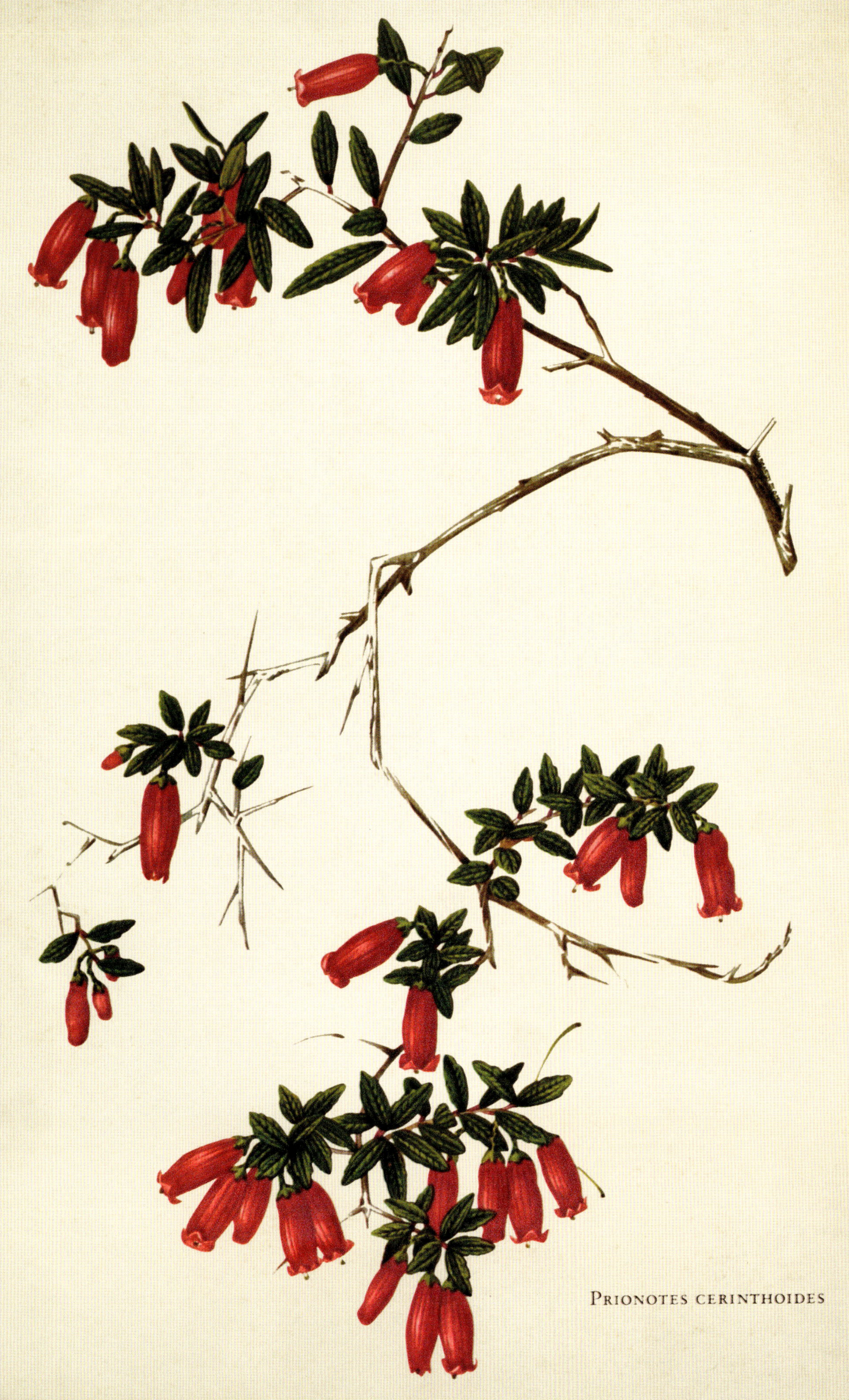

Prionotes cerinthoides

Modern Florilegia

Both botanical art in general and the concept of the florilegium (*see* page 34) have seen something of a renaissance in recent years, and Anne-Marie Evans, who developed a botanical illustration course at the English Gardening School in London, and with her husband wrote a well-received book entitled *An Approach to Botanical Painting* (1993), can take some of the credit for this, as artists trained by her have contributed to two significant recent projects.

The first of these, entitled *Flower Paintings from the Apothecaries' Garden*, is a collection of contemporary botanical illustrations, produced by the Chelsea Physic Garden Florilegium Society. This group of artists, founded in 1995, is in the process of recording the plants currently grown in Britain's second oldest botanic garden, and one work from each of the 56 artists involved is included in the book, which was published in 2005. A wide range of plants is illustrated, in a variety of styles, and both Dr Andrew Brown, then chairman of the Society, and Gillian Barlow, vice-chairman, contributed paintings.

The Highgrove Florilegium is a much larger and more costly work, consisting of two large folio volumes, published as a limited edition of 175 numbered sets, each signed by HRH Prince Charles, in 2008 and 2009. This book illustrates a selection of the plants grown in the Prince of Wales's garden at Highgrove, in Gloucestershire, and 71 artists combined to produce watercolours for the 124 coloured plates that were reproduced at their original size by a recently introduced process known as stochastic lithography. Unlike conventional lithography, which uses evenly spaced half-tone dots of various sizes, this process uses smaller, unevenly spaced, microdots designed to enhance the detail and colour contrast of the image.

An eclectic range of plants is illustrated, from modest ferns and bulbs through to perennials, shrubs and large trees, with a sprinkling of fruit and vegetables, including even well-known cultivars such as lettuce "Little Gem". In addition to the paintings, the hand-made paper, marbled paper boards, decorative endpapers, dark red goatskin binding embossed with gold leaf, and handmade felt book cover are all beautifully executed, with great attention to detail. The text was written by Professor Christopher Humphries and Dr Frederick J. Rumsey of the Natural History Museum. Royalties from the sale of this book have been covenanted to The Prince's Charities Foundation, which supports a wide range of causes.

Other florilegium societies have been founded in the twenty-first century, including one at Hampton Court, set up in 2004 by another group of artists trained by Anne-Marie Evans. The artists aim to establish an archive of botanical paintings depicting the plants grown in the garden and glasshouses of the palace.

In addition, florilegium societies have been established at Brooklyn Botanic Garden (2000) in New York and at Sheffield Botanical Gardens, UK (2002). A similar society has been established in Australia at the Royal Botanic Gardens, Sydney, where a group of 32 artists is currently engaged in illustrating a selection of the most significant plants growing in the estates of the gardens and Domain Trust; the objective is to publish a florilegium in 2016 to mark the bicentennial celebration of Sydney Botanic Garden.

OPPOSITE: Magnolia sieboldii *by Mayumi Hashi from* The Highgrove Florilegium.

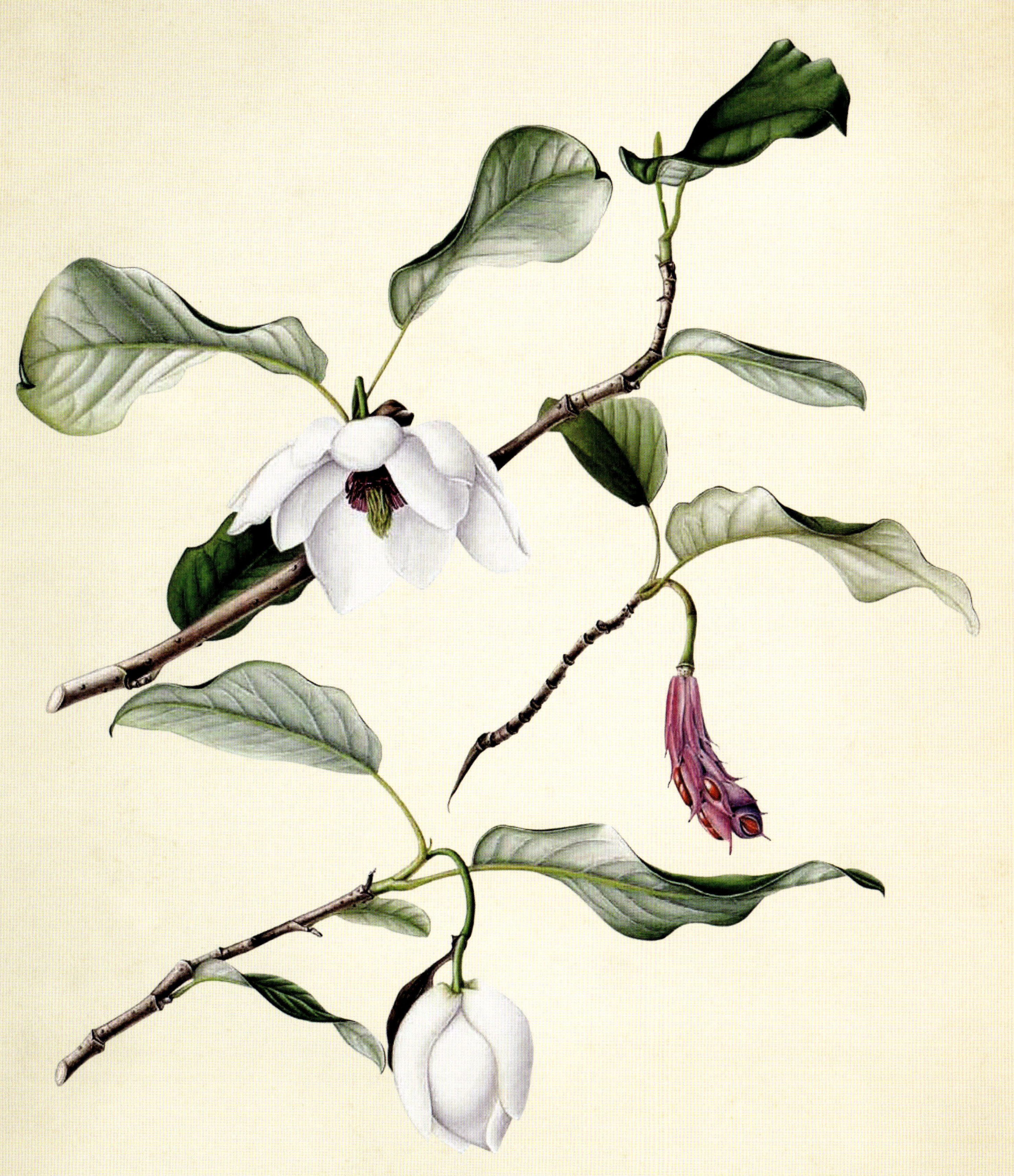

17

The Flowers of War and Beyond

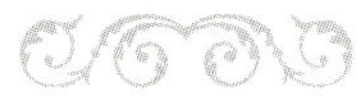

Geoffrey Herklots and Paul Furse

Geoffrey Alton Craig Herklots (1902–86) was a marine biologist, ornithologist and botanist. He was appointed Reader in Biology at Hong Kong University in 1928, and in 1951 wrote an excellent naturalist's guide entitled *The Hong Kong Countryside*.

During the Pacific Campaign of the Second World War, along with other foreign civilians, Herklots was separated from his wife and young children, and interned in a prison camp. The conditions were dreadful, and the prisoners nearly starving, so Geoffrey did all he could to teach the other internees about which wild fruits, roots and berries they could eat, and to grow vegetables to supplement the rations.

Just before the surrender of Hong Kong to the Japanese, he had published *Vegetable Cultivation in Hong Kong*, and during his time in prison he rewrote it and improved it. In 1972, he produced a new and exceptionally useful book entitled *Vegetables in South-East Asia*, with numerous line drawings, mostly by himself, but with about 20 contributed by Tang Ying-Wei, artist to the Agriculture and Fisheries Department in Hong Kong.

After the war, during which many of his drawings had been lost, Herklots became Secretary for Development in Hong Kong, with responsibility for the reorganization and modernization of the fishing and farming industries. He later visited many other tropical countries, including Trinidad, where he was Principal of the Imperial College of Tropical Agriculture from 1953 to 1960, and spent two years in Nepal, where he established a botanic garden.

Herklots wrote on a number of subjects, including orchids and other tropical plants. His books are illustrated with his own line drawings, which are wonderfully accurate and models of clarity, being annotated with details of collection and habitat which makes them most useful to botanists.

He wrote and illustrated, with line drawings and 16 colour plates, *Flowering Tropical Climbers* (1976) and continued working up to the time of his death on 14 January 1986; four pages of his line drawings of maples appear in the *European Garden Flora* (Vol. 5) published in 1997. He was awarded the prestigious Victoria Medal of Honour in 1980.

John Paul Wellington Furse (1904–78) was a submariner and champion boxer, an expert on natural history, especially lilies, fritillaries and other bulbs. He was from an artistic background: his father, Charles Furse, was a fashionable Edwardian portrait painter, and his mother Dame Katherine Furse was founder of the VAD and Director of the Wrens; Marianne North (*see* pages 210–15) was his great aunt. After retiring from the Navy in 1960, Admiral Furse, accompanied by his wife Polly, made four major expeditions to Turkey, Iran and Afghanistan in a Land Rover in search of bulbous plants, which were grown in botanic gardens and at the Royal Horticultural Society's garden at Wisley. Travel in Iran and Afghanistan was then primitive, but not dangerous, and the Land Rover, famous for its strength and reliability, made such a contribution to the botany of the area that it has a plant named after it, *Scrophularia landroveri*, a tough figwort with dingy greenish flowers. They drove from England in mid-winter, aiming to arrive in Iraq and Iran in early February, to see the flowers before the nomads' sheep and goats had grazed them to the ground. They camped in the wilds, as far as possible from any habitation, collecting herbarium specimens, bulbs and seeds, and painting specimens to show the range of flower colours.

Paul Furse painted flowers in a strong, bold style with gouache, while Polly did more delicate landscapes. He also grew many of his collections in his garden at Smarden in Kent, and painted examples from there. His paintings are now in the Lindley Library and at the Royal Botanic Gardens, Kew, and some were reproduced in Patrick

ABOVE: A study of an Acer *leaf by Geoffrey Herklots for* The European Garden Flora.

OPPOSITE: Fritillaria pinardii *painted by Admiral Paul Furse in 1957.*

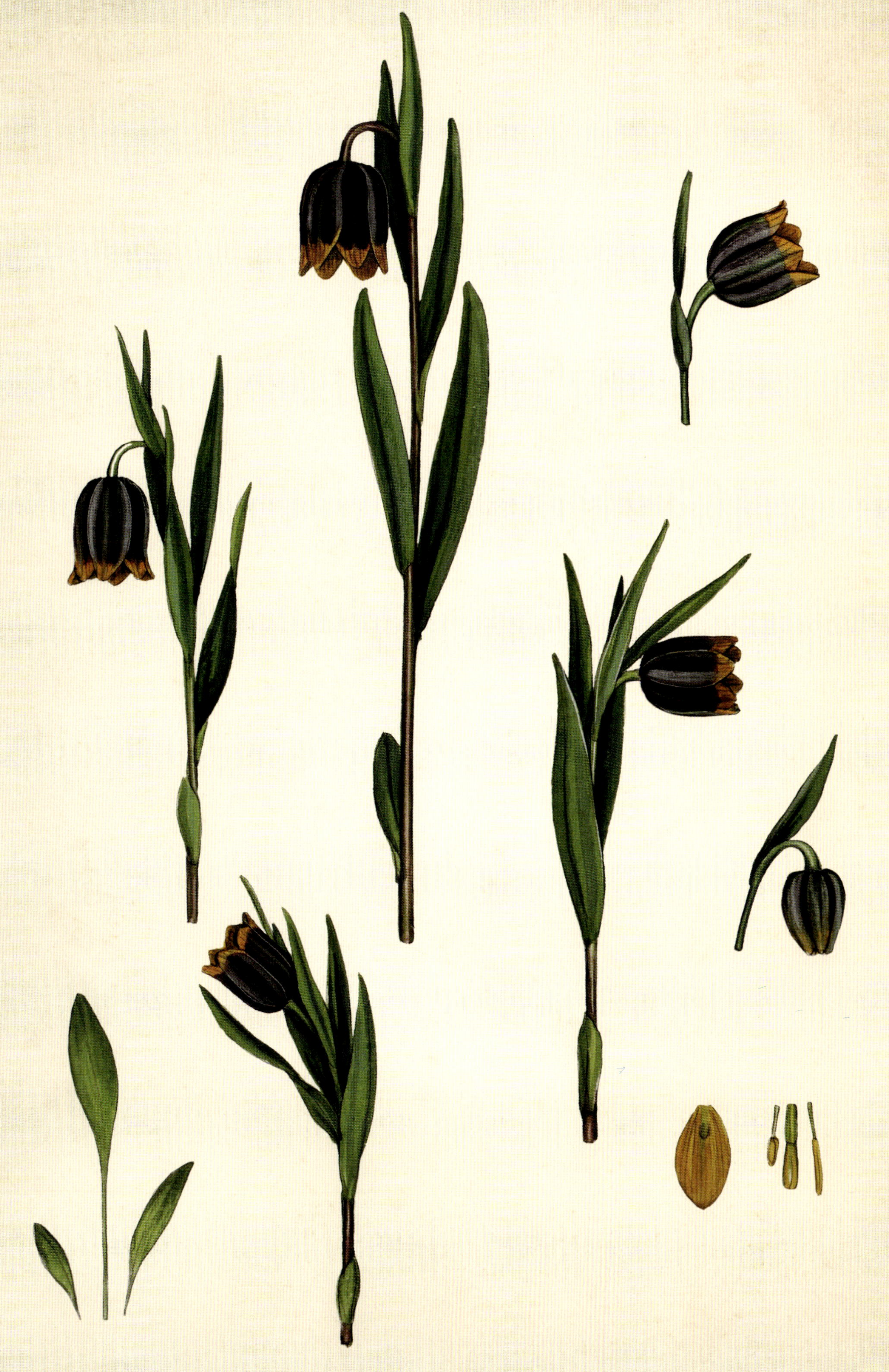

MARGARET MEE

Margaret Brown was born on 22 May 1909 in Chesham, Buckinghamshire to George Brown, a keen amateur naturalist, and his wife, Isabella. Margaret, her two sisters and her brother had an idyllic country childhood, and the girls were at first educated at home by an aunt, who was an illustrator of children's books. Margaret went on to Dr Challoner's Grammar School in Amersham, where she was fortunate in having an art master who encouraged his pupils to collect and paint flowers.

Margaret always enjoyed drawing and painting, and was also a voracious reader. The children were also regaled with travel tales by their maternal grandfather, and these probably introduced Margaret to the fascination of foreign lands. In 1926, she enrolled at the School of Art, Science and Commerce in Watford, but did not complete the course, leaving to work in London and Liverpool, before travelling to Germany. She had become interested in left-wing politics and wanted to see at first hand the effect of Nazism; she witnessed the burning of the Reichstag in February 1933, and watched, horrified, as Jews were arrested.

She returned to England, became involved in the trade-union movement and married a trade unionist, Reg Bartlett. After the breakdown of her marriage, she took a job in the drawing office of the de Havilland aircraft factory, where she stayed for the duration of the Second World War, leaving afterwards to pursue part-time studies at St Martin's School of Art in London. Here, she met her second husband, Greville Mee, with whom she enjoyed a long and happy marriage. Margaret became a full-time student at Camberwell College of Art, where she was taught by Victor Pasmore, one of the founders of the Euston Road School, a group of artists who favoured naturalism and realism. Pasmore was apparently a good, if demanding, teacher, and particularly keen to promote a good and observant eye for detail, a quality evident in Mee's work. Margaret learnt about proportion and depth in painting, as well as the technique of gouache, the medium she later favoured for her plant paintings.

In 1952, Margaret flew out to Brazil at short notice to see her sister Catherine, who was unwell, leaving Greville to follow by sea. This errand of mercy turned into a stay lasting many years, and Margaret took a job teaching at the British School in Sao Paulo, while her husband built up his career as a commercial artist. The pair liked to escape at weekends into the countryside, and it was while out walking that Margaret started to paint the flowers that she saw. Despite the death of her sister, Margaret and Greville decided to stay on in Brazil, and she soon became friendly with another teacher who agreed to accompany her on her first trip up the Amazon, during the school holidays in January 1956. This expedition involved a flight of over 2000 miles, followed by a slow train journey and then riverboat, before joining a smaller boat that sailed up and down the river trading goods. The two ladies finally transferred to a dug-out canoe paddled by an Indian boy.

On her return to Sao Paulo, Margaret worked at improving her painting technique, and was encouraged by the botanists at the Sao Paolo Institute, which arranged an exhibition of her work in the city. This led to a larger exhibition in Rio de Janeiro, and in 1960, another at the Royal Horticultural Society in London, where she was awarded a Grenfell Medal for her painting. By now, Margaret was in thrall to the lure of the tropical forest, and planned to return at the earliest opportunity. She also became friendly with

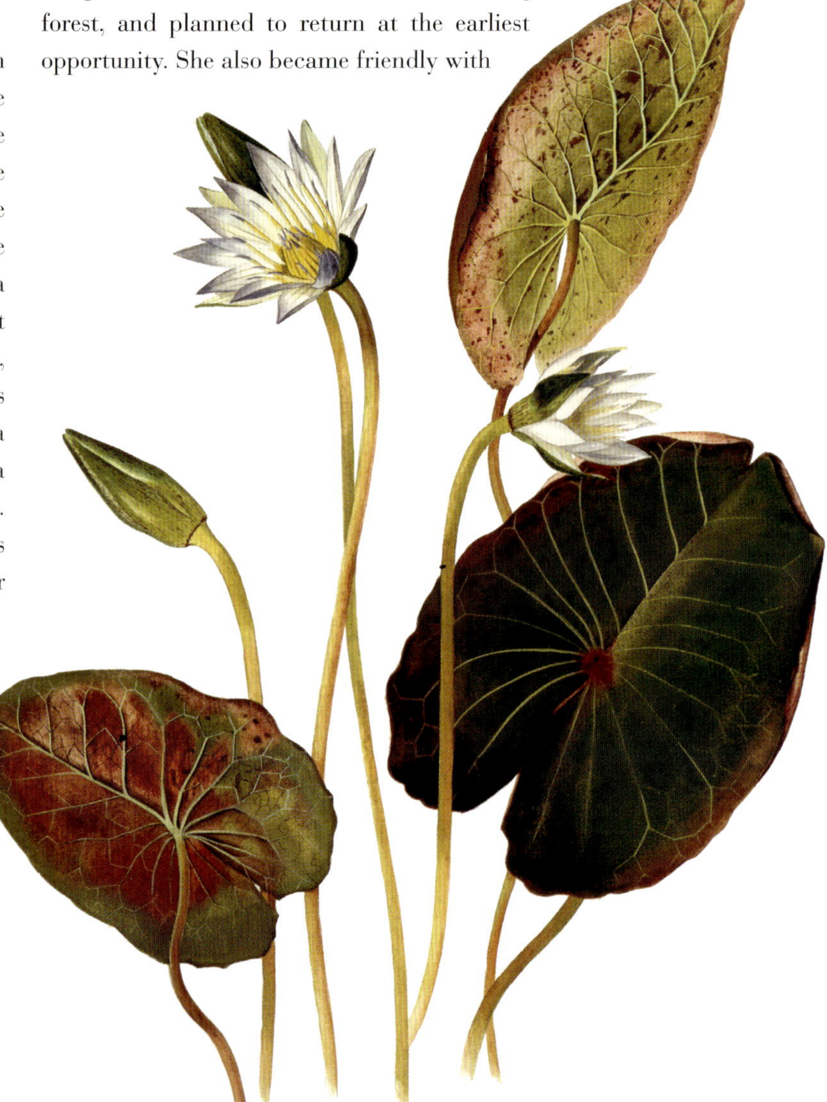

OPPOSITE: Gustavia pulchra, *flowering in the Rio Negro forest in Amazonia, painted by Margaret Mee.*

RIGHT: Nymphaea *species painted in the Amazon by Margaret Mee.*

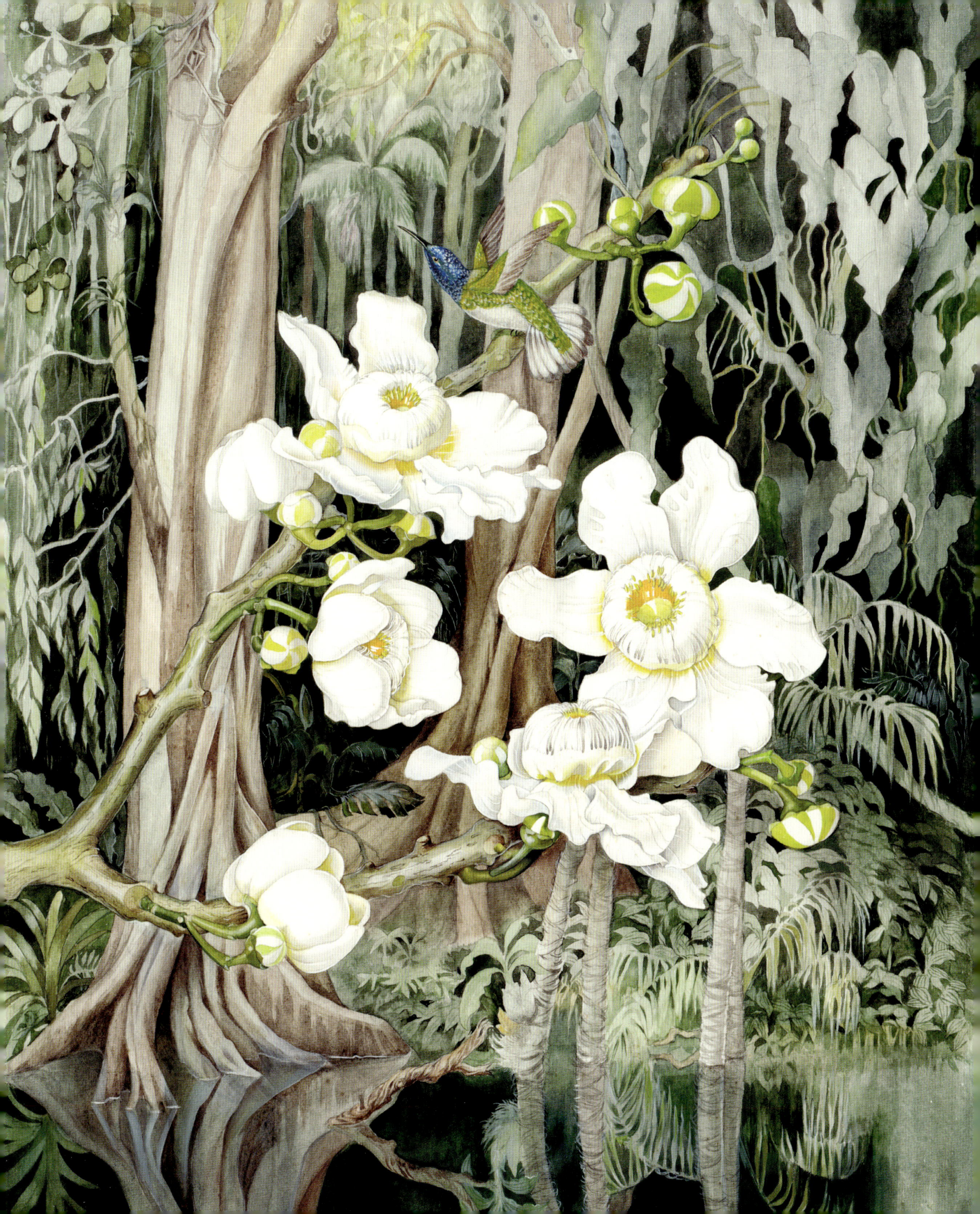

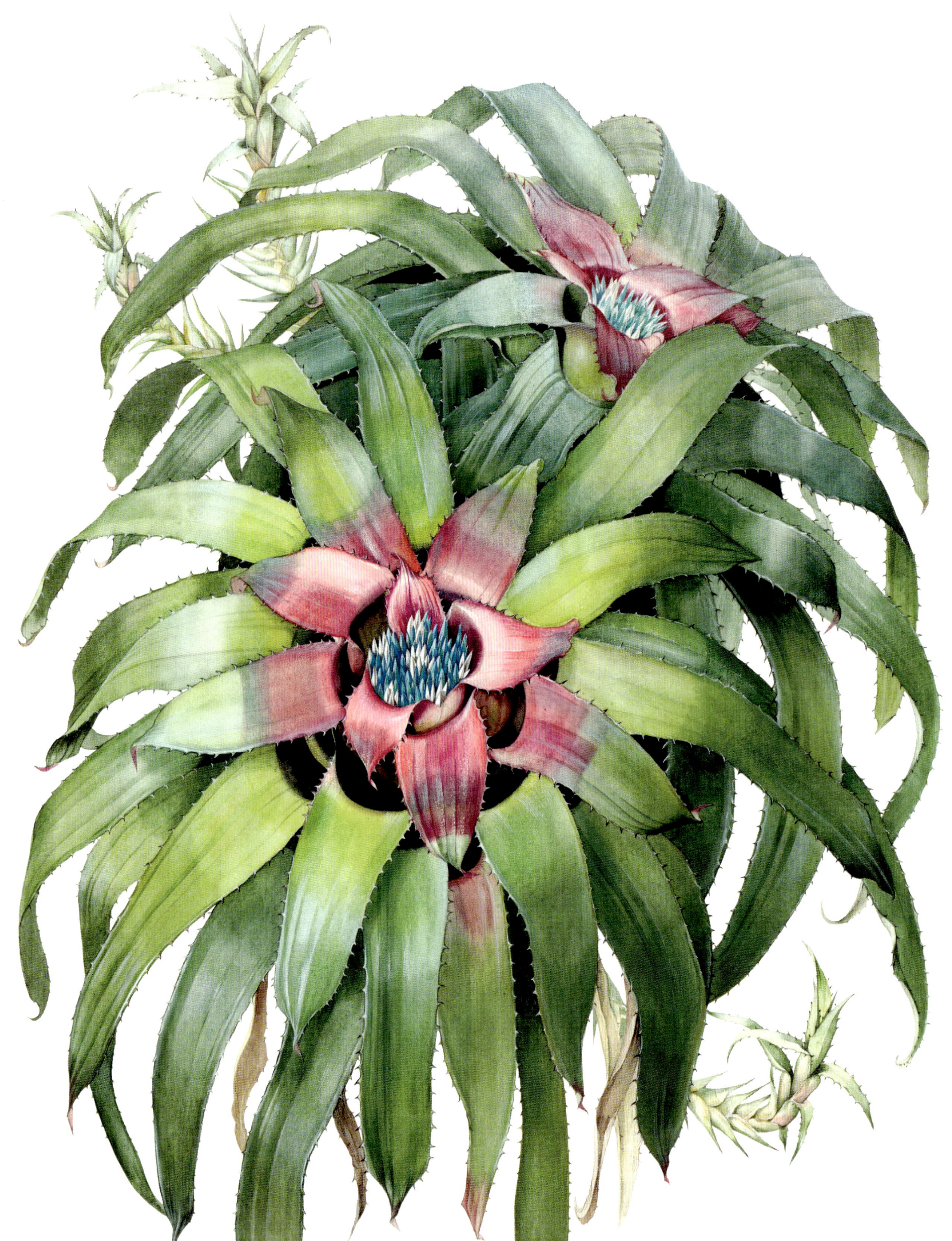

OPPOSITE: Neoregelia margaretae *collected in the Amazon and painted by Margaret Mee in 1979, from a plant cultivated by Roberto Burle Marx.*

RIGHT: Selencereus wittei, *a night-blooming cactus in fruit, painted by the Rio Negro by Margaret Mee in 1988.*

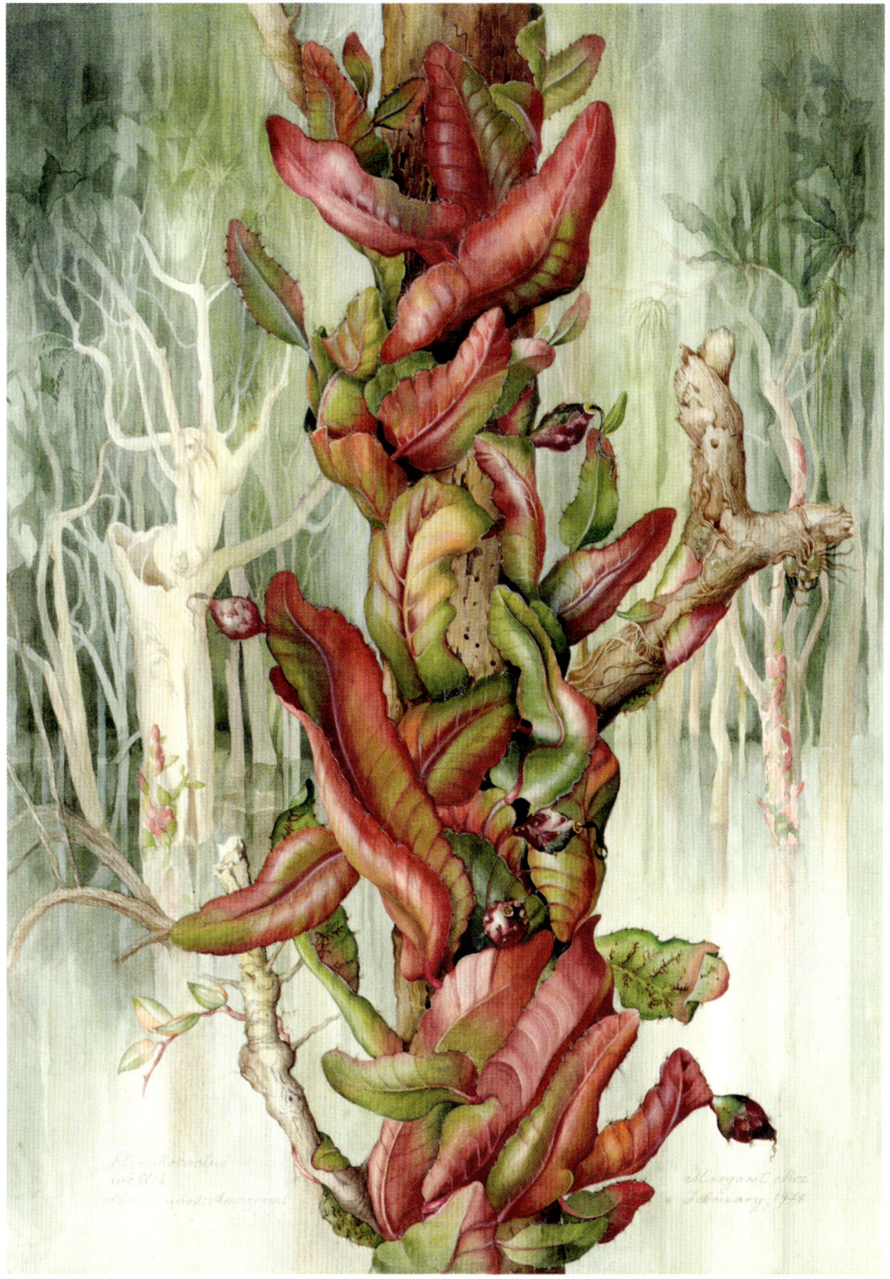

the Brazilian artist, Roberto Burle Marx (1909–94), who had become increasingly interested in plants and gardens, and also worked as a garden and landscape designer.

During her excursions up the rivers, Margaret made detailed drawings in the field and gouache sketches, an exceedingly difficult task, as she described in her journal:

> We reached the Rio Negro where I found, for the first time, flowers of the magnificent *Gustavia pulchra*. There was no time to lose, for the wind was battering the delicate petals as I drew. Will the exquisite beauty of this species save it from extinction?

The resulting painting of this beautiful plant, complete with hummingbird, against a background of rainforest tree trunks and foliage, is particularly striking. In 1967, her book of 31 folio paintings, entitled *Flowers of the Brazilian Forest*, was published by the Tryon Gallery in London.

She was especially fascinated by the elusive flowers of night-flowering plants, and in 1978 painted a waterlily, *Nymphaea rudgeana*, whose scented white flowers open at dusk: "Does this beautiful lake where I found *Nymphaea rudgeana* in flower still exist? They were putting the surrounding primary forest to axe and fire...." In 1981, she painted another nocturnal flowerer, the cactus *Selenicereus wittii*, but had to make do with faded flowers from the previous night. Seven years later, by maintaining an all-night vigil on a riverboat, she was finally able to observe the same species in flower, and sketched it by the light of a torch; this was one of her last paintings.

Margaret's work continued to attract attention, and in 1976 she was awarded the MBE for services to botany in Brazil. In 1988, an educational charity, the Margaret Mee Amazon Trust (MMAT), based in Britain, was established to raise awareness of the importance of the conservation of Brazil's forests, particularly in Amazonia, and to create a scholarship scheme. The official launch took place at the autumn preview of an exhibition at Kew of 60 of Margaret's paintings of the Amazon, months before her untimely death in a car crash. After her death, friends and supporters set up a sister organization in Rio de Janeiro, entitled the Fundação Botânica Margaret Mee (FBMM), which promoted local scholarships. Since then, young botanical artists have come from Brazil to Kew to study botanical painting under the supervision of Christabel King. The collection of Amazonian paintings by Margaret, her sketchbooks and several notebooks, as well as other memorabilia are stored at Kew.

Margaret Mee was a brave and outspoken woman, determined to draw attention to the shameful destruction of the rainforest, and her paintings stand as a fitting memorial to her.

BARBARA EVERARD AND GRAHAM STUART THOMAS

Barbara Everard (1910–80) was born at Telscombe, near Lewes in Sussex. She started work in the reproduction antique business, and later moved to Kent, where she met and married Raymond Wallace Everard. The couple moved to work in Malaya, but after the fall of Singapore to the Japanese in 1942, Barbara, who by now had a young child, returned to England, while her husband, who was taken prisoner, was forced to work on the notorious Burma railway. After a spell of recuperation, Raymond obtained a job managing a rubber plantation in Malaya and he, Barbara and their son returned there in 1946. Barbara began to collect and paint local plants, both wild and cultivated.

In 1952, the Everards returned to England, and Barbara exhibited a group of orchid paintings at the Royal Horticultural Society, for which she was awarded a Grenfell Gold Medal. This was the first of many exhibitions, and she continued as a freelance artist, working to commission and also illustrating several books, including *Flowers of the World* (1970), with text by botanist Dr Brian D Morley. For this, Barbara produced 92 watercolour plates, containing over 1000 plants from different geographical regions. She also contributed line drawings to several of the excellent Field Guides produced by Oleg Polunin and Anthony Huxley in the 1970s and 80s.

Barbara returned to Malaya in 1975 and worked on illustrations of endangered plants, and became increasingly interested in conservation; she was particularly interested in orchids, and later established a Conservation Trust with the Orchid Society of Great Britain. Barbara drew other subjects in pencil, ink and charcoal and also did a number of good landscapes in oil, but it is for her plant paintings that she is best known; 250 of her paintings and drawings are held in the library at Kew.

Graham Stuart Thomas was one of the most influential garden designers and plantsmen of the twentieth century, as well as being an amateur artist. He was born in Cambridge on 3 April 1909 and from an early age enjoyed drawing. He trained as a gardener at the University Botanic Garden, Cambridge and spent the major part of his working life as a nurseryman, garden designer and author, who not only wrote, but also illustrated, his own books.

It was while working at Hilling's nursery in Sunningdale that he met Gertrude Jekyll, who encouraged his interest in perennials and roses. He was the first Gardens Advisor to the National Trust, where he made an impact with his interest in historically accurate planting schemes, and subtle use of herbaceous plants; he also wrote, and did pencil drawings for a book on the Trust's gardens, published in 1979. Graham was an early devotee of old roses, and did much, through the collection at Mottisfont in Hampshire, and his writings on this subject, including three books, *The Old Shrub Roses* (1955), *Shrub Roses of Today* (1962) and *Climbing Roses Old and New* (1965), to encourage a resurgence of interest in these beautiful shrubs.

Graham wrote a number of other excellent books, notably on perennials, and in 1987 published a volume containing all his flower paintings and drawings, a valuable record of the plants grown in twentieth-century British gardens. He was awarded an OBE in 1975 and the Victoria Medal of Honour by the RHS in 1968. He died in 2003.

LEFT: Rosa multibracteata *and its hybrid,* 'Cerise Bouquet' *painted by Graham Stuart Thomas for his* Shrub Roses of Today *(1962).*

OPPOSITE: The Venus's fly trap, Dionaea muscipula, *painted by Barbara Everard.*

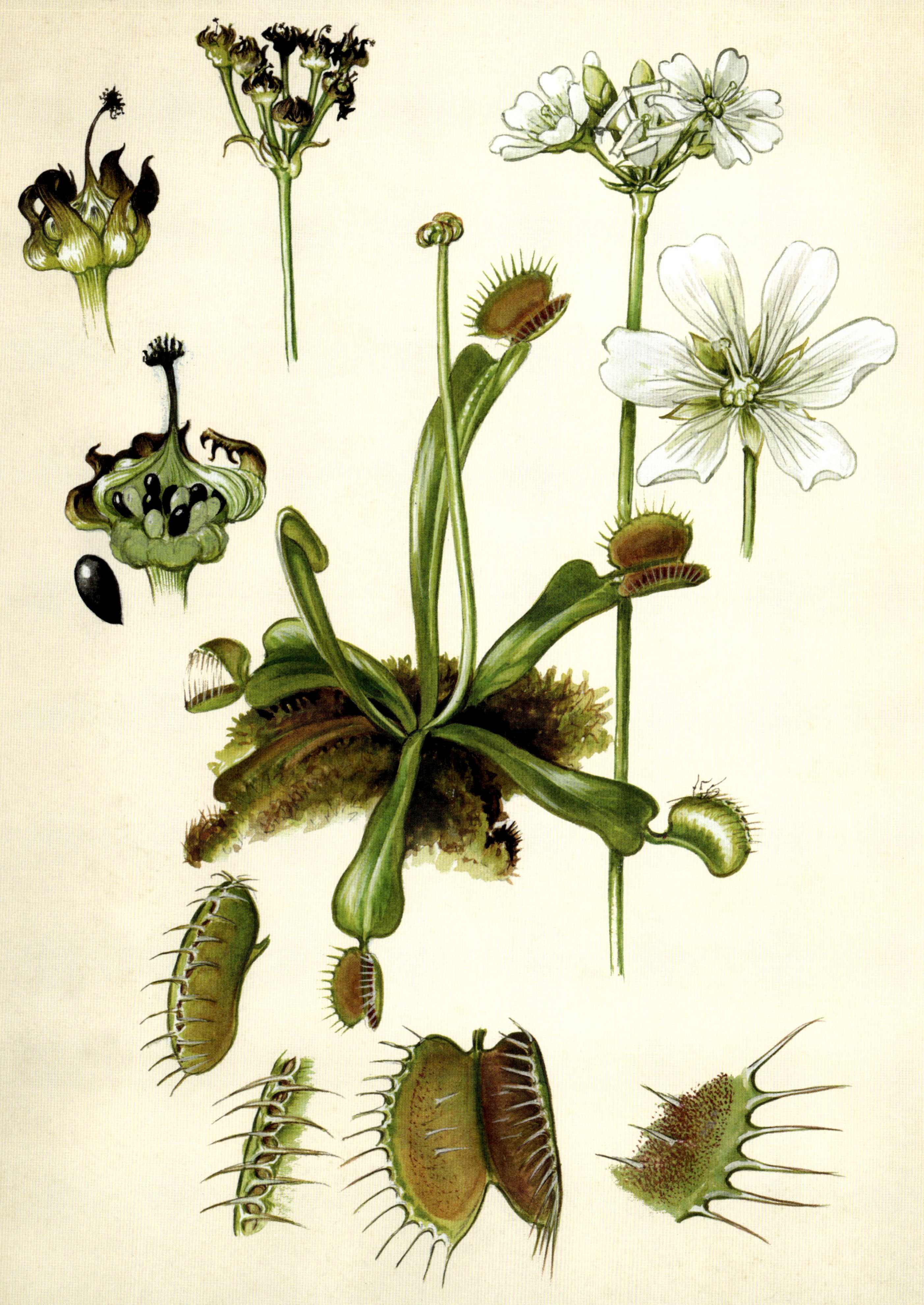

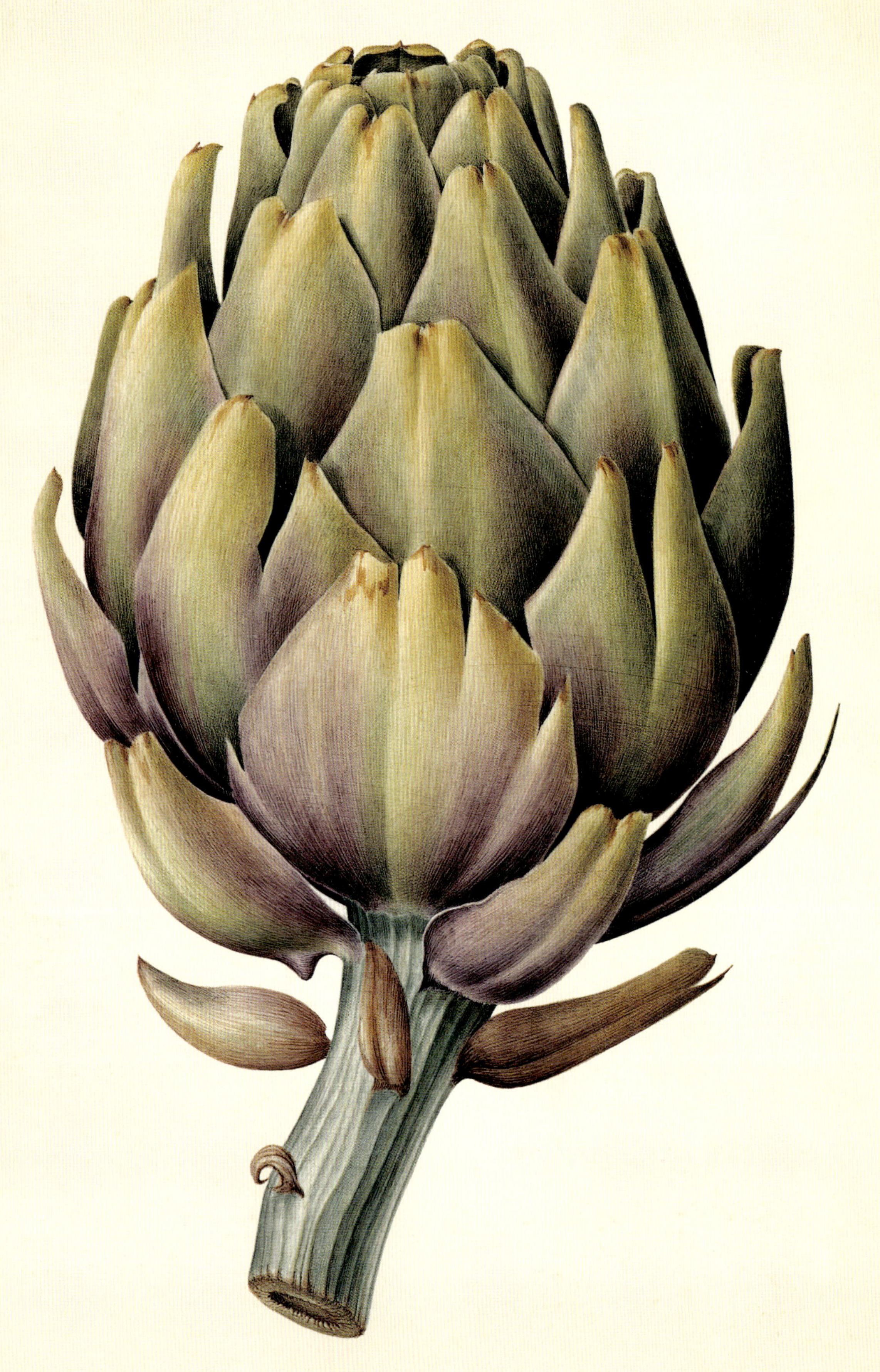

RORY McEWEN AND RAYMOND BOOTH

Rory (Roderick) McEwen (1932–82) was a talented folk singer and brilliant flower painter, whose work has had a lasting influence on later generations of botanical artists. He was a pupil of Wilfrid Blunt at Eton, at the time when Blunt was preparing *The Art of Botanical Illustration* (1950), and so became familiar with the work of Aubriet, Ehret and Redouté. As a child, Rory was taught to paint flowers by his French governess, and after school and National Service, he started painting again, finding, in his own words, that "my hand had unknowingly educated itself…". While at Cambridge, through friendship with Sacheverell Sitwell, he made paintings for C. Oscar Moreton's *Old Carnations and Pinks* (1955), and in 1964 for Moreton's *The Auricula*, in both of these the roots have as much importance in the design as the flowers and leaves.

A third major commission was for illustrations for a limited edition of *Tulips and Tulipomania* by Wilfrid Blunt (1979), including a set of prints of tulip flowers hugely enlarged. These three showed McEwen's delight in complicated flowers, with stripes, flames and complicated overlays of colour, all represented with an accuracy and crispness which surpassed what could be achieved by photography. His other, and at the time revolutionary, idea was to paint vegetables as works of art, not only beautiful specimens such as flowering globe artichokes, but series of rotting onions, recording their peeling skins and the spread of moulds. His later works, all painted on Italian vellum, concentrated on simple forms, fallen leaves or dying violets, "True Facts from Nature", painted with such concentration that they are seen in a new light.

Raymond Booth's botanical art has a very individual, independent style. Like Marianne North, he works in oils on paper or board, and paints landscapes as well as birds, plants and other natural history subjects. He is a keen gardener, with a deep knowledge of the plants he paints. He was born in Leeds, Yorkshire in 1929 and, at the age of 16, won a scholarship to Leeds College of Art, but his training was interrupted by two years of National Service, which he spent mostly in Egypt.

After returning to England he was found to have tuberculosis and was treated in a sanatorium, where he was fortunately free to paint. His work was exhibited at the RHS, and at the Walker Gallery in London, and from 1962 at the Fine Art Gallery.

His style is very highly finished, with great attention to detail and he paints plants with thick leaves and shining surfaces, with consummate skill. His paintings, along with those by the Australian Paul Jones, were used for a large folio on *Camellia* published in two volumes in 1956 and 1960. His most important published work is *Japonica Magnifica*, published in 1992, with text by Don Elick, an American living in Japan. It is a folio, with colour plates showing the plants at full size, and often in their natural settings. He grew almost all the species in the book himself, and in addition to being a beautiful book, it is valuable for helping the gardener to grow some beautiful and temperamental Japanese native plants. There are also landscape drawings and paintings. In addition to flowers, he paints landscapes, and these are represented in collections in England and in the USA.

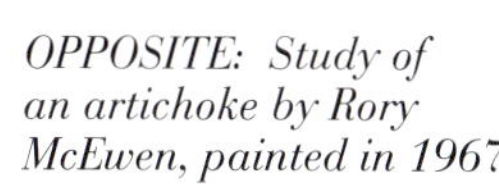

OPPOSITE: Study of an artichoke by Rory McEwen, painted in 1967.

LEFT: A camellia, painted by Raymond Booth.

Exhibiting Botanical Watercolours

Wilfrid Blunt, the foremost critic of botanical illustration, wrote that for the botanical artists there is always a conflict between art and science: how much should the specimen be manipulated or "improved" in the service of art without compromising the accuracy and the science. To achieve a balance, the artist must also study or have sufficient knowledge of the plant to know which characters are typical of the species and which unique to the actual specimen to be painted. A true scientific botanical illustration should not only represent the model, but also the species as a whole.

Many botanical artists wish to exhibit their paintings, either to sell them or to have them judged against similar work. One of the organizations which holds regular exhibitions is the Royal Horticultural Society, in London. There have been exhibits of flower paintings at the society's flower shows since 1913, and a judging panel has assessed them since circa 1926. The intention of these exhibits has been to encourage the accurate painting of flowers, both wild and cultivated, without being too prescriptive of the exact styles or format used. This is in contrast to illustrations for *Curtis's Botanical Magazine*, published for the Royal Botanic Gardens, Kew, where the format, size and style of the painting must conform to certain standards in order to maintain consistency in this long-running journal.

Other organizations, such as The Society of Botanical Artists, also hold regular exhibitions in London. In New York, the American Society of Botanical Artists at the New York Botanical Garden holds shows and has a newsletter, which acts as a meeting point for artists from around the world. Brooklyn Botanic Garden also has a thriving botanical art group, the BBG Florilegium Society (*see* page 228), and there are popular groups in Australia and South America.

These shows give new artists a chance to compare their work with that of the experts and enable them to meet collectors who may buy their work. Many botanical libraries, as well as private collectors, use such exhibitions to find new talent and to add to their collections. Even those artists who show for the first time regularly sell all their exhibits, particularly if their work is in an unusual style or happens to feature botanically interesting subjects.

OPPOSITE: A study of a rose 'Bouquet d'or' *by Regine Hagedorn, painted in 2004.*

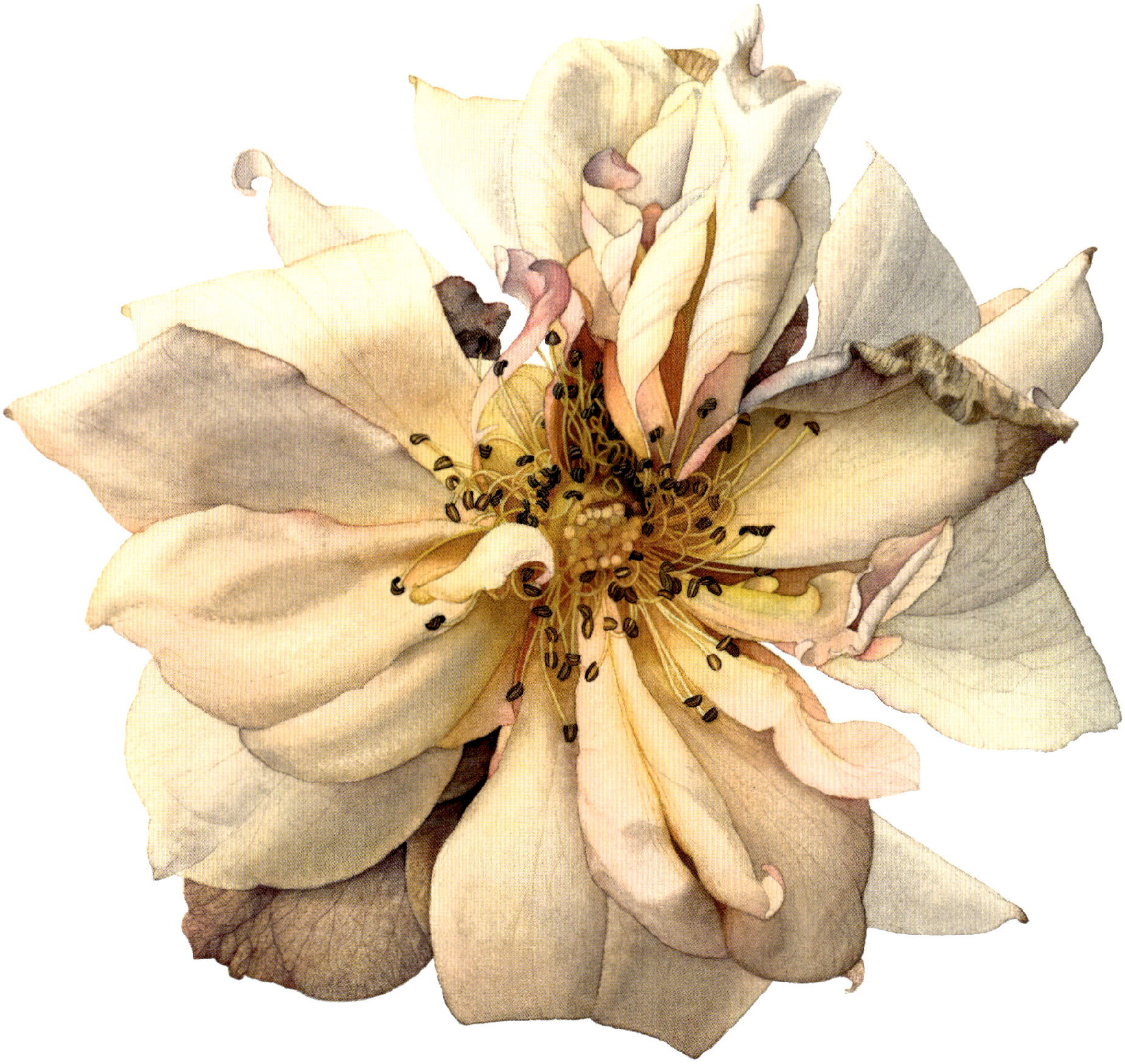

ROSIER NOISETTE 1872

"BOUQUET D'OR" II

200 %.

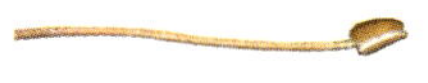

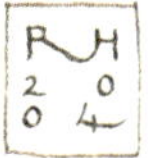

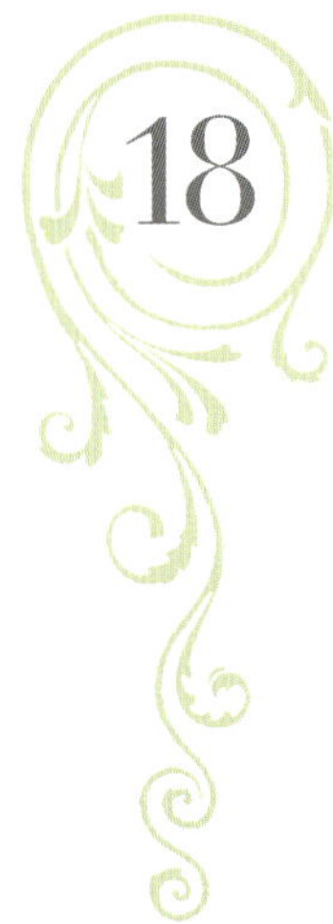

18

Carrying on the Tradition

The most exciting development in the world of botanical art in the last ten years has been the opening of the Shirley Sherwood Gallery of Botanical Art at Kew. In 1990, Dr Shirley Sherwood, who read Botany at Oxford, began collecting contemporary botanical paintings, encouraged by her husband, James who was a ship owner and founder of Orient-Express Hotels.

Not only has Dr Sherwood bought paintings from all the leading botanical artists, but she has supported and encouraged them, organized exhibitions of their work and classes in botanical painting in the company's hotels around the world. She commissioned the building of the special gallery at Kew, which opened in 2008, and is the first purpose-built botanical art gallery in the world. This chapter mentions only a few of those whose work may be seen in the gallery, or in recent botanical publications.

As can be seen from this book, Kew has long been home to the cream of botanical artists, and this continues to the present day. Christabel King (1950–) is the present chief artist, contributing most of the paintings in *Curtis's Botanical Magazine*, and teaching botanical illustration to Brazilian students who come to Kew under the Margaret Mee Foundation. In 1988, she was the artist on a Kew expedition to the Mountains of the Moon, and her paintings from there were published by Elm Tree in 1989. She has made a speciality of illustrating *Cactaceae*, handling the painting of the spines and hairs with particular skill. Her paintings have also appeared in other Kew publications such as *The Genus Roscoea* (2007) and *Hardy Heathers* (2011).

Mary Grierson (1912–2011) worked at Kew for many years until her retirement. After working as a cartographer in the war, she learnt botanical painting at Flatford Mill under John Nash, and came to work at Kew in 1960. She illustrated many books, including *Orchidaceae* by David Hunt

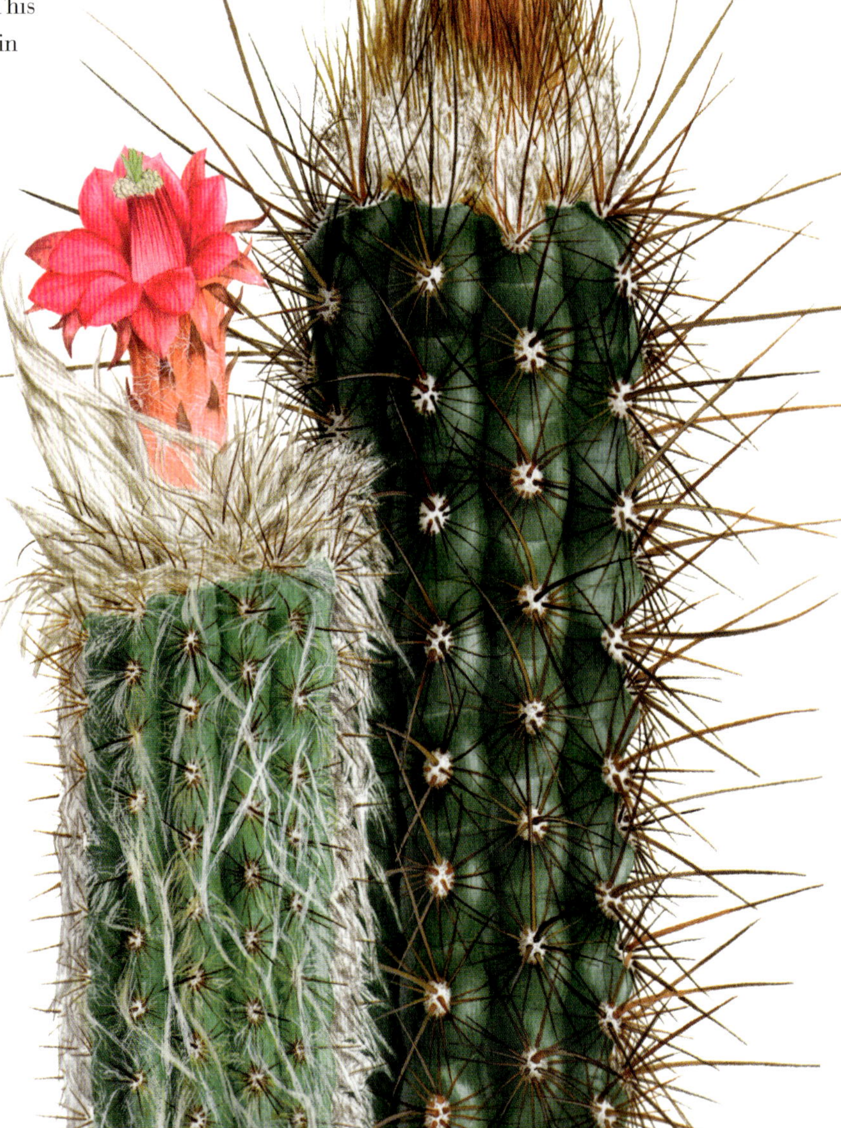

RIGHT: Christobel King's study of Oreocereus doelzianus.

OPPOSITE: Mary Grierson's study of Doryanthes palmeri *made from a specimen in the Australia house at the Royal Botanic Gardens, Kew.*

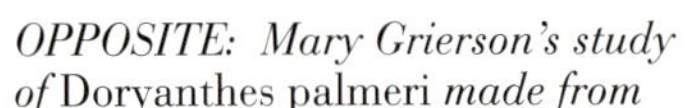

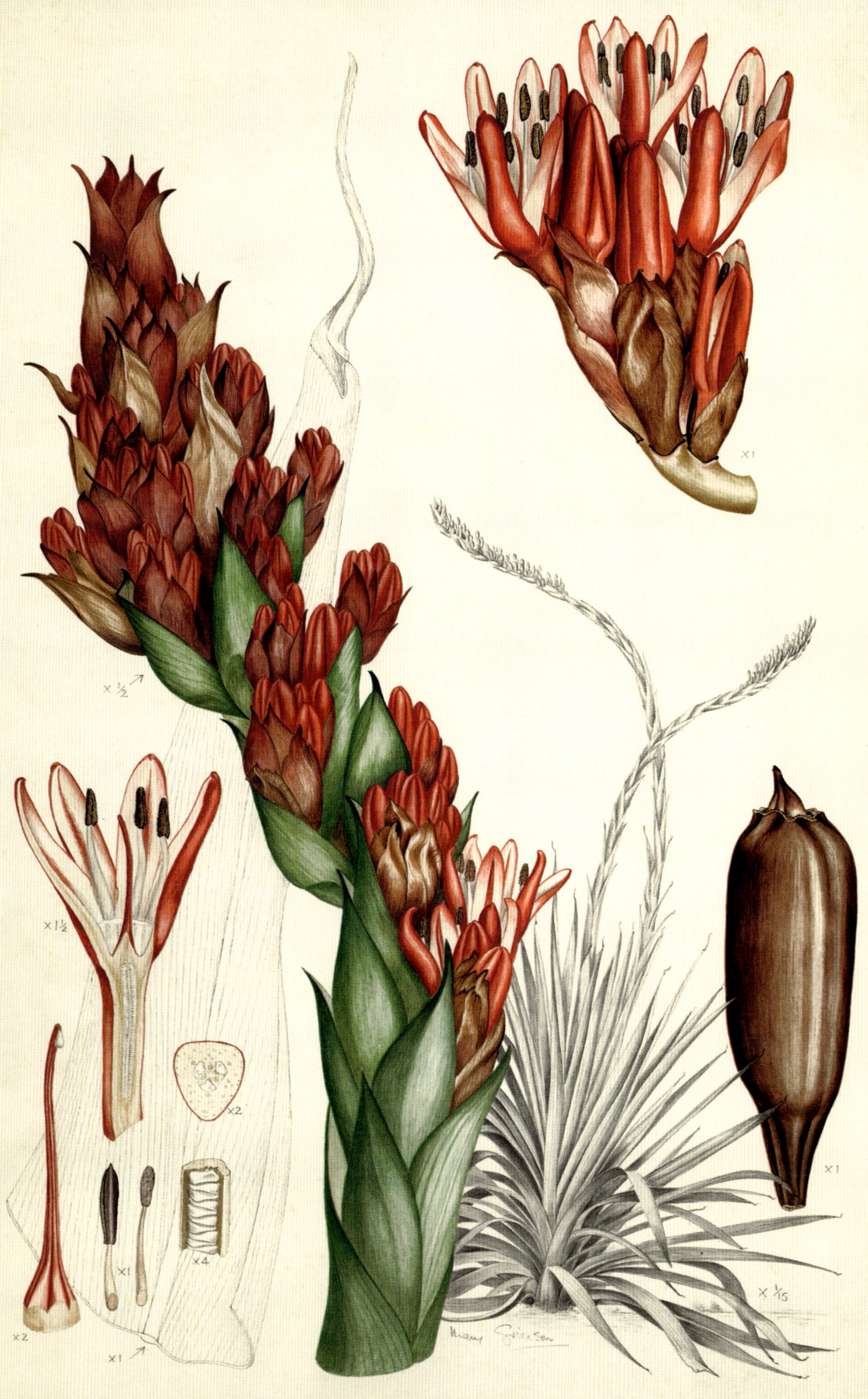

DORYANTHES PALMERI   Australia House, Kew.

and *Hellebores* by Brian Mathew (1989), as well as providing numerous plates for other Kew publications. She often combined pencil drawing of plant habit or locality with watercolour for flower details on the same painting.

Wendy Walsh (1912–) was born in England, but has spent most of her life in Ireland where her husband was Agent at Trinity College, Dublin. She started painting flowers when she visited her son in the Gilbert and Ellis Islands, now Kirabati, in the Pacific, and was fascinated by the exotic vegetation of the islands. On returning to Ireland, she continued to paint, publishing several books on Irish wild and garden flowers with Charles Nelson, then at the National Botanic Gardens in Glasnevin. Her style is delicate, using transparent watercolour, and particularly suited to wild flowers and often incorporating pencil scenes of the landscape. This work also appeared on a set of Irish postage stamps, and her work has recently been published at Kew in Nelson's *Hardy Heathers* (2011).

It was the work of Pandora Sellars (1936–) that inspired Shirley Sherwood to begin collecting botanical painting. Sellars is a teacher and freelance botanical artist, who often worked at Kew or for Kew botanists, producing wonderful paintings. Her particular style is to build up a series of different coloured leaves with contrasting shapes into a decorative page, like a dense tropical jungle, or painting flowers with a backing of leaves rather than white paper. When faced with a simpler subject the precision of her work and her handling of the shiny texture of a leaf surface is outstanding. This was well shown in her paintings for *The Genus Arum*, by Peter Boyce (1993), and her paintings of *Erythronium* species will shortly be published.

Both in Australia and South Africa, botanical painting thrives as an art form and as a part of scientific publishing, and several artists in both countries regularly produce wonderful records of their native flora.

Philippa Nikulinsky, born in Kalgoorlie in 1942, has made a speciality of painting the shrubby flora of the arid areas of Western Australia, often camping out in the bush to paint the flowers. She produced a beautiful series of *Lechenaultia* paintings for Kew and illustrated a folio entitled *Soul of the Desert*

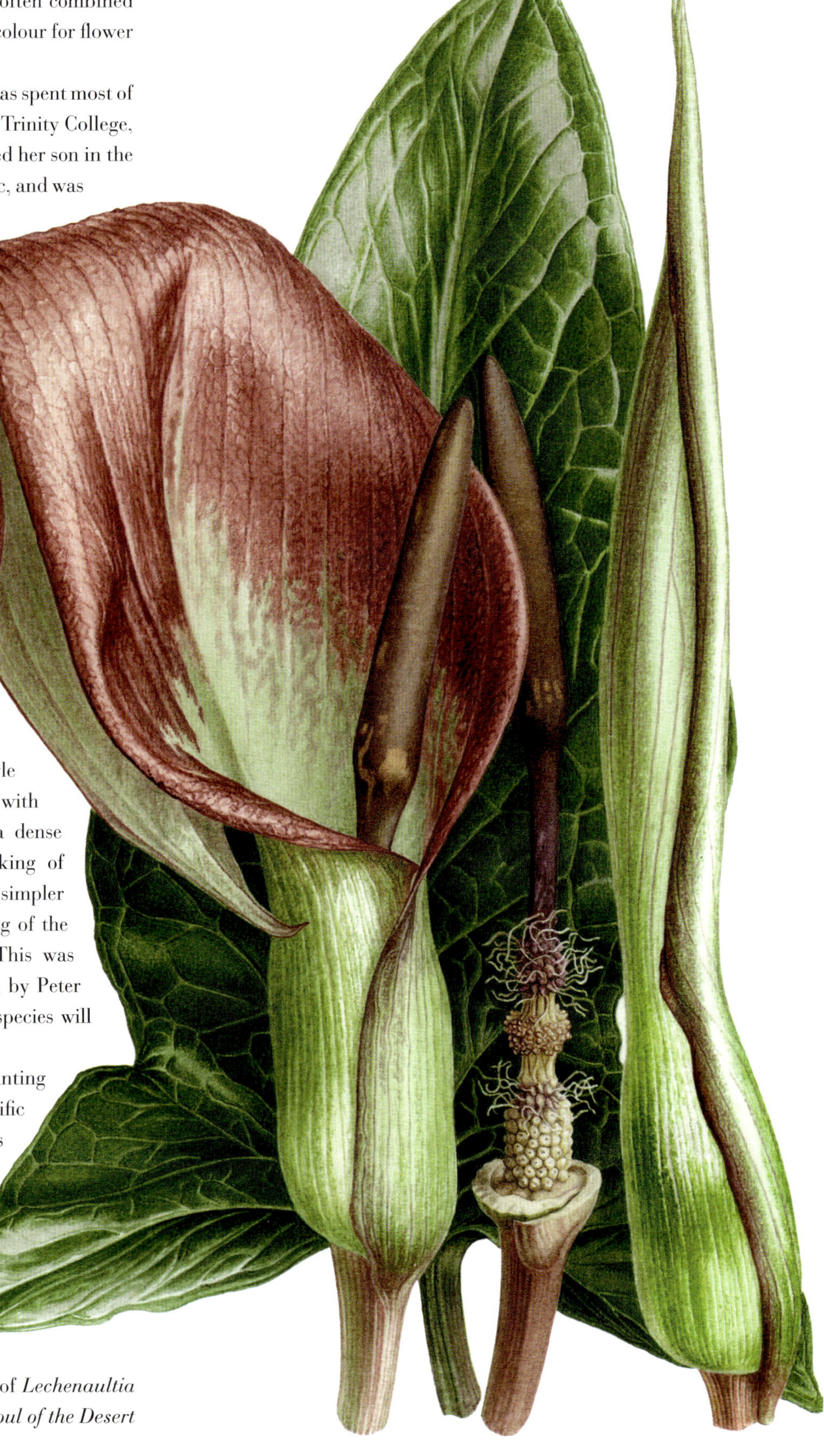

(2000). Celia Rosser was born in Melbourne in 1930, and was based at Monash University. Her main work is a monumental monograph on the genus *Banksia*, *The Banksias*, with text by Alex George, published in three volumes between 1981 and 2000. For this, they travelled all over Australia, studying the plants and painting all the species. This book has achieved the highest praise as one of the best botanical monographs published this century. Lucy T. Smith, who was born in Sydney in 1968, and studied in Queensland, illustrated palms for the Townsville Palmetum in 1995, and now works for the palms research group at Kew and teaches in London. She is a regular illustrator for Kew publications, and is working on a huge project, illustrating the palms of New Guinea.

There have probably been more beautifully illustrated botanical monographs published in South Africa in recent years, than in all the rest of the world. Its uniquely rich and beautiful flora has inspired botanists, artists and gardeners to study, paint and grow these wild flowers, and at the same time generous subscribers and supporters have enabled these books to be published. *Ericas in Southern Africa* was published in 1967, with paintings by Irma von Below, Cynthia Letty and Fay Anderson. Anderson produced the paintings for Peter Goldblatt's revision of *Moraea* (1986), and Ellaphie Ward-Hilhorst (1920–94) illustrated the three volumes of Van der Walt and Vorster's *Pelargonium* monograph (1977–88). The paintings for Olive Hilliard and B. L. Burtt's monograph of *Dierama* by Auriol Batten are particularly attractive, as they have pencil drawings of the veldt and the rocky Drakensberg Mountains behind the graceful arching stems of the flowers. Accurate and beautiful paintings of flowers are particularly significant in countries such as South Africa where there are small populations of very rare species that may become extinct, so a painting may be a lasting monument to a lost species.

Botanical painting in Japan and China developed in a completely different direction from Europe, losing the realism that was so well-developed by the twelfth century, and becoming more stylized, concentrating on simplicity and the clever use of the paint brush, with a whole school of flower painting using only black ink. This did not really change until the mid twentieth century. There are now many Japanese botanical artists producing work in the European style, but with a distinctly oriental elegance and sense of design. Mieko Ishikawa (1950–) has illustrated several books and exhibited her work at RHS shows in London; she is famous for her paintings of wild cherry blossom and of pine branches, both particularly Japanese subjects; she has painted a wonderful

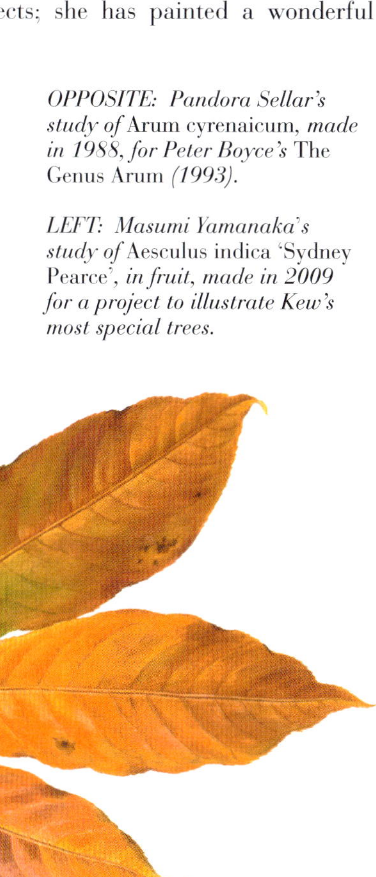

OPPOSITE: Pandora Sellar's study of Arum cyrenaicum, *made in 1988, for Peter Boyce's* The Genus Arum *(1993).*

LEFT: Masumi Yamanaka's study of Aesculus indica 'Sydney Pearce', *in fruit, made in 2009 for a project to illustrate Kew's most special trees.*

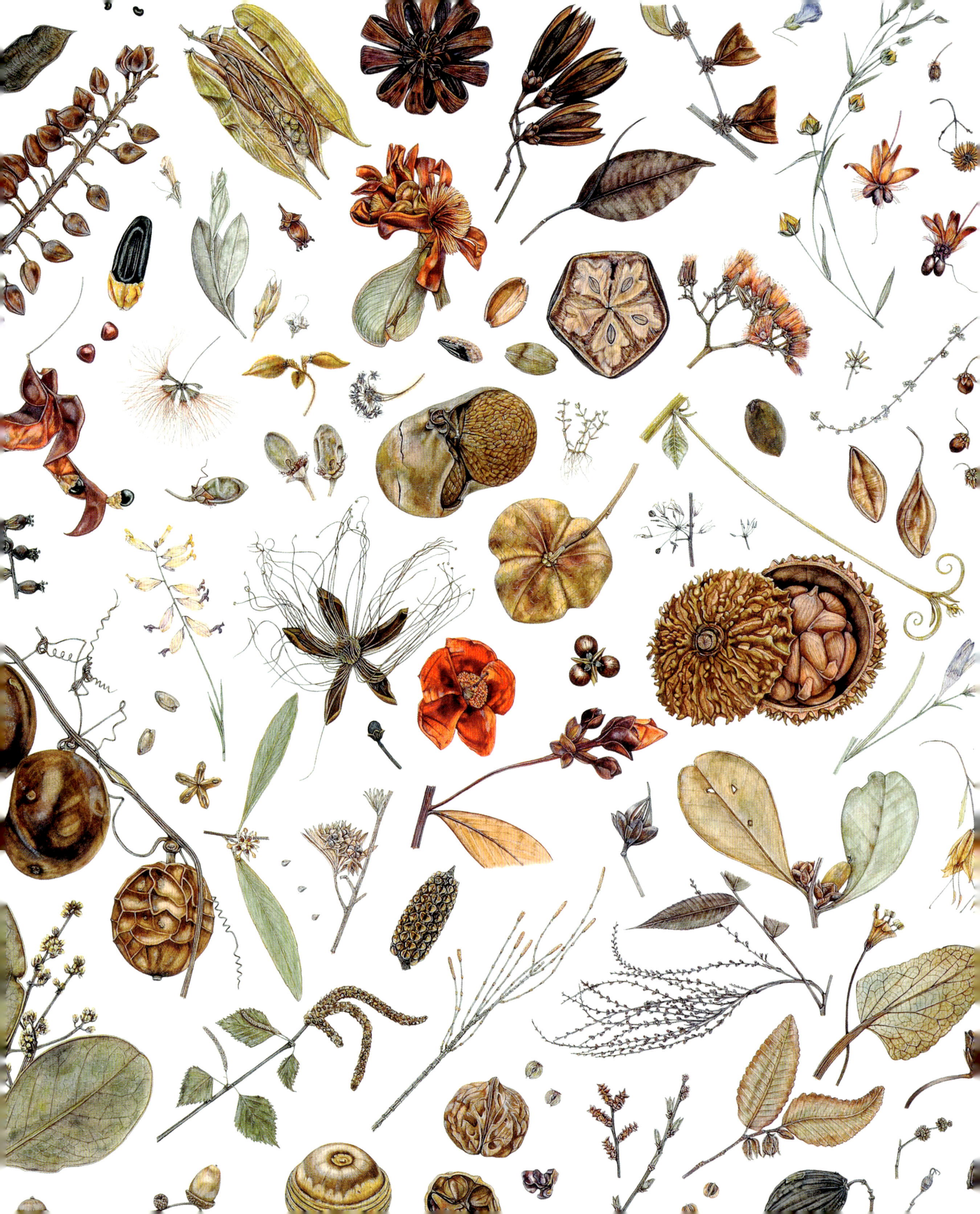

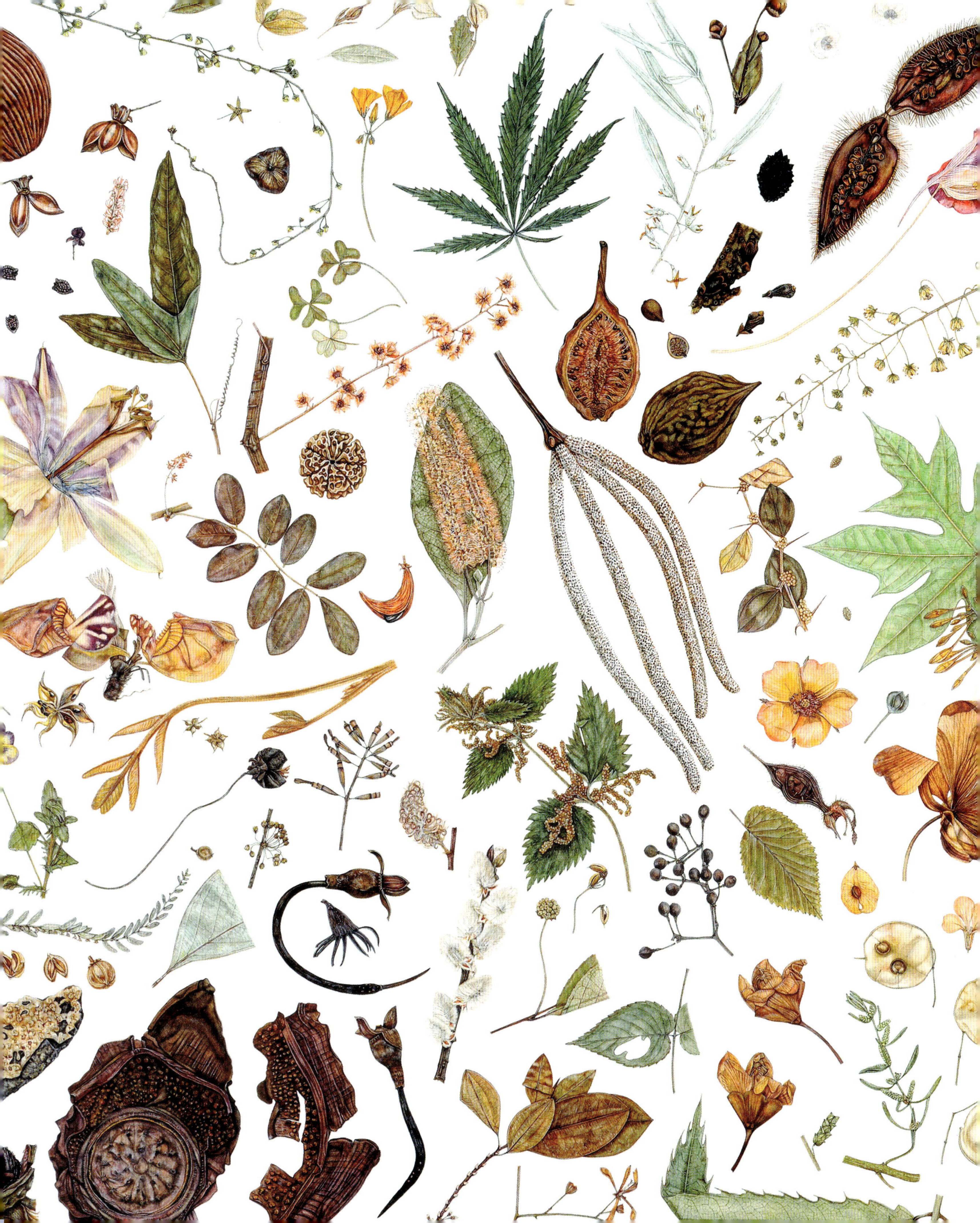

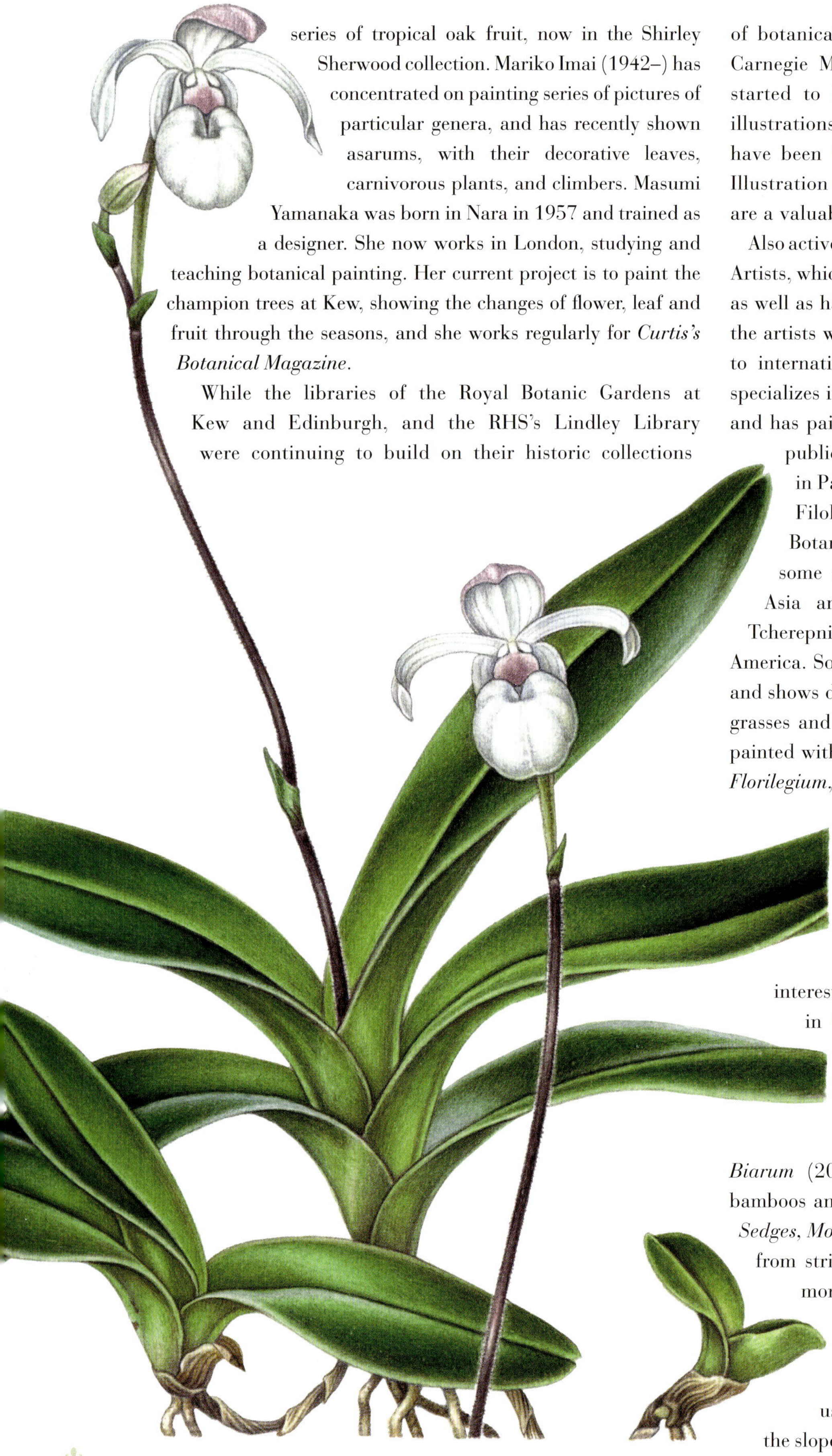

series of tropical oak fruit, now in the Shirley Sherwood collection. Mariko Imai (1942–) has concentrated on painting series of pictures of particular genera, and has recently shown asarums, with their decorative leaves, carnivorous plants, and climbers. Masumi Yamanaka was born in Nara in 1957 and trained as a designer. She now works in London, studying and teaching botanical painting. Her current project is to paint the champion trees at Kew, showing the changes of flower, leaf and fruit through the seasons, and she works regularly for *Curtis's Botanical Magazine*.

While the libraries of the Royal Botanic Gardens at Kew and Edinburgh, and the RHS's Lindley Library were continuing to build on their historic collections of botanical illustrations, the Hunt Botanical Library, of the Carnegie Mellon University in Pittsburgh, founded in 1961, started to build up a representative collection of botanical illustrations and now has over 30,000 items. Since 1968, there have been 13 International Exhibitions of Botanical Art and Illustration featuring contemporary artists and their catalogues are a valuable record of artists of the twentieth century.

Also active in North America is the American Society of Botanical Artists, which holds exhibitions and publishes a quarterly journal, as well as having affiliated groups throughout the USA. Many of the artists who belong to this also exhibit abroad and contribute to international journals. New Yorker Carol Woodin (1956–) specializes in painting orchids, particularly from South America, and has painted new slipper orchids (*Phragmipedium*) for Kew publications. Catherine M. Watters was born and grew up in Paris, but is now based in California. She teaches at the Filoli Center at Woodside, and is associated with Quarryhill Botanical Garden in Sonoma Valley; she has illustrated some of the new plants that have been introduced from Asia and grown there. Both Kate Nessler and Jessica Tcherepnine regularly exhibit in London, though they live in America. Some of Nessler's recent work has been on raw vellum, and shows dry, autumnal plants as well as flowering clumps with grasses and moss. Tcherepnine favours strong subjects that are painted with precision and verve. Her work is in *The Highgrove Florilegium*, the Shirley Sherwood Collection and in libraries around the world.

While working in the service of the science of botany, botanical illustrators must show the features of a particular species, preferably in great detail, but at other times they may please their own fancies, and use botanical images to create interesting works of art. Ann Farrer was born in Melbourne in 1950, but has spent most of her life in England, working and teaching at Kew and travelling to the Himalayas like her relative Reginald. She has produced paintings for many different Kew publications, and has illustrated monographs of *Biarum* (2008), *Tropaeolum* (2010) and many paintings of bamboos and other grasses, including *Collins Guide to Grasses, Sedges, Mosses and Ferns* (1984). Recently, she has moved away from strict botanical illustration, and her work has become more abstract, while still using detailed parts of plants as her subject matter, revelling in the patterns they can make. Her exhibition in 2011 at the Park Walk Gallery in London, "The Earth Beneath my Feet", uses tufts of grass and moss, pulled up by sheep from the slopes of Ingleborough as subjects. Each grass stem or leaf and every scrap of moss is precisely painted and can be identified, and she explores the patterns they make on the page.

The work of Regine Hagerdorn, born in Gottingen in 1952, surprised and delighted those who saw it at an RHS exhibition in 2002. She had taken twigs of different roses and painted them in exquisite detail, showing the different skin colours, thorns and buds. Equally detailed treatment was given to a plate of flowers, hips and attendant beetles, images of the stamens of *Rosa* 'Mermaid' and enlarged single rosebuds.

Rachel Pedder-Smith has taken the detailed representation of botanical specimens a stage further towards abstract art. She studied at Leeds Metropolitan University and then at the Royal College of Art where she took a Ph.D. During this time she studied at Kew, and made drawings for *Legumes of the World* (2005). While doing this, she was fascinated by the diversity of legume pods and seeds, and began to paint them in great detail. An open pod of the tropical tree *Afzelia africana*, each side mirroring its other half, formed a striking double image in her exhibition in the Park Walk Gallery in 2004. Her legume work also provided the subject in 2004 for her major painting of dry beans and peas, showing the amazing diversity in the family. This page of over 200 different plant parts is a study in reds and browns, and is both informative and decorative as a work of art and was exhibited in the exhibition "Watercolour" in Tate Britain in 2011. Now Pedder-Smith has taken this idea further, showing one dried herbarium specimen of each family of flowering plants, in their evolutionary sequence, as revealed by study of their DNA. This major work shows hundreds of different specimens from the Kew herbarium, and is a study in muted browns, with touches of reds, greens and purples, the delicate flowers becoming translucent as their cells dry out. It takes the largest sheets of watercolour paper available, runs to over 5 m (16½ ft) long, and was the subject of a special exhibition in the Shirley Sherwood Gallery of Botanical Art at Kew in April and May 2012.

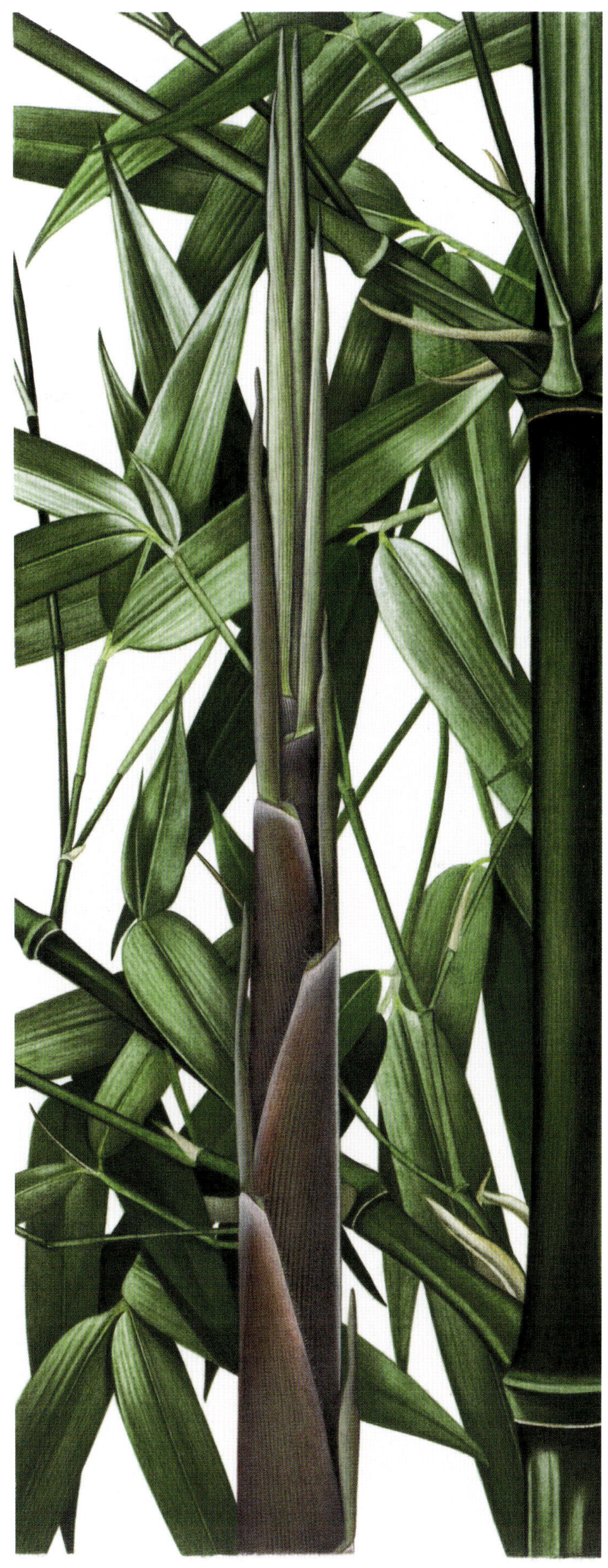

OPPOSITE: Carol Woodin's 1993 study of the Mexican slipper orchid, Mexipedium xerophyticum.

RIGHT: Ann Farrer's study of Phyllostachys heteroclada, *made in 1996.*

PREVIOUS PAGES: Rachel Pedder-Smith's painting of specimens of different plant families from the Herbarium at the Royal Botanic Gardens, Kew.

INDEX

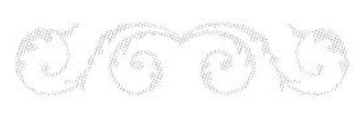

Page numbers in *italics* denote an illustration
Page numbers in **bold** denote a key entry

A
Abelia biflora *169*
Abricot noir á feuilles de pêcher *141*
Acer *230*
A. palmatum var. *dissectum* *84*
Achillea *29*
Aconitum ferox. *177*
Adianthus asarifolium *200*
Adiantum capillus-veneris *111*
A. muscosum *55*
Adoration of the Mystic Lamb *18–19*
Aesculus hippocastanum 25, *61*
A. indica 'Sydney Pearce' *245*
Agapanthus umbellatus *133*
Aiton, William 100
Aiton, William Townsend (W. T.) **100**, 110, 189
Akrotiri fresco *12*
Albizia lebbeck *178–79*
Album de Redoute 136
Aldovandri 32
Almanach de flore... 138
Almeidea rubra *81*
Aloe Africana *45*
alpines 222
Alstroemeria pelegrina *132*
A. salsilla *118*
Altaica, Flora 80
altarpieces, flowers on 16–17
"Amande sultane" *140*
Amaranthus cristatus *56*
American Society of Botanical Artists 240, 248
Amoenitates Exoticae 83
ancient herbals 14–15
Anderson, Fay 245
Andrews, Henry C. 148, **156–57**
Aniciae Iulianae Codex **14**, 68
Anisotome latifolia *193*
Annunciation, The 20
Antarctica 192
Antarctica, Flora *192, 193,* 194, 200
Aponogeton distachyon *152*
Approach to Botanical Painting, An 228
Arabis verna *190, 191*
Aristolchia *71*
Armeniaca percicaefolia 140
Arnold Arboretum 219
Art of Botanical Drawing *208*
Artemisia 28
artichoke *238*
Artocarpus incisa *206*
Arum *36*
Genus Arum, The 244
Arum cyrenaicum *244*
A. dioscoridis 71
A. maculatum 28
Asia, Central 76, 79–80
Asiatic Magnolias in Cultivation *220, 221*
Aster calendulaefolius *138*
Astragalus *66*
A. hispidus *67*
Atlantica, Flora 67, 132
Aubriet, Claude 59, **62–63**, 86
Auer, Alois 110
Augusta, Princess 100, 147, 161
Auricula *43*
Auricula, The 239
auriculas *137*
Australian plants 1700–1830, discovering 88–97

B
Bailey, Captain F. M. 226
Baker, J. G. 216
Ball, John 204
banana 31
Banister, John 48
Banks' Florilegium 92
Banks, Sir Joseph 47, 59, 90–92, 95, **98–99**, *99*, 103–04, 112, 161, 188
Banksias, The 245
Bartolozzi, Francesco 131
Barton, Benjamin Smith 52
Bartram, John **52**, 55, 56
Bartram, William **52**, 55, 56
basil *28*
Basseporte, Madeleine Françoise 59, 86, *87*, 128
Bateman, James 124
Batten, Auriol 245
Baudin 55
Bauer, Ferdinand 68, **92**, *94*, **95**, 104, 107, 117, 121
Bauer, Ferdinand, pictures by *94, 95, 96, 105*
Bauer, Francis (Franz) 47, 68, 73, **104**, 107, 117, 150, 200
Bauer, Francis, pictures by *102, 104, 114*
Bayer, Christoph 186
Beauharnais, Vicomte Alexandre François Marie de 144
Beberstein, Baron Friedrich August Marschall von 79
Bellini, Giovanni 22, 26
Bentham, George 194, 197, 207
Bentinck, Margaret, Duchess of Portland 112
Besler, Basilius 34, 37, **38**
Bessa, Pancrace 55, **128**, 136, 138
Bessa, Pancrace, pictures by *138, 139*
Biarum 248
Blunt, Wilfred 32, 92, 128, 239, 240
Boccius, Father Norbert 68
Boehmeria macrophyllum *182*
Bomaria salsilla *118*
Bombax ceiba *113, 183*
B. malabaricum *183*
Bommer, E. 80
Bond, George 110
Bonpland, Aimé 123, 132, 138, *144*
Book of Flowers, The 74
books of hours, flowers in 16, 17
Booth, Raymond 239, *239*
borage (*Borago officinalis*) *24*
boreali-americana, Flora 55
Bosschaert, Ambrosius 46
Botanic Garden 150, 153, *153*
Botanical and horticultural illustrated journals 146–57
Botanical Cabinet 150
Botanical Illustrations 200
botanical libraries 248
Botanical Miscellany 148
Botanical Register, The 107, 109, **148**, *148, 149*, 150, 162
Botanist, The 153
Botanist's Guide 188
Botanist's Repository, The 148, 156
Botany Bay (Sydney) 91, 99
Bourdichon, Jean 17
Boyton House 107
Bradbury, Henry 110
brasiliensis, Flora 124
Brassavola nodosa *117*
brassica *24*
Brazil 212, 232
Brazilian wildflowers *213*
Bringing China to Europe 218–29
British India, Flora of 197
Brompton Botanic Garden 147
Brown, Capability 100
Brown, Robert 68, 72, 88, 92, 95, 107, 190
Brueghel, Jan the Elder 46
Brueghel, Jan the Elder, pictures by *46*
Brunfels, Otto 24
Brunonia sericea 96
Buchan, Alexander 90
Bulbocodium vernum *152*
bulbs 26
Bulstrode 112
Burbridge, Frederick William 208
Burbridge, Frederick William, picture by *208*
Burmann, Professor 83
Burtt, B. L. 245
Bury, Priscilla 153
Busbecq, Ogier 25, **26**, 62
Bute, Lord (John Stuart, Earl of) 100, 147

C
Calanthe versicolor *109*
Calcutta Botanic Garden 173, 176, 183
Calendarium 37
Callistephus chinensis *138*
Camellia cuspidata *222*
C. japonica tricolor *239*
camellias *162, 239*
Campbell, Dr Archibald 183
Cape, The 83, 92, 95, 104, 156
Capparis cynophallophora *115*
Captain Cook's florilegium 92
Caragana sinica *101*
Cardamine pratensis *146*
Cardopatium corymbosum *70*
Caribbean, The 114
Carlos III, King of Spain 118
carnations 11, *16*, 26
Carrara herbal (Sloane MS 2020) 14
Caspian Sea 80
Castanea sativa *67*
Catesby, Mark **51**, 56
Catesby, Mark, pictures by *49, 50, 51*
Cathcart, J. F. 183
Catherine the Great 76, 79, 186
Cattley, William 107
Cattleya skinneri *125*
Caucasus 78
Cecil, William (Lord Burghley) 31
Cels, J.-M. 132
Centauria plantarum rariorum Rossiae meridionalis *78*, *79*, 79
Cephalotus follicularis *92*, 95
Chambers, William 100, 161
Chamisso, Adelbert 79
Champion, Major John George 207
Charles, HRH Prince of Wales 34, 228
Charles, I King of England 42
Charlotte, Queen 104, 112, 126
Chatsworth 124, 153, 162
Cheirostemon pentadactylon *123*
Chelsea Physic Garden 59, 80, 146
cherry, black *50*
Chien Lung, Emperor 161
Chile
China 121
Bringing China to Europe 218–29
Chinese plant painting 11, 26, 158, 245
Chinese plant painting for Europeans 159–61
Early Chinese Plant Drawings 158–69
European paintings of Chinese plants 162–67
Choix des Plantes: dont la plupart sont cultivées dans le jardin de Cels 138
Choix des plus belles fleurs 136, *137*
Choris, Login Andrewitsch 79
Chrysanthemum *164–65*
C. mawii *203*
Ciccimara, P., picture by *221*
Cinchona condaminea *122*
Clifford, George 59, 60
Climbing Roses, Old and New 236
Clusia odorata *208*
Clusius (Charles d'Escluse) 26, 29, 42
coconut palm *212*
Codex Vindobonensis, pictures from *14, 15*
Codonopsis gracilis *199*
C. inflata *199*
C. javanica *199*
Collaert, Aedriaen 34
collage, flower 112, *112, 113*
Collecteana Botanica 107
Collins Guide to Bulbs 230
Collins Guide to Grasses, Sedges, Mosses and Ferns 248
Collinson, Peter 52, 56, 59, 66
Collybia peronata *186*
Coloured Engravings of Heaths 156
columbines *16*
Colutea galegifolia *139*
Commelin, Caspar 44
Commelin, Jan 44
Commentarii, in libros sex Pedacii Dioscoridis de medica material 25, *25*
Companion to the Botanical Magazine, A 147
Company School in India, The **170–87**, 207

Compton, Henry 42, 48
Cook, Captain James 89, *89*, 98–99, 103
coral tree flowers, Brazil *210*
cornflowers *16*
Corona imperialis *42*
Cotton, Barbara 150
Cours d'horticulture 142
Cratevus 14
Crete 12
Cronier, Félice, picture by *143*
Crown Imperial fritillary *42*
Cruÿdeboeck 29
crysanthemums 158, 161, *164–65*
currants, red *35*
Curtis, Charles M. 176
Curtis, John 107
Curtis, William 146
Curtis's Botanical Magazine 67, 95, 109, 110, 126, **146–55**, 166, 190, 194, 200, 221, 240
Cymbidium ensifolium *166*
Cypripedium pubescens *58*
Cytinus hypocistus *72*

D
dahlia *53*
"Dahlia double" *136*
Dalbergia scandens 174
Dampier, William 88, *88*, 95
dandelion *187*
Danica, Flora (1648, Simon Paulii) 186
Daphne jasminea 72
D. petraea *222*, *223*
Darwin, Charles 188, **192**, 197, 204, 212
Das große Rasenstück 22
Datura arborea *211*
David, Père and the French missionaries 168–69
Davidia involucrata var. *Vilmoriniana* *218*
de Berry, Duchesse 136, 138, 142
de Candolle, Augustin 73, 132, 141
de Heem, Jan Davidsz 46
De Historia Stirpium 25, *25*
de Jussieu, Bernard 55, 59, 60
de la Billadière, J. J. H. 67, 92
de Lahaie, Felix (Delahaye, Felix) 92
de L'Obel, Mathias **28–29**, 31
De metamorphosibus Insectorum Surinamensium *74*
de Pannemaeker, P. 154
de Passe, Crispin van 40
Decaisne, Joseph 154
Decandria monogynia *207*
Delanica, Flora 112
Delany, Mrs, and her paper mosaicks **112–13**
Delavay, Père Jean Marie 168
Delile, Eulalia 178
Delineations of Exotick plants cultivated in the Royal Gardens at Kew *103*, 104
Delphinium pelegrinum *100*
Denmark 186–87
D'Entrecasteaux, Admiral Bruni 92
Derris scandens *174*
d'Escluse, Charles, *see* Clusius
Description des plantes de l'Amérique 55
Description des plantes rares cultiviées á Malmaison et á Navarre 132, 138, *144*
description of the genus Cinchona, A 104
description of the genus Pinus, A 104, *105*
Descriptions of the East 66
Desfontaines, René-Louiche 67, 132
Deshima Island 82
Designs of Chinese buildings 161
Dierama 245
Dietsch, Johann Christoph 186
Dillman, John 100
Dionaea musipula 236
Dioscorides, Pedanios 11, **14**, 24, 28, 68, 71
Dissertatio de generatione et metamorphosibus insectorum 75
Dodoens, Rembert 29
Dombey, Joseph 118, 121
Dombeya wallichii *180*
d'Orléans, Gaston 35, 86
Doryanthes excelsa 95, *97*
D. palmeri *243*
Douce, Francis 16, 17
Draba mawii *200*
Dracunculus vulgaris 14
dragon arum 14, *127*
Drake, Miss Sarah Anne 109, 124, 150, 176
Drake, Miss Sarah Anne, pictures by *108*, *109*
Drawings of British Plants 226
Drimia maritima *15*
Dryas octopetala *76*, 78
Dryopteris dilatata *110*
Dürer, Albrecht 11, 16, **22–3**
Dürer, Albrecht, pictures by *22*, *23*
Duhamel du Monceau, Henri-Louis 138, 142
Duncanson, Thomas 110
Duperrey, Loius-Isidore 138
Dutch East India Company 83
Dutch flower paintings, seventeenth century 46–47
Dykes, W. R. 225

E
Early Chinese Plant Drawings 158–69
early works of the sixteenth century 22–33
East India Company 159, 161, 166, 168, **170–78**, 207
Edwards, Sydenham Teast 126, 147, 148, 150
Edwards, Sydenham Teast, picture by *149*
Egerton MS 747 14
Ehret, Georg Dionysius 32, 51, 52, **56–9**, 60, 66, *66*, 100, 112
Ehret, Georg Dionysius, pictures by *56*, *57*, *58*, *59*, *61*
Eichler, August 124
Eichstätt garden 38
Elements of Botany 52
Elwes and the genus *Lilium* 216–17
Elwes, Henry John **216–17**, 221, 222
Empress Joséphine 131, 132, 135, 138, **144–145**, 166
Endeavour 89, 98
Endemic Flora of Tasmania, The 226, *227*
Endlicher, S. L. 124
England, the golden age in 100–111
English herbals, sixteenth century 28–29
English Rock Garden, The 222
Englishmen in Levant 66–67
Engravings of Heaths 156
Enkianthus quinqueflorus *163*
Erica conspicua *156*
E. massoni *102*
Ericas in Southern Africa 245
Eriobotrya japonica *167*
Erysimum tenuifolium *152*
Euphorbia dendroides *33*
European Garden Flora (Vol. 5) 230
European paintings of Chinese plants 162–67
Evans, Anne-Marie 228
Evans, Sir Arthur 12
Everard, Barbara 236
Everard, Barbara, picture by *237*
exhibiting botanical watercolours 240–41
Exotic flora 190
Eyre, General John 207
Eyre, General John, picture by *207*

F
Farges, Paul Guillaume 169
Farrer, Ann 248
Farrer, Ann, picture by *249*
Farrer, Reginald 222
Farrer, Reginald, picture by *223*
ferns
nature-printed British Ferns, The 110
Ferns of Great Britain and Ireland, The *109*, 110, *110*
Feuillée, Louis 114
Figures of the most beautiful, useful and uncommon Plants described in the Gardener's Dictionary 52
Filices Exoticae *201*, *202*
fine editions, seventeenth century 44–45
Fiori Diversi 35
Fischer, F. E. L. Von 80
Fitch, W. H. (Walter Hood) 183, 190, **200–03**, 216
Fitch, W. H. (Walter Hood), pictures by *192*, *193*, *195*, *198*, *199*, *200*, *201*, *202*, *203*, *205*
Fleurs dessinées d'après nature 128
Fleurs et fruits 138
Flinders, Mathew 92, 95, 99
Flora Altaica 80
Flora Antarctica *192*, *193*, 194, 200
Flora Atlantica 67, 132
Flora boreali-americana 55
Flora brasiliensis 124
Flora Danica (1648, Simon Paulii) 186
Flora Danica, the story of 186–87
Flora Delanica 112
Flora Graeca **68–73**, 188
Flora Graeca 67, **68–73**, *68–73*,
Flora Hongkongensis 207
Flora Japonica (Carl Peter Thunberg) 83, *84*
Flora Japonica (Siebold And Zuccarini) 83, *83*, *85*
Flora Londinensis 146–47, *146*, *147*
Flora of British India 197
Flora of Louisiana 226
Flora Parisiensis secundum systema sexuale disposita... 138
Flora Peruviana, et Chilensis 121
Flora Rossica *76*, *77*, 78
Flora Sibirica 76
Flora Taurico-Caucasica 79
Floral Cabinet and Magazine of Exotic Beauty, The 153
Flore des Serres et des Jardins de l'Europe 154
Flore parisienne, contenant la description des plantes qui croissent naturellement aux environs de Paris 141–42
florilegia 11
Florilegium (Aedriaen Collaert) 34
Florilegium (Alexander Marshal) **42–43**, 48
Florilegium Imperiale 117
Florilegium of Japanese Plants 85
Flower Paintings from the Apothecaries' Garden 228
Flowering Tropical Climbers 230
flowers in religious paintings 21
flowers in Renaissance painting 16–19
Flowers of the Brazilian Rainforest 235
Flowers of the World 236
flowers of war and beyond 230–41
Forrest, George 221, 225
Forsythia suspensa *83*
Fortune, Robert 162, 166
Fothergill, John 52, 59, 100, 112
Fragmenta botanica 117
France 55
France, the golden age in 128–45
Franchet 168
Francis I, Emperor of Austria 117
Fritillaria assyriana *231*
F. caucasica *79*
F. pinardii *231*
Fructus theobromae *123*
Fruitist, The: a treatise on orchard and garden fruits, their description, history and management *153*
Fuchs, Leonhart 25, *25*
Fuci 188
Fugosia digitata *209*
Fulham Palace (London) 42, 48
fumitory *14*
Furse, Paul 230
Furse, Paul, picture by *231*

G
Galeandra devoniana *108*
Galen 25
Galves, Isidor 121
Garden of Flowers, A 40
Garden, Alexander 56
Gardener's Chronicle, The 150, 153
Gardenia jasminoides *56*
Gartenflora 154, *154*, *155*
Genera and Species of Orchidaceous Plants, The 107
Genera filicum; or Illustrations of the Ferns, and other allied Genera 109, 200
Genera Plantarum 194
Genus Arum, The 244
Genus Crocus, The 204–05
Genus Iris, The 225
Genus Lilium, The 216, *216*, *217*
Genus Rosa, The 225
George III 104, 112
Geranium heteritrichon *176*
G. microphylum *192*
Geraniums 157
Geranologia *131*, 132
Gerard, John **31**,
Gerard's herbal, sixteenth century 30–31
Gerardus emaculatus 31
Germany, seventeenth century 37–39
Ghent 154
Ghent Altarpiece 16, *18–19*
Gleichenia flabellate *202*
Gmelin, Johann Georg 76
Godman, Alice 216

Goldblatt, Peter 245
Gomphrena globosa *44*
Good, Peter 92
Gorichaud 176
Govindoo 181
Graeca, Flora 67, **68–73**, *68–73*, 188
Granville, Mary, *see* Delany, Mrs
Gray, Asa 188, 197
Greece 11, 12
Grenfell Medals 232, 236
Grete Herball, The 28
Greville, Robert K. 95
Grevillea banksii *94*, 95
Grierson, Mary 242
Grierson, Mary, picture by *243*
Grove, Alfred 216
Guatemala 124
Guillemin, J. B. Antoine 132
Guirlande de Julie 35
Gustavia pulchra *233*, 235

H
Hagerdorn, Regine 249
Hamelia patens *117*
Hance, H. F. 207
Handbook of the British Flora 197
Hardy Heathers 244
Hare (Dürer) 22, *22*
Hartlib, Samuel 42
Hartweg, Theodore 124
Hashi, Mayumi, picture by *229*
Hastings Hours, The 17
Hawkins, John 68, 71, 72
heathers *156*, 244, 245
Ericas in Southern Africa 245
Hardy Heathers 244
heathery, The; or, A monograph of the genus Erica 156, *156*, *157*
Heliotropium americanum *34*
Hellebores *244*
Henderson, Peter 126
Henry VIII 28
Henry, Augustine 217, **218**, 221
Henslow, John Stevens 153
Herball or generall Historie of Plantes, The *30*, 31
herbals 11
Herbarium Apulei 24
Herbarium Vivae Eicones 24, *24*
herbs 11
Herklots, Geoffrey 230
Herolt, Johanna Helena 44
Herreyns, Willem Jacob 128
Hervier général de l'amateur 138
Hibiscus rosa-sinensis *129*
H. tiliaceus *159*
Highgrove Florilegium, The 34, **228**, *229*, 248
Hilliard, Olive 245
Histoire des arbres Forestiers de l'Amérique Septentrionale 55, 138
Histoire des Chênes d'Amérique *54*, 55, 138
Histoire naturelle des oranges 142
Historia naturalis plamarum 123
Historia Plantarum 48
Holden, S. 153
Hong Kong Countryside, The 230
Hongkongensis, Flora 207
Hooker, Joseph Dalton 148, 181, 188, **192–203**, *192*, 204, 212
Hooker, Joseph Dalton, picture by 194
Hooker, William 107, 109, 150
Hooker, William Jackson 107, 109, 148, 150, 178, 181, **188–90**, *188*, 200, 210
Hooker, William Jackson, pictures by *189*, *190*
Horti medici amstelodamensis rariorum plantarum descriptio et icons 44, *44–45*
Horticultural Society, *see* Royal Horticultural Society
Hortus Brittano-Americanus *50*, 51
Hortus Cliffortianus 59, 60
Hortus Europae americanus 51
Hortus Eystettensis 34, **37**, *38*, *39*, 44, 48
Hortus Floridus 40, *40*, *41*
Hortus Itzteinensis 44
Hortus Kewensis 100, 110
Hortus Medicus (Amsterdam) 44
Hortus Woburnensis 156
House of the Vettii 12
Huizong, Emperor 158
Humboldt, Alexander von **123**, 138, 141
Humphries, Professor Christopher 228
Hunt Botanical Library 248
Hunt, David 242
Hutton, Mrs Janet *207*
Hutton, Mrs Janet, pictures by *178–79*, *183*, *206*
Huxley, T. H. 188
hyacinths *26*
Hyacinthus *37*
Hydrangea petiolaris *83*

I
Icones Plantarum (Jakob Theodor) 31
Icones Plantarum (W. J. Hooker) 200
Icones plantarum Garlliae rariorum 141
Icones plantarum Indiae Orientalis: or figures of Indian Plants 181
Icones plantarum Japonicarum 83, *84*
Icones Plantarum novarum vel imperfecte cognitarum floram Rossicam, imprimis Altaicam illustrantes 80, 80
Icones plantarum rariorum 68
Icones plantarum rariorum (1781–93) 117
Icones plantarum rariorum horti caeseri Schoenbrunnensis Descriptiones et icones (1797–1804) 117
Icones plantarum Syriae rariorum 67, *67*
Icones Selectae Plantarum quas in Japoniae collegit et delineavit Englebertus Kaempfer 82, *82*
Idstein (Frankfurt am Main) 44
Illustration Horticole 154
Illustrationes florae Novae Hollandiae *94*, 95, *96*
Illustrations of Himalayan Plants (Joseph Hooker) **183**, *184*, 194, 196, 197, 216
Illustrations of Indian Botany 181
Illustrations of orchidaceous plants, by Francis Bauer 107
Illustrations of the botany and other branches of natural history of Hymalayan mountains and the flora of Cashmere 176
Illustrations of the botany of Captain Cook's voyage round the world in H.M.S. Endeavour in 1768–71 92
Imai, Mariko 248
Imperial Martagon lily *42*
Index Kewensis 197
India
British India, Flora of 97
Company School in India, The **170–87**, 207
Dutch East India Company 83
East India Company 159, 161, 166, 168, **170–78**, 207
Magnoliaceae of British India, The 183
Instituto Botanico del Nuevo Reino de Granada 118
Investigator, H. M. S. 92, 95, 99
Iris germanica 22
I. kolpakowskiana Rgl 155
Irises 11, *16*, 21
Genus Iris, The 225
Ishikawa, Mieko 245
Islam, sixteenth century 26–27
Iznik ceramics 26, *27*

J
Jacquemont, Victor 178
Jahangir, Mogul Emperor 170
Japan 82–85, 245
Japonica Magnifica 239
Japonica, Flora (Carl Peter Thunberg) 83, *84*
Japonica, Flora (Siebold And Zuccarini) 83, *83*, *85*
Jardin de St Pétersbourg 80
Jardin des Plantes, *see also* Jardin du Roi 131, 138
Jardin du Roi 55, 59, 62, 86, 118, 128, 131
Jardin Fleuriste 154
Jardine, Hyacinth 197
Jensen, Johan Laurentz 128
Johann Konrad von Gemmingen 37, 38
Johnson, Thomas 31, 48
Johnstone, G. H. 221
Joubert, Jean 86
Journal des observations physiques, mathématiques et botaniques 114
Journal of a tour in Marocco and the Great Atlas *204*
Journals of Hipólito Ruiz, The 121

K
Kaempfer, Englebert 82–83
Kaempferia galanda *173*
K. rotunda Linn. *171*
Keiga, Kawahara 85
Kennedy, John 156
Kerr, William 161, 166
Kew, a new era at 188–205
Kew, the Royal Gardens at 52, 98, **100–111**, 161, 166, **188–205**, 235, 240, 242–49
Kilburn, William 147
King, Christabel 235, 242
King, Christabel, picture by *242*
King, Sir George 182
Kingdon-Ward, Frank 225, 226
Kishmush grape 150
Knappe, Karl Friedrich 79
Knight, T. A. 150
Knossos 11, 12, *13*
Knowles, G. B. 153
Koberger, Anton 22
Komarov Botanical Institute 80
Kunth, C. S. 123

L
La Pomologie française *128*, 129, *140*, *141*, 142
Lachenalia aloides *131*
Lack, Hans Walter 117
Ladies' Botany 109
Lagerstroemeria speciosa *162–63*
Lambert, Aylmer Bourke 104, 107, 121
Larix decidua *105*
L. kaempferi *85*
Laryx griffithii 197
Lathrea squamaria *25*
Lavandula × intermedia *17*
le Clerc, Sebastien 35
Le Grand Herbier 28
Le Jardin du très Chrétien Henri IV, Roi de France et de Navarre... 34
Le Jardin fruitier 138
Le Jardin fruitier du Muséum 154
Le Moyne, Jacques 17, *17*
Le Voyage aux Régions équinocxiales du Nouveau Continent 123, 138
Lee & Kennedy Nursery 92, 156
Legrand, P. F. 128
Legumes of the World 249
Leiden Botanical Garden 59
Lemaire, Charles 154
Leonardo Da Vinci **20–1**
Leonardo Da Vinci, pictures by *20*, *21*
Les Jardins de la Malmaison 132, 144
Les Liliacées *131*, 132, *132*, *133*, 136, 145
Les Roses *134*, 135, *135*, 136, 145, *145*
Les Vélins du Muséum 86–87
Lettsom, John 147
Letty, Cynthia 245
Levant, travellers to 62–75
L'Héritier de Brutelle, Charles-Louis 100, 103, **131–32**
Ligozzi, Jacopo 11, 22, **32–3**
lilies 10, 11, 12, 16, 21, 216–17
Genus Lilium, The 216, *216*, *217*
Lilies (Patrick Synge) 230
Lilium auratum 198
L. candidum 12
L. chalcedonicum 12, 26
L. persicum *42*
L. regale *219*
L. superbum *126*, *217*
L. washingtonianum *216*
Lily Prince, Knossos *13*
lily, martagon *37*, *42*
Lindley, John 68, 73, 107, 109, 150, 153, 161, 162, 190
Linnaeus and plant classification **60–61**
Linnaeus, Carolus 55, 59, **60–61**, *60*, 72, 83, 126
Linnean Society 98, 146, 217
Liparia splendens *144*
Liverpool Botanic Garden 174
Lobb, Thomas 207
Lobb, William 207
Loddiges, Conrad 150
Loddiges, George 124
Londinensis, Flora 146–47, *146*, *147*
London, seventeenth century 42–43
lotus flower 158
Louis XII, King of France 17
Louis XIII, King of France 35, 86
Louis XVI, King of France 128
Louis XVIII, King of France 136
Louisiana, Flora 226
Lugard, Charlotte 208
Lugard, Charlotte, picture by *209*
Lugard, Edward James 208
Lyte, Henry 29

M
Macartney, Lord 161
Madonna in a Garland of Flowers 46
Madonna of the Carnation 21
Madonna of the Rose Garden (altarpiece) 16
Madonna with the Iris, The 23
Magnolia campbellii subsp. *Mollicomata 221*
M. grandiflora 51
M. nitida 220
M. sieboldii 229
Magnoliaceae of British India, The 183
magnolias 220–21
Margaret Mee Amazon Trust (Foundation) 235, 242
Marianne North Gallery, Kew 212–13
Marica caerulea 153
Marie-Antoinette 131, 136
Maries, Charles 208
Marshal, Alexander **42**,
Martagon imperiale 42
Martius, Friedrich Philipp von 123
Martyn, Thomas 126
Mary in a Garden with Cherry Tree 17
Masson, Francis 83, 99, **103–04**
Masson, Francis, pictures by *98, 99, 103*
Master Mary of Burgundy 17
Mathew, Brian 244
Matiz, Francisco Javier 118
Matthioli, Pier Andrea 25
Maund, Benjamin 150, 153
Maund, Eliza 153
Maund, Sarah 153
Maw, George **204–05**, *204*
Maxillaria tetragona 189
McEwen, Rory 239
McEwen, Rory, picture by 238
Meconopsis simplicifolia 185
Meconopsis villosa 184
Medici Dukes 32, 174
medicinal plants 34
medicinal plants 10–11
Medinilla rubicunda 181
Mee, Margaret 232–35
Mee, Margaret, pictures by *232, 233, 234, 235*
Meen, Margaret 52
Meen, Margaret, picture by *53, 67*
Melanthium 41
Melocactus 31
Mengjiang, Zhao 158
Mengjiang, Zhao, picture by *158*
Menzies, Arichibald 99
Merian, Maria Sybilla 44, **74–75**
Merian, Maria Sybilla, pictures by *74, 75*
Messerschmidt, Daniel 76
Mexico 48, 52, 123, 124
Mexipedium xerophyticum 248
Meyer, Albrecht 25
Michaux, André **55**, 136, 138
Michaux, François **55**, 136, 138
Michel, Etienne 138
Miller, Philip 52, 56, 59, 66
Mills, Miss R. 153
Minoan Civilization 10–12, *11*
Mirabilis jalapa 32
Missions Etrangères 168
Mitella diphylla 86
Modern florilegia 228–29
Mondandrian Plants of the order Scimataceae 174
Mongolia 80
Moninckx, Jan 44
Moninckx, Maria 44
monocotyledons 29
Monographie de Melastomacées 138
Moore, Thomas 110
Moore, Thomas, picture by *110*
Morin, Louis 62
Morina persica 63
Morocco 204
mosaic, flower 112, *112, 113*
Murray, John 148
Musa × paradisiaca l. 87
Muséum nationale d'Histoire naturelle 141, 168
Museum Tradescantianum 48
Mutis, José Celestino **118, 121**, 123
Mutis, José Celestino, picture by *118*
Mutisia clematis 119
names of plants in Greek, Latin, Englishe, Duche & Frenche, with the commune names that herbaries and apothecaries use, The 28

N
Napoleon 135, 138, 144–45
Narcissus 35, 41
narcissus *26*
Narcissus tazetta 11, 46, 158, *158*
nasturtiums (*Tropaeolum*) 121
Natural History of Aleppo, The 59, **66**, *66*, 173
Natural History of Carolina, Georgia, Florida and the Bahama Islands, The 48,49, 51, *51*
naturalism, Chinese 158
nature-printed British Ferns, The 110
Nelson, Charles 244
Nerine serniensis 42
Neomarica caerulea 153
Nessler, Kate 248
Netherlands, the, seventeenth century 40–41
Neues Blumenbuch 74
New Herball, A 28
New Illustration of the sexual system of Carolus von Linnaeus 126
New Kreüterbuch 25, 28
Nievre Herball, A 29
Nikulinsky, Philippa 244
Nodder, Frederick, picture by *93*
Noeregelia margaretae 234
Noisette, Louis Claude 138
Noltie, Henry J. 178
North American plants 48–61
North, Marianne **210–15**, 230, 239
North, Marianne, pictures by *210, 211, 212, 213, 214, 215*
Norton, John 31
Novae Hollandiae plantarum specimen 92
Nymphaea 232
N. lotus 215

O
oak *54*
Oeder, Georg Christian 186
Old Carnations and Pinks 239
Old Shrub Roses, The 236
Ophrys bombyliflora 31
Opuntia 38
orchids *31, 58*, 107, 124236
Genera and Species of Orchidaceous Plants, The 107
Orchidaceae 242
Orchidaceae of Mexico and Guatemala 124
Orchids of the Sikkim-Himalaya, The 183
Sertum orchidaceum, a wreath of the most beautiful orchidaceous flowers 109
Orchis mascula 30
Orchus purpurea 147
Oreocereus doelzianus 242
Origins of botanical art 10–21
Ornithogalum angustifolium 21
Ottoman Empire 62
Ottoman Turkey 26

P
Pachira marginata 214
Paeonia (monograph) 221
Paeonia 160
Pallas, Peter Simon 76
Pancratium maritimum 12
Pantling, Robert 183
Papaver orientalis 63
Paradisi in sole paradisus terrestris 42
Paradisus Londinensis, The 150
Parisiensis, Flora 138
Parkinson, John 42
Parkinson, Sydney 90–92, *90, 93*, 107
Parkinson, Sydney, picture by 91
Parsons, Alfred 225
Passiflora 57, 149
P. laurifolia 149
P. quadrangularis 116
Paulii, Simon 186
Pavón Jiminez, Don José 118, 121, 144
Paxton, Joseph 153
Paxton's Flower Garden 153
Paxton's Magazine of Botany, and Register of Flowering Plants 153, *153*
peach *151*
Pedder-Smith, Rachel 249
Pedder-Smith, Rachel, pictures by *246–47*
pelargoniums 157, 245
Pelargonium cullatum subsp. *strigifolium 130*
Pena, Pierre **28–29**
peonies 11, 16, 22, 161
Paeonia (monograph) 221
Paeonia moutan cv 160
Pericallis lenata 148
Persian lily 42
Peru 121
Peruviana, et Chilensis, Flora 121
Petasites japonicus 83
Peter the Great 76, 80
Pether, Abraham 126
Phleum pratense 186
Phyllostachys heteroclada 249
Phormium tenax 93
Physalis alkekengi 112
pineapple 48
Pinguicula grandiflora 152
pitcher-plant *92*, 95
plant classification and Linnaeus 60–61
Plantae aequinoctiales 122, 123
Plantae Asiaticae rariores 176
Plantae Davidiana 168
Plantae Delavayanae 169, *169*
Plantarum novarum vel minus cognitarum quas in itinere Caspio-Caucasico observavit 80, *80*
Plantarum succulentarum Historia 132
Plantin, Christophe 26, 29
Plants from the coast of Coromandel 173
Pliny the Elder 12, 14, 24
Plumier, Charles 55
Pococke, Richard 66
Poiteau, Pierre-Antoine 67, 123, 128, 138, **141–42**
Poiteau, Pierre-Antoine, pictures by *129, 140, 141, 142*
polyanthus *75*
Pomona Herefordiensis 150
Pomona Londinensis 150
Pompeii 12
Portinari Altarpiece 16, *16*
potato 31
pre-classical era 12–13
Prêtre, Jean Gabriel 166
Prêtre, Jean Gabriel, picture by *167*
primroses *75*
primulas 222
printing 24
Prionotes cerinthoides 227
Prodromus (Peru and Chile) 121
Prodromus Flora Graeca 72
Prodromus florae Novae Hollandiae et Insulae Van-Diemen 95
Prodromus Florae Peninsulae Indiae Orientalis 181
Protea canaliculata 157
Prunus × dasycarpa Ehrh. *141*
Pseudoiris lutea 29
Psychotria *124*
Pteris vittata 173

Q
Quercus alba 54
Quian Xuan *10–11*

R
Rabel, Daniel 35, 86
Rabel, Daniel, picture by *35*
Raffles, Sir Thomas Stamford 161
Rafflesia arnoldii 161
Rage for Rock Gardening, A 222
Rariorum Plantarum Historia 26
Ray, John 48, 51
Razumovsky, Alexei Kirillovich 80
Real Jardín Botánico 121
Recollections of a Happy Life 215
Recueil de Plantes 35
Redouté brothers 55, 67, 138
Redouté, Antoine-Ferdinand 131
Redouté, Henri-Joseph 128, 138
Redouté, Pierre-Joseph 46, 55, 86, 103, 121, 128, **130–45**, *131*, 157
Redouté, Pierre-Joseph, pictures by *54, 130, 131, 132, 133, 134, 135, 136, 137, 144, 145*
Redouté, Pierre-Joseph, pictures in style of *143*
Reeves, John 161, 162, 166
Reeves, John, picture by *159*
Regel, Eduard August von 80, **154**
Reinagle, Philip 126
Revue Horticole 154
Rheum compactum (R. Nutans) 77
rhododendrons 194, 222
Rhododendron dalhousiae 194, 195
R. lacteum 224
R. leucaspsis 226
Rhododendrons of Sikkim-Himalaya, The 194
Ricci, Matteo 159
Riocreux, Alfred 154, 178
Risso, Antoine 142
Rizo Blanco, Salvador, picture by *119*
Robert, Nicolas 35, **86**, *86*
Robert, Nicolas, pictures by *34, 86*

Robin, Jean 34
Robin, Vespasien 34
Robins, Thomas the Elder 47
Robins, Thomas the elder, picture by *47*
Roe, Sir Thomas 170
Roemeria hybrida 69
Rosa 'Celestial' *225*
R. '*Bouquet d'or*' *241*
R. gallica 12
R. gallica 'Officinalis' 26
R. kamtschatica 134
R. multibracteata 236
R. oxyodon 78
R. redutea glauca 145
Rosarum Monographica: or a botanical history of roses 107
Roscoe, William 174
roses 11, 12, 236
Climbing Roses, Old and New 236
Genus Rosa, The 225
Old Shrub Roses, The 236
Roses (Andrews) 157
Shrub Roses of Today 236
Ross-Craig, Stella 221, 222, **226**
Ross-Craig, Stella, picture by *226*
Rosser, Celia 245
Rossica, Flora 76, 77, 78
Rössler, Martin 186
Rössler, Michael 186
Round, Frank 225
Roxburgh Collection 170–75
Roxburgh, William 173, 176, 178, 181
Roxburgh, William, pictures commissioned by *170, 171, 172, 173*
Royal Botanic Garden, Edinburgh 222
Royal Botanic Gardens, Kew, *see* Kew Gardens
Royal Botanical Institution (Denmark) 186
Royal Horticultural Society (RHS) 98, 99, 146, 150, 161, 162, 210, 240
Royal Society, The 98
Royle, John Forbes 176
Rubens, Pieter Paul 46
Rubens, Pieter Paul, picture by *46*
Rubus crataegifolius 154
Ruiz López, Don Hipólito 118, 121, 144
Rumsey, Dr Frederick J. 228
Rungia 181
Russell, Alexander 59, **66**, 170
Russell, Patrick 170, 173
Russia, Exploration of 76–81
Ruysch, Rachel 46, 131

S
Sacabiosa africana frutescens 44
saffron crocus 10, 12
Saffron Gatherers, The 12
Saint Jerome in the Wilderness 22
Saint-Hilaire, Henri Jaume 128
Salisbury, R. A. 150
Salvia forskaelei 68
Sanguisorba alpina 80
Sansom, Francis 148, 150
Santorini 12
Sarracenia purpurea 49
sassafras *50*
Scabiosa africana frutescens 44
Scharf, Johann 117
Schedel, Sebastian 37
Schedel, Sebastian, pictures by *36, 37*
Scheidweiler, M. 154
Schizanthus retusus 152
Schmutzer, Mathias 117
Schönbrunn Palace, Vienna 68, 114
Schongauer, Martin 16, 22
Schongauer, Martin, school of 17
Second World War 230, 232, 236
Sedelmayer, Martin 117
Sedum acre 147
S. dasphyllum var. *glanduliferum 205*
Selectarum stirpium Americanarum historia 114, *114, 115, 116, 117*
Selenipedium lindleyanum 124
S. vittatum 124
Sellars, Pandora 244
Selencereus wittei 235
Sertum Anglicum 103, 131
Sertum orchidaceum, a wreath of the most beautiful orchidaceous flowers 109
Sertum Petropolitanum 80
seventeenth century florilegia 34–47
Severeyns, G. 154
Sharawadgi 161
Sherard, William 51, 68
Sherwood, Dr Shirley 242, 244
Shirley Sherwood Gallery 242, 248, 249
Shrub Roses of Today 236
Siberia 78
Sibirica, Flora 76
Sibthorp, Humphrey 56, 68
Sibthorp, John **68–73**, 92, 107, 108
Siebold, Philipp Franz von 83
Sims, John 148
Sinclair, George 156
Skinner, George Ure 124
Sloane, Sir Hans 56, 59, 82
Smith, Edwin 153
Smith, F. W. 153
Smith, James Edward (J. E.) 68, 72, 126, 188
Smith, Lucy T. 245
Smith, Matilda 197, **221**
Smith, Matilda, pictures by *218, 219*
Snelling, Lilian 217, 221, **222**, 225, 226
Snelling, Lilian, pictures by *222, 224*
Solander, Daniel 90, *90*, 92, 112
Some Further Recollections of a Happy Life 215
Soulié, Jean-André 169
South Africa 156–57, 245
South America, exploring 114–25
Sowerby family 73
Sowerby, James 103, 131, 147, 148, 188
Species Plantarum 60, *60*, 89
Spicilegium Neilgherrense 181
Spöring, Herman 90, 91
St Bavo's Cathedral 16
St Petersburg Botanical Garden 80, *80*
Stachelina arborescens 71
Stapelia grandiflora 99
S. pedunculata 103
S. reticulata 98
S. novae 98, 99, **103**
Star of Bethlehem *20*, 21
Stearn, William 92
Steller, Georg W. 76
Stern, F. C. 221
Still Life of Flowers 47
Stirpes novae 101, 131
Stirpium adversaria nova 28
Stones, Margaret 217, **226**
Stones, Margaret, picture by *227*
Strelizia depicta 104
S. reginae 104
Sturt's Desert Pea (*Wildampia* A. S. George) 88, *89*
sugar maple *52*
Suleiman 26
Sultan Ahmet III 62, 64
sun spurge *20*, 21
Sutherlandia furtescens 42
Swainsona galegifolia 139

T
Tabernaemontanus 31
Talauma hodgsoni 196
Talbot, Milo 226
Taraxicum officinale 187
Taurico-Caucasica, Flora 79
Taylor, Simon 100, 103
Taylor, Simon, picture by *100*
Tcherepnine, Jessica 248
Tcherepnine, Jessica, picture by *249*
Temple of Flora, The, or Garden of Nature 126
Theatrum Florae 35, *35*
Theodor, Jakob 31
Theophrastus 11
Thistleton-Dyer, Harriet 221
Thomas, Graham Stuart 236
Thomas, Graham Stuart, picture by *236*
Thornton, Robert John 126
Thornton's *The Temple of Flora* 126–27
Thory, Claude-Antoine *135*
Thunberg, Carl Peter **83**, 103
toothwort *24*
Tournefort, Joseph Pitton De **62–63**, *62*, 66, 72, 86
Tradescant, John the Younger 42, **48**, *48*
Traité des Arbres et Arbustes que l'on cultive en France en plain terre 138, 142
Traité du Citrionier 138
Transactions of the Horticultural Society of London 150, *150*
Transactions of the Linnean Society 161
Trees of Great Britain and Ireland, The 217
Trew, Dr 56
Triana, José Jerónimo, picture by *120*
Tropaeolum 121, 248
Tropaeolum dekerianum 120
Tulipa praecox rubra 26
T. sibthoriana 71
Tulips 26, *26*, 40, 46, *59*, **64**, *64, 65, 143*
Tulips and Tulipomania 239
tulips from Ottoman Empire 64–65
tupelo *50*
Turkish Empire, sixteenth century 26–27
Turner 28
Turner, Dawson 188
Turpin, Pierre Jean Françios 67, 123, 138, **141–42**

U
Urban, Ignatius 124

V
Vahl, Martin 186
Vallet, Pierre 34, 35
van de Borcht, Pierr 29
van der Goes, Hugo 16, *16*
Van Eyck, Hubert 16,
Van Eyck, Hubert, picture by *18–19*
Van Eyck, Jan 16,
Van Eyck, Jan, picture by *18–19*
van Houtte, Loius 154
van Huysum, Jan 46, 128, 131
van Kouwenhoorn, Pieter 40
van Os, Jan 46
van Spaëndonck, Cornelis 128
van Spaëndonck, Gerard 46, 86, **128–29**, 131, 136, 138, 141
van Spaëndonck, Gerard, picture by *129*
Veitch, James and Sons Nursery 207, 208, 218, 219
Veitch, John Gould 194
vélins 35, 59 **86–87**, 128, 131, 132, 136
Ventenat, Etienne-Pierre 132, 138, **144**
Venus's fly trap (Dionaea muscipola) 237
Verelst, Simon, picture by *65*
Vesuvius 12
Victoria amazonica 106–07
Victoria Regia 107
Victorian travellers 206–
Vigna adenantha 91
Viola 39
violets *16*, 21, 39
Virgin of the Rocks, The 21, *21*
Vishnupersaud 176
von Below, Irma 245
von Eichwald, Karl Eduard 80
von Jacquin, Joseph Franz 117
von Jacquin, Nikolaus Joseph 68, 104, **114, 117**
von Ledebour, Karl Frederick 79–80
Voyage autour du monde 138
Voyage dans l'Inde 178
Voyage into the Levant 62, *63*
Voyage to New Holland, &c. In the Year 1699 89
Voyage to Terra Australis, A 95

W
Wallich, Nathaniel 176, 178
Wallich, Nathaniel, pictures commissioned by *176, 177*
Walsh, Wendy 244
Walther, Johann the Elder 34, **44**
Ward-Hilhorst, Ellaphie 245
Watters, Catherine M.
Webster, A. V., picture by *220*
Weddell, E., picture by *97*
Wedgwood, John 99
Weiditz, Hans 17, 24
Westcott, F. 153
Wheeler-Cuffe, Lady Charlotte 208
White, Gilbert 146
White, John 48
White, John, picture by *48*
Wight, Robert 178, **181**
Wilmott, Ellen 225
Wilson, E. H. (Ernest Henry) 208, **218–19**, 221
Withers, Mrs Augusta 124, 150, 153
Wolgemut, Michael 22
wood anemones *20*, 21
woodcuts, sixteenth century 24–25
Woodin, Carol 248
Woodin, Carol, picture by *248*
Worring, Andreas 110

Y
Yamanaka, Masumi 248
Yamanaka, Masumi, picture by *245*
Ying-Wei, Tang 230

Z
Zingiber cassumunar R. 175
Z. elatum 172
Z. montanum 175
Z. rubens 170
Zubov, Count Valerian 79
Zuccarini *83*, 85

BIBLIOGRAPHY

It would be impossible to write a book such as this, which spans several centuries and continents, without recourse to numerous reference works. The list below includes the main sources consulted. In addition, numerous original letters and papers have been consulted in the Library and Archives of The Royal Botanic Gardens, Kew.

Aiton, W., *Hortus Kewensis*, London, 1805–14

Andrews, H., *Coloured Engravings of Heaths*, London, 1802–09

Bauer, F., *Delineations of Exotick plants cultivated in the Royal Garden at Kew* (3 parts), London, 1796–1801

Baumann, H., *Greek Wild Flowers* (trans.), London, 1993

Baytop, T., *Istanbul Lalesi*, Ankara, 1992

Besler, B., *Hortus Eystettensis…*, Nuremberg, 1613

Blunt, W. & Stearn, W.T., *The Art of Botanical Illustration*, London, 1950

Bonpland, A., *Descriptions des plantes rare cultivées à Malmaison et à Navarre*, Paris, 1813

Brown, A., Cribb, P. & Barlow, G., *Flower Paintings from the Apothecaries' Garden*, Woodbridge, 2005

Calmann, G., *Ehret, Flower Painter Extraordinary*, Oxford, 1977

Collins, M., *Medieval Herbals: The Illustrative Traditions*, London & Toronto, 2000

Commelin, J., *Catalogus plantarum Horti Medici Amstelodamensis*, Amsterdam, 1697–1701

Cullen, J. *et al.* eds, *The European Garden Flora*: Vol. 5., Cambridge, 1997

Curtis's Botanical Magazine (passim)

Desmond, R., *The History of the Royal Botanic Gardens, Kew*, London, 2007

Desmond, R. & Ellwood, C., *Dictionary of British and Irish Botanists and Horticulturists*, London, 1994

Edwards, S. & Lindley, J. eds, *The Botanical Register*, London, 1815–47

Elick, D. & Booth, R., *Japonica Magnifica*, Stroud, UK & Portland, Oregon, 1992

Elwes, H., *A monograph of the genus Lilium*, London, 1880

Franchet, A., *Plantae Davidianae ex Sinarum Imperio*, 2 vols, Paris, 1884–88

Hayden, R., *Mrs Delany: Her Life and her Flowers*, London, 1992

Henrey, B., *British Botanical and Horticultural Literature Before 1800*, Oxford, 1975

Herklots, G., *Flowering Tropical Climbers*, Folkestone, 1976

Herklots, G., *The Hong Kong Countryside*, London, 1972

Herklots, G., *Vegetables in South-East Asia*, London, 1972

Hooker, J., *Flora Tasmaniae*, London, 1855–60

Hooker, J., *Rhododendrons of the Sikkim Himalaya*, London, 1849–51

Hulton, P. & Smith, L., *Flowers in Art from East and West*, London, 1979

Humphries, C., Rumsey, F. & Crouch, H., *The Highgrove Florilegium*, 2 vols, London, 2008–09

Jellicoe, G. & S., Goode, P. & Lancaster, M. eds, *The Oxford Companion to Gardens*, Oxford, 1986

Lack, H. W., *Florilegium Imperiale*, Munich, 2006

Lambert, A., *Description of the Genus Pinus*, 2 vols, 1803–24

Lancaster, R., *Plantsman's Paradise, Travels in China*, Woodbridge, 2008

Leith-Ross, P. & McBurney, H., *The Florilegium of Alexander Marshal at Windsor*, London, 2000

Lewis, J., *Walter Hood Fitch: A Celebration*, London, 1992

L'Héritier de Brutelle, C.-L., *Sertum Anglicum…*, Paris, 1788

Lindley, J., *Rosarum Monographia; or A Botanical History of Roses*, 1820

Mabberley, D. J., *Mabberley's Plant Book*, Cambridge, 2008

Magee, J., *The Art and Science of William Bartram*, Pennsylvania & London, 2007

Masson, F., *Stapeliae novae…*, London, 1796

Maw, G., *A monograph of the genus Crocus*, London, 1886

Morley, B. D. & Everard, B., *Wild Flowers of the World*, London, 1970

Morrison, T. ed., *Margaret Mee. In search of Flowers of the Amazon Forests*, Woodbridge, 1988

Needham, J., *Science and Civilisation in China*, Vol. 6. Pt 1., Cambridge, 1986

Nissen. C., *Die Botanische Buchillustration*, 2nd ed., Stuttgart, 1966

Noltie, H., *Indian Botanical Drawings* 1793–1868, Edinburgh, 1999

Noltie, H., *Robert Wight and the Botanical Drawings of Rungiah & Govindoo*, Edinburgh, 2007

North, M., *Recollections of a Happy Life*, 2 vols, London, 1892

O'Neill, J. & McLean, E.P., *Peter Collinson and the Eighteenth-Century Natural History Exchange*, Philadelphia, 2008

Parkinson, J., *Paradisi in sole paradisus terrestris*, London, 1629

Rabel, D., *Theatrum Florae*, Paris, 1633

Redouté, P.-J., *Choix de plus belles fleurs*, Paris, 1827–33

Redouté, P.-J., *Les Liliacées*, Paris, 1802–16

Redouté, P.-J. & Thory, A., *Les Roses*, Paris, 1817–24

Robinson, T., *William Roxburgh*, Chichester & Edinburgh, 2008

Schultes, R. & von Thenen de Jaramillo-Arango, M. (trans.), *The Journals of Hipolito Ruiz*, Portland, 1998

Shulman, N., *A Rage for Rock Gardening: The Story of Reginald Farrer*, London, 2002

Sibthorp, J., *Flora Graeca: sive Plantarum rariorum historia…*, 10 vols, London, 1806–40

Sloan, K., *A Noble Art: Amateur Artists and Drawing Masters c. 1600–1800*, London, 2000

Stearn, W. ed., *John Lindley 1799–1865*, Woodbridge, 1999

Stearn, W. et al., *Siebold's Florilegium of Japanese Plants*, Vol. II, Tokyo, 1994

Thornton, R., *Temple of Flora*. London, 1812

Van de Passe, C., *Hortus Floridus…*, Utrecht, 1614

Veitch, J., *Hortus Veitchii*, London, 1906

Ventenat, E., *Choix de Plantes*, Paris, 1803

Ventenat, E., *Jardin de la Malmaison*, Paris, 1803–04

Watson, Wed., *A Magnificent Collection of Botanical Books* (Sale catalogue), London, 1987

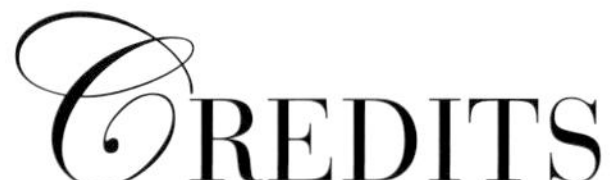

CREDITS

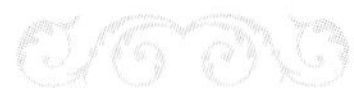

Publishers' Acknowledgements

The Publishers would like to thank Gina Fullerlove, Head of Publishing at The Royal Botanic Gardens, Kew for her help and support in the production of this book. Special thanks to the following members of staff at the Library and Archives at Kew for their enthusiasm and patience in researching material and providing information throughout the project: Lynn Parker, Trishya Long, Paul Little and Julia Buckley.

Picture Credits

The publishers would like to thank the following sources for their kind permission to reproduce the pictures in this book.

Key: t=Top, b=Bottom, c=Centre, l=Left and r=Right

Addison Publications Ltd.: /Magnolia sieboldii by Mayumi Hashi from The Highgrove Florilegium © A G Carrick Ltd., 229

Archivo del Real Jardin Botánico, CSIC, Madrid: 95

The Bodleian Libraries, University of Oxford: (MS. Douce 219, fol. 37r): 16r

The Bridgeman Art Library: /Alte Pinakothek, Munich, Germany: 46, /© Bristol City Museum and Art Gallery, UK: 129, /Photo © Christie's Images: 64, / Cincinnati Art Museum, Ohio, USA / John J. Emery Endowment: 10-11, /Photo © The Fine Art Society, London, UK: 239, /Fitzwilliam Museum, University of Cambridge, UK: 47, /Galleria degli Uffizi, Florence, Italy /Alinari: 16l, /Musee National de la Renaissance, Ecouen, France /Giraudon: 27, /National Gallery, London, UK: 21, 23, /Private Collection: 22, /© The Right Hon. Earl of Derby: 52, /The Royal Collection © 2011 Her Majesty Queen Elizabeth II: 20, 43, /St. Bavo Cathedral, Ghent, Belgium: 18-19, /Victoria & Albert Museum, London, UK: 17

© The Trustees of the British Museum: 48b, 112, 113

Barbara Everard: 236

Ann Farrer: 249

Paul Furse: 231

Getty Images: /De Agostini: 11, 12

Mary Grierson: 243

Regine Hagedorn: 241

Christabel King: 242

Rory McEwen: 238

The Natural History Museum: 90–91, 93, 159

Rachel Pedder-Smith: 246–247

Private Collection: 60b

Réunion des Musées Nationaux: /© Muséum national d'Histoire naturelle, Dist. RMN /image du MNHN, bibliothèque central: 87

The Royal Horticultural Society: 236

Pandora Sellars: 244

Photo Scala, Florence: /Courtesy of the Ministero Beni e Att. Culturali: 32, 33, / Image copyright The Metropolitan Museum of Art /Art Resource: 158

Margaret Stones: 227

Topfoto.co.uk: 13

Carol Woodin: 248

Masumi Yamanaka: 245

Every effort has been made to acknowledge correctly and contact the source and/ or copyright holder of each picture and Carlton Books Limited apologises for any unintentional errors or omissions, which will be corrected in future editions of this book.

Publishers' Credits

Editorial Manager: Vanessa Daubney
Additional Editorial Work: Lesley Malkin & Catherine Rubinstein
Art Director & Cover Design: Katie Baxendale
Design: Sally Bond
Photography: Russell Porter
Picture Research: Jenny Meredith
Production Controller: Maria Petalidou